1.15 miles	≅ 1 minute (1′) of latitude ≅ 1 nautical mile
1.6449	= coefficient for 90% standard deviation
3.141 592 654	= π
*6 miles	= length, width, of normal township
*10 sq. chains (ch²)	= (Gunter's) 1 acre
*15° longitude	= width of one time zone = 360°/24 hr
15°F	changes length of 100-ft steel tape by 0.01 ft
*16½ ft	= 1 rod = 1 pole = 1 perch = ¼ ch (Gunter's)
*20°C	= standard temperature (Celsius) in taping = 68°F
23°26½′	= maximum declination of sun at solstices
$23^h56^m04.091^s$	= length of sidereal day in mean solar time, and $3^m55.909^s$ solar time short of mean solar day; also $3^m56.555^s$ sidereal time short of mean solar day
*24 hr	= 360° of longitude
*25.4 mm	= 1 in. (U.S. standard foot of 1959)
*36	= number of sections in normal township
*50	= Beaman arc reading for 0° vertical angle (30 in old arcs)
57°17′44.8″	= 1 radian (rad) = 57.295 779 51°
*66 ft	= length of Gunter's chain = 100 links (lk)
69.1 miles	≅ 1° latitude
*80 ch	= (Gunter's) 1 mile
100	= usual stadia ratio
101 ft	≅ 1 second (1″) of latitude
300	≅ stadia ratio in some precise levels
333⅓	≅ stadia ratio in some precise levels, for yard rods
*400 grads	= 360°
*480 ch	= width and length of normal township
490 lb/ft³	= density of steel for tape computations
*640 acres	= one normal section of 1 mile²
6076.10 ft	= 1 nautical mile
*3600/3937	= ratio U.S. yd/m for old legal (1866) and surveyor's foot
4,046.9 m²	= 1 acre
*6400 mils	= 360°
5,729.577 951 ft	= radius of 1° curve, arc definition
5,729.650 686 ft	= radius of 1° curve, chord definition
10,000 km	= distance from equator to pole (basis for length of meter)
*43,560 ft²	= 1 acre
206,264.806 25 sec	= 1 radian = cot 1 sec = 180°/π in sec
299,792.5 km/sec	= speed of light, and other electromagnetic waves, in vacuum
1,650,763.73	= wave lengths of krypton gas in vacuum, 1960 meter length
6,356,583.8 m	= earth's polar semi-axis (Clarke ellipsoid 1866)
6,378,206.4 m	= earth's equatorial semi-axis (Clarke ellipsoid 1866)
20,906,000 ft	= mean radius of earth = 3960 miles
29,000,000 lb/in.²	= Young's modulus of elasticity for steel

* Denotes exact value. All others correct to figures shown.

SEVENTH EDITION

ELEMENTARY SURVEYING

RUSSELL C. BRINKER
Adjunct Professor of Civil Engineering,
New Mexico State University
PAUL R. WOLF
Professor, Civil and Environmental Engineering
University of Wisconsin at Madison

HARPER & ROW, PUBLISHERS, New York
Cambridge, Philadelphia, San Francisco,
London, Mexico City, São Paulo, Sydney

Cover design: **DanielsDesign**

Sponsoring Editor: Cliff Robichaud
Project Editor: Robert Greiner
Designer: Michel Craig
Production Manager: Marion Palen
Compositor: Syntax International Pte. Ltd.
Printer and Binder: The Murray Printing Company
Art Studio: Vantage Art, Inc.

ELEMENTARY SURVEYING, Seventh Edition

Library of Congress Cataloging in Publication Data

Brinker, Russell C. (Russell Charles), 1908–
Elementary surveying.

Includes bibliographies and index.
1. Surveying. I. Wolf, Paul R. II. Title.
TA545.B86 1984 526.9 83-18462
ISBN 0-06-040982-7

ACKNOWLEDGMENTS

The authors wish to acknowledge the use of helpful suggestions, assistance, or pertinent material for this or previous editions, or both, given by Professors A. S. Cutler, O. S. Zelner, and L. F. Boon; C. B. Andrews; P. P. Rice; D. F. Griffin; A. S. Chase; L. Perez; Lt. Col. W. L. Baxter; E. G. Rich; J. P. Rastroni; D. V. Smith; E. C. Wagner, H. E. Kallsen, J. L. Clapp, R. B. Buckner, and S. D. Johnson; G. B. Lyon; W. A. Wintz, Jr.; D. C. McKee; J. M. DeMarche; C. F. Meyer; C. H. Drown; J. R. Coltharp; P. W. McDonnell, Jr.; J. O. Eichler; D. C. McNeese; D. A. Tyler; D. S. Turner; J. O. Meadows; E. F. Kuhlan; R. E. Hauck; K. S. Curtis; P. E. Borgo; P. B. Newlin; A. C. Kellie; C. E. Balleisen; W. I. Strong; E. F. Burkholder; H. Z. Lewis; A. P. Vonderohe, and D. F. Mezera. Also, W. C. Wattles, R. B. Irwin, T. E. Henderson, F. A. Sieker, F. P. Thomack, R. H. Holdridge, R. Minnick, J. M. Kessler, B. A. Dewitt, J. W. Schoonmaker, J. D. Henry, E. Gammon, R. J. Fish, and E. Zimmerman. Special thanks are given to Louise Shafer for her many contributions to this book.

Illustrative material and other help has been freely given by the U.S. Bureau of Land Management, U.S. Geological Survey, National Geodetic Survey, U.S. Soil Conservation Service, Defense Mapping Agency, and Technical Advisors, Inc.

Manufacturers of surveying equipment who provided photographs include the Keuffel & Esser Company; Kern Instruments, Inc.; Lietz Company; American Paulin System; Carl Zeiss Oberkochen; Wild Heerbrugg Instruments, Inc.; Lenker Manufacturing Company; Warren Knight Company;

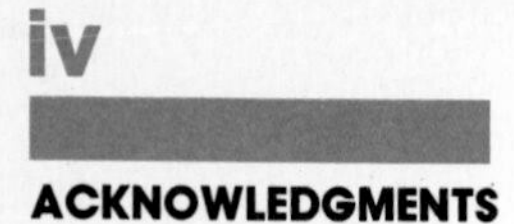

AGA Geodimeter, Inc.; Hewlett Packard, Inc.; Telludist, Inc.; Dietzgen, Inc.; W. & L. E. Gurley; Numonics Corporation; Magnivox; Owen Ayres & Associates, Inc.; Bausch & Lomb, Inc.; Shell Canada Resources, Ltd.; Kelsh Instrument Division, Danko Arlington, Inc.; and Benchmark Company.

CONTENTS

PREFACE

This Seventh Edition of *Elementary Surveying* follows the approach of previous editions in providing an easy-reading textbook containing basic theory and practical material for use in field and classroom. Inclusion of pertinent, less well-known facts, and emphasis on the professional aspects of surveying should stimulate interest in this historic profession. Discussions of advances in technology leading to improved methods and equipment were principal motivations for revising this book.

Chapters 1 through 17 provide material for a one-course program (project)—production of a topographic map—following the complete path through theory of errors, notekeeping, use of equipment, field methods, computations, and drafting procedures.

Chapters 18 through 28 fill out a second course of more advanced and special subjects, while giving one-course terminal students a view of what lies ahead. Material in the text, but not covered in the classroom, will still be valuable for future use, because surveying and engineering majors are likely to retain their college textbooks for later reference in professional work and to prepare for registration examinations.

Past editions, and this one, have benefitted from the ideas and reviews of numerous teachers and practitioners as indicated by the long list of names in the Acknowledgments section.

Among the many changes that improve and update the book are the following:

1. Three chapters (Linear Measurements, Leveling, and Transits and Theodolites) have been broken into two new chapters each in order to improve organization and to produce moderate subject lengths.
2. The chapters on Astronomical Observations, Construction Surveys, Circular Curves, Parabolic Curves, and Volumes have been substantially changed.
3. More emphasis is placed on newer types of equipment, such as "total station" instruments and digital theodolites with automatic readout. Photos and instrument properties have been updated.
4. Use of EDM instruments (EDMIs) in a "tracking mode" for construction stakeout is discussed.
5. Discussion of Inertial and Satellite Doppler systems has been increased, and material on the UTM Coordinate System introduced.
6. Three computer programs, written in BASIC, for traverse computations, and reduction of azimuth data for Polaris and Solar observations, are included in the Appendix, and example problems illustrate their use.
7. A new chapter on Coordinate Geometry in surveying computations has been added to the Appendix.
8. The new adjustments of the nation's horizontal and vertical datums, being made by the National Geodetic Survey, are described.
9. New sections in the Circular Curves chapter are Curve Layout by Deflection Angles with Electronic Tacheometers ("total station instruments"), and Intersection of Two Circular Curves.
10. An excellent computer-driven plot of a subdivision demonstrates the advances in this area since the Sixth Edition was published.
11. Recent data on government surveys and mapping projects are noted, and clarifying points on Public Lands Surveys added.
12. Methods of reducing EDM distances, to correct for a vertical offset of the EDM instrument (EDMI) and theodolite upon which it is mounted, have been added.
13. A brief section on observations for Latitude and Longitude extends coverage in the chapter on astronomy.
14. A new two-page Isogonic Chart replaces the old one.
15. A few more suggestions on recording notes were added to an already lengthy list.
16. Several more contour characteristics have been added to increase that comprehensive tabulation.
17. The recriprocal leveling noteform has been revised.
18. A list of selected relevant references has been added after each chapter.

Material on older-type instruments has been somewhat reduced but not eliminated. Steel tapes, dumpy levels, transits and planetables are still being manufactured, sold, and used by students in their first courses at many schools, and by people in the field. They illustrate basic fundamentals of surveying, such as the theory of errors, by introducing "hands-on" measurements. Judgment of beginners is enhanced by reading a vernier and tape, rather than pushing a button and getting the answer automatically selected and recorded.

Stress in the text continues to be placed on the theory of errors, and on correlation of theory and practical field methods. Nearly 900 end-of-chapter problems are included, and answers to approximately a fourth of these are given in the back of the book to assist students in self-study.

Engineers, architects, geologists, and foresters must be capable of making measurements and analyzing the precision and accuracy of results obtained by other people. They should be qualified to properly locate and set machinery, lay out buildings and other common structures, and understand and prepare simple topographic maps. Each of these areas is discussed, and proper field procedures to obtain a desired precision noted.

A few references are made to costs so that students will learn early in their college work to associate the three bases of surveying and engineering practice: theory, applications, and costs. All surveying is a constant struggle to reduce or isolate errors and mistakes. At the end of most chapters, students are reminded of this point through lists of typical errors and mistakes.

Although all seven editions retain the title *Elementary Surveying* (thereby avoiding going back to the "first" edition again), the contents go beyond the elementary stage in length and scope. The large number of chapters, however, permit inclusion or omission of subjects to correspond with the class time available for students in surveying, civil engineering, other engineering curricula, architecture, geology, agriculture, and forestry.

Chapters are arranged in the order found most convenient at numerous colleges. Fundamental material is collected in the first seventeen chapters. Theory and use of the fundamental ground surveying instruments—the tape, EDM equipment, level, transit, theodolite, and planetable—are described in detail, and new types of equipment noted. Any chapter following Chapter 17 can be omitted without loss of continuity, although several are short enough to be suitable for a single assignment.

Limited coverage of such subjects as field astronomy, boundary surveys and photogrammetry is given to fit various programs offered. For example, the brief chapter on boundary surveys is intended to make students aware of a few problems involved in the survey and transfer of property, and the legal requirements of professional registration. Some instructors give broad survey-type courses and want their students to get an overall view of the many surveying functions. It is believed that the arrangement and scope of material presented herein will meet that need also.

Taping, electronic distance measurement, leveling, and use of the transit and theodolite are taken up in order because students find it easier to acquire some facility with the equipment in this sequence. Also, this arrangement permits the start and continuation of field work with a minimum of preliminary lecture time, and makes it possible to begin effective computation and drafting after just a few periods in the field if bad weather is encountered.

The difficulty in getting through all the background material (basic concepts of the profession, history, theory of errors, and notekeeping methods) before starting field work in the first week (other than pacing) is recognized.

Nevertheless, the authors believe these topics should precede the theory and use of instruments

The subject of noteforms—an important part of surveying and engineering—is discussed in a separate chapter. Most sample noteforms are collected in Appendix D, rather than scattered throughout the book so that they may be found easily.

PART ONE

1 INTRODUCTION

1-1. DEFINITION OF SURVEYING. Surveying has traditionally been defined as the science and art of determining relative positions of points above, on, or beneath the surface of the earth, or establishing such points. In a more general sense, however, surveying can be regarded as that discipline which encompasses all methods for gathering and processing information about the physical earth and environment. Conventional ground systems are now supplemented by aerial and satellite surveying methods, which evolved through the defense and space programs.

In general, the work of a surveyor can be divided into five parts:

1. *Research analysis and decision making.* Selecting the survey method, equipment, most likely corner locations, and so on.
2. *Field work or data acquisition.* Making measurements and recording data in the field.
3. *Computing or data processing.* Performing calculations based on the recorded data to determine locations, areas, volumes, and so on.
4. *Mapping or data representation.* Plotting measurements or computed values to produce a map, plat, or chart, or portraying the data in a numerical or computer format.
5. *Stakeout.* Setting monuments and stakes to delineate boundaries or guide construction operations.

Descriptions of the variety of instruments and field and office procedures used by surveyors in accomplishing these tasks are given in this book.

1-2. IMPORTANCE OF SURVEYING. Surveying is one of the oldest and most important arts practiced by man because from the earliest times it has been necessary to mark boundaries and divide land. Surveying has now become indispensable to our modern way of life. The results of today's surveys are being used to (a) map the earth above and below sea level; (b) prepare navigational charts for use in the air, on land, and at sea; (c) establish property boundaries of private and public lands; (d) develop data banks of land-use and natural resource information which aid in managing our environment; (e) determine facts on the size, shape, gravity, and magnetic fields of the earth; and (f) prepare charts of our moon and planets.

Surveying continues to play an extremely important role in many branches of engineering. For example, surveys are required to plan, construct, and maintain highways, railroads, rapid-transit systems, buildings, bridges, missile ranges, launching sites, tracking stations, tunnels, canals, irrigation ditches, dams, drainage works, urban land subdivisions, water supply and sewage systems, pipelines, and mine shafts. Surveying or surveying methods are commonly employed in laying out assembly lines and jigs, fabricating and placing large equipment, providing control for aerial photography, and in many related tasks in agronomy, archeology, astronomy, forestry, geography, geology, geophysics, landscape architecture, meteorology, paleontology, and seismology, but particularly in military and civil engineering. Optical alignment is an application of surveying in shop practice (installation of machinery, fabrication of airplanes, etc.).

All engineers must know the limits of accuracy possible in construction, plant design and layout, and manufacturing processes, even though someone else may do the actual surveying. In particular, surveyors and civil engineers who are called upon to design and plan surveys must have a thorough understanding of the methods and instruments used, including their capabilites and limitations. This knowledge is best obtained by making measurements with the kinds of equipment used in practice to get a true concept of the theory of errors, and the small but recognizable differences that occur in observed quantities.

In addition to stressing the need for reasonable limits of accuracy, surveying emphasizes the value of significant figures. Surveyors and engineers must know when to work to hundredths of a foot instead of to tenths or thousandths, or perhaps the nearest foot, and what precision in field data is necessary to justify carrying out computations to the desired number of decimal places. With experience, they learn how available equipment and personnel govern procedures and results.

Neat sketches and computations are the mark of an orderly mind, which in turn is an index of sound engineering background and competence. Taking field notes under all sorts of conditions is excellent preparation for the kind of recording and sketching expected of all engineers. Additional training that has a carry-over value is obtained in arranging computations properly.

Engineers who design buildings, bridges, equipment, and so on are fortunate if their estimates of loads to be carried are correct within 5%. Then a factor of safety of 2 or more is applied. But except for topographic work, only exceedingly small errors can be tolerated in surveying, and there is no factor

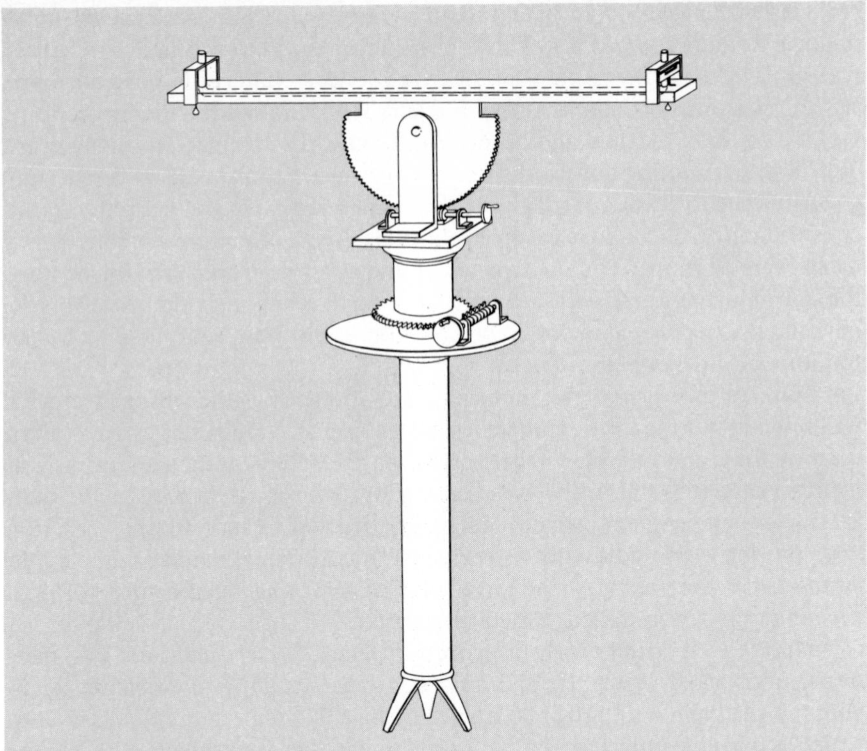

Figure 1-1. The diopter.

of safety. Traditionally, therefore, surveying stresses both manual and computational precision.

1-3. HISTORY OF SURVEYING. The oldest historical records in existence today which bear directly on the subject of surveying state that this science had its beginning in Egypt. Herodotus says Sesostris (about 1400 B.C.) divided the land of Egypt into plots for the purpose of taxation. Annual floods of the Nile River swept away portions of these plots, and surveyors were appointed to replace the bounds. These early surveyors were called *rope-stretchers*, since their measurements were made with ropes having markers at unit distances.

As a consequence of this work, early Greek thinkers developed the science of geometry. Their advance, however, was chiefly along the lines of pure science. Heron stands out prominently for applying science to surveying in about 120 B.C. He was the author of several important treatises of interest to surveyors, including *The Dioptra*, which related the methods of surveying a field, drawing a plan, and making calculations. It also described one of the first pieces of surveying equipment recorded, the *diopter* (Figure 1-1). For many years Heron's work was the most authoritative among Greek and Egyptian surveyors.

Significant development in the art of surveying came from the practical-minded Romans, whose best-known writing on surveying was by Frontinus. Although the original manuscript disappeared, copied portions have been preserved. This noted Roman engineer and surveyor, who lived in the first century, was a pioneer in the field and his essay remained the standard for many years.

The engineering ability of the Romans was demonstrated by their extensive construction work throughout the empire. Surveying necessary for this construction resulted in the organization of a surveyors' guild. Ingenious instruments were developed and used. Among these were the *groma*, used for sighting; the *libella*, an A frame with a plumb bob, for leveling; and the *chorobates*, a horizontal straightedge about 20 ft long with supporting legs and a groove on top for water to serve as a level.

One of the oldest Latin manuscripts in existence is the *Codex Acerianus*, written in about the sixth century. It contains an account of surveying as practiced by the Romans and includes several pages from Frontinus's treatise. The manuscript was found in the 10th century by Gerbert and served as the basis for his text on geometry, which was largely devoted to surveying.

During the Middle Ages, Greek and Roman science was kept alive by the Arabs. Little progress was made in the art of surveying, and the only writings pertaining to it were called "practical geometry."

In the 13th century Von Piso wrote *Practica Geometria*, which contained instructions on surveying. He also authored *Liber Quadratorum*, dealing chiefly with the *quadrans*, a square brass frame having a 90° angle and other graduated scales. A movable pointer was used for sighting. Other instruments of the period were the *astrolabe*, a metal circle with a pointer hinged at its center and held by a ring at the top, and the *cross staff*, a wooden rod about 4 ft long with an adjustable cross arm at right angles to it. The known lengths of the arms of the cross staff permitted distances to be measured by proportion and angles.

Early civilizations assumed the earth to be a flat surface, but by noting the earth's circular shadow on the moon during lunar eclipses and watching ships gradually disappear as they sailed toward the horizon, it was slowly deduced that the planet actually curved in all directions.

Determining the true size and shape of the earth has intrigued humans for centuries. History records that a Greek named Eratosthenes, about 220 B.C., first attempted to compute its dimensions. He ascertained the angle subtending the meridian arc between Syene and Alexandria in Egypt by measuring shadows cast by the sun at these cities. The arc length was found by multiplying the number of caravan days between Syene and Alexandria by the average daily distance traveled. From the angle and arc measurements, applying elementary geometry, Eratosthenes calculated the earth's circumference to be about 25,000 mi. Subsequent precise geodetic measurements using better instruments and a technique equivalent geometrically to Eratosthenes's have shown his value, though slightly too large, to be amazingly close to the current accepted one. Actually, of course, the earth approximates an oblate spheroid having an equatorial radius about $13\frac{1}{2}$ miles longer than the polar radius.

In the 18th and 19th centuries the art of surveying advanced more rapidly. The need for maps and location of national boundaries caused England and France to make extensive surveys requiring accurate triangulation; thus geodetic surveying began. The U.S. Coast and Geodetic Survey (now the National Geodetic Survey of the U.S. Department of Commerce) was established by an act of Congress in 1807. Initially its charge was to perform hydrographic surveys and prepare nautical charts. Later its activities were expanded to include establishment of control monuments throughout the country.

Increased land values and the importance of exact boundaries, along with the demand for public improvements in the canal, turnpike, and railroad eras, brought surveying into a prominent position. More recently, the large volume of general construction, numerous land subdivisions with better records required, and demands posed by the fields of exploration and ecology have entailed an augmented surveying program. Surveying is still the sign of progress in the development and use of the earth's resources.

During World Wars I and II, and the Korean and Vietnam conflicts, surveying in its many phases played an important role because of the stimulus provided to improve instruments and methods used to make measurements and maps. Progress continued into the space program where new equipment and systems were needed to supply precise control for missile alignment and moon mapping of proposed landing sites. Electronic distance-measuring (EDM) equipment, laser devices, north-seeking gyroscopes, improved aerial cameras, helicopters, inertial and doppler surveying systems, remote sensors, and various-size computers are but a few products of today's technology now being directly applied in modern surveying with terrific impact. Landsat spacecraft now provide images of global coverage every 18 days for down-to-earth projects such as land-cover inventories, natural resource mapping, water quality assessment, and flood control.

Traditional surveying instruments—transit, level, and steel tape—are now frequently supplanted by the theodolite, automatic level, electronic distance-measuring equipment (Figure 1-2), and aerial camera (see Figure 28-1). In the mapping field, except for small areas, photogrammetry has replaced ground surveys on many kinds of projects. But conventional ground methods are still essential for establishing locations of horizontal and vertical control points, property corners, and construction layout.

1-4. GEODETIC AND PLANE SURVEYS. Two general classifications of surveys are *geodetic* and *plane*. They differ principally in the assumptions on which the computations are based, although field measurements for geodetic surveys are sometimes performed to a higher order of accuracy than those for plane surveys.

In geodetic surveying, the curved surface of the earth is considered by performing the computations on a *spheroid* (curved surface approximating the size and shape of the earth). It is now becoming common to do geodetic computations in a three-dimensional, earth-centered Cartesian coordinate system. The calculations involve solving equations derived from spherical trigonometry and calculus. Geodetic methods are employed to determine relative positions of widely spaced monuments and compute lengths and directions of the long lines between them. These monuments serve as the basis for referencing other subordinate surveys of lesser extent.

In the past, field measurements for geodetic surveys consisted primarily of angles observed using ground-based theodolites and distances measured with

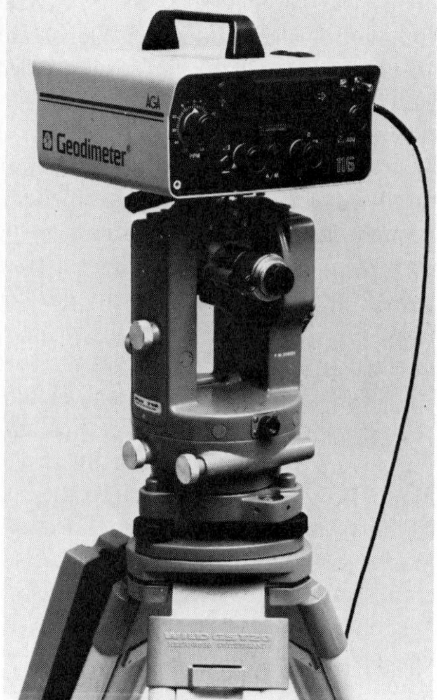

Figure 1-2. Wild T-2 theodolite (**left**), Geodimeter 116 electronic distance measurement instrument (**right**), and Lietz B2C automatic level (**below**). (Courtesy Wild Heerbrugg Instruments, Inc.; AGA Geodimeter, Inc.; and Lietz Company.)

tapes or electronic devices. Although these types of measurements are still used, recently some rather exotic equipment has been developed to supplement geodetic surveys. This includes doppler systems, which measure changes in radio signals broadcast from satellite-borne transmitters; and inertial systems that are carried in helicopters or ground vehicles and employ gyroscopes and accelerometers to measure positional changes in northing, easting, and elevation. These new systems, described in more detail in Chapter 20, are capable of achieving high accuracies in shorter time periods.

In plane surveying, except for leveling, the reference base for field work and computations is assumed to be a flat horizontal surface. The direction of a plumb line (and thus gravity) is considered parallel throughout the survey region, and all measured angles are presumed to be plane angles. For areas of limited size, the surface of our vast spheroid is actually very nearly flat. On a line 5 mi long, the spheroid arc and chord lengths differ by only about 0.02 ft. A plane surface tangent to the spheroid has departed only about 8 in at 1 mi from the point of tangency. In a triangle having an area of 75 mi^2, the difference between the sum of the three spheroidal angles and three plane angles is about 1 sec. It is evident, therefore, that except in surveys covering extensive areas, the earth's surface can be approximated as a plane, thus simplifying computations and techniques. In general, algebra, plane and analytical geometry, and plane trigonometry are used. Even for very large areas, such as those involved in state plane coordinate systems described in Chapter 21, plane surveying can be used, with some allowances being made in computations. Use of state plane coordinates is a shortcut to accomplishing geodetic computations. This book concentrates primarily on methods of plane surveying, an approach that satisfies the requirements of most projects.

1-5. SPECIALIZED SURVEYS. Many types of surveys are so specialized that a person proficient in a particular discipline may have little contact with the other areas. Persons seeking careers in surveying and mapping, however, should be knowledgeable in every phase, since all are closely related in modern practice. Some important classifications are described briefly here.

Control surveys establish a network of horizontal and vertical monuments that serve as a reference framework for other surveys.

Topographic surveys determine locations of natural and artificial features and elevations used in map making.

Land, boundary, and *cadastral surveys* are (usually) closed surveys to establish property lines and corners. The term *cadastral* is now generally applied to surveys of the Public Lands Systems. There are three major categories: *original surveys* to establish new section corners in unsurveyed areas that still exist in Alaska and several western states; *retracement surveys*, which recover previously established boundary lines; and *subdivision surveys* to establish monuments and delineate new parcels of ownership.

Hydrographic surveys define shorelines and depths of lakes, streams, oceans, reservoirs, and other bodies of water. *Sea surveying* is associated with port and offshore industries and the marine environment, including measurements and marine investigations made by shipborne personnel.

Route surveys are made to plan, design, and construct railroads, highways, pipelines, and other linear projects. They normally begin at a control point and progress to another control point in the most direct manner permitted by field construction.

Construction surveys, run while construction is in progress, control elevations, horizontal positions, dimensions, and configurations. Also, they secure essential data for computing construction pay quantities.

As-built surveys provide exact final location and layout of engineering works, their positional verification, and records that include design changes.

Mine surveys are performed above and below ground to guide tunneling and other operations associated with mining, including geophysical surveys for mineral and energy resource exploration.

Solar surveys map property boundaries, solar access easements, position obstructions and collectors according to sun angles, and meet other requirements of zoning boards and title insurance companies.

Optical tooling (also referred to as *industrial surveying* or *optical alignment*) is a method of making extremely accurate measurements for manufacturing processes where small tolerances are required.

Except for control surveys, most other types described are usually performed using plane surveying procedures, but geodetic methods may be employed on the others if a survey covers an extensive area and/or requires extreme accuracy.

Ground and aerial (*or photogrammetric*) *surveys* are broad classifications sometimes used. Ground surveys utilize measurements made with ground-based equipment such as tapes, electronic distance-measuring devices, levels, and theodolites. Aerial or photogrammetric surveys employ cameras and other sensors carried in aircraft to get data for studies and mapping. Procedures for securing and reducing aerial data are described in Chapter 28. Aerial surveys have been used in all the specialized types of surveying listed except for optical tooling, and in this area *terrestrial* (ground-based) photographs are often used.

1-6 PRESENT STATUS OF SURVEYING. There is an increasing demand for good maps in the United States, and various government agencies are attempting to provide them though handicapped by insufficient funds and personnel. A common misconception is that the entire country has been adequately mapped by now. Actually, about 95% of the United States is covered by mapping at scales of 1 mi to the inch or larger. At the present rate of new map production and periodic revision of satisfactorily mapped areas, complete coverage will be attained in the early 1980s. The primary basic map series is on a $\frac{1}{24,000}$ scale.

Accurate mapping cannot be done without good *basic control.* National survey nets for geodetic horizontal and vertical control are continually being extended over the country, primarily by the U.S. National Geodetic Survey, for the control of nautical charts and topographic maps, and to provide coordinated position data for all surveyors. The horizontal-control survey net consists of arcs of first-order and second-order triangulation, lines of first-order and second-order traverse, and first- and second-order trilateration (see Chapter 20). The data from this net are coordinated and correlated on the North American datum of 1927 extending through Alaska and Central America.

The vertical-control survey net consists of bench marks of known elevations located throughout the country. Elevations are referred to the National Geodetic Vertical Datum (NGVD) of 1929.

Four U.S. government agencies do surveying and mapping on a large scale.

1. The Coast and Geodetic Survey, now National Geodetic Survey (NGS) and part of the National Ocean Survey (NOS), was organized to map the coast.

Its activities include triangulation, traverse, trilateration and precise leveling to extend control, preparation of nautical and aeronautical charts, photogrammetric surveys, tide and current studies, collection of magnetic data, gravimetric surveys, and worldwide control survey operations that involve satellites. The basic control points established by this organization are the foundation for all large-area surveying.

2. The General Land Office, now Bureau of Land Management (BLM), established in 1812, directs the public lands surveys. Lines and corners have been set for most public lands in the conterminous United States, but much work remains in Alaska and is proceeding with "modern techniques."

3. The U.S. Geological Survey (USGS), established in 1879, will ultimately map the entire country. Its standard $7\frac{1}{2}$-min and 15-min quadrangle maps show topographic and cultural features, and are suitable for general use plus a variety of engineering and scientific purposes. More than 9.5 million copies are distributed each year.[1]

4. The Defense Mapping Agency (DMA) prepares maps, associated products, and provides services for the Department of Defense and all land combat forces. It is divided into the following military mapping groups: Aerospace Center, Defense Mapping School, Hydrographic Center, Inter-American Geodetic Survey, and Topographic Center. The DMA Topographic Center fulfills a key mission in an era when accurate mapping, charting, and geodesy products are essential to realize the complete potential of new weapons. Technological advances in weaponry demand corresponding improvements in mapping, charting, and geodesy to obtain accuracies that were just dreams only a few years ago.

In addition, units of the Corps of Engineers, U.S. Army, have made extensive surveys for emergency and military purposes. Some of these surveys provide data for engineering projects, such as those connected with flood control.

Extensive surveys have also been conducted for special purposes by nearly 40 federal agencies, including the Forest Service, National Park Service, International Boundary Commission, Bureau of Reclamation, Tennessee Valley Authority, Mississippi River Commission, U.S. Lake Survey, and Department of Transportation. Likewise, many cities, counties, and states have had extensive surveying programs, as have various utilities.

1-7. THE SURVEYING PROFESSION. Land or boundary surveying is classified as a learned profession because the modern practitioner needs a wide background of technical training and experience and must exercise independent judgment. Registered (licensed) professional surveyors must have a thorough knowledge of mathematics—particularly geometry and trigonometry with

[1] As of this writing, mail orders for maps covering areas east of the Mississippi River should be addressed to the Branch of Distribution, U.S. Geological Survey, 1200 South Eads Street, Arlington, Virginia 22202, and for locations west of the Mississippi River to Branch of Distribution, USGS, Federal Center, Denver, Colorado 80225.

some calculus; a solid understanding of surveying theory, instruments, and methods in the areas of geodesy, photogrammetry, remote sensing, cartography, and computers; some competence in economics (including office management), geography, geology, astronomy, and dendrology; and a familiarity with laws pertaining to land and boundaries. They should be knowledgeable in field operations and computations and able to do neat drafting. Above all, they are governed by a professional code of ethics and are expected to charge reasonable fees for their work.

The personal qualifications of surveyors are as important as their technical ability in dealing with the public. They must be patient and tactful with clients and their sometimes hostile neighbors. Few people are aware of the painstaking research of old records required before field work is started. Diligent, time-consuming effort may be needed to locate corners on nearby tracts for checking purposes as well as to find corners for the property in question.

Permission to trespass on private property or to cut obstructing tree branches and shrubbery must be obtained through a proper approach. Such privileges are not conveyed by a surveying license or by employment in a state highway department (but a court order can be secured if a landowner objects to necessary surveys).

All 50 states, Guam, and Puerto Rico have registration laws for professional surveyors and engineers (as do the provinces of Canada). Some states presently have separate licensing boards for surveyors. In general, a surveyor's license is required to make property surveys, but not for construction, topographic, or route work unless boundary corners are set.

To qualify for registration as either a professional Land Surveyor (LS) or an Engineer (PE) it is necessary to have an appropriate college degree, although some states allow relevant experience in lieu of formal education. In addition, candidates take a Surveyor-in-Training (SIT) or Engineer-in-Training (EIT) test, acquire 2 or more years of additional practical experience, and then pass a 2-day written examination. Within a few years, an acceptable technical degree is expected to become a prerequisite for new registrants, and Continuing Education (CE) credits are required for registration renewal.

A $1\frac{1}{2}$-day national LS examination on fundamentals, principles, and the practice of land surveying is now used by most states. The remaining $\frac{1}{2}$ day is devoted to local legal customs and aspects. Thus, transfer of registration from one state to another has become easier.

Typical state laws require an LS to sign all plats, assume responsibility for any liability claims, and be an *active part* of the field survey. Depending on various conditions, an LS may be able to supervise only two or three parties. Thus, more surveyors with management ability will have to become registered.

A 1981 estimate of the number of surveyors in private practice in the United States was 62,000, with 42,000 of them registered. These are small numbers for a country of 3.6 million mi^2 and over 220 million people!

1-8. FUTURE CHALLENGES IN SURVEYING. Surveying is in the early stages of a revolution in the way data are stored, retrieved, and shared. This is due to developments in computer technology. The demands on surveyors will be very different in a few years from what they are now.

The National Geodetic control network must be maintained, new adjustments made (as is being done presently), and supplemented to meet the requirements of high-order future surveys. New topographic maps with larger scales are necessary for better planning and design. Existing maps of our rapidly expanding urban areas need revision and updating to reflect changes. Long-range planning and assessment of environment impacts of proposed construction projects call for maps and/or data banks that contain a variety of land information such as ownership, location, acreage, soil types, land uses, and natural resources. Cadastral surveys of the yet unsurveyed public lands are essential. Monuments set many years ago by the original surveyors have to be recovered and remonumented for preservation of property boundaries. Appropriate surveys with very demanding accuracies are necessary to position drilling rigs as mineral and oil explorations press farther offshore. And in the space program, the desire for maps of neighboring planets will continue.

These and other opportunities offer professionally rewarding indoor or outdoor life, or both, for numerous people with suitable training in the various branches of surveying.

NOTE: Answers for these problems, and some in later chapters, can be obtained by consulting a dictionary, the bibliographies, later chapters, and professional surveyors.

PROBLEMS

1-1. In what other contexts and occupations is the term *survey* employed?
1-2. List 10 uses for surveying other than property and construction surveys.
1-3. Describe some surveying applications in forestry and mining.
1-4. Name two uses for lasers other than in surveying.
1-5. How is surveying used in modern "dry" farming and contour plowing?
1-6. What methods and instruments are used to control locations of offshore oil leases and platforms?
1-7. How did the early Romans control elevations for the long, large aqueducts built to bring water to their cities?
1-8. Why is it necessary to make accurate surveys of underground mines?
1-9. What are patent surveys? Reconnaissance surveys?
1-10. How are surveying methods (optical tooling) applied in fabricating huge ships and planes?
1-11. List at least five weather conditions that affect survey accuracies.
1-12. What astronomical observations are made by surveyors?
1-13. Why should a surveyor have a working knowledge of geology? Of dendrology?
1-14. Which kinds of surveys discussed in Section 1–5 would you classify as "engineering" surveys?
1-15. Five parts of surveying are listed in Section 1-1. Which do you think requires the most experience? Why?
1-16. Suggest a simple method for leveling a garage floor 20 × 22 ft prior to concreting if a transit or level is not available.
1-17. What surveying measurements does a contractor need to lay a 36-in.-diameter sewer line that are not required for an 8-in. water main?
1-18. Describe the purpose of horizontal lines in some binoculars.
1-19. What organizations in your state will furnish maps and surveying reference data to surveyors and engineers?
1-20. Why should the purchaser of a farm, city lot, or home demand a survey before making final payment?

1-21. Do the subdivision laws of your community specify the accuracy required for surveys made to lay out a subdivision? If so, what limits are set?
1-22. Can mechanics' liens be used by surveyors and civil engineers? Explain.
1-23. List the legal requirements for registration as a land surveyor in your state.
1-24. How does every surveyor benefit from the U.S. satellite program?
1-25. Explain why an aerial photograph is not a map. How is it converted to a map?

BIBLIOGRAPHY

American Society of Civil Engineers and American Congress on Surveying and Mapping. 1978. *Definitions of Surveying and Mapping Terms*, manual no. 34.

Brinker, R. C. 1978. 4567 *Review Questions for Surveyors*, 11th ed. R. C. Brinker, Box 1399, Sun City, AZ 85372.

Brown, C. M. 1971. "Surveyors Service to Society." *Surveying and Mapping* 31(no. 3): 439.

Frederick, D. G. 1981. "The Mandate of the Public Agencies in the United States." *Surveying and Mapping* 41(no. 1):47.

Knudsen, J. D. 1980. "Alaska Mechanic's Lien Law." *Surveying and Mapping* 40(no. 1): 69.

Laferriere, J. A. 1971. "Importance of Mapping in a Modern Society." *Surveying and Mapping* 31(no. 4):581.

Peyton, H. J., Jr. 1951. "Early Development of Horizontal Angle-Measuring Surveying Instruments." *Surveying and Mapping* 11(no. 4):381.

Pryor, W. T. 1975. "Metrication." *Surveying and Mapping* 35(no. 3):229.

Quinn, A. O. 1981. "A Report of the Activities of the Private Sector in the Surveying and Mapping Profession." *Surveying and Mapping* 41(no. 1):55.

Ridgeway, H. H. 1982. "Surveying: The Profession and Its Requirements." *ASCE Journal of the Surveying and Mapping Division* 108(SU1):18.

Rutscheidt, E., and C. Andregg. 1971. "Instruments and Methods for Surveying and Mapping." *Surveying and Mapping* 31(no. 2):271.

Southard, R. B. 1980. "The Changing Scene in Surveying and Mapping." *Surveying and Mapping* 40(no. 4):397.

"Surveyors Speak Out." 1982. *Point of Beginning*, Aug.–Sept., p. 42.

Whalen, C. T. 1982. "The New Adjustment of the North American Vertical Datum." *American Congress of Surveying and Mapping Bulletin* 78:39.

2 THEORY OF MEASUREMENTS AND ERRORS

2-1. INTRODUCTION. Making measurements and the subsequent computations utilizing them are fundamental tasks of surveyors. The process requires a combination of human skill and mechanical equipment applied with the utmost judgment. Experience and good physical conditions improve the human factor; superior equipment enables good operators to do better work with more consistent results and in less time. Measurements are never exact, however, and no matter how carefully made, they will always contain errors.

Surveyors, whose work must be performed to exacting standards, should therefore thoroughly understand the different kinds of errors, their sources and expected magnitudes under varying conditions, and manner of propagation. Only then can they select instruments and procedures necessary to reduce error sizes to within tolerable limits.

Of equal importance, surveyors must be capable of assessing the magnitudes of errors in their measurements so either they can be accounted for in calculations or, if necessary, new ones taken. The design of measurement programs, comparable to other engineering design, is now practiced. Matrix algebra and the electronic computer are two tools used to plan measurement projects and investigate and distribute errors after results have been obtained.

2-2. TYPES OF MEASUREMENTS IN SURVEYING. Five kinds of measurements illustrated in Figure 2-1 form the basis of plane surveying: (1) horizontal angles, (2) horizontal distances, (3) vertical angles, (4) vertical distances, and (5) slope distances. Horizontal angles, as angle *AOB*, and horizontal distances, *OA* and

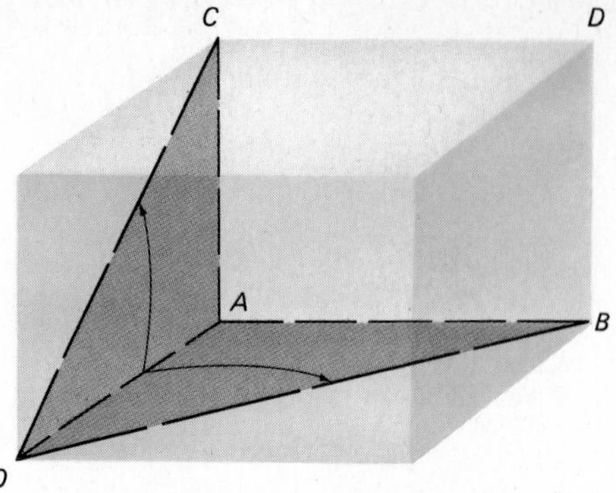

Figure 2-1. Kinds of measurements in surveying.

OB, are measured in horizontal planes; vertical angles, like *AOC*, in vertical planes. Vertical lines, *AC* and *BD*, are measured in the direction of gravity; slope distances, *OC*, determined along inclined planes. Using combinations of these basic measurements, relative positions between any points can be computed.

2-3. UNITS OF MEASUREMENT. The units of measurement in surveying are those for *length*, *angle*, *area*, and *volume*. The *English system* length unit of *foot* has been most commonly used in the United States. It bears a definite relation to the *meter*, the basis of the *metric system*. Originally the meter was defined as $\frac{1}{10,000,000}$ of the earth's meridional quadrant. When the metric system was legalized for use in the United States in 1866, a meter was defined as the interval under certain physical conditions between lines on an International Prototype bar made of 90% platinum and 10% iridium, and accepted as equal to 39.37 in. A copy of the bar is held by the U.S. Bureau of Standards and compared periodically with the international standard stored in France.

In October 1960, at the General Conference on Weights and Measures (CGPM), the United States and 35 other nations agreed to redefine the meter in terms of the wavelength of a certain kind of light. Now a meter is equal to the length of 1,650,763.73 waves of the orange-red light produced by burning the element krypton (Kr-86). The new definition permits industries to make more accurate measurements and check their own instruments without recourse to the standard meter-bar in Washington. The wavelength of orange-red krypton light is a true constant, whereas there is a risk of instability in the metal meter-bar. If the CGPM had been one year later, the laser might well have become the standard instead of krypton light.

Surveys in the United States are most commonly made in linear units of feet and *decimals* of a foot, although use of the metric system is gradually increasing. In construction, on the other hand, feet and *inches* are often used. Because surveyors provide measurements for developing construction plans and guiding building operations, they must understand all the various systems of units, and be capable of making conversions between them. Caution must always be exercised to ensure that measurements are recorded in their proper units and conversions correctly made.

Units of length used in past and present surveys in the United States include the following:

1 foot = 12 inches
1 yard = 3 feet
1 meter = 39.37 inches = 3.2808 feet[1]
1 rod = 1 pole = 1 perch = $16\frac{1}{2}$ feet
1 vara = approximately 33 inches (an old Spanish unit often encountered in the southwest United States)
1 Gunter's chain = 66 feet = 100 links (lk) = 4 rods
1 engineer's chain = 100 feet = 100 links (lk)
1 mile = 5280 feet = 80 Gunter's chains (ch)
1 nautical mile = 6076.10 feet
1 fathom = 6 feet

A common unit of area is the *acre.* Ten square chains (Gunter's) equal 1 acre. Thus an acre contains 43,560 ft^2. In the metric system, area is commonly given in *hectares.* One hectare is equal to 10,000 m^2 or 2.471 acres. The *arpent* (equal approximately to 0.85 acre but varies somewhat in different states) was used in land grants of the French crown; when employed as a linear term, it refers to the length of a side of 1 square arpent.

The unit of angle used in surveying is the *degree,* defined as $\frac{1}{360}$th of a circle. One degree (1°) equals 60 min, and 1 min equals 60 sec. Divisions of seconds are given in tenths, hundredths, and thousandths. Other methods have also been used to subdivide a circle, for example, 400 *grad* (with 100 *centesimal min/grad* and 100 *centesimal sec/min*) and 6400 *mil.*

The radian is an angle subtended by an arc of a circle having a length equal to the radius of the circle. Obviously, 2π rad = 360°, 1 rad = 57°17′44.8″ = 57.2958°, and 0.01745 rad = 1°.

2-4. INTERNATIONAL SYSTEM OF UNITS (SI).[2]

A worldwide movement is now under way to adopt the *International System of Units,* generally known as *SI.* The system, which involves no changes in dimensions or values, is being advocated as a means of standardizing and simplifying units of measurement throughout the world. Units in the SI system of major concern to surveyors (standard symbols given in parentheses) are the *meter* (m) for distances, *radian* (rad) for plane angles, *square meter* (m^2) for areas, and *cubic meter* (m^3) for volumes.

[1] The foot originally adopted in the United States was based on 39.37 in being equal exactly to 1 m, and thus 1 ft equal to 0.3048006 m. In 1959, however, the United States officially adopted the inch as equal to exactly 2.54 cm, so under this standard 1 ft equals 0.3048000 m. This difference in standards amounts to one part in 500,000, or 1 ft in 99.6969 . . . mi. Since all surveying prior to 1959 was done on the original standard, it would have been confusing to change. Therefore the original basis, now called the *U.S. Survey Foot,* has been officially adopted for surveying.

[2] A pamphlet entitled "International Standard ISO 1000," which describes the SI system in detail, is available at nominal cost from the American National Standards Institute, 1430 Broadway, New York, NY 10018.

Subdivisions of the meter are the *millimeter* (mm), *centimeter* (cm), and *decimeter* (dm), equal to 0.001 m, 0.01 m, and 0.1 m, respectively. A *kilometer* (km) equals 1000 m, or approximately five-eighths of a mile. Degrees, minutes, and seconds are also acceptable subdivisions of the radian for plane angles measured in SI.

Transition from English to SI units has begun in the United States, but will require a gradual and lengthy process to complete because the public must be educated to the system and become accustomed to it. The English system can never be completely replaced because our archives will always contain volumes of valuable data recorded in these units. Original land descriptions for title insurance investigations must continue giving lengths in feet, just as some still carry distances in chains and links.

For surveyors the transition will likewise be a long one because many surveying instruments are graduated in English units, and changes must be made at a pace equal to the acceptance rate by the profession, clients, and general public. Although conversion will be difficult, and at times confusing, surveyors presently burdened with theory and computations in the archaic and awkward yard, foot, and inch units should welcome the change.

This book uses both English and SI units in discussion and example problems.

2-5. SIGNIFICANT FIGURES. In recording measurements, an indication of the accuracy attained is the number of digits (significant figures) recorded. By definition, the number of significant figures in any value includes the positive (certain) digits plus one (*only one*) digit that is estimated and, therefore, questionable. For example, a distance recorded as 873.52 ft is said to have five significant figures; in this case the first four digits are certain and the last digit is questionable. To be consistent with the theory of errors, it is essential that data be recorded with the correct number of significant figures. If a significant figure is dropped off in recording a value, the time spent in acquiring certain accuracy has been wasted. On the other hand, if data are recorded with more figures than those that are significant, false accuracy will be implied and time may be wasted in making computations.

The number of significant figures is often confused with the number of decimal places. Decimal places may have to be used to maintain the correct number of significant figures, but in themselves they do not indicate significant figures. Some examples follow:

Two significant figures: 24, 2.4, 0.24, 0.0024, 0.020
Three significant figures: 364, 36.4 0.000364, 0.0240
Four significant figures: 7621, 76.21, 0.0007621, 24.00

Zeros at the end of an integral value may cause difficulty because they may or may not indicate significant figures. In the value 2400 it is not known how many figures are significant; there may be two, three, or four. One method of eliminating this uncertainty is to place a bar over the last significant figure, as in $240\bar{0}$, $24\bar{0}0$, or $2\bar{4}00$. Another method is to express the value in terms of powers of 10; the significant figures in the measurement are then written as a number between 1 and 10, including the correct number of zeros at the end, and the decimal point is placed by annexing a power of 10. As an example, 2400 becomes $2.400 \times (10)^3$ if both zeros are significant, $2.40 \times (10)^3$ if one is, and $2.4 \times (10)^3$ if there are only two significant figures.

In engineering computations it is imperative that calculations be consistent with the measured values. For addition or subtraction, the answer is rounded off, retaining the digit found in the *rightmost column* full of significant figures as the rightmost significant figure in the answer. Two examples are shown.

46.4012	57.301
1.02	1.48
375.0	629.
422.4	688.

In multiplication the percentage error of a product is equal to the sum of the percentage errors of all factors. This percentage error can be maintained by having the number of significant figures in the answer equal to the least number of significant figures in any of the factors. For example, 362.56×2.13 will give 772.2528 when multiplied out, but there should be only three significant figures in the answer, which becomes 772. Likewise, in division the quotient should be rounded off to contain as many significant figures as the least number of significant figures in either the divisor or the dividend.

In surveying, four types of problems relating to significant figures are encountered.

1. The field measurements are given to some specific number of significant figures, thus dictating that a corresponding number should be shown in a computed value. In an intermediate calculation it is common practice to carry at least one more digit than required and then round off the answer to the correct number of significant figures. If natural trigonometric functions are used, they should always have one more place than the number of significant figures desired held in the answer.

2. There may be an implied number of significant figures. For instance, the length of a football field might be specified as 100 yards. But in laying out the field, such a distance would probably be measured to the nearest hundredth of a foot, not the nearest half-yard.

3. Each factor may not cause an equal variation. For example, if a steel tape 100.00 ft long is to be corrected for a change in temperature of 15°F, one of these numbers has five significant figures while the other has only two. A 15° variation in temperature changes the tape length by only 0.01 ft, however. Therefore an adjusted tape length to five significant figures is warranted for this type of data. Another example is the computation of a slope distance from horizontal and vertical distances, as in Figure 2-2. The vertical distance V is given to two significant figures, and the horizontal distance H measured to five significant figures. From these data the slope distance S can be computed

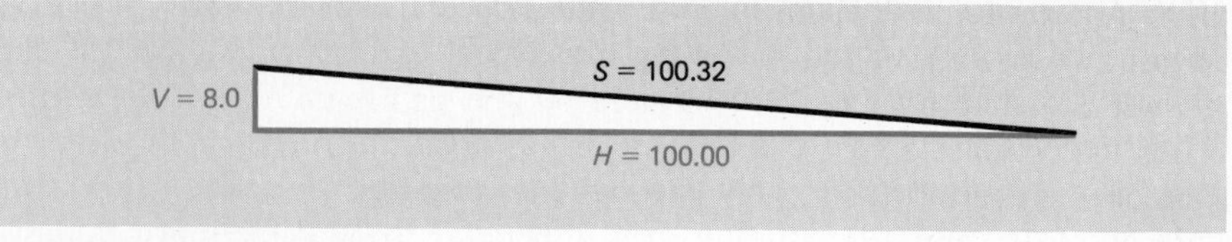

Figure 2-2. Slope correction.

to five significant figures. For small angles of slope, a considerable change in the vertical distance produces a relatively small change in the difference between slope and horizontal distances.

4. Measurements are recorded in one system of units but must be converted to another. A good rule to follow in making these conversions is to retain in the answer a number of significant figures equal to those in the measured value. As an example, to convert 178 ft $6\frac{3}{8}$ in to meters, the number of significant figures in the measured value would first be determined by expressing it in its smallest units of $\frac{1}{8}$ in, or $178 \times 12 \times 8 + 6 \times 8 + 3 = 17{,}139$. Thus, the measurement contains five significant figures and the answer $17{,}139 \div (8 \times 39.37 \text{ in/m}) = 54.416$ m is properly expressed with five significant figures. (Note that 39.37 used in the conversion is an exact constant and does not limit the number of significant figures.)

2-6. ROUNDING OFF NUMBERS. Rounding off a number is the process of dropping one or more digits so that the answer contains only those digits that are significant or necessary in subsequent computations. In rounding off numbers to any required degree of accuracy in this text, the following procedures will be observed:

1. When the digit to be dropped is less than 5, the number is written *without* the digit. Thus, 78.374 becomes 78.37.
2. When the digit to be dropped is exactly 5, the nearest *even* number is used for the preceding digit. Thus, 78,375 becomes 78.38, and 78.385 is also rounded to 78.38.
3. When the digit to be dropped is greater than 5, the number is written with the preceding digit *increased* by one. Thus, 78.376 becomes 78.38.

The procedures in 1 and 3 are standard practice. When rounding off the value 78.375 in 2, however, some people always take the next higher hundredth, whereas others invariably use the next lower hundredth. Using the nearest even digit, however, produces better balanced results in a series of computations.

2-7. DIRECT AND INDIRECT MEASUREMENTS. Measurements may be made directly or indirectly. Examples of *direct measurements* are applying a tape to line, fitting a protractor to an angle, or turning an angle with a transit or theodolite.

An *indirect measurement* is secured when it is not possible to apply a measuring instrument directly to the distance or angle to be measured. The answer is therefore determined by its relationship to some other known value. Thus the distance across a river can be found by measuring the length of a line on one side, the angle at each end of this line to a point on the other side, and then computing by one of the standard trigonometric formulas. Since many indirect measurements are made in surveying, a thorough knowledge of geometry and trigonometry is essential.

2-8. ERRORS IN MEASUREMENTS. It can be unconditionally stated that (1) *no measurement is exact*, (2) *every measurement contains errors*, (3) *the true value of a measurement is never known*, and therefore (4) *the exact error present is always unknown.* These facts are demonstrated by the following: When a distance is scaled with a rule divided into tenths of an inch, the distance can be read only to hundredths (by interpolation). If a better rule graduated in hundredths of an inch is available, however, the same distance might be estimated to thousandths of an inch. And with a rule graduated in thousandths of an inch, a reading to ten-thousandths is possible. Obviously, accuracy of measurements depends on the division size, reliability of equipment used, and human limitations in interpolating closer than about one-tenth of a scale division. As better equipment is developed, measurements will more closely approach their true values. Note that *measurements*, not *counts* (of cars, bolts, buildings, or other objects), are under consideration here.

Mistakes are caused by a misunderstanding of the problem, carelessness, or poor judgment. Large mistakes are often referred to as *blunders* and are not considered in the succeeding discussion of errors. They are detected by systematic checking of all work, and eliminated by redoing part of the job or even all of it. It is very difficult to detect small mistakes because they merge with errors. When not exposed, these little mistakes must therefore be treated as errors and will contaminate the various types of errors.

2-9. SOURCES OF ERRORS IN MAKING MEASUREMENTS. Errors in measurements stem from three sources and are classified accordingly.

Natural Errors. These are caused by variations in wind, temperature, humidity, refraction, gravity, and magnetic declination. For example, the length of a steel tape varies with changes in temperature.

Instrumental Errors. These result from any imperfection in the construction or adjustment of instruments and from the movement of individual parts. For example, the painted graduations on a rod may not be perfectly spaced, or the rod may be warped. The effect of many instrumental errors can be reduced or even eliminated by adopting proper surveying procedures or applying computed corrections.

Personal Errors. These arise principally from limitations of the human senses of sight and touch. For example, there is a small error in the measured value of an angle when the vertical cross hair in a transit is not aligned perfectly on target, or if the top of a rod is out of plumb when sighted.

2-10. TYPES OF ERRORS. Errors in measurements are of two types: systematic and random.

Systematic Errors. These errors conform to mathematical and physical laws. Their magnitude may be constant or variable, depending on conditions. Systematic errors, also known as *cumulative errors*, can be computed and their effects eliminated by applying corrections. For example, a 100-ft steel tape that is 0.02 ft too long introduces a 0.02-ft error each time it is used if a correction is not made. The change in length of a steel tape resulting from a given temperature differential can be computed by a simple formula, and the correction easily applied.

Random Errors. These are the errors that remain after mistakes and systematic errors have been eliminated. They are caused by factors beyond the control of the observer, obey the law of probability, and are sometimes called *accidental errors*. They are present in all surveying measurements.

The magnitudes and algebraic signs of random errors are matters of chance. There is no absolute way to compute or eliminate them. Random errors are also known as *compensating errors*, since they tend to partially cancel themselves in a series of measurements. For example, a person interpolating to hundredths of a foot on a tape graduated only to tenths, or reading a level rod marked in hundredths, will presumably estimate too high on some values and too low on others. Individual personal characteristics may nullify such partial compensation, however, since some people are inclined to interpolate high, others interpolate low, and many favor certain digits—for example, 7 instead of 6 or 8, 3 instead of 2 or 4, and particularly 0 instead of 9 or 1.

2-11. MAGNITUDE OF ERRORS. A *discrepancy* is the difference between two measured values of the same quantity. A small discrepancy indicates there are probably no mistakes and random errors are small. Small discrepancies do not preclude the presence of systematic errors, however.

Precision refers to the degree of refinement or consistency of a group of measurements. If multiple measurements are made of the same quantity and small discrepancies result, this indicates high precision. The degree of precision attainable is dependent on equipment sensitivity and observer skill. In surveying, precision should not be confused with *accuracy*, which denotes the absolute nearness of measured quantities to their true values. A survey may be precise without being accurate. To illustrate, if refined methods are employed and readings taken carefully, say to 0.001 ft, but there are instrumental errors in the measuring device and corrections are not made for them, the survey cannot be accurate. As a numerical example, two measurements of a distance with a tape assumed to be 100.000 ft long that is actually 100.020 ft might give results of 453.270 and 453.272 ft. These values are precise but they are not accurate, since there is a systematic error of approximately 0.090 ft in each. The *apparent precision* obtained would be expressed as 0.002/453.271 = 1/220,000, which is excellent, but accuracy of the distance is only 0.09/453.71 = 1 part in 5000. Also, a survey may appear to be accurate when rough measurements have been taken. For example, the angles of a traverse may be read with a compass to only the nearest $\frac{1}{4}°$ and yet produce a zero misclosure error. On good surveys, precision and accuracy are consistent throughout.

2-12. ELIMINATING MISTAKES AND SYSTEMATIC ERRORS. All field operations and office computations are governed by the constant effort to eliminate mistakes and systematic errors. Mistakes can be corrected only if discovered. Comparing several measurements of the same quantity is one of the best ways to isolate mistakes. Making a commonsense estimate and analysis is another. Assume that five measurements of a line are recorded as follows: 567.91, 576.95, 567.88, 567.90, and 567.93. The second value disagrees with the others, apparently because of a transposition of figures in reading or recording. This mistake can be eradicated by (a) repeating the measurement or (b) casting out the doubtful value.

When a mistake is detected it is usually best to repeat the measurement. If, however, a sufficient number of other measurements of the quantity are available and in agreement, as in the foregoing example, the widely divergent result may be discarded. Serious consideration must be given to the effect on an average before discarding a value. It is seldom safe to change a recorded number, even though there appears to be a simple transposition of figures. Tampering with physical data is always bad practice and will certainly cause trouble, even though done infrequently.

Systematic errors can be calculated and proper corrections applied to the measurements, or a field procedure used that automatically eliminates the errors. For example, the error due to sag of a tape supported at the ends only can be computed and subtracted from each measurement. If, however, the tape is supported throughout its length or at short intervals, the sag error is zero or negligible. A leveling instrument out of adjustment causes incorrect readings, but if all sights are made the same length, the errors cancel in differential leveling.

2-13. PROBABILITY. At one time or another, everyone has had an experience with games of chance, such as coin flipping, card games, or dice, which involve probability. In basic mathematics courses, laws of combinations and permutations are introduced. It is shown that things which happen randomly or by chance are governed by mathematical principles referred to as probability. These theories are applicable in many sociological and scientific measurements. In Section 2-10 it was pointed out that random errors exist in all surveying work. Their frequency and magnitude are governed by the same general principles of probability.

For convenience, the term *error* will be used to mean only random error for the remainder of this chapter. It will be assumed that all mistakes and systematic errors have been eliminated before random errors are considered. Ways to compute and correct for systematic errors are discussed later.

TABLE 2-1. **PROBABILITY FOR TWO MEASUREMENTS**

VALUE OF ERROR	NUMBER OF POSSIBILITIES	PROBABILITY	PROBABILITY, AS DECIMAL
−0.10	1	$\frac{1}{121}$	0.0083
−0.09	2	$\frac{2}{121}$	0.0165
−0.08	3	$\frac{3}{121}$	0.0248
−0.07	4	$\frac{4}{121}$	0.0331
−0.06	5	$\frac{5}{121}$	0.0413
−0.05	6	$\frac{6}{121}$	0.0496
−0.04	7	$\frac{7}{121}$	0.0579
−0.03	8	$\frac{8}{121}$	0.0661
−0.02	9	$\frac{9}{121}$	0.0744
−0.01	10	$\frac{10}{121}$	0.0826
0.00	11	$\frac{11}{121}$	0.0909
0.01	10	$\frac{10}{121}$	0.0826
0.02	9	$\frac{9}{121}$	0.0744
0.03	8	$\frac{8}{121}$	0.0661
0.04	7	$\frac{7}{121}$	0.0579
0.05	6	$\frac{6}{121}$	0.0496
0.06	5	$\frac{5}{121}$	0.0413
0.07	4	$\frac{4}{121}$	0.0331
0.08	3	$\frac{3}{121}$	0.0248
0.09	2	$\frac{2}{121}$	0.0165
0.10	1	$\frac{1}{121}$	0.0083
		Total:	1.000

2-14. OCCURRENCE OF RANDOM ERRORS. When making physical measurements, it is necessary to record values read from scales, dials, gauges, or similar equipment. It is characteristic of a measurement that *it cannot be made exactly* so it will always contain random errors. The sizes of these errors can be reduced by refining the equipment and procedures used.

To develop the principle of how random errors occur, suppose that a distance measurement of 10.4 in is made with a scale on which a reading can be estimated to 0.1 and is correct to ± 0.05. In this case the true value of the measurement is between 10.35 and 10.45; and to the nearest hundredth, it may be 10.35, 10.36, 10.37, 10.38, 10.39, 10.40, 10.41, 10.42, 10.43, 10.44, or 10.45. Thus there are 11 possible values for the correct answer. For this discussion it is assumed that all these readings have the same possibility of being correct. The probability of any one answer being correct is therefore $\frac{1}{11}$ or 0.0909.

Consider a line requiring two adjacent measurements made with this scale, each having the same possible error. The answer, a sum of two measurements, can be the total of any pair of 11 possibilities for each separate measurement, all having an equal chance of being correct. From mathematics, if one event can happen *n* ways and another can occur *r* ways, the two events together can happen *nr* ways. For the assumed conditions, there are $(11)(11) = 121$ possibilities. The difference between the sum of the measurements and the true

TABLE 2-2. **PROBABILITY FOR THREE MEASUREMENTS**

VALUE OF ERROR	NUMBER OF POSSIBILITIES		PROBABILITY
±0.15	1 each, or	2	0.0008*
±0.14	3 each, or	6	0.0023
±0.13	6 each, or	12	0.0045
±0.12	10 each, or	20	0.0075
±0.11	15 each, or	30	0.0113
±0.10	21 each, or	42	0.0158
±0.09	28 each, or	56	0.0210
±0.08	36 each, or	72	0.0270
±0.07	45 each, or	90	0.0338
±0.06	55 each, or	110	0.0413
±0.05	66 each, or	132	0.0496
±0.04	75 each, or	150	0.0563
±0.03	82 each, or	164	0.0616
±0.02	87 each, or	174	0.0654
±0.01	90 each, or	180	0.0676
0.00	91	91	0.0684
	Sum:	1331	1.0000**

* $\frac{1}{1331} = 0.0008$
** This sum equals 2 times the first 15 numbers in column 3, plus the last number.

value will be between −0.10 and +0.10. Only one pair of possible values can give a difference of −0.10; that is the pair for which the difference in each measurement is −0.05. An error of −0.09 can be obtained in two ways; there may be a difference of −0.05 in the first reading and a difference of −0.04 in the second, or a difference of −0.04 in the first reading and a difference of −0.05 in the second. This analysis can be continued to obtain the results shown in Table 2-1.

If three adjacent measurements are added in the same manner, with a maximum difference of −0.05 (and +0.05), all three would have to be off by −0.05 (or +0.05) to get an error of −0.15 (or +0.15). Also, by principles of mathematics, the total number of possibilities is $(11)(11)(11) = (11)^3 = 1331$. The complete development is shown in Table 2-2.

Values in the first and third columns of Table 2-2 are plotted as a bar graph called a *histogram* in Figure 2-3 and show graphically how the errors are distributed. Each bar of the figure has a width, or *class interval*, equal to an error increment of 0.01, and a height representing the probability of an error corresponding to its abscissa value; for example, the bar for an error of 0.00 has a height of 0.0684. If the number of errors being considered is large, as in this case, by connecting the centers of plotted bar tops, a smooth and continuous *probability curve* is obtained. The bell shape of this curve is characteristic of a *normally distributed* group of errors and thus often referred to as the *normal error distribution curve*. Statisticians frequently call it the *normal density curve*, since it shows the densities of errors of various sizes. In surveying, normal or very nearly normal error distributions almost always occur, and henceforth in this book that condition will be assumed.

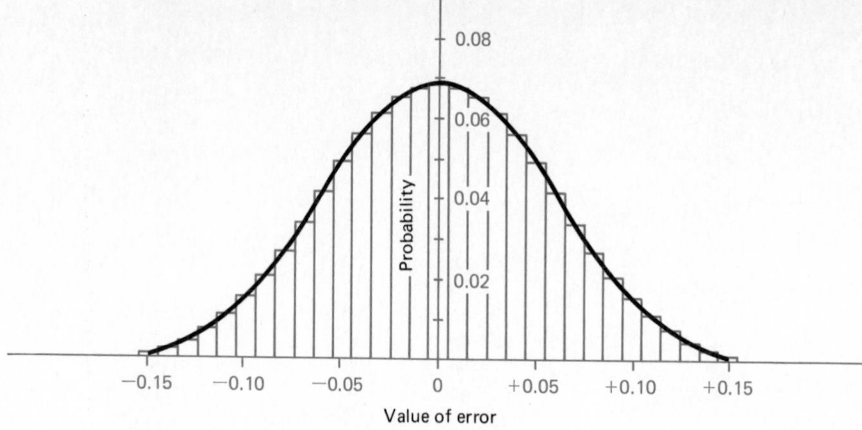

Figure 2-3. Histogram and probability curve of errors for three measurements, with maximum error of ± 0.05 units in each.

The area under the entire probability curve represents the sum of all probabilities in the third column of Table 2-2, or *1*. Also, the total area under the curve between any two ordinates equals the probability that the true error size in any three combined measurements will actually occur between errors represented at the selected ordinates.

If the same measurements of the preceding example had been taken with a smaller possible error (that is, more precisely), the probability curve would be similar to Figure 2-4. This curve shows that a greater percentage of values have small errors, and fewer measurements contain big ones. For readings

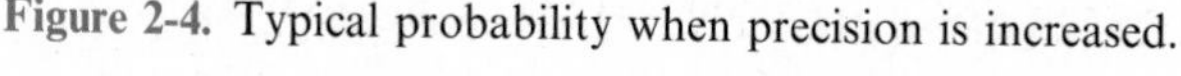

Figure 2-4. Typical probability when precision is increased.

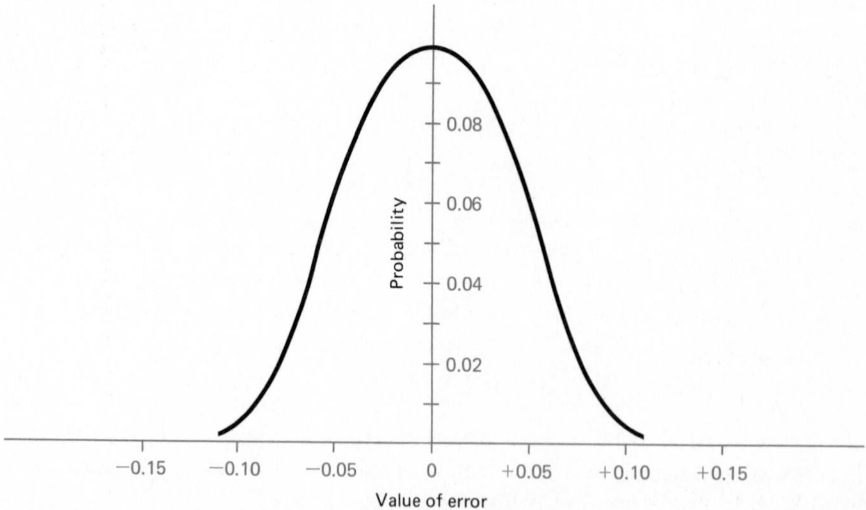

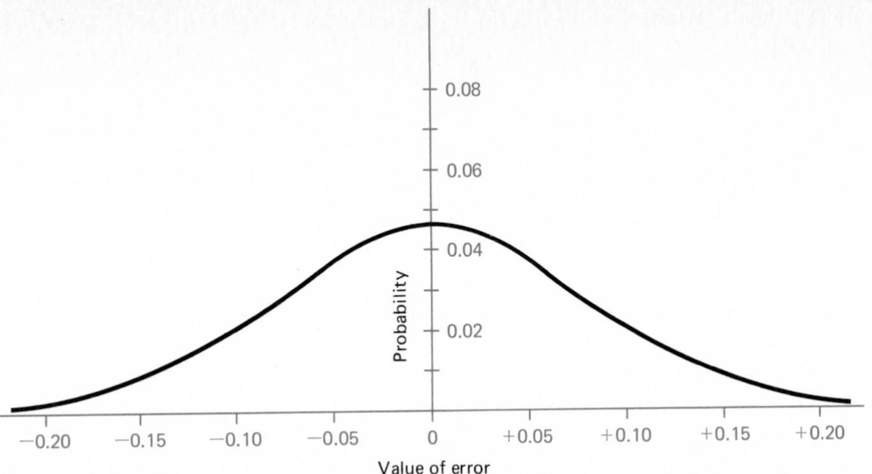

Figure 2-5. Typical probability when precision is decreased.

taken less precisely, the opposite effect is produced in Figure 2-5, which shows that a considerable percentage have large errors. In all three cases, however, the curve maintained its bell-shaped characteristic.

In this section, the equipment used was assumed to be simple and the errors primarily due to lack of precision in estimating readings. Of course, there are many additional sources of random errors, but the probability curve remains fundamentally the same.

2-15. GENERAL LAWS OF PROBABILITY. From an analysis of the data in the preceding section and the curves in Figures 2-3, 2-4, and 2-5, some general laws of probability can be stated:

1. Small errors occur more often than large ones; that is, they are more probable.
2. Large errors happen infrequently and are therefore less probable; for normally distributed errors, unusually large ones may be *mistakes* rather than random errors.
3. Positive and negative errors of the same size happen with equal frequency; that is, they are equally probable.

2-16. MOST PROBABLE VALUE. In physical measurements, the true value of any quantity is never known. Its *most probable value* can be calculated, however, if redundant measurements have been made. Redundant measurements are observations in excess of the minimum needed to determine a quantity. For a single unknown, such as a line length, that has been directly and independently measured a number of times using the same equipment and procedures,[3] the first

[3] The significance of using the same equipment and procedures is that the measurements are of equal reliability or *weight*. The subject of unequal weights is discussed in Section 2-26.

measurement establishes a value for the quantity and all additional observations are redundant. The most probable value in this case is simply the arithmetic mean, or

$$\bar{M} = \frac{\sum M}{n} \tag{2-1}$$

where $\bar{M}$ is the most probable value of the quantity, $\sum M$ the sum of the individual measurements M, and n the total number of observations. Equation (2-1) can be derived using the principle of least squares, which is based on the theory of probability. However, it can also be deduced intuitively according to the general law of probability stated in Section 2-15, *that positive and negative errors of the same magnitude occur with equal frequency.*

In more complicated problems where the observations are not made with the same instruments and procedures, or if several interrelated quantities are being determined through indirect measurements, most probable values are calculated by employing least squares methods. Procedures for least squares computation are beyond the scope of this text, and the treatment here relates to multiple direct observations of the same quantity using the same equipment.

2-17. RESIDUALS. Having determined the most probable value of a quantity, it is possible to calculate *residuals.* A residual is simply the difference between any measured value of a quantity and its most probable value, or

$$v = M - \bar{M} \tag{2-2}$$

where v is the residual in any measurement M, and $\bar{M}$ is the most probable value for the quantity. Residuals are theoretically identical to errors with the exception that residuals can be calculated whereas errors cannot because true values are never known. Thus, residuals rather than errors are the values to deal with in analysis and adjustment of survey measurements.

2-18. MEASURES OF PRECISION. Referring again to Figures 2-3, 2-4, and 2-5, although the curves have similar shapes, there are significant differences in their error spreads or *dispersions*; that is, their abscissa widths differ. The magnitude of dispersion is an indication of the precision of the measurements. *Standard deviation* (often interchangeably called *standard error*) and *variance* are statistical terms commonly used for expressing precisions of groups measurements. The equation for standard deviation is

$$\sigma = \pm\sqrt{\frac{\sum v^2}{n-1}} \tag{2-3}$$

In Eq. (2-3), σ is the standard deviation of a group of measurements of the same quantity, v the residual of an individual observation, $\sum v^2$ the sum of squares of the individual residuals, and n the number of observations.

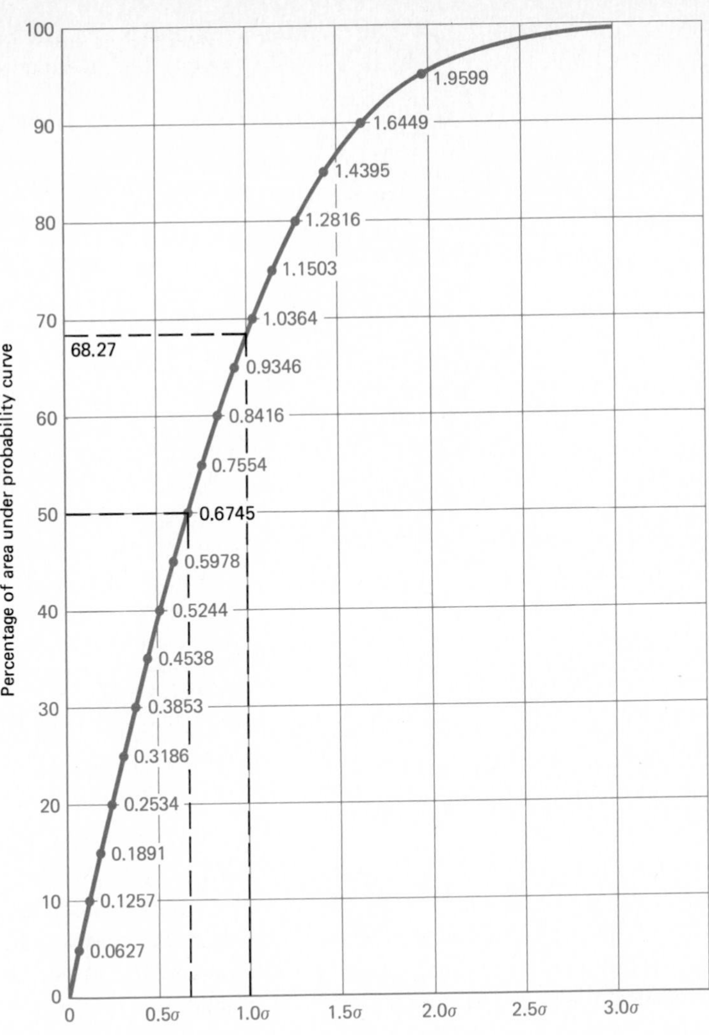

Figure 2-6. Relation between error and percentage of area under probability curve.

Variance is equal to σ^2, *the square of standard deviation.* When the mean is an exact value, as it may be for some data other than physical measurements, then n is substituted for the term $(n - 1)$. This also may be done when n is very large. Since neither of these conditions tends to exist in surveying, the term $(n - 1)$ is used.

In surveying, a deviation is thought of as an error, and therefore the term *standard error* will be used instead of standard deviation in this book.

Figure 2-6 is a graph showing the percentage of the area of a probability curve corresponding to the range of error between equal positive and negative values. Thus the area between errors of $+\sigma$ and $-\sigma$ represents 68.27% of the total area under the probability curve, and hence gives the limits of errors

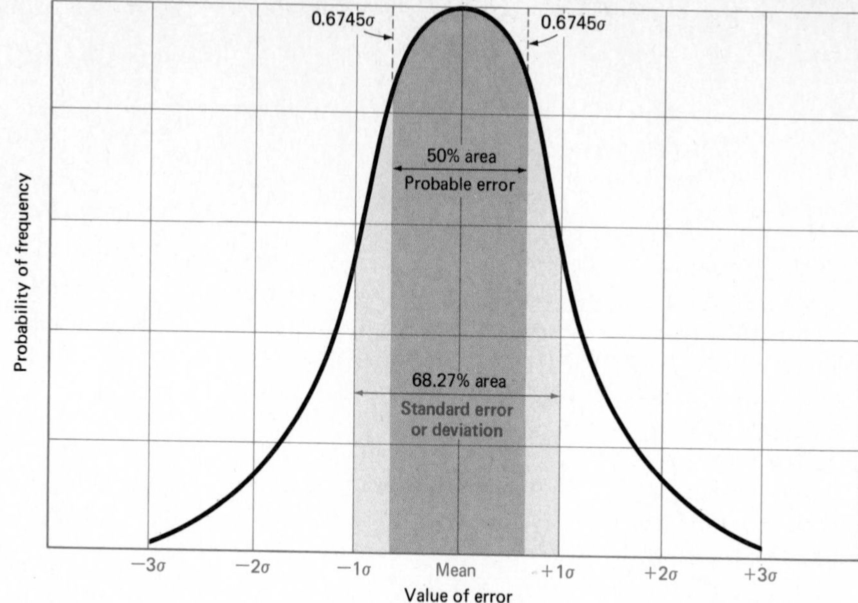

Figure 2-7. Typical probability curve showing standard error and probable error.

which can be expected to occur 68.27% of the time. This relation is shown more clearly on the typical probability curve in Figure 2-7. In a corresponding manner the percentage of error for any proportion of the probability curve can be obtained from Figure 2-6. Once the standard error has been found, the entire probability curve can be evaluated.

2-19. INTERPRETATION OF STANDARD ERROR. It has been shown that the standard error establishes the limits within which measurements are expected to fall 68.27% of the time. In other words, if a measurement was repeated ten times, it would be expected that approximately seven of the results would fall within limits established by the standard error, and conversely about three of them would fall anywhere outside these limits. Another interpretation is that one additional measurement would have a 68.27% chance of falling within the limits set by the standard error. A third deduction is that the true value has a 68.27% probability of falling within the standard error limits.

2-20. THE 50, 90, AND 95% ERRORS. From the data given in Figure 2-6, the probability of an error of any percentage likelihood can be determined. The general equation is

$$E_p = C_p \sigma \tag{2-4}$$

where E_p is the percentage error and C_p the numerical factor taken from Figure 2-6.

Using Eq. (2-4) the following are expressions for errors that have a 50, 90, and 95% chance of occurring:

$$E_{50} = 0.6745\sigma \tag{2-5}$$

$$E_{90} = 1.6449\sigma \tag{2-6}$$

$$E_{95} = 1.9599\sigma \tag{2-7}$$

The 50% error, or E_{50}, is the so-called *probable error*. It establishes limits within which the measurements will fall 50% of the time. In other words, a measurement will have the same chance of coming within these limits as it will have of falling outside of them. The 50% limits are shown graphically in Figure 2-7. In the past, probable error was employed extensively in discussing random errors, but now it is seldom used.

The x axis is an asymptote of the probability curve, so the 100% error cannot be evaluated. This means no matter what error is found, a larger one is possible. For this reason either the 90 or 95% error is often considered to be the maximum practical value.

EXAMPLE 2-1

To clarify definitions and use the equations given in Sections 2-16 through 2-20, suppose that a line has been measured 10 times with the results shown in column (1) of the following table. It is assumed these measurements have already been corrected for all systematic errors. Compute the most probable value for the line length, its standard error, and errors having 50, 90, and 95% probability.

(1) LENGTH (ft)	(2) RESIDUAL v (ft)	(3) v^2
1000.57	+0.12	0.0144
1000.39	−0.06	0.0036
1000.37	−0.08	0.0064
1000.39	−0.06	0.0036
1000.48	+0.03	0.0009
1000.49	+0.04	0.0016
1000.32	−0.13	0.0169
1000.46	+0.01	0.0001
1000.47	+0.02	0.0004
1000.55	+0.10	0.0100
$\sum = 10{,}004.49$	$\sum = -0.01$	$\sum v^2 = 0.0579$

By Eq. (2-1), $\bar{M} = \dfrac{10{,}004.49}{10} = 1000.45$ ft

By Eq. (2-2), the residuals are calculated. These are tabulated in column (2) and their squares listed in column (3).

By Eq. (2-3), $\sigma = \pm\sqrt{\dfrac{\sum v^2}{n-1}} = \pm\sqrt{\dfrac{0.0579}{9}} = \pm 0.08$ ft

By Eq. (2-5), $E_{50} = \pm 0.6745\sigma = \pm 0.6745(0.08) = \pm 0.05$ ft

By Eq. (2-6), $E_{90} = \pm 1.6449(0.08) = \pm 0.13$ ft

By Eq. (2-7), $E_{95} = \pm 1.9599(0.08) = \pm 0.15$ ft

The following conclusions can be drawn concerning this example:

1. The most probable length is 1000.45 ft.
2. The standard error of a single measurement is ± 0.08 ft.
3. The normal expectation is that 68% of the time, a recorded length would lie between 1000.37 and 1000.53 ft; that is, about seven values would lie within these limits. (Actually seven of them do.)
4. The probable error (E_{50}) is ± 0.05 ft. Therefore it can be anticipated that half, or five, of the measurements will fall in the interval 1000.40 to 1000.50 ft. (Four values do.)
5. Ninety percent of the time a measured length would not contain an error larger than ± 0.13 ft, and its value would be within the range 1000.32 to 1000.58 ft.
6. The 95% error would be ± 0.15 ft, and the length would lie between 1000.30 and 1000.60 ft 95% of the time. (Note that all measurements indeed are within the limits of both the 90 and 95% errors.)

2-21. ERROR OF A SUM. The following formula, derived from the general law of propagation of random errors, gives the error in the sum of quantities that each contain random errors:

$$E_{\text{sum}} = \sqrt{E_a^2 + E_b^2 + E_c^2 + \cdots} \tag{2-8}$$

In Eq. (2-8), E represents any specified error, and a, b, and c are the separate measurements.

EXAMPLE 2-2

Assume that a line is measured in three sections, with errors in the individual parts equal to ± 0.012, ± 0.028, and ± 0.020 ft, respectively. Then the error of the total length is

$$E_{\text{sum}} = \pm\sqrt{0.012^2 + 0.028^2 + 0.020^2} = \pm 0.036 \text{ ft}$$

A similar computation applies to the error of any product, and, thus, to the error of an area. In Figure 2-8 the error in side A of a rectangular parcel of land is E_a; in side B it is E_b. Therefore, the error in area caused by E_a is BE_a, and that due to E_b is AE_b. The equation for the error in area, product AB, which again can be derived from the general law of propagation of random errors, is

$$E_{\text{prod}} = \sqrt{A^2E_b^2 + B^2E_a^2} \tag{2-9}$$

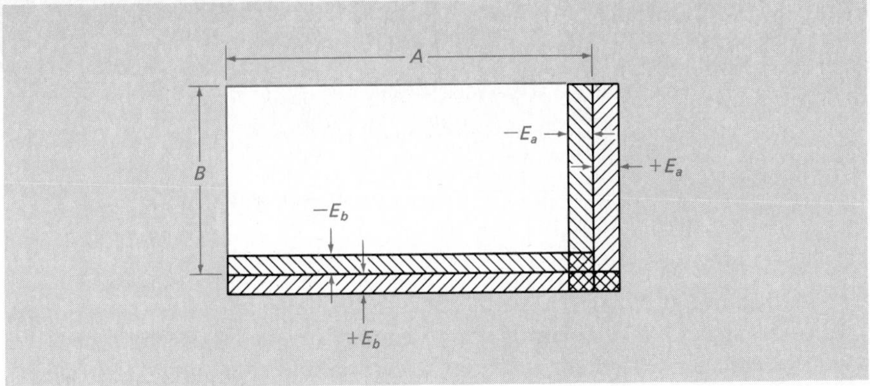

Figure 2-8. Error of area.

EXAMPLE 2-3

For a rectangular lot $50.00 \pm 0.01 \times 100.00 \pm 0.02$ ft, calculate the error in the area.

By Eq. (2-9),

$$E = \pm\sqrt{50^2(0.02)^2 + 100^2(0.01)^2} = \pm 1.41 \text{ ft}^2$$

The area as computed might be $(50.00)(100.00) = 5000.0000$ ft^2. However, the rule for significant figures (see Section 2-5) states that there cannot be more significant figures in the answer than in any of the individual factors used. Accordingly, the area should be rounded off to $500\bar{0}$ ft^2 (four significant figures). From the calculation, the error is ± 1.41 ft^2, the last digit in 5000 is questionable, and the number of significant figures to be used in the answer (four) is verified.

2-22. ERROR OF A SERIES. Sometimes a series of similar quantities, such as the angles of a traverse, are read with each measurement being in error by about the same amount. The total error in the sum of all measured quantities of such a series is called the *error of the series*, designated as E_{series}. If the same error E in each measurement is assumed and Eq. (2-8) applied, the series error is

$$E_{\text{series}} = \sqrt{E^2 + E^2 + E^2 + \cdots} = \sqrt{nE^2} = E\sqrt{n} \qquad (2\text{-}10)$$

where E represents the error in any individual measurement and n the number of measurements.

This equation shows that when the same operation is repeated, the random errors tend to balance out and the remaining error of the series is proportional to the square root of the number of observations. This equation has extensive use—for instance, to determine the allowable misclosure error for angles of a traverse, to be discussed in Chapter 12.

EXAMPLE 2-4

Assume that any distance of 100 ft can be taped with an error of ± 0.02 ft if certain techniques are employed, and the error in taping 5000 ft using these skills is needed.

Since the number of 100-ft lengths in 5000 ft is 50, then by Eq. (2-10):

$$E_{\text{series}} = \pm E\sqrt{n} = \pm 0.02\sqrt{50} = \pm 0.14 \text{ ft}$$

EXAMPLE 2-5

A distance of 1000 ft is to be taped with an error of not more than ± 0.10 ft. It is desired to determine how accurately each 100-ft length must be measured to ensure the error will not exceed a permissible limit.

Since $E_{\text{series}} = E\sqrt{n}$ and $n = 10$, the allowable error E in 100 ft is

$$E = \frac{E_{\text{series}}}{\sqrt{n}} = \frac{0.10}{\sqrt{10}} = \pm 0.03 \text{ ft}$$

EXAMPLE 2-6

Suppose it is required to tape a length of 2500 ft with an error of not more than ± 0.10 ft.

If 100 ft is again considered the unit length, $n = 25$ and by Eq. (2-10), the allowable error E in 100 ft is

$$E = \frac{0.10}{\sqrt{25}} = \pm 0.02 \text{ ft}$$

This analysis shows that a larger number of possibilities provides a greater chance for errors to cancel out.

2-23. ERROR OF THE MEAN. Section 2-16 stated that the most probable value of a group of repeated measurements is the arithmetic mean, and the mean itself is subject to error. Nevertheless, in many types of surveys, error of the mean is commonly used for comparisons.

By applying Eq. (2-10), which is a special form of Eq. (2-8), it is possible to find the error for the sum of a series of measurements where each one has the same error. Since the sum divided by the number of measurements gives an average value, the error of the mean may be found by the relation

$$E_m = \frac{E_{\text{sum}}}{n}$$

from which

$$E_m = \frac{E\sqrt{n}}{n} = \frac{E}{\sqrt{n}} \tag{2-11}$$

where E is the specified error of a single measurement, E_m some specified error of the mean, and n the number of observations.

The error of the mean at any percentage probability can be determined and applied to all criteria that have been developed. The standard error of the mean, $(E_{68})_m$ or σ_m, is

$$(E_{68})_m = \sigma_m = \frac{\sigma}{\sqrt{n}} = \sqrt{\frac{\sum v^2}{n(n-1)}} \tag{2-12}$$

The 90% error of the mean is

$$(E_{90})_m = \frac{E_{90}}{\sqrt{n}} = 1.6449\sqrt{\frac{\sum v^2}{n(n-1)}} \tag{2-13}$$

These equations show that *the error of the mean varies inversely as the square root of the number of repetitions.* Thus, to double the accuracy—that is, to reduce the error by one-half—four times as many measurements must be made.

EXAMPLE 2-7

Calculate the standard error of the mean and the 90% error of the mean for the measurements of Example 2-1.

By Eq. (2-12):

$$\sigma_m = \frac{\sigma}{\sqrt{n}} = \pm\frac{0.08}{\sqrt{10}} = \pm 0.025 \text{ ft}$$

Also, by Eq. (2-13):

$$(E_{90}) = (1.6449)\pm 0.025 = \pm 0.04 \text{ ft}$$

These values show the error limits of 68 and 90% probability for the line's length. It can be said that the exact length of line has a 68% chance of being within ± 0.025 of the mean and a 90% likelihood of falling not farther than ± 0.04 ft from the mean.

2-24. APPLICATIONS. In the example problems it has been shown that the equations of probability are applied in two ways:

1. To analyze measurements that have already been made, for comparison with other results or with specification requirements.
2. To establish procedures and specifications in order that the required results will be obtained.

Application of the various probability equations must be tempered with judgment and caution. Remember, a basic assumption made was that an infinite number of errors is being considered. Frequently, in surveying, only a few observations—often from 4 to 10—are taken. If these results are typical and representative, then the answer obtained using probability equations will be reliable; if they are not, the conclusions can be misleading.

2-25. ADJUSTMENT OF MEASUREMENTS. In Section 2-8 it was emphasized that the true value of any measured quantity is never known. In some types of problems, however, the sum of several measurements must equal a fixed value; for example, the sum of the interior angles of a polygon has to total $(n - 2)(180°)$. In practice, therefore, the measured angles of a polygon are adjusted to make them add to the required total. Correspondingly, distances—either horizontal or vertical—may be altered slightly to meet certain requirements. The methods employed will be explained in later chapters where the operations are taken up in detail. In making these adjustments, the principles of probability are most important.

2-26. WEIGHTS OF MEASUREMENTS. It is evident that some measurements are more precise than others because of better equipment, improved techniques, and superior field conditions. In making adjustments, it is consequently desirable to assign *relative weights* to individual observations. It can be shown that relative weights are inversely proportional to variances, or

$$W_a \propto \frac{1}{\sigma_a^2} \tag{2-14}$$

In Eq. (2-14), W_a is the weight of an observation a, which has a variance σ_a^2. Thus, the higher the precision (the smaller the variance), the larger should be the relative weight of the measured value being adjusted. In some cases variances are unknown originally and weights must be assigned to measured values based on estimates of their relative precisions. If a quantity is measured repeatedly and the individual observations have varying weights, the weighted mean can be computed from the expression:

$$\bar{M}_w = \frac{\sum WM}{\sum W} \tag{2-15}$$

where $\bar{M}_w$ is the weighted mean, $\sum WM$ the sum of the individual weights times their corresponding observations, and $\sum W$ the sum of the weights.

EXAMPLE 2-8

Suppose four measurements of a distance are recorded as 482.16, 482.17, 482.20, and 482.18 and given weights of 1, 2, 2, and 4, respectively, by the survey-party chief. Determine the weighted mean.

By Eq. (2-15):

$$\bar{M}_w = \frac{482.16 + 482.17(2) + 482.20(2) + 482.18(4)}{1 + 2 + 2 + 4} = 482.18 \text{ ft}$$

In computing adjustments involving unequally weighted measurements, corrections applied to observed values should be made inversely proportional to the relative weights.

EXAMPLE 2-9

Assume the measured angles of a certain triangle are: $A = 49°51'15''$, wt. 1; $B = 60°32'08''$, wt. 2; and $C = 69°36'33''$, wt. 3. Perform a weighted adjustment of the angles.

Angle adjustments are made inversely proportional to their weights as in the accompanying tabulation. Angle C with the greatest weight (3) gets the smallest correction, $2x$; B receives $3x$; and A, $6x$.

	ANGLE	WT.	CORRECTION	NUMERICAL CORR.	ROUNDED CORR.	ADJUSTED ANGLES
A	49°51′15″	1	$6x$	+2.18″	+2.2″(or 2″)	49°51′17″
B	60°32′08″	2	$3x$	+1.09″	+1.1″(or 1″)	60°32′09″
C	69°36′33″	3	$2x$	+0.73″	+0.7″(or 1″)	69°36′34″
Sum	179°59′56″	6	$11x$	+4.00″	+4.0″(or 4″)	180°00′00″

$11x = 4''$ and $x = +0.36''$

It must be emphasized again that adjustment computations based on the theory of probability are valid only if systematic errors and mistakes have been eliminated by employing proper procedures, equipment, and calculations.

PROBLEMS

2-1. Convert the following distances given in meters to feet:
(a) 4129.57 m (b) 738.29 m (c) 1129.30 m

2-2. Convert the following distances given in feet to meters:
(a) 537.52 ft (b) 13,280.75 ft (c) 4132.8 ft

2-3. Compute the lengths in feet corresponding to the following distances measured with a Gunter's chain:
(a) 15 ch 45 lk (b) 74 ch 23 lk (c) 41 ch 5 lk

2-4. Express 174,538 ft^2 in:
(a) acres (b) hectares (c) square Gunter's chains

2-5. Convert 4.753 hectares to:
(a) acres (b) square Gunter's chains

2-6. What are the lengths in feet and decimals for the following distances shown on a building blueprint:
(a) 27 ft $4\frac{3}{8}$ in (b) 52 ft $1\frac{3}{4}$ in

2-7. What is the area in acres of a rectangular parcel of land measured with a Gunter's chain if the recorded sides are as follows:
(a) 79 ch 17 lk and 51 ch 39 lk (b) 16 ch 10 lk and 12 ch 82 lk

2-8. Compute the area in acres of triangular lots shown on a plat having the following recorded right-angle sides:
(a) 153.72 ft and 438.50 ft (b) 3 ch 8 lk and 1 ch 95 lk

2-9. A distance is expressed as 135,285.19 U.S. survey feet. What is the length in U.S. standard feet?

2-10. What are the radian and degree-minute-second equivalents for the following angles given in grads:
(a) 57 grads (b) 36.2 grads (c) 163.71 grads

2-11. Give answers to the following problems in the correct number of significant figures:
(a) sum of 91.72, 0.00154, 156, and 9.7
(b) sum of 1.2354, 0.052, 1130, and 483.6

(c) product of 1128.95 and 1.29
(d) quotient of 4930.27 divided by 5.9

2-12. Express the value or answer in powers of 10 to the correct number of significant figures:
(a) 11,432
(b) 4520
(c) square of 11,291
(d) sum of (11.285 + 0.5 + 146.1) divided by 7.2

2-13. Convert the adjusted angles of a triangle to radians and show a computational check:
(a) 48°27′13″, 81°11′48″, and 50°20′59″
(b) 29°58′04″, 64°32′00″, and 85°29′56″

2-14. Explain the difference between systematic and random errors.

2-15. Discuss the difference between precision and accuracy.

A distance *AB* is measured repeatedly and the results, in feet, are listed in Problems 2–16 through 2–19. Calculate the (a) most probable length of the line, (b) standard error of a single measurement, and (c) standard error of the mean for each set of results.

2-16. 728.89, 728.92, 728.91, 728.87, 728.90, 728.95, 729.04, 728.86, 728.92, and 728.79.

2-17. Same as Problem 2-16 except discard one measurement, 729.04.

2-18. Same as Problem 2-16 except discard two measurements, 729.04 and 728.79.

2-19. Same as Problem 2-16 except include two additional measurements, 728.86 and 728.88.

In Problems 2-20 through 2-23, determine the range within which measurements should fall (a) 50% of the time and (b) 90% of the time. List the percentage of values that actually fit within these ranges.

2-20. For the data of Problem 2-16.

2-21. For the data of Problem 2-17.

2-22. For the data of Problem 2-18.

2-23. For the data of Problem 2-19.

In Problems 2-24 through 2-26, calculate the angle's most probable value, the standard error of a single measurement, and standard error of the mean.

2-24. 49°23′10″, 49°23′00″, 49°23′30″, 49°22′50″, 49°23′20″, and 49°23′10″.

2-25. Same as Problem 2-24 with two additional measurements of 49°23′10″ and 49°23′20″.

2-26. Same as Problem 2-24 with four additional measurements of 49°23′10″, 49°23′20″, 49°23′20″, and 49°23′10″.

2-27. A field party is capable of making taping measurements with a standard error of ±0.015 ft per 100-ft tape length. What standard error would be expected if a distance of 5000 ft was taped by this party?

2-28. Repeat Problem 2-27, except the standard error per 20-m tape length is ±0.008 m and a distance of 1200 m is taped. What is the expected 90% error in 1200 m?

2-29. A distance of 4000 ft must be taped in a manner to ensure a standard error less than ±0.10 ft. What must be the standard error per 100-ft tape length to achieve the desired precision?

2-30. Lines of levels were run requiring n instrument setups. If each backsight and foresight rod reading has a standard error σ, what are the standard errors in each of the following level lines?
(a) $n = 24$, $\sigma = \pm 0.005$ ft
(b) $n = 18$, $\sigma = \pm 0.5$ mm

2-31. Line *AD* is measured in three sections, *AB*, *BC*, and *CD*, with lengths and standard errors as listed below. What are the standard errors in the total length *AD*?
(a) $AB = 835.21$, ±0.06 ft; $BC = 1278.43$, ±0.13 ft; $CD = 492.87$, ±0.05 ft
(b) $AB = 129.856$ m, ±0.021 m; $BC = 295.300$ m, ±0.290 m; $CD = 246.205$ m, ±0.025 m

2-32. A distance AB was measured four times as 577.83, 577.81, 577.85, and 577.84. The measurements were given weights of 2, 3, 1, and 2, respectively, by the observer. (a) Calculate the weighted mean for distance AB. (b) What difference results if later judgment revises the weights to 3, 2, 1, and 3?

2-33. Determine the weighted mean for the following angles and weights:
(a) 59°21′48″, wt 2; 59°21′40″, wt 1; 59°21′51″, wt 3; 59°21′42″, wt 1
(b) 65°38′20″, wt 2; 65°38′16″, wt 1; 65°38′23″, wt 3; 65°38′21″, wt 2

2-34. Specifications for measuring angles of an n-sided figure limit the total angular closure to E. How accurately must each angle be measured for the following values of n and E?
(a) $n = 6$, $E = 30$ sec
(b) $n = 12$, $E = 1$ min

2-35. What is the area of a rectangular field and its error for the following recorded values:
(a) 468.10, ±0.08 ft by 620.56, ±0.10 ft
(b) 86.25, ±0.012 m by 140.80, ±0.020 m

2-36. Adjust the angles of triangle ABC for the following angular values and weights:
(a) $A = 49°27'31''$, wt. 2; $B = 61°42'18''$, wt. 1; $C = 68°50'19''$, wt. 4
(b) $A = 80°14'05''$, wt. 1; $B = 38°37'47''$, wt. 2; $C = 61°07'58''$, wt. 3

2-37. Determine the relative weights and perform a weighted adjustment (to the nearest second) for angles A, B, and C of a plane triangle, given the following observations:

ANGLE A OBSERVATION NO.	VALUE	ANGLE B OBSERVATION NO.	VALUE	ANGLE C OBSERVATION NO.	VALUE
1	52°12′	1	67°20′	1	60°22′
2	52°13′	2	67°18′	2	60°21′
3	52°13′	3	67°24′	3	60°23′
4	52°12′	4	67°22′	4	60°24′

2-38. A line of levels was run from bench mark A to B, B to C, and C to D. Elevation differences obtained between bench marks, with their standard errors, are listed below. What is the difference in elevation from bench mark A to D and the standard error of that elevation difference?
(a) BM A to BM B = +73.87, ±0.09 ft; BM B to BM C = −113.05, ±0.17 ft; and BM C to BM D = −48.90, ±0.12 ft.
(b) BM A to BM B = −30.821, ±0.015 m; BM B to BM C = +49.378, ±0.022 m; and BM C to BM D = +61.805, ±0.018 m

BIBLIOGRAPHY

Aguilar, A. M. 1973. "Principles of Survey Error Analysis and Adjustment." *ASCE Journal of the Surveying and Mapping Division* 99(no. SU1):107.

Barry, B. A. 1978. *Errors in Practical Measurement in Science, Engineering, and Technology*. New York: Wiley.

Mikhail, E. 1976. *Observations and Least Squares*. New York: Harper & Row.

Mikhail, E. M., and G. Gracie. 1981. *Analysis and Adjustment of Survey Measurements*. New York: Van Nostrand.

Uotila, U. A. 1973. "Useful Statistics for Surveyors." *Surveying and Mapping* 33(no. 1):67.

Whitten, C. A. 1974. "The Metric System and the Land Surveyor." *Bulletin, American Congress on Surveying and Mapping*, no. 44, p. 9.

Whitten, C. A., et al. 1980. "Planning for Metrication for Surveying and Mapping." *Bulletin, American Congress on Surveying and Mapping*, no. 70, p. 37.

Wolf, P. R. 1980. *Adjustment Computations: Practical Least Squares for Surveyors*, 2nd ed. Madison, Wis.: P.B.L. Publishers.

3 SURVEYING FIELD NOTES

3-1. INTRODUCTION. Surveying field notes, whether in books or on an electronic data recorder, are the only permanent record of work done in the field. If they are incomplete, incorrect, lost, or destroyed, much or all the time and money invested in making accurate records has been wasted. Hence the note-keeper's job is frequently the most important and difficult one in a party. A field book containing information gathered over a period of weeks is worth many thousands of dollars because of the high costs to maintain a party of two, three, or more persons in the field. Therefore it should be carefully guarded and have the owner's name and address lettered with India ink on the cover and inside.

Data in field notes are normally used by office personnel to make drawings or computations. Accordingly, it is essential that notes be intelligible to anyone without verbal explanations. The Reinhardt system of slope lettering is generally employed for clarity and speed; it requires a minimum number of simple strokes to form any letter.

Property surveys are subject to court review under some conditions, so field notes become an important factor in litigation. Also, because they may be used as references in land transactions for generations, it is necessary to index and preserve them properly. The salable "goodwill" of a surveyor's business depends largely on the office library of field books. Cash receipts may be kept in an unlocked desk drawer, but field books are stored in a fireproof safe!

Original notes are those taken at the same time measurements are being made. All other sets are copies and must be so marked. Copied notes may not be accepted in court because they are open to question concerning possible mistakes, such as interchanging numbers, and omissions. The value of a distance or an angle placed in the field book from memory, 10 min after the observation, is definitely unreliable.

Students are tempted to scribble notes on scrap sheets of paper for later transfer in neater form to the regular field book. This practice may result in loss of some or all of the original data and defeats one purpose of a surveying course—to provide experience in taking notes under job conditions. In a real job situation, a surveyor is not likely to spend any time at night transcribing scribbled notes. Certainly an employer will not pay for this evidence of incompetence.

Notes should be lettered with a sharp pencil of at least 3-H hardness so an indentation is made in the paper. Books so prepared will withstand damp weather in the field (or even a soaking) and still be legible, whereas graphite from a soft pencil, or ink from a pen, leaves an undecipherable smudge under such circumstances.

Erasures of observed data are not permitted in field books. If a number has been recorded incorrectly, a line is run through it without destroying the number's legibility, and the proper value is noted above (see Plate D-3, left page[1]). If an entire page is to be deleted, diagonal lines are drawn through opposite corners and VOID is lettered prominently with the reasons.

3-2. REQUIREMENTS OF GOOD NOTES. The following points are considered in appraising a set of field notes.

Accuracy. This is the most important quality in all surveying operations.

Integrity. A single omitted measurement or detail can nullify use of the notes for plotting or computing. If the project was far from the office, it is time-consuming and expensive to return for a missing measurement. Notes should be checked carefully for completeness before leaving the survey site, and never "fudged" to improve closures.

Legibility. Notes can be used only if they are legible. A professional-looking set of notes is likely to be professional in quality.

Arrangement. Noteforms appropriate to the particular survey contribute to accuracy, integrity, and legibility.

Clarity. Advance planning and proper field procedures are necessary to ensure clarity of sketches and tabulations and to make mistakes and omissions more evident. Avoid crowding notes; paper is relatively cheap. Costly mistakes in drafting and computing are the end results of ambiguous notes.

3-3. TYPES OF FIELD BOOKS. Since field books contain valuable data, suffer hard wear, and must be permanent in nature, it is good economy to use only the best for practical work. Various kinds of field books are available, but bound and loose-leaf types are the most common.

[1] See Appendix D for typical field note plates.

The bound book, a standard for many years, has a sewed binding, a hard cover of leatherite, polyethylene, or covered hardboard, and contains 80 leaves. Its use ensures maximum testimony acceptability for property survey records in courtrooms.

Bound duplicating books permit copies of notes to be made through carbon paper in the field. Alternate pages are perforated for easy removal.

Loose-leaf books have come into wide use because of many advantages, which include (a) assurance of a flat working surface, (b) simplicity of filing individual project notes, (c) ready transfer of partial sets of notes between field and office, (d) provision for holding pages of printed tables, diagrams, formulas, and sample forms, (e) the possibility of using different rulings in the same book, and (f) a saving in sheets (since none are wasted by filing partially filled books) and thus lower total cost. Disadvantages include possible loss of loose sheets and potential use of cheaper, poor-quality paper.

Stapled, sewed, or spiral-bound books are not suitable for practical work. They may be satisfactory for abbreviated surveying courses that have only a few field periods, because of limited service required and low cost.

Special column and page rulings provide for particular needs in leveling, transit and theodolite work, topographic surveying, cross sectioning, and so on.

New automatic reading and recording systems have been developed for surveying, but field sketches and some other handwritten information must still be recorded. Electronic theodolites, distance-measuring units, and "total-station" systems provide visually displayed digital readings. These can be recorded on tape cassettes or microprocessors with various capacities, and interfaced with different instruments and computers to go through the complete process from field measurements to printouts and map plotters. The cost of data collector and interface units may be comparable to or exceed that for some precise theodolites or EDM instruments.

A helpful notekeeping "instrument" is a camera. A moderately priced, reliable, lightweight unit can produce records of monuments set or found and other admissible field evidence.

3-4. KINDS OF NOTES. Four types of notes are kept in practice: (1) sketches, (2) tabulations, (3) descriptions, and (4) combinations of these. The most common type is a combination form, but an experienced recorder selects the version best fitted to the job at hand. Appendix D contains typical noteforms illustrating some field problems covered in this text.

For a simple survey, such as measuring the distances between hubs on a series of lines, a sketch showing the lengths is sufficient. In measuring the length of a line forward and backward, a tabulation properly arranged in columns is adequate, as in Plate D-2 in Appendix D. The proverb about one picture being worth 10,000 words might well have been written for notekeepers.

The location of a reference point may be difficult to identify without a sketch, but often a few lines of description are enough. Bench marks usually are so described, as in Plate D-3.

In notekeeping this axiom is always pertinent: When in doubt about the need for any information, include it and make a sketch. It is better to have too many data than not enough.

3-5. ARRANGEMENTS OF NOTES. Note styles and arrangement depend on departmental standards and individual preference. Highway departments, mapping agencies, and other organizations engaged in surveying furnish their field personnel with sample noteforms, similar to those in Appendix D, to aid in preparing uniform and complete records that can be checked quickly.

It is desirable for students to have an expertly designed set of noteforms covering their first field work as guides, to set high standards and save time. The noteforms shown in Appendix D are a composite of several models. They stress the open style, especially helpful for beginners, in which some lines or spaces are skipped for clarity. Thus angles measured at a point *A* (see Plate D-8) are placed opposite *A* on the page, but distances measured between hubs *A* and *B* on the ground are recorded on the line between *A* and *B* in the field book.

Left- and right-hand pages are practically always used in pairs and therefore carry the same number. A complete title should be lettered across the top of the left page and may be extended over the right one. Titles may be abbreviated on succeeding pages for the same survey project. Location and type of work are placed beneath the title. Some surveyors prefer to confine the title on the left page and keep the top of the right one free for date, party, weather, and other items. This design is revised if the entire right page has to be reserved for sketches and bench-mark descriptions. Arrangements shown in Appendix D are eminently satisfactory and demonstrate to students the flexibility of noteforms.

The left page is generally ruled in six columns designed for tabulation only. Column headings are placed between the first two horizontal lines at the page top and follow from left to right in the anticipated order of reading and recording. The upper part of the left or right page must contain four items.

1. *Project name, location, date, time of day* (A.M. *or* P.M.), *and starting and finishing times.* These entries are necessary to document the notes and furnish a timetable, as well as correlate different surveys. Precision, troubles encountered, and other facts may be gleaned from the time required for a survey.

2. *Weather.* Wind velocity, temperature, and adverse weather conditions such as rain, snow, sunshine, and fog have a decided effect on accuracy in surveying operations. Surveyors are unlikely to do the best possible work at temperatures of 15°F or with rain pouring down their necks. Hence, weather details are important in reviewing field notes, applying corrections to tape lengths due to temperature variations, and for other purposes.

3. *Party.* The names and initials of party members and their duties are required for documentation and future reference. Jobs can be described by symbols, such as ⊼ for instrument operator, ϕ for rodperson, N for notekeeper, and HT for head tapeperson. The party chief is frequently the notekeeper.

4. *Instrument type and number.* The type of instrument used (with its make and serial number) and the degree of adjustment affect the accuracy of a survey. Identification of the specific equipment employed may aid in isolating some errors—for example, a particular tape with an actual length that is later found to disagree with the distance recorded between its end graduations.

To permit ready location of desired data, each field book must have a table of contents that is kept current daily. In practice, surveyors cross-index their notes on days when field work is impossible.

3-6. SUGGESTIONS FOR RECORDING NOTES. Observing the suggestions listed here will eliminate some common mistakes in recording notes.

1. Use the Reinhardt system of lettering.[2] Reserve uppercase letters for emphasis.
2. Letter the notebook owner's name and address on the cover and first inside page in India ink. Number all field books for record purposes.
3. Use a hard pencil, at least 3-H or 4-H, and keep it sharp.
4. Begin a new day's work on a new page. For property surveys having complicated sketches, this rule may be waived.
5. Employ any orderly, standard, familiar noteform type, but if necessary, design a special arrangement to fit the project.
6. Immediately after a measurement, always record it directly in the field book rather than on a sheet of scrap paper for copying later.
7. Include explanatory statements, details, and additional measurements if they might clarify the notes for field and office personnel.
8. Record what is read without performing any mental arithmetic. Put down what you read!
9. Do not erase recorded data. Run a single line through an incorrect value (but retain its legibility) and place the correct figure above or below it. To *void* an entire page, run diagonal lines to the page corners. State the reason for doing so.
10. Carry a straightedge for ruling lines and a small protractor to lay off angles.
11. Run notes down the page, except in route surveys, where they usually progress upward to conform with sketches made while looking in the forward direction. (See Plate D-13.)
12. Use sketches instead of tabulations when in doubt.
13. Make drawings to general proportions rather than to exact scale or without plan, and recognize that the usual preliminary estimate of space required is too small. Letter parallel with or perpendicular to the appropriate feature, showing clearly to what they apply. Dimension lines, as used in machine graphics, are seldom necessary.
14. Exaggerate details or sketches if clarity is thereby improved, or prepare separate diagrams.
15. Line up descriptions and drawings with corresponding numerical data. For example, the beginning of a bench-mark description should be placed on the same line as its elevation, as in Plate D-3.
16. Avoid crowding. If it is helpful to do so, use several right-hand pages of descriptions and sketches for a single left-hand sheet of tabulation. Similarly, use any number of pages of tabulation for a single drawing.

[2] A single-stroke style of lettering described in engineering drawing textbooks and used in the noteforms of Appendix D.

17. Paper is cheap compared with the value of time that might be wasted in misinterpretation by office personnel of compressed field notes, or in returning to the field for clarification.
18. Use explanatory notes when they are pertinent, always keeping in mind the purpose of the survey and needs of the office force. Put these notes in open spaces to avoid conflict with other parts of the sketch.
19. Employ conventional symbols and signs for compactness.
20. Have north at the top, or left side, of all sketches if possible. A meridian arrow is vital.
21. Keep tabulated figures inside of and off column rulings, with decimal points and digits in line vertically.
22. Make a mental estimate of all measurements before receiving and recording them in order to eliminate large mistakes.
23. Repeat aloud values given for recording. For example, before putting down a distance of 124.68, call out "one, two, four, point six, eight" for verification by the tapeperson who submitted the measurement.
24. Place a zero before the decimal point for numbers less than 1; that is, record 0.37 instead of .37.
25. Show the precision of measurements by means of significant figures. For example, record 3.80 instead of 3.8 only if the reading was actually determined to hundredths.
26. Do not superimpose one number over another or on lines of sketches, and do not try to change one figure to another, as a 3 to a 5.
27. Make all possible arithmetic checks on the notes and record them before leaving the field.
28. Compare all misclosures and error ratios while in the field. On large projects where daily assignments are made for several parties, completed work is shown by satisfactory closures.
29. Arrange essential computations made in the field so they can be checked later.
30. Title, index, and cross-reference each new job or continuation of a previous one by client's organization, property owner, and description.
31. Sign surname and initials in the lower right-hand corner of the right page on all original notes. This places responsibility just as signing a check does.
32. Letter COPY in large size diagonally across pages of nonoriginal notes, but do not obscure or touch a sketch or any figure in so doing.

PROBLEMS

3-1. What information should normally be included in a good set of field notes?
3-2. Why should a pen not be used in field notekeeping?
3-3. Explain the reason for item 26 in Section 3-6 when recording notes.
3-4. What erasures, if any, are permitted in a field book?
3-5. Why should a zero be placed before a decimal point in tabulating values, such as 0.73?
3-6. State your idea of a sixth point (see Section 3-2) worth considering in appraising a set of field notes.

3-7. Why are stapled and spiral-bound field books not satisfactory for practical work?

3-8. Give two reasons why ruled vertical and horizontal lines are necessary on field book pages.

3-9. What disadvantage might an electronic note recorder have for a land surveyor running a small office?

3-10. Why are sketches in field books usually not drawn to exact scale?

3-11. When should sketches be made instead of just recording data?

3-12. If, as frequently happens, a sketch becomes too small, what should be done?

3-13. Give an example of inconsistency in recording elevations on a sketch.

3-14. Why should lettering be kept parallel with, or at right angles to, the pertinent feature?

3-15. North is generally placed at the top or left side, and sketches arranged to read from the bottom and right side of pages. Why?

3-16. Discuss two problems that might arise from omission in a field book of the first names and initials of a surveying field party.

3-17. Justify the requirement to list in a field book the make and serial numbers of tapes, transits, theodolites, levels, and EDMIs used on a survey.

3-18. What other information or notes, in addition to the four major items listed in Section 3-5, might be helpful on the right-hand field book page?

3-19. A resurvey was made of a lot in a 50-year-old subdivision but the date was omitted by mistake in the field book. What effect might this have on you as the purchaser of that lot?

3-20. If some feature is knowingly left out of field notes (such as an irregular old fence), what should be done?

3-21. What information should be included in surveying notes for fences, roads, trees, and buildings?

3-22. Differentiate between notes that should run down the page, and those that traditionally run up the page.

3-23. A theodolite with digital readout of angles is used in an isolated area. What advantages and disadvantages might result?

3-24. What similarity exists between the signer of a set of field notes and the endorser of a check?

3-25. Some states require the signer of a land survey plat to have taken an "active" part in the field survey. Why?

3-26. If the accuracy of a distance measurement (say to 0.01 ft) is questionable, how should this be noted?

3-27. When additions or insertions to notes must be made at a later date, how is this situation handled?

3-28. List the type, or types, of field notes used in each of Plates D-1 through D-13 in Appendix D.

BIBLIOGRAPHY

Brinker, R. C., B. A. Barry, and R. Minnick. 1980. *Noteforms for Surveying Measurements*, 2nd ed. Landmark Enterprises, Rancho Cordova, CA 95670.

Pafford, F. W. 1962. *Handbook of Survey Notekeeping*. New York: Wiley.

4 DISTANCE MEASUREMENT; TAPING

PART I
METHODS OF LINEAR MEASUREMENT

4-1. INTRODUCTION. Distance measurement is the basis of all surveying. Even though angles may be read precisely with elaborate equipment, the length of at least one line must be measured to supplement the angles in locating points.

In plane surveying the distance between two points means the horizontal distance. If the points are at different elevations, the distance is the horizontal length between plumb lines at the points.

Lengths of lines may be specified in different units. The unit generally used in plane surveying in the United States is the foot, decimally divided. In architectural and machine work and on some construction projects, the unit is a foot divided into inches and fractions of an inch. Geodetic surveying usually employs the meter. Chains, varas, rods, and other units have been, and still are, utilized in some localities, and for special purposes.

4-2. METHODS OF MEASURING HORIZONTAL DISTANCES. In surveying, linear measurements are obtained by (a) pacing, (b) odometer readings, (c) optical rangefinders, (d) tacheometry (stadia), (e) subtense bars, (f) taping, and (g) electronic distance measurement (EDM). Of these methods, taping and EDM are most commonly used by surveyors. Methods (a) through (e) are briefly discussed in the following sections. Taping is described in detail in Part II of this chapter and EDM is covered in Chapter 5. Distances can also be estimated, a technique useful in making field note sketches and checking measurements for mistakes.

Triangulation is a method for determining positions of points from which horizontal distances can be computed (see Chapter 20). In this procedure, lengths of lines are computed trigonometrically from measured base lines and angles. A variation of triangulation is the *airborne control* (ABC) system. It uses a helicopter with an optical plumb bob (*hoversight*). Electronic distance measurements (and/or angles) are taken from known ground control points to the helicopter as it hovers at a measured height over a ground station of unknown position. Other methods for determining positions of points from which horizontal distances can be calculated utilize *inertial* and *satellite doppler* surveying systems. These devices, described in Chapter 20, are currently used chiefly to position widely spaced points in control surveys. *Photogrammetry* can also be used to obtain horizontal distances. This topic is covered in Chapter 28.

4-3. PACING. Distances obtained by pacing are sufficiently accurate for many purposes in surveying, engineering, geology, agriculture, forestry, and military field sketching. Pacing is also used to detect blunders that may occur in taping or stadia readings.

Pacing consists of counting the number of steps or paces in a required distance. The length of an individual's pace must first be determined. This is best done by walking with natural steps back and forth over a measured level course at least 300 ft long, and dividing the known distance by the average number of steps. For short distances the length of each pace is needed, but the number of steps taken per 100 ft is desirable for checking long lines. Plate D-1 shows the notes for a field problem on pacing.

It is possible to adjust one's pace to an even 3 ft, but a person of average height finds such a step tiring if maintained for very long. The length of an individual's pace varies when going uphill or downhill and changes with age. For long distances, a pocket instrument called a *pedometer* can be carried to register the number of paces, or a *passometer* attached to the body or leg counts the steps. Some surveyors prefer to count *strides*, a stride being two paces.

Pacing is one of the most valuable things learned in surveying, since it has practical applications for everybody and requires no equipment. Experienced pacers can measure distances of 100 ft or longer with an accuracy of $\frac{1}{50}$ to $\frac{1}{100}$ if the terrain is open and reasonably level.

4-4. ODOMETER READINGS. An odometer converts the number of revolutions of a wheel of known circumference to a distance. Lengths measured by an odometer on a vehicle are suitable for some preliminary surveys in route-location work. They also serve as a rough check on measurements made by other methods. A precision of approximately $\frac{1}{200}$ is reasonable. Other types of measuring wheels are available and useful for determining short distances, particularly on curved lines. Odometers give surface distances which should be corrected to horizontal if the ground slopes severely (see Section 4-13).

4-5. OPTICAL RANGEFINDERS. These instruments operate on the same principle as rangefinders on single-lens reflex cameras. Basically, when focused, they

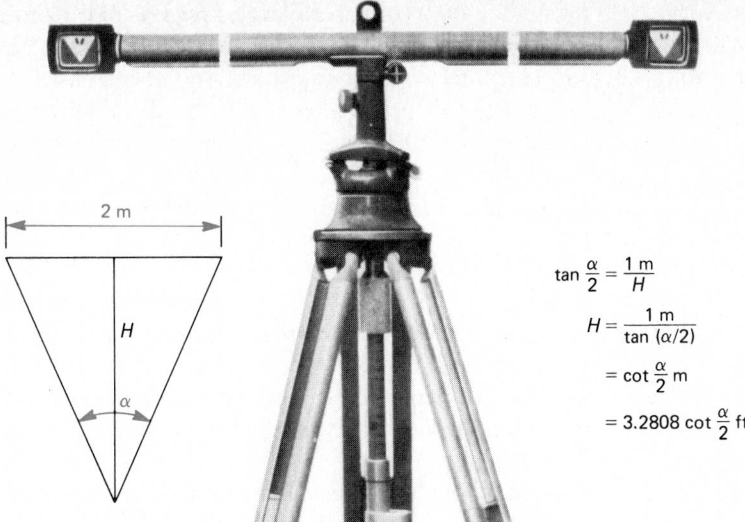

Figure 4-1. Subtense bar. (Courtesy Kern Instruments, Inc.)

solve for object distance f_2 in Eq. (6-6) where focal length f and image distance f_1 are known. An operator looks through the lens and adjusts the focus until a distant object viewed is focused in coincidence, whereupon a distance reading is obtained. These instruments are capable of accuracies of 1 part in 50 at distances up to 150 ft, but accuracy diminishes as length increases. They are suitable for reconnaissance, sketching, or checking more accurate measurements for mistakes.

4-6. TACHEOMETRY. Tacheometry (*stadia* is the more common term in the United States) is a surveying method used to quickly determine the horizontal distance to, and elevation of, a point. Stadia measurements are obtained by sighting through a telescope equipped with two or more horizontal cross hairs at a known spacing. The apparent intercepted length between the top and bottom hairs is read on a graduated rod held vertically at the desired point. The distance from telescope to rod is found by proportional relationships in similar triangles. A precision of $\frac{1}{500}$ is achieved with reasonable care. A detailed explanation of the method is given in Chapter 15.

4-7. SUBTENSE BAR. Several other optical methods of determining distances indirectly have been developed in which the angle subtended by a known distance between endmarks on a horizontal rod or bar (such as a subtense bar) is read on a precise transit or theodolite.

The Invar subtense bar shown in Figure 4-1 (along with a geometric diagram) is set on a tripod and aligned perpendicular to the survey line by means of a sighting device on top of the bar.[1] Fixed targets near the bar ends are

[1] Invar is a metal that has a low coefficient of thermal expansion and thus precisely maintains its true length despite temperature variations (see Section 4-15.2).

precisely 2 m apart. The horizontal angle α between targets is measured with a theodolite reading to 1 sec or less, and the horizontal distance is computed. Referring to Figure 4-1, the horizontal distance is given by:

$$H = \cot\frac{\alpha}{2} \quad \text{(meters)} \tag{4-1}$$

or

$$H = 3.2808 \cot\frac{\alpha}{2} \quad \text{(feet)} \tag{4-2}$$

where H is the horizontal distance and $\alpha/2$ half the measured angle subtended by the targets.

An important characteristic of the subtense-bar method is that *horizontal distances always result, even though inclined sights are taken, because α is measured in a horizontal plane.*

For sights of 500 ft (150 m) or less, and using a 1-sec theodolite, an accuracy of 1 part in 3000 can be achieved. Accuracy diminishes with increased line length, but this can be offset somewhat by repeating the angles several times, or taking readings from both ends of the line and averaging. The subtense-bar method of distance measurement was often used in the past to obtain distances over inaccessible courses—for example, over bodies of water. EDM devices have now almost totally replaced this procedure.

PART II
DISTANCE MEASUREMENTS BY TAPING

4-8. INTRODUCTION TO TAPING. Measurement of horizontal distance by taping consists of applying the known length of a graduated tape directly to a line a number of times. Two types of problems arise: (1) measuring a distance between fixed points, such as two stakes in the ground, and (2) laying out a distance with only the starting mark in place.

Taping is performed in six steps: (1) lining in, (2) applying tension, (3) plumbing, (4) marking tape lengths, (5) reading the tape, and (6) recording the distance. Application of these steps in taping on level and sloping ground is detailed in Sections 4-11 and 4-12.

4-9. TAPING EQUIPMENT. Various types of equipment used for taping in the United States, past and present, are described in this section.

4-9.1. HISTORICAL EQUIPMENT

Poles. Early surveyors struggled with braced timber panels and wood and metal poles. These devices resulted in the term *pole* as a unit of measure. Its length was $16\frac{1}{2}$ ft, the same as the *rod.*

Gunter's Chain. A Gunter's chain was the best measuring device available to surveyors for many years in the United States and is referred to in old field

Figure 4-2. Gunter's chain.

notes and deeds. It was 66 ft (4 poles) long and had 100 links, each link equal to 0.66 ft or 7.92 in. The links were made of heavy wire, had a loop at each end, and were joined together by three rings (Figure 4-2). The outside ends of the handles fastened to the end links were the 0 and 66-ft marks. Successive tags had one, two, three, or four teeth to mark every tenth link from each end. The center tag was round. With 600 or 800 connecting link and ring surfaces subject to frictional wear, hard use elongated the chain, and its length had to be adjusted by means of bolts in the handles.

Distances measured with chains were recorded either in chains and links or in chains and decimals of chains—for example, 7 ch 94.5 lk or 7.945 ch. Decimal parts of links were estimated. The 66-ft length of the Gunter's chain was selected because of its relevance to the mile and the relationship of a square chain to an acre. Thus, 1 ch $= \frac{1}{80}$ mile, and 10 ch^2 $= 10 \times 66^2 = 43{,}560$ ft^2 $=$ 1 acre.

Engineer's Chain. An engineer's chain had the same construction as a Gunter's chain but was 100 ft long and each of its 100 links had a length of 1 ft.

Chains are seldom, if ever, used today, although a steel tape graduated like a Gunter's chain is manufactured. Nevertheless, the many chain surveys on record oblige the modern practitioner to understand the limits of accuracy possible with this equipment and conversion of distances recorded in chains and links to feet. The term *chaining* continues to be used interchangeably with *taping*, even though tapes are employed exclusively.

Wires. Before thin flat steel, now used in modern tapes, could be produced efficiently, wires were utilized for measuring lengths. They still are practical in special cases—for example, hydrographic surveys.

4-9.2. TAPES IN CURRENT USE

Surveyor's and Engineer's Tapes. These tapes are made of steel $\frac{1}{4}$ to $\frac{3}{8}$ in wide and weigh 2 to 3 lb/100 ft. Lengths of 100, 200, 300, and 500 ft, and 30, 50, 60, and 100 m are standard. The 100-ft tape is most common. All can be wound on a reel, or done up in loops 5 ft long to make a figure 8, then *thrown* into a circle having a diameter of about $9\frac{1}{2}$ in. For long-distance taping (300 to 500 ft), *band chains* (also known as *chain tapes*) are available with smaller cross sections ($\frac{1}{8}$ to $\frac{5}{16}$ in by 0.016 to 0.025 in).

Tapes are graduated at every foot and marked from 0 to 100. Some tapes have only the last foot at each end subdivided into tenths, or tenths and hundredths, of a foot. Others are graduated in feet, tenths, and hundredths throughout. Still others (*adding tapes*) have an extra graduated foot beyond the zero mark. In all cases, a metal ring or loop at each end allows a handle or leather thong to be attached.

Special-Purpose Tapes. Tapes having suitable cross sections, lengths, composition, and graduation arrangements are manufactured for special purposes, such as base-line measurement, city engineering work, oil riggers and gaugers' use, and topographic surveys.

Builder's Tapes. Builder's tapes have smaller cross sections and are lighter in weight than surveyor's tapes. Since most building plans prepared by engineers and architects carry dimensions in feet and inches, the builder's tape is graduated in those units.

Invar Tapes. Invar tapes are made of a special nickel steel (35% nickel and 65% steel) to reduce length variations caused by differences in temperature. The thermal coefficient of expansion and contraction is only about $\frac{1}{30}$ to $\frac{1}{60}$ that of an ordinary steel tape. The metal is soft and somewhat unstable. This weakness of Invar tapes, along with their cost of perhaps 10 times that of ordinary tapes, makes them suitable only for precise geodetic work and as a standard for comparison with working tapes.

Lovar Tapes. A somewhat newer version, the Lovar tape, has properties and a cost between those of steel and Invar tapes.

Cloth Tapes. Cloth (or metallic) tapes are actually made of high-grade linen $\frac{5}{8}$ in wide with fine copper wires running lengthwise to give additional strength and prevent excessive elongation. Metallic tapes commonly used are 50, 100, and 200 ft long and come in enclosed reels. Although not suitable for precise work, metallic tapes are convenient and practical for many purposes but should *not* be used around electrical units.

Glass-Fiber Tapes. Glass-fiber tapes can be employed for the same types of work as metallic tapes and are safe around electrical equipment.

4-9.3. TAPING ACCESSORIES

Chaining Pins or Taping Pins. Sometimes called surveyor's arrows, they are used to mark tape lengths. Most taping pins are made of number 12 steel wire, sharply pointed at one end, have a round loop at the other end, and are painted with alternate red and white bands. Sets of 11 pins carried on a steel ring are standard.

Hand Level. This simple instrument, described in Section 6-16, is used to keep the tape ends at equal elevations when measuring over rough terrain.

Tension Handles. These facilitate the application of an exact standard or known tension. A complete unit consists of a wire handle, a clip to fit the end ring of the tape, and a spring balance reading up to 30 lb in $\frac{1}{2}$-lb calibrations.

Clamp Handles. These holders are used to apply tension by a positive, quick grip using a scissors-type action on any part of a steel tape without damage to the tape or injury to hands.

Pocket Thermometer. Thermometers for field use are about 5 in long, graduated from perhaps -30 to $+120°F$ in 1 or 2° divisions, and kept in protective metal cases.

Tape Repair Kits. A tape repair kit contains sleeve splices to be placed over the two parts of a broken tape, hammered down, and fastened with eyelets by a combined hand puncher and riveter.

Poles. Range poles (flags or lining rods) made of wood, steel, or aluminum are about 1 in thick and 6 to 10 ft long. They are round or hexagonal in cross section and marked with alternate red and white bands 1 ft long which can be used for rough measurements. A wooden range pole has a metal shoe at the base. The main utility of range poles is to mark alignment.

Plumb Bobs. Plumb bobs for taping should weigh a minimum of 8 oz and have a fine point. At least 6 ft of fish-line cord, free of knots, is necessary. Plumb-bob points are now standardized to simplify replacement.

Full equipment for a taping party consists of one 100-ft steel tape, one 50-ft metallic tape, two range poles, 11 chaining pins on a ring, two plumb bobs, one hand level or clinometer, keel (colored lumber crayon), and a field book. Some of this equipment is shown in Figure 4-3.

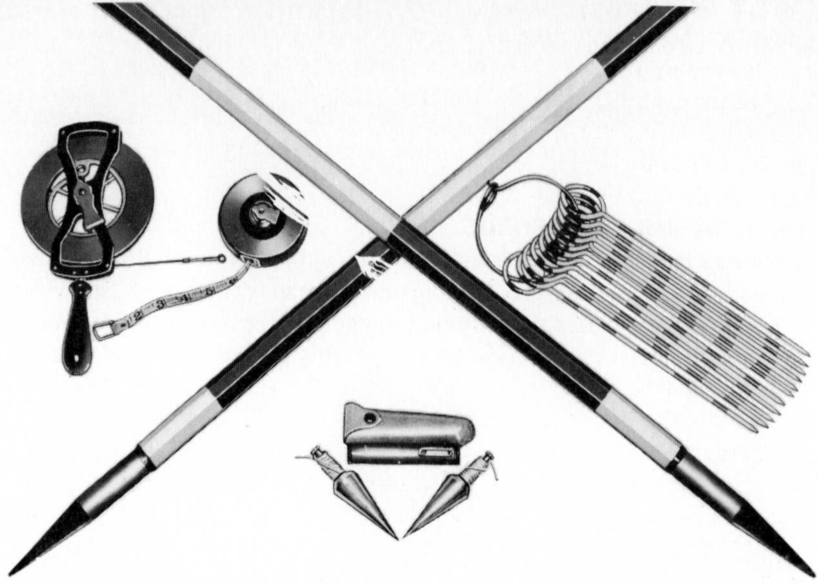

Figure 4-3. Taping equipment for a field party. (Courtesy W. & L. E. Gurley.)

4-10. CARE OF TAPING EQUIPMENT. The following points are pertinent in the care of tapes and range poles:

1. Considering the cross-sectional area of the average surveyor's steel tape and its permissible stress, a pull of 100 lb will do no damage. If the tape is kinked, however, a pull of less than 1 lb will break it. Therefore, always check to be certain that any loops and kinks are eliminated before tension is applied.
2. If a tape gets wet, wipe it first with a dry cloth, then with an oily rag.
3. Tapes should be either kept on a reel or "thrown" into circular loops, but not handled both ways.
4. Each tape should have an individual number or tag to identify it.
5. Broken tapes can be mended by riveting and/or applying a sleeve device, but a mended tape should not be used on important work.
6. Range poles are made with the metal shoe and point in line with the section above. This alignment may be lost if the pole is used improperly.

4-11. TAPING ON LEVEL GROUND. The subsections that follow describe the six steps in taping on level ground.

4-11.1. LINING IN. The line to be measured should be definitely marked at both ends, and at intermediate points where necessary, to ensure unobstructed sight lines. Range poles are ideal for this purpose. The forward tapeperson is lined in by the rear tapeperson (or by a transit or theodolite for greater accuracy). Directions are given by vocal or hand signals.

4-11.2. APPLYING TENSION. The 100-ft end of a tape is held over the first (rear) point by the rear tapeperson, while the forward tapeperson, holding the zero end, is lined in. For accurate results the tape must be straight and the two ends held at the same elevation. A specified tension, generally 10, 12, 15, 20, or 25 lb, is applied. To maintain a steady pull, tapepersons wrap the leather thong at the tape's end around one hand, keep forearms against their bodies, and face at right angles to the line. In this position, they are off the line of sight. Also, the body need only be tilted to hold, decrease, or increase the pull. Sustaining a constant tension with *outstretched* arms is difficult, if not impossible, for a pull of 15 lb or more. Good communication between head and rear tapepersons will avoid jerking the tape, save time, and get better results.

4-11.3. PLUMBING. Weeds, brush, obstacles, and surface irregularities may make it undesirable to lay a tape on the ground. Instead, the tape is held above ground in a horizontal position. Each end point on the tape is marked by placing the plumb-bob string over the proper tape graduation and securing it with one thumb. The rear tapeperson continues to hold a plumb bob over the fixed point, while the forward tapeperson marks the length. In measuring a distance shorter than a full tape length, the forward tapeperson moves the plumb-bob string to a point on the tape over the ground mark.

4-11.4. MARKING TAPE LENGTHS. When the tape has been lined in properly, tension has been applied, and the rear tapeperson is over the point, "stick" is called out. The forward tapeperson then places a pin exactly opposite the zero mark of the tape and calls "stuck." If the point is being plumbed over soft ground, the bob is released by raising the thumb, and a chaining pin carefully set in the hole made by the plumb-bob point. The pin should form a right angle with the tape but approximately a 45° angle with the ground. The point where the pin enters the ground is checked by repeating the measurement until certainty of its correct location is assured.

After checking the measurement, the forward tapeperson signals that the point is OK, the rear tapeperson pulls up the rear pin, and they move ahead. The forward tapeperson drags the tape, paces roughly 100 ft, and stops. Just before the 100-ft end reaches the pin that has been set, the rear tapeperson calls "tape" to notify the forward tapeperson that they have gone 100 ft. The process is repeated until a partial tape length is needed at the end of the line.

When a surveyor is working on pavement, the plumb bob is eased to the surface, and the point's position marked by a scratch, a spike, keel, a nail in a bottle cap, or other means.

4-11.5. READING THE TAPE. There are two common styles of graduations on surveyor's tapes. *It is necessary to identify the type being used before starting work* to avoid making 1-ft mistakes repeatedly.

The more common type of tape is calibrated from 0 to 100 by full feet in one direction, and has an additional foot beyond the zero end graduated from 0 to 1 ft in tenths (and perhaps hundredths) in the other direction, making the complete tape 101 ft long. With a full-foot graduation held by the rear tapeperson at the last pin set [like the 87-ft mark in Figure 4-4(a)], the graduations

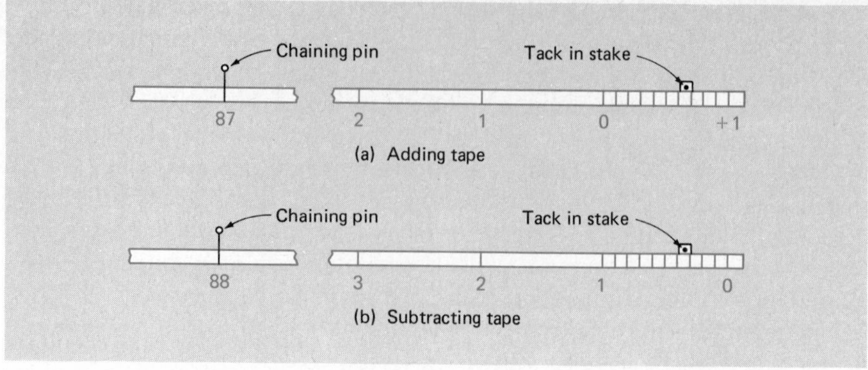

Figure 4-4. Reading partial tape lengths.

between zero and the tape end should straddle the closing point. The head tapeperson reads the additional length of 0.68 ft beyond the zero mark. To ensure correct recording, the rear tapeperson calls "87." The head tapeperson repeats and adds the partial foot reading, calling "87.68." Since part of a foot has been added, this type of tape is known as an *adding tape.*

The other kind of tape found in practice is calibrated from 0 to 100 by full feet, and the first foot at each end (from 0 to 1 and from 99 to 100) is graduated in tenths (and perhaps hundredths). Thus the complete tape is 100 ft long. With a full-foot graduation held at the last chaining pin set, the graduated section of the tape between the zero mark and the 1-ft mark should straddle the closing point, as indicated in Figure 4-4(b), where the 88-ft mark is being held on the last chaining pin and the tack marking the end of the line is opposite 0.32 ft read from the zero end. The partial tape length is then 88.00 − 0.32 = 87.68 ft. The quantity 0.32 ft is said to be *cut off*, and this type of tape is called a *subtracting* or *cut tape.* To ensure subtraction of a foot from the number at the full-foot graduation used, the following field procedure and calls are recommended: Rear tapeperson calls "88"; forward tapeperson says "cut point three-two"; rear tapeperson answers "eighty seven point six eight"; forward tapeperson replies "check."

Subtraction of the decimal of a foot is avoided if the forward tapeperson reads (counts) 0.68 ft backward from the 1-ft graduation. Calls of "88," "0.68," "87.68," and "check" are made in this procedure. The only justification for a subtracting tape, if there is any, appears to be its use in "getting off a plus" (described in Section 4-14) on route surveys. These tapes afford reduced chance of an arithmetic error or making the mistake of using 101 ft for a full tape length.

The same routine should be used throughout all taping by a party and the results tested in every possible way. A single failure to subtract 1 ft in the procedure just described when using a cut tape will destroy the precision of a hundred other measurements. For this reason, the adding tape is more nearly foolproof. The greatest danger arises when changing from one style to the other.

It is customary to have the 100-ft end of the subtracting tape ahead in route surveys, where stationing along the line is continuous. Some surveyors prefer the arrangement in other work also, when setting intermediate points or measuring partial tape lengths.

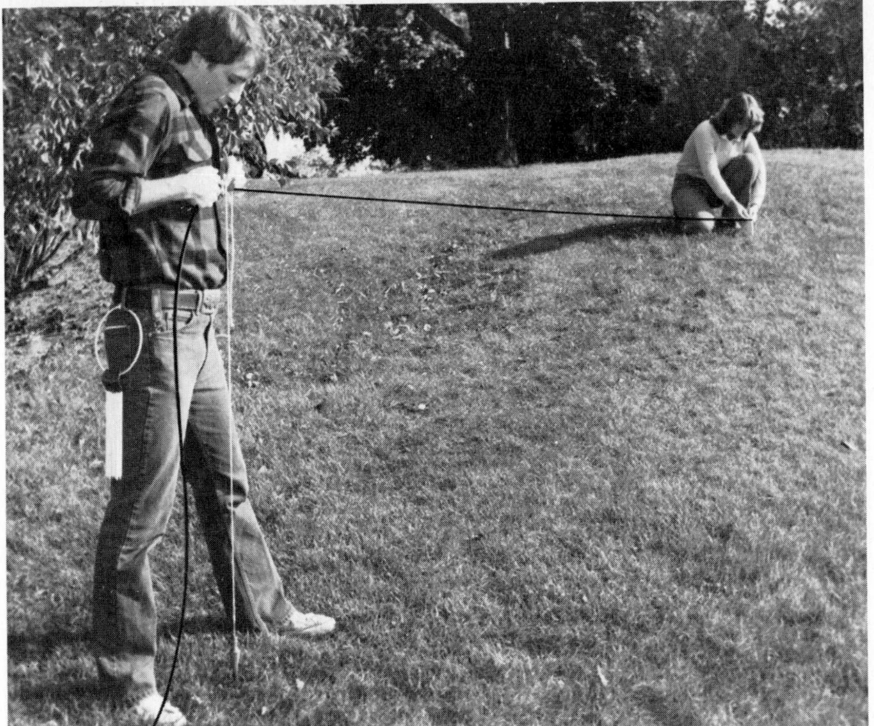

Figure 4-5. Breaking tape.

4-11.6. RECORDING THE DISTANCE. Accurate field work may be canceled by careless recording. After the partial tape length is obtained at the end of a line, the rear tapeperson determines the number of full 100-ft tape lengths by counting the pins collected from the original set of 11. For distances longer than 1000 ft, a notation is made in the field book when the rear tapeperson has 10 pins and one remains in the ground. This signifies a tally of 10 full tape lengths and is traditionally called an "out." The forward tapeperson starts out again with 10 pins and the process is repeated.

Taping is a skill that can best be taught and learned by field demonstrations and practice.

4-12. HORIZONTAL MEASUREMENTS ON UNEVEN GROUND. In taping on uneven or sloping ground, it is standard practice to hold the tape horizontal and use a plumb bob at one or perhaps both ends. It is difficult to keep the plumb line steady for heights above the chest. Wind exaggerates this problem and may make accurate work impossible.

Where a 100-ft length cannot be held horizontal without plumbing from above shoulder level, shorter distances are measured and accumulated to total a full tape length. This procedure, called *breaking tape*, is illustrated in Figure 4-5.

As an example of this operation, assume that when the 100-ft end of the tape is held at the rear point, the forward tapeperson can advance only 30 ft

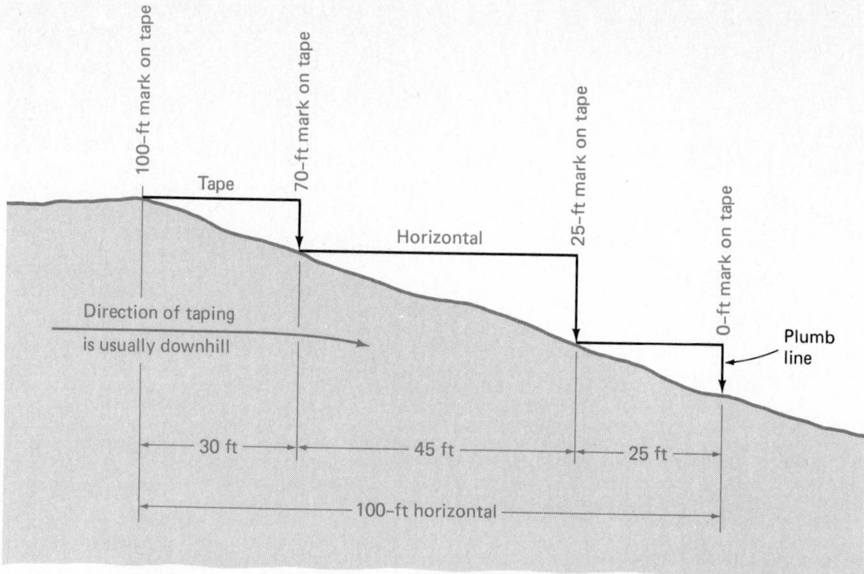

Figure 4-6. Procedure for breaking tape (when tape is not in a box or on a reel).

without being forced to plumb from above the chest. A pin is therefore set beneath the 70-ft mark, as in Figure 4-6. The rear tapeperson moves ahead to this pin and holds the 70-ft graduation while another pin is set at, say, the 25-ft mark. Then, with the 25-ft graduation over the second pin, the full 100-ft distance is marked at the zero point.

To avoid kinking the tape, the full length is pulled ahead by the forward tapeperson, which does waste some time in the process of walking ahead and then back. But the partial tape lengths are added mechanically to make a full 100 ft by holding the proper graduations. No mental arithmetic is required. The rear tapeperson returns the pins set at the intermediate points to the forward tapeperson to keep the tally clear on the number of full tape lengths established. In all cases the tape is leveled by eye or hand level, with the tapepersons keeping in mind the natural tendency to have the downhill end of a tape too low. Practice will improve the knack of holding a tape at right angles to the plumb-bob string.

In an alternate procedure, only the partial length of 30 ft (see Figure 4-6) is pulled ahead, a point marked, another 45 ft advanced, and finally, after adding the partial values (which step is eliminated in the other method), a 25-ft length establishes the full station.

Taping downhill is preferable to measuring uphill, since in taping downhill the rear point is held steady on a fixed object while the other end is plumbed. In taping uphill, the forward point must be set while the other end is wavering somewhat.

4-13. SLOPE MEASUREMENTS. In measuring the distance between two points on a steep slope, it may be desirable to tape along the slope and determine the angle of inclination α, or the difference in elevation d (Figure 4-7), rather than

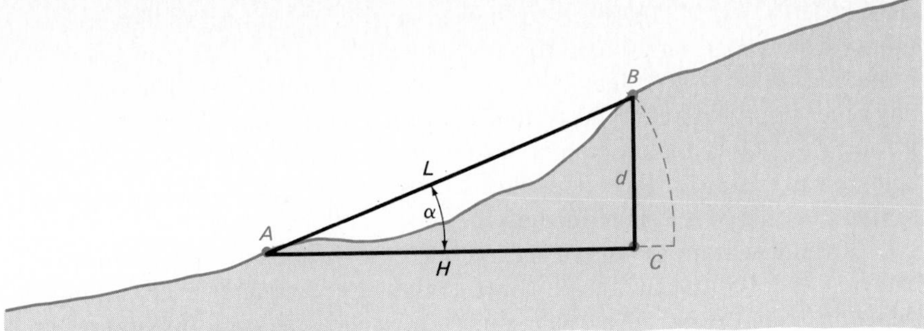

Figure 4-7. Slope measurement.

break tape every few feet. Long tapes (200 to 500 ft) are advantageous for measuring along slopes (as well as across rivers and ravines), and in some military operations.

In Figure 4-7, if angle α is determined, the horizontal distance between points A and B can be computed from the relation

$$H = L \cos \alpha \tag{4-3}$$

where H is the horizontal distance between points, L the slope length separating them, and α the vertical angle from the horizontal, usually obtained with an Abney hand level, clinometer (see Figure 6-16), transit, or theodolite (see Chapter 10).

In a variation of slope taping, the difference in elevation d between the ends of the tape is found by leveling, and the horizontal distance is computed using the following expression derived from the Pythagorean theorem:

$$H = \sqrt{L^2 - d^2} \tag{4-4}$$

Another approximate formula, obtained from the first term of a binomial expansion of the Pythagorean theorem, may be used to reduce slope distances to horizontal:

$$H = L - \frac{d^2}{2L} \quad \text{(approx.)} \tag{4-5}$$

As demonstrated in Eq. (4-5), the term $-d^2/2L$ equals C in Figure 4-7, and is a correction to be subtracted from the measured slope length to obtain the horizontal distance.

The error in using the approximate formula for a 100-ft length grows with increasing slope, but for inclinations up to 10% the answer is correct to the nearest 0.001 ft. More precise results are obtained for slopes steeper than 10% by including the second term of the binomial expansion; thus,

$$H = L - \left(\frac{d^2}{2L} + \frac{d^4}{8L^3}\right) \tag{4-6}$$

4-14. STATIONING. In route surveying, stationing is carried along continuously from a starting point designated as station 0 + 00. The term *full station* is applied to each 100-ft length, where a stake is normally set. The position of any other point is given by its total distance from the point of beginning. Thus, station 7 + 84.9 is a unique point 784.9 ft from the starting mark, this distance being measured along the survey line. The partial length beyond a *full* station, in this example 84.9 ft, is termed a *plus*.

Taping in stations (with a subtracting tape) is done most conveniently by carrying the 100-ft end ahead. Since stakes are driven at every change in direction (*angle point*) of a route survey as well as at each full station, it is necessary to follow each plus station with a full one. *To determine the plus station* of an angle point, for example, at station 7 + 84.9, the 100-ft end of the tape is pulled beyond the angle point by the forward tapeperson, who then walks back and holds a full-foot graduation on the stake (in this case, the 85-ft mark). Meanwhile the rear tapeperson reads the number of tenths of a foot from the 1-ft mark (in this example 0.9 ft).

To get off a plus and establish the next full station, 8 + 00, a rear tapeperson holds the 84-ft graduation at the plus station, and a forward tapeperson sets a pin 0.9 ft back from the 100-ft mark. Note that all subtractions are eliminated by holding the foot mark corresponding to the plus at that station and reading the decimal part of the plus from the 100-ft graduation. This method, like others to be discussed, exemplifies the advantage of systematizing field procedures to reduce the possibility of mistakes.

4-15. SOURCES OF ERROR IN TAPING. There are three fundamental sources of error in taping.

1. *Instrumental errors.* A tape may differ in actual length from its nominal length because of a defect in manufacture or repair, or as a result of kinks.
2. *Natural errors.* The horizontal distance between end graduations of a tape varies because of the effects of temperature, wind, and weight of the tape itself.
3. *Personal errors.* Tapepersons may be careless in setting pins, reading the tape, or manipulating the equipment.

Common types of errors in taping are given under the following nine headings:

1. Incorrect length of tape.
2. Temperature other than standard.
3. Inconsistent pull.
4. Sag.
5. Poor alignment.
6. Tape not horizontal.
7. Improper plumbing.
8. Faulty marking.
9. Incorrect reading or interpolation.

Some of these classifications produce systematic errors; others, random errors. They are described in the following sections.

4-15.1. INCORRECT LENGTH OF TAPE. Incorrect length of a tape is one of the most important errors. It is systematic. Tape manufacturers do not guarantee steel tapes to be exactly their nominal length—for example, 100.00 ft—or provide a standardization certificate unless requested and paid for as an extra. The true length is obtained by comparing it with a standard tape or distance. The National Bureau of Standards in Gaithersburg, Maryland, will make such a comparison for a nominal fee and certify the exact distance between end graduations under given conditions of temperature, tension, and manner of support.

A 100-ft steel tape usually is standardized for each of two sets of conditions—for example, 68°F, a 12-lb pull, with the tape fully supported throughout; and 68°F, a 20-lb pull, with the tape supported at the ends only. Schools and surveying offices normally have at least one standardized tape that is used only to check other tapes subjected to wear.

An error due to incorrect length of a tape occurs each time the tape is used. If the true length, known by standardization, is not exactly equal to its nominal value of 100.00 ft recorded for every full length, the correction can be determined and applied from the formulas:

$$C_l = \left(\frac{l - l'}{l'}\right) L \tag{4-7}$$

and

$$\bar{L} = L + C_l \tag{4-8}$$

where C_l is the correction to be applied to the measured (recorded) length of a line to obtain the true length, l the actual tape length, l' the nominal tape length, L the measured (recorded) length of line, and $\bar{L}$ the corrected length of line.

EXAMPLE 4-1

A 100-foot steel tape when compared with a standard is actually 100.02 ft long. What is the corrected length of the line measured with this tape and found to be 565.75 ft?

By Eq. (4-7):

$$C_l = \left(\frac{100.02 - 100.00}{100.00}\right) 565.75 = +0.11 \text{ ft}$$

By Eq. (4-8):

$$\bar{L} = 565.75 + 0.11 = 565.86 \text{ ft}$$

The example illustrates that in measuring unknown distances with a tape that is too long, a correction must be added. Conversely, if the tape is too short, the correction will be minus, resulting in a decrease (C_l is still added but has a negative sign).

An alternate method of making corrections for an incorrect tape length is to compute the amount by which a full tape length is too long or too short and then multiply it by the number of tape lengths in the line. Thus, in Example 4-1, the tape is 0.02 ft too long and the total correction is 5.6575 × 0.02 = 0.11 ft. This value is then added to the measured length of 565.75 to get 565.86 ft.

From a practical standpoint, the effect of any error is to make the tape length incorrect. Note that the true (actual) distance equals the measured distance plus a correction, and the proper algebraic sign for Eq. (4-8) is "built in." This is also true for the corrections discussed in succeeding sections. However, students should still try to reason whether a certain condition "makes" a tape too long or too short and apply the correction accordingly.

4-15.2. TEMPERATURE OTHER THAN STANDARD. Steel tapes are standardized for 68°F (20°C) in the United States. A temperature higher or lower than this value causes a change in length that must be considered.

The coefficient of thermal expansion and contraction of steel used in ordinary tapes is approximately 0.0000065 per unit length per degree Fahrenheit, and 0.0000116 per unit length per degree Celsius. For any tape, the correction for temperature can be computed and applied using the formulas:

$$C_t = k(T_1 - T)L \tag{4-9}$$

and

$$\bar{L} = L + C_t \tag{4-10}$$

where C_t is the correction in the length of a line due to nonstandard temperature, k the coefficient of thermal expansion and contraction of the tape, T_1 the tape temperature at time of measurement, T the tape temperature when it has standard length, L the measured (recorded) length of line, and $\bar{L}$ the corrected length of the line.

Errors due to temperature change may be practically eliminated by either (1) measuring temperature and making corrections according to Eqs. (4-9) and (4-10), or (2) using an *Invar* tape manufactured from a nickel-steel alloy. Temperature effects on the length of such tapes are negligible for most practical work. *Lovar* tapes with coefficients approximately $\frac{1}{3}$ that of steel can also be used to reduce temperature effects. Invar and Lovar tapes are fragile and lose calibration more readily than steel tapes, especially if mistreated.

EXAMPLE 4-2

The recorded length of a line measured at 30.5°F with a steel tape that is 100.00 ft long at 68°F was 872.54 ft. What is the line's corrected length?

By Eq. (4-9):

$$C_t = 0.0000065(30.5 - 68)872.54 = -0.21 \text{ ft}$$

By Eq. (4-10):

$$\bar{L} = 872.54 - 0.21 = 872.33 \text{ ft}$$

Errors due to temperature changes are systematic and have the same sign if the temperature is always above 68°F or always below that standard. When the temperature is above 68°F during part of the time occupied in measuring a long line, and below 68°F for the remainder of the time, the errors tend to partially balance each other, but corrections should still be computed and applied.

Temperature effects are difficult to assess in taping. The air temperature read from a thermometer may be quite different from that of the tape to which it is attached. Sunshine, shade, wind, evaporation from a wet tape, and other conditions make the tape temperature uncertain. Field experiments prove that temperatures on the ground or in the grass may be 10 to 25° higher or lower than those at shoulder height because of a 6-in "layer of weather" (microclimate) on top of the ground. Since a temperature difference of 15°F produces a change of 0.01 ft per tape length, the importance of large variations is obvious.

Shop measurements made with steel scales and other devices likewise are subject to temperature effects. The precision required in fabricating a large airplane or ship can be lost by this one cause alone.

4-15.3. INCONSISTENT PULL. When a steel tape is pulled with a tension greater than standard, it elongates in an elastic manner. The modulus of elasticity of a material is the ratio of unit stress to unit elongation, or

$$E = \frac{\text{unit stress}}{\text{elongation per unit length}} = \frac{P/A}{e/L}$$

Elongation e is the correction for pull and is computed and applied using the following formulas:

$$C_p = (P_1 - P)\frac{L}{AE} \tag{4-11}$$

and

$$\bar{L} = L + C_p \tag{4-12}$$

where C_p is the total elongation in tape length due to pull, in feet; P_1 the pull applied to the tape, in pounds; P the standard pull for the tape, in pounds; A the cross-sectional area, in square inches; E the modulus of elasticity of steel, in pounds per square inch; L the measured (recorded) length of line; and $\bar{L}$ the corrected length. An average value of E is 29,000,000 lb/in^2 for the kind of steel used in tapes. The cross-sectional area of a steel tape can be obtained from the manufacturer, by measuring its width and thickness with calipers, or by dividing the total tape weight by its length (in feet) times the unit weight of steel (490 lb/ft^3), and multiplying by 144 to convert square feet to square inches.

Equations (4-11) and (4-12) also apply when the metric system is used. In that case, to produce the correction C_p in meters, comparable units are P and P_1 in kilograms, L and $\bar{L}$ in meters, A in square centimeters, and E in kilograms per square centimeter. An average value of E for steel in these units is approximately 2,000,000 kg/cm^2.

Errors resulting from incorrect tension can be eliminated by (1) using a spring balance to measure and maintain the standard pull, or (2) applying a pull other than standard and making corrections for the deviation from standard according to Eqs. (4-11) and (4-12).

EXAMPLE 4-3

A steel tape that is 100.000 ft long under a pull of 12.0 lb when supported throughout, and has a cross-sectional area of 0.005 in^2, is applied fully supported with a 20-lb pull to measure a line whose recorded length is 686.79 ft. What is the corrected length of the line?

By Eq. (4-11):

$$C_p = \frac{(20 - 12)686.79}{0.005(29{,}000{,}000)} = +0.038 \text{ ft}$$

By Eq. (4-12):

$$\bar{L} = 686.79 + 0.038 = 686.83 \text{ ft}$$

Errors due to incorrect pull may be either systematic or random. The pull applied by even an experienced tapeperson is sometimes greater or less than the desired value. An inexperienced person, particularly one who has not used a spring balance on a tape, is likely to apply less than the standard tension consistently.

4-15.4. SAG. A steel tape not supported along its entire length sags in the form of a catenary, a good example being the cable of a suspension bridge. Sag shortens the horizontal (chord) distance between end graduations, because the tape length remains the same (Figure 4-8). Sag can be diminished (by greater tension) but not eliminated unless the tape is supported throughout.

The actual sag of a tape (for example, 6 in below the horizontal) is not important. The reduced chord distance between the end graduations is the critical factor.

For a small deflection v at the center of a tape length, the equation of a parabola can be used to investigate the sag effect. If C_s is the correction for sag (difference between length of curve and straight line from one support to the next), in feet; L_s the unsupported length of the tape, in feet; d the chord distance between supports, in feet; w the weight of the tape per foot of length, in pounds; both W and wL_s the total tape weight between supports, in pounds; and P_1 the pull on the tape, in pounds; then

$$L_s - d = \frac{8v^2}{3d}$$

0 Foot mark

Foot mark 100

(a) Tape supported throughout

(b) Tape supported at ends only

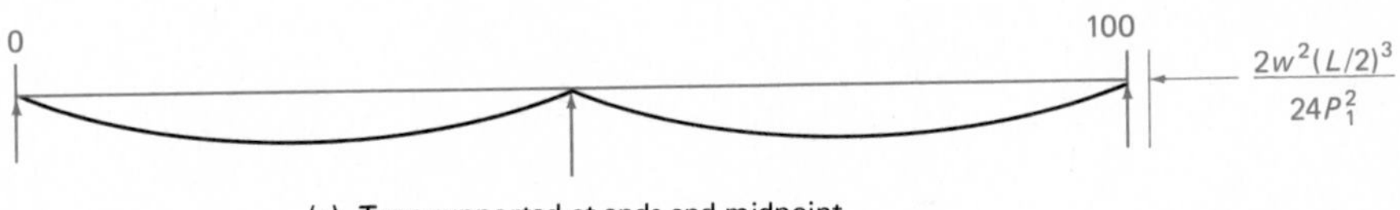

(c) Tape supported at ends and midpoint

Figure 4-8. Effect of sag.

Also, by mechanics,

$$P_1 v = \frac{wd^2}{8}$$

Combining these two relations and assuming $L_s = d$ to simplify results, the following equation is obtained:

$$C_s = -\frac{L_s}{24}\left(\frac{W}{P_1}\right)^2 = -\frac{W^2 L_s}{24P_1^2} = -\frac{w^2 L_s^3}{24P_1^2} \qquad \text{(4-13)}$$

Theoretically a term $\cos^2 \alpha$ (where α is the vertical angle) should be included on the right side of Eq. (4-13). Practically, however, it has little significance except for steep slopes and long tape lengths. Equation (4-13) gives the sag correction, which is always negative, for each measurement taken with the tape unsupported. English system units for Eq. (4-13) are pounds for W, w, and P_1 and feet for L_s. In the metric system, kilograms are used for W, w, and P_1 and meters for L_s. Having measured a line in several segments, and having calculated a sag correction for each segment, the corrected length is given by

$$\bar{L} = L + \sum C_s \qquad \text{(4-14)}$$

where L is the recorded length of a line, $\sum C_s$ the sum of the individual sag corrections, and $\bar{L}$ the corrected length of the line.

The effects of errors due to sag can be eliminated by (1) supporting the tape at short intervals or throughout, or (2) computing a sag correction for each unsupported segment and applying the total to the recorded length according to Eqs. (4-13) and (4-14).

EXAMPLE 4-4

A steel tape 100.000 ft long weighs 1.50 lb (0.015 lb/ft) and is used supported at the ends only as in Figure 4-8(b). A line in three segments is measured using a 12-lb pull and recorded as 250.52 ft. What is the length of the line corrected for sag?

By Eq. (4-13), for each 100-ft segment

$$C_s = \frac{-(1.50)^2(100)}{24(12)^2} = -0.065 \text{ ft}$$

and for the 50.52-ft segment

$$C_s = \frac{-(0.015)^2(50.52)^3}{24(12)^2} = -0.008 \text{ ft}$$

The corrected distance is

$$\bar{L} = 250.52 - 2(0.065) - 0.008 = 250.38 \text{ ft}$$

Note in this example the dramatic decrease in sag correction to a practically negligible figure for the 50.52-ft tape length. Thus, the effects of sag can be almost eliminated by supporting the tape at its midpoint when full tape lengths are measured.

As previously stated, sag corrections are always negative, whereas corrections for pull are positive if the tension applied exceeds the standard pull. These factors can be regulated to offset each other. By setting Eqs. (4-11) and (4-13) equal to each other and rearranging, the following formula, which can be solved by trial, is obtained for a pull that produces offsetting corrections:

$$P_1 = \frac{0.2W\sqrt{AE}}{\sqrt{P_1 - P}} \qquad \text{(4-15)}$$

where P_1 is the total pull of the tape, in pounds or kilograms; P the pull for a standardized tape (supported throughout), in pounds or kilograms; W the tape weight, in pounds or kilograms; A the tape cross-sectional area, in square inches or square centimeters; and E the modulus of elasticity of steel, in pounds per square inch or kilograms per square centimeter.

The pull required to balance the sag for a tape that has a standard tension of 12.0 lb, a cross-sectional area of 0.0050 in^2, and a weight of 1.7 lb is found by trial to be 30.3 lb. Thus,

$$30.3 = \frac{0.2(1.7)\sqrt{0.0050 \times 29{,}000{,}000}}{\sqrt{30.3 - 12.0}}$$

The pull required to make the distance between end graduations exactly equal to nominal length (for example 100.000 ft or 30.0000 m) with the tape unsupported is called the *normal tension.* Normal tension is not commonly used because it may be too great for convenient application and it changes with temperature variations. Theoretically a tension that also offsets temperature can be computed, but in cold weather an impractical pull often results.

4-15.5. POOR ALIGNMENT. If one end of a tape is off-line or the tape is snagged on an obstruction, a systematic error is introduced. The correction for offset from alignment, C_a, can be calculated by Eqs. (4-4) through (4-6) with both d and L in the horizontal plane; that is, d is the distance the tape is off-line and L the length of tape involved.

When a pin marking the end of a 100-ft length is set 1.4 ft off-line, by Eq. (4-5) the error in that measurement is $1.4^2/200 = -0.01$ ft. A similar error enters the next tape length when the succeeding pin is put correctly on-line.

If the center of a 100-ft tape catches on brush and is 1.0 ft off-line, the error produced in the two 50-ft lengths is $2(1.0^2/100) = -0.02$ ft.

Errors resulting from poor alignment are systematic in effect and always make the recorded length longer than the true distance. They may be reduced (but never eliminated) by care in setting pins, lining in properly, and keeping the tape straight. Snapping the tape while applying tension will straighten it. A moderate amount of field practice enables a rear tapeperson to keep the forward tapeperson (who sights along marks on the back line) within much less than a foot off the correct course.

4-15.6. TAPE NOT HORIZONTAL. The error caused by a tape being inclined in the vertical plane is the same as that resulting from it being off-line in the horizontal plane. Corrected lengths can also be determined by Eqs. (4-4) through (4-6), where d is the difference in elevation between the tape ends and L the tape length.

Errors due to the tape not being horizontal are systematic and always make the recorded length longer than the true length. They are reduced by using a hand level to check elevations of the tape ends, or by running differential levels (see Section 7-4) over the taping points. The errors cannot be completely eliminated, since the tape is certain to be out of level on some measurements despite the best efforts of a tapeperson.

4-15.7. IMPROPER PLUMBING. Practice and steady nerves are necessary to hold a plumb bob still long enough to mark a point or permit an instrument sight. The plumb bob moves around, even in calm weather. On very light slopes and on smooth surfaces such as pavements, inexperienced tapepersons obtain better results by laying the tape on the ground instead of plumbing. Experienced tapepersons plumb most measurements.

Errors due to improper plumbing are random, since they may make distances either too long or too short. The errors would be systematic, however, when taping directly against or in the direction of a strong wind.

Touching the plumb bob on the ground, or steadying it with one foot, decreases its swing. Practice in plumbing will reduce errors.

TABLE 4-1. TYPES OF ERRORS

TYPE AND CLASS OF ERROR		SYSTEMATIC (S) OR RANDOM (R)	DEPARTURE FROM NORMAL TO PRODUCE 0.01-ft ERROR FOR A 100-ft TAPE
Tape length	I	S	0.01 ft
Temperature	N	S or R	15°F
Pull	P	S or R	15 lb
Sag	N, P	S	0.6 ft at center for 100-ft tape standardized by support throughout
Alignment	P	S	1.4 ft at one end of 100-ft tape or 0.7 ft at midpoint
Tape not level	P	S	1.4 ft
Plumbing	P	R	0.01 ft
Marking	P	R	0.01 ft
Interpolation	P	R	0.01 ft

4-15.8. FAULTY MARKING. Chaining pins should be set perpendicular to the taped line but inclined 45° to the ground. This position permits plumbing to the point where the pin enters the ground without interference from the loop.

Brush, stones, and roots deflect a chaining pin and may increase the effect of incorrect marking. Errors from these sources tend to be random and are kept small by carefully locating a point, then checking it.

4-15.9. INCORRECT READING OR INTERPOLATION. The process of reading to hundredths on tapes graduated only to tenths is called *interpolation.* This process is readily learned and can be applied in many areas of surveying and engineering.

Errors due to interpolation are random over the length of a line. They can be reduced by care in reading, by using a small scale to determine the last figure, and by correcting any disposition toward particular values. Tabulating the number of times each digit from 0 through 9 is interpolated in work covering a period of several days, and plotting a polar graph of the results, will expose any predilection for a few numerals.

4-15.10. SUMMARY OF EFFECTS OF TAPING ERRORS. An error of 0.01 ft is significant in many surveying measurements. Table 4-1 lists the nine types of errors; classifies them as instrumental (I), natural (N), or personal (P), and systematic (S) or random (R); and gives the departure from normal that produces an error of 0.01 ft in a 100-ft length. The summary verifies practical experience that recorded lengths of lines are more often too long than too short.

The accepted method of reducing errors on precise work is to make several measurements of the same line with various tapes, at different times of day, and in opposite directions. An accuracy of $\frac{1}{10,000}$ can be obtained by careful attention to details.

4-16. TAPE PROBLEMS. All tape problems develop from the fact that a nominal 100-ft tape is longer or shorter than 100.00 ft because of manufacture,

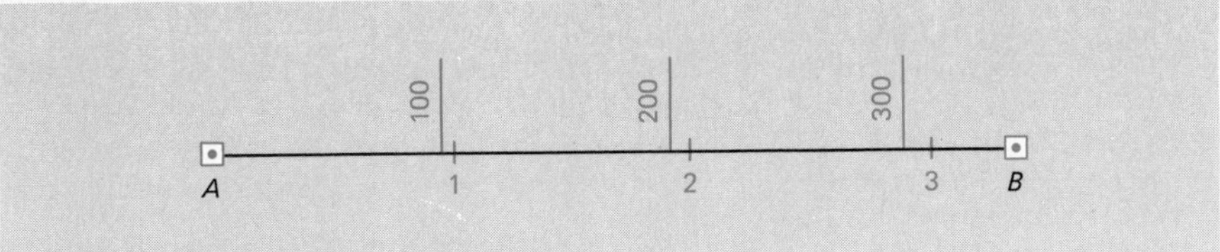

Figure 4-9. Taping between fixed points, tape too long.

temperature changes, tension applied, or some other reason. There are only four versions of the problem: A line can be *measured* between two fixed points, or a distance *laid off* from one fixed point, with a tape that is either too long or too short. The solution of a particular problem is always simplified and verified by drawing a sketch.

Assume that the fixed distance *AB* in Figure 4-9 is measured with a tape that is later found to be 100.03 ft long. Then (the conditions in the figure are greatly exaggerated) the first tape length would extend to point 1; the next, to point 2; and the third, to point 3. Since the distance remaining from 3 to *B* is less than the correct distance from the 300-ft mark to *B*, the *recorded* length *AB* is too small and must be increased by a correction. If the tape had been too short, the *recorded* distance would be too large, and the correction must be subtracted.

In laying out a required distance from one fixed point, the reverse is true. The correction must be subtracted from the desired length for tapes longer than the nominal value and added for tapes shorter than the assumed length. A simple sketch like Figure 4-9 makes clear whether the correction should be added or subtracted for any of the four cases.

4-17. COMBINED CORRECTIONS IN A TAPING PROBLEM.

In taping linear distance, several types of systematic errors often occur simultaneously. An example is precise taping using *chaining bucks* (tripods) to suspend the ends of the tape (usually at different elevations). The tape may not be exactly 100.000 ft long, and the taping is, of course, performed during ambient temperatures that are not standard. For convenience, a uniform pull may be applied that is other than standard, and because the tape is suspended at the ends only, a sag error results. On uneven ground the taping buck tops will not be at the same elevations, but these elevation differences can be measured. In this instance corrections for five conditions would be applied: (1) tape length, (2) temperature, (3) pull, (4) sag, and (5) slope.

EXAMPLE 4-5

A steel tape standardized at 68°F and supported throughout under a tension of 20 lb was found to be 100.012 ft long. The tape had a cross-sectional area of 0.0078 in^2 and a weight of 0.0266 lb/ft. This tape was used supported at the ends only with a constant tension of 15 lb to measure a line from *A* to *B* in nine segments. The data given in the following table were recorded. Apply corrections for tape length, temperature, pull, sag, and tape not horizontal to determine the correct length of the line.

SECTION	MEASURED (RECORDED) DISTANCE (ft)	TEMPERATURE (°F)	DIFFERENCE IN ELEVATION (ft)	INCLINATION CORRECTION (ft)
A–1	100.000	58	1.26	0.008
1–2	100.000	58	0.98	0.005
2–3	100.000	59	0.60	0.002
3–4	100.000	59	0.81	0.003
4–5	100.000	59	1.22	0.007
5–6	100.000	60	2.06	0.021
6–7	100.000	60	2.54	0.032
7–8	100.000	60	2.68	0.036
8–B	70.564	61	1.87	0.025
	Sum = 870.564			0.139

SOLUTION

a. The tape length correction by Eq. (4-7) is

$$C_\ell = \left(\frac{100.012 - 100.000}{100.000}\right)870.564 = +0.104 \text{ ft}$$

b. The temperature corrections by Eq. (4-9) are (Note: Separate corrections are required for distances measured at each different temperature):

$$C_{t_1} = 0.0000065(58 - 68)200.000 = -0.013 \text{ ft}$$
$$C_{t_2} = 0.0000065(59 - 68)300.000 = -0.018 \text{ ft}$$
$$C_{t_3} = 0.0000065(60 - 68)300.000 = -0.016 \text{ ft}$$
$$C_{t_4} = 0.0000065(61 - 68)70.564 = -0.003 \text{ ft}$$
$$\sum C_t = -0.050 \text{ ft}$$

c. The pull correction by Eq. (4-11) is

$$C_p = \frac{(15 - 20)870.564}{0.0078(29{,}000{,}000)} = -0.019 \text{ ft}$$

d. The sag corrections by Eq. (4-13) are (Note: Separate corrections are required for the two different suspended lengths):

$$C_{s_1} = -8\left[\frac{(0.0266)^2(100.000)^3}{24(15)^2}\right] = -1.048 \text{ ft}$$

$$C_{s_2} = -\left[\frac{(0.0266)^2(70.564)^3}{24(15)^2}\right] = -0.046 \text{ ft}$$

$$\sum C_s = -1.094 \text{ ft}$$

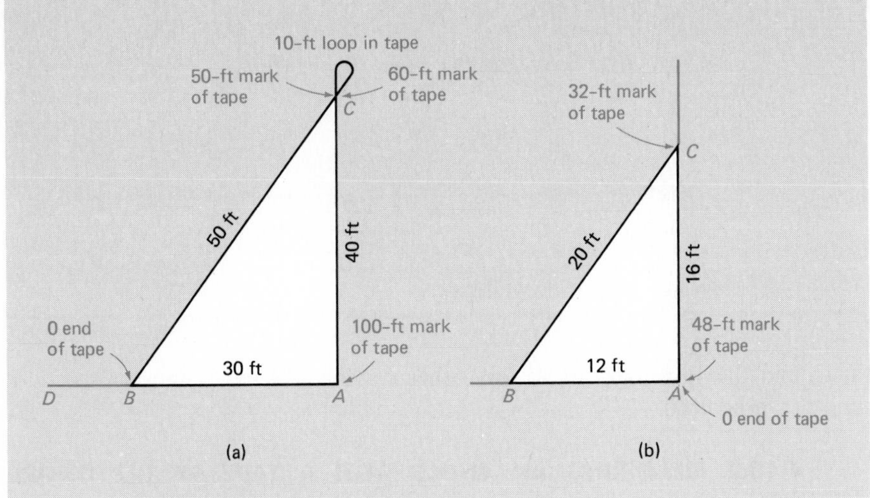

Figure 4-10. Laying out right angle with tape.

e. Corrections for slope have been calculated using the correction term $-d^2/2L$ of Eq. (4-5) and entered in the right-hand column of the table. To clarify, the first entry was obtained as

$$C = \frac{(1.26)^2}{2(100.000)} = -0.008 \text{ ft}$$

From the table, the sum of slope corrections is -0.139 ft.

f. Finally, the corrected distance AB is obtained by adding all corrections to the measured distance, or

$$\begin{aligned} AB &= 870.564 + 0.104 - 0.050 - 0.019 - 1.094 - 0.139 \\ &= 869.366 \text{ ft} \end{aligned}$$

Field data and corrections for this example have been carried out to thousandths of a foot. Ordinary taping does not justify working to this precision, but the procedure is the same.

4-18. SPECIAL FIELD OPERATIONS USING A TAPE. Many problems arising in the field can be solved by taping. Some examples follow.

4-18.1. LAYING OUT A RIGHT ANGLE WITH A TAPE. A right angle is readily laid out by the 3-4-5 method. In Figure 4-10(a), to erect a perpendicular to AD at A, measure 30 ft along AD and set point B. Then with the tape's zero graduation at B and the 100-ft mark at A, form a loop in the tape by bringing the 50- and 60-ft graduations together and pull each part of the tape taut to locate C. One person can make the layout alone by tying the tape thongs to stakes beyond A and B.

If a 50-ft metallic tape is used, a possible procedure is indicated in Figure 4-10(b). The zero mark is held at A, the 12-ft mark at B, the 32-ft mark

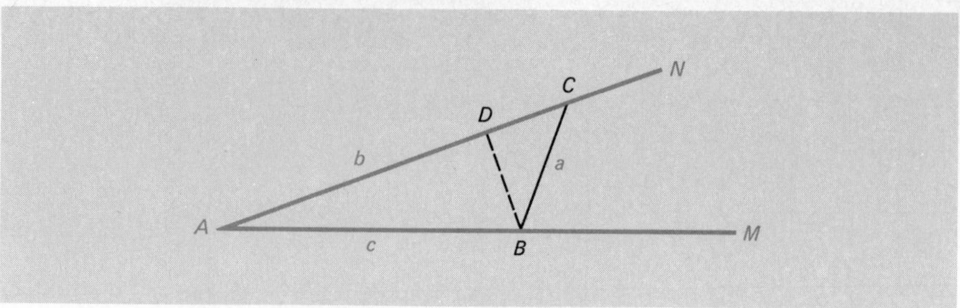

Figure 4-11. Measuring an angle with tape.

at C, and the 48-ft mark at A. Any other distances in the proportions of 3, 4, and 5 can be used.

4-18.2. MEASURING AN ANGLE WITH A TAPE BY THE CHORD METHOD. If all three sides of a triangle are known, the angles can be computed. To find angle A of Figure 4-11, measure any definite lengths along AM and AN, such as AB and AC. Also measure BC. Then

$$\sin \tfrac{1}{2}A = \sqrt{\frac{(s-b)(s-c)}{bc}} \tag{4-16a}$$

or

$$\cos A = \frac{b^2 + c^2 - a^2}{2bc} \tag{4-16b}$$

where a, b, and c are the sides of triangle ABC and $s = \frac{1}{2}(a + b + c)$.

For $b = 30.0$ ft, $c = 25.0$ ft, and $a = 12.5$ ft, angle A is calculated equal to 24°09′.

An isosceles triangle may be formed by making AB equal to AC. Then

$$\sin \tfrac{1}{2}A = \frac{a}{2c} \tag{4-17}$$

Selecting a value of 50 ft for AB and AC simplifies the arithmetic. Thus if $AB = AC = 50.0$ ft and BC measures 20.90 ft, then $\sin \frac{1}{2}A = 0.2090$ and angle $A = 24°08'$.

4-18.3. MEASURING AN ANGLE WITH A TAPE BY THE TANGENT METHOD. If AD and a perpendicular BD are measured (see Figure 4-11), $\tan A = BD/AD$. By making AD equal to 50 or 100 ft, the tangent is easily computed. To illustrate, if $AD = 100.00$ ft and BD measures 44.80 ft, then $\tan A = 0.4480$ and angle $A = 24°08'$. This procedure is not as convenient as the chord method because it requires establishing a perpendicular at D.

4-18.4. LAYING OFF ANGLES. An angle can be laid off by reversing the tangent method just described. Along the initial side of the angle, a unit distance of 10, 20, 50, or 100 ft is laid off, like AB in Figure 4-12. A perpendicular BC is erected and if $AB = 100$, its length made equal to 100 times the natural tangent

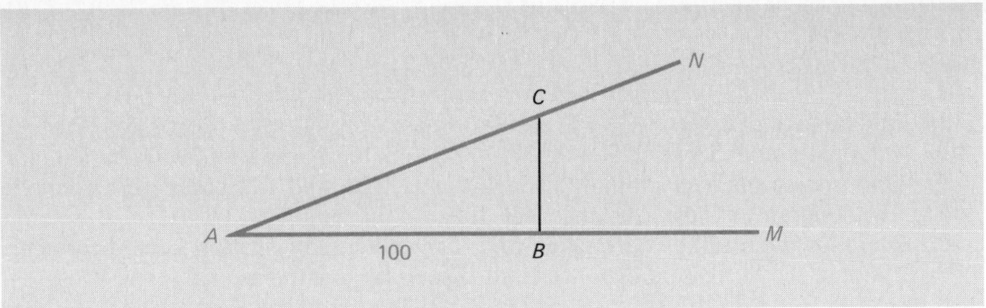

Figure 4-12. Laying off angle by tangent method.

of the desired angle. Points *A* and *C* are connected to give the required angle at *A*. This method is used by draftsmen as well as surveyors in the field.

4-18.5. TAPE SURVEY OF A FIELD. A field may be completely surveyed by taping. In fact, this was the only method available before instruments for measuring angles were built. Now EDM equipment makes the method useful again.

The procedure consists of dividing the area into a series of triangles and measuring the sides of each one. For small areas, one corner of the field is selected as the apex, and the distances to all other corners and the perimeter length are measured. In Figure 4-13, if corner *G* is chosen as the reference point, distances *GA*, *AB*, *BC*, *CD*, *DE*, *EF*, and *FG* along the perimeter, and diagonal lengths *GB*, *GC*, *GD*, and *GE* locate all corners of the field.

For larger areas it is better to establish a central point, such as *P* in Figure 4-14, and measure the perimeter and all lines radiating from *P* to the corners. The field can be plotted and the area determined from these data. The central-point method may appear to require more work, but the short lengths of the interior distances compensate for their greater number. Also, all corners are more likely to be visible from a selected point within the field.

Figure 4-13. Survey of field by tape.

Figure 4-14. Survey of field using central point.

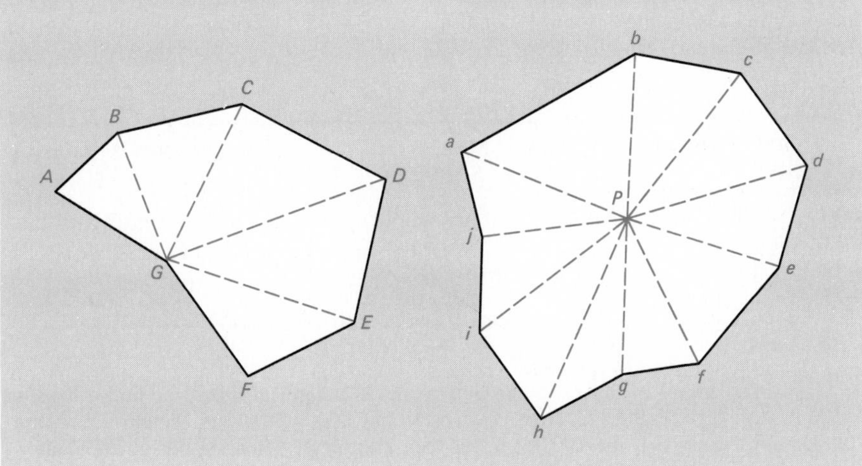

PROBLEMS

4-1. List six methods of measuring horizontal distances. Give an advantage and a disadvantage of each.

4-2. A student counted 172, 173, 172, 171, 174, and 172 paces while walking along a 500-ft course on level ground. Then 121, 122, 120, and 122 paces were counted in walking an unknown distance *AB*. What is the length of *AB*?

4-3. The following readings were taken on a 2-m subtense bar with a 1-sec theodolite. Compute the horizontal distance from theodolite to subtense bar.
(a) 0°28′14″, 0°28′16″, 0°28′15″, 0°28′15″
(b) 0°13′35″, 0°13′34″, 0°13′35″, 0°13′34″

4-4. For the data given, compute the horizontal distance for a recorded slope distance *AB*.
(a) $AB = 429.37$ ft, slope angle = 4°35′
(b) $AB = 258.69$ m, difference in elevation *A* to $B = 10.7$ m
(c) $AB = 651.45$ ft, grade = 4.5%

4-5. Compute the horizontal distance between the ends of a 100-ft tape using approximate Eq. (4-5) and exact Eq. (4-4) for differences in elevation of 5, 10, 15, 20, and 25 ft. Carry out the computations far enough to show differences in results and tabulate your answers.

A 100-ft steel tape, NBS 420, cross-sectional area 0.0030 in^2, weight 1 lb, and standardized at 68°F, is 100.014 ft between endmarks when supported throughout under 12-lb pull, and 99.998 ft when supported at the ends only with a 15-lb pull. What is the true length of a recorded distance *AB* for the conditions given in Problems 4-6 through 4-9? (Assume all full tape lengths except the last.)

	RECORDED DISTANCE *AB*	AVERAGE TEMP.	MEANS OF SUPPORT	TENSION
4-6.	242.90	68°F	Throughout	12 lb
4-7.	584.77	68°F	Ends only	15 lb
4-8.	420.31	55°F	Throughout	12 lb
4-9.	669.55	92°F	Ends only	15 lb

For the NBS 420 tape of Problems 4-6 through 4-9, determine the true length of the recorded distance *BC* for the conditions shown in Problems 4-10 through 4-13. (Assume all full tape lengths except the last.)

	RECORDED DISTANCE *BC*	AVERAGE TEMP.	MEANS OF SUPPORT	TENSION	ELEV. DIFFERENCE PER 100 FT.
4-10.	200.00	104°F	Throughout	21 lb	2.7 ft
4-11.	649.20	90°F	Ends only	18 lb	3.9 ft
4-12.	576.18	46°F	Throughout	19 lb	2.3 ft
4-13.	837.65	34°F	Ends only	20 lb	2.2 ft

In Problems 4-14 through 4-18 determine the length of *CD* to be laid out using a 100-ft steel tape, NBS 422, with cross-sectional area 0.0060 in^2, weight 2.0 lb, and standardized at 68°F to be 100.011 ft between endmarks when supported throughout with a 12-lb pull, and 99.948 ft if supported at the ends only under 16-lb tension.

	REQUIRED DISTANCE *CD*	AVERAGE TEMP.	MEANS OF SUPPORT	TENSION	SLOPE
4-14.	78.00 ft	68°F	Throughout	12 lb	0
4-15.	87.68 ft	97°F	Ends only	16 lb	1.5 ft/100 ft
4-16.	248.62 ft*	76°F	Throughout	15 lb	3.9 ft/100 ft
4-17.	97.00 ft	35°F	Ends only	20 lb	2.8% grade
4-18.	622.85 ft*	104°F	Ends only	18 lb	3° slope

* Assume all full tape lengths except the last one.

A 30-m steel tape measured 30.0150 m when standardized fully supported under a 7-kg pull at a temperature of 20°C. The tape weighed 0.90 kg and had a cross-sectional area of 0.028 cm^2. What is the true length of a recorded distance *AB* for the conditions given in Problems 4-19 through 4-21? (Assume all full tape lengths except in the last one.)

	RECORDED DISTANCE *AB*	AVERAGE TEMP.	MEANS OF SUPPORT	TENSION	ELEV. DIFFERENCE PER 100 M
4-19.	51.375 m	28°C	Throughout	9 kg	3.0 m
4-20.	82.480 m	10°C	Ends only	10 kg	Horizontal
4-21.	114.095 m	12°C	Ends only	10 kg	2.5 m

A 20-m steel tape measured 19.9895 m when standardized fully supported under a 7-kg pull at a temperature of 20°C. The tape weighed 0.65 kg and had a cross-sectional area of 0.030 cm^2. Determine length *BC* to be laid out using this tape for the conditions given in Problems 4-22 and 4-23. (Assume all full tape lengths except the last.)

	REQUIRED DISTANCE *BC*	AVERAGE TEMP.	MEANS OF SUPPORT	TENSION	SLOPE
4-22.	40.000 m	11°C	Throughout	10 kg	4.0% grade
4-23.	156.527 m	39°C	Ends only	10 kg	2°30′ slope

4-24. What difference in temperature from standard, if neglected in the use of a steel tape, will cause an error of 1 part in 3000? One part in 5000? One part in 10,000?

4-25. When measuring a distance *AB*, the first taping pin was placed 1.5 ft to the right of line *AB* and the second pin set 1 ft left of line *AB*. The recorded distance was 251.57 ft. Calculate the corrected distance.

4-26. In taping from *A* to *B*, a tree on-line necessitated setting an intermediate point *C* offset 6 ft to the side of line *AB*. Line *AC* was then measured as 386.29 ft along a uniform 4% slope. Line *CB* on horizontal ground was measured as 185.10 ft. Find the length of *AB*.

4-27. Chaining pins for a line measured with a 100-ft steel (adding) tape having an extra graduated foot are set by mistake at the 101-ft mark. The length is recorded as exactly 800 ft. What is the correct distance?

4-28. The distance between two fixed points on a construction site was measured at a temperature of 110°F with a steel tape that was 100.000 ft long at 68°F. It was recorded as 2508.06 ft. Compute the distance corrected for temperature. What distance would have been recorded if the temperature at the time of measurement had been 28°F?

4-29. A tape with a cross section of $\frac{5}{16} \times 0.025$ in is 100.045 ft long when supported throughout at a tension of 15 lb. What is the length between end graduations if the same tension is applied, but the tape is supported at the ends and midpoint? If it is supported at the ends and quarter-points?

4-30. A 100-ft steel tape having a cross-sectional area of 0.0044 in^2 is exactly 100.000 ft long at 68°F when fully supported under a pull of 12 lb. What is the normal tension for this tape?

4-31. What pull (normal tension) is required to make a tape exactly 100.000 ft between its ends when used in an unsupported mode, if the tape has a cross-sectional area of 0.0056 in^2 and its length is 99.990 ft when supported under a 15-lb pull?

4-32. What error in a measured distance results from the conditions noted?
(a) One end of a 30-ft length of tape is off-line by 1.0 ft.
(b) One end of an 80-ft length of tape is too high by 2.4 ft.
(c) One end of a 100-ft tape is off-line by 1.5 ft and too low by 1.7 ft.
(d) One end of a 30-m tape is off-line by 25 cm and too high by 15 cm.

4-33. To determine the angle AOB between two intersecting fences without setting up a transit, convenient distances $OA = 100.00$ ft and $OB = 80.00$ ft are measured from the intersection along the fence lines. If the distance AB is 48.90 ft, what is the intersection angle?

4-34. In a seven-sided figure similar to Figure 4-13, the following lengths were obtained in measuring the sides: $AB = 265.83$, $BC = 421.71$, $CD = 524.09$, $DE = 453.16$, $EF = 375.87$, $FG = 449.98$, $GA = 413.70$, $GB = 432.47$, $GC = 550.26$, $GD = 763.11$, and $GE = 648.52$ ft. Compute the angles at each corner.

4-35. A base line taped on chaining bucks has a recorded length of 1483.295 ft. The average temperature was 87.8°F, pull applied 16 lb, and total inclination correction for the line −0.523 ft. The tape weighed 1.50 lb and its length between the 0- and 100-ft marks was 100.017 ft when supported throughout with a pull of 16 lb at 68°F. Make all corrections and compute the sea-level length of the line assuming its average elevation is 4975 ft above MSL. [Mean sea level (MSL) reduction is discussed in Section 21-5.]

4-36. Determine the most probable length of a line AB, the standard deviation, and the 90% error of a single measurement for the following series of measurements made under the same conditions: 648.29, 648.33, 648.27, 648.36, 648.32, 648.29.

4-37. The standard deviation of taping a 600-ft distance is ±0.08 ft. Using the same procedures, what should it be for an 1800-ft distance?

4-38. An irregular field is measured with a 100-ft steel tape that is 100.06 ft long, and the area is erroneously found to be 27.956 acres. What is the true area?

4-39. In measuring with a 100-ft standardized steel tape having a cross section of 0.25 × 0.025 in, if the tape is supported throughout, which of the following errors is most serious? (a) A constant temperature difference of + 10°F from standard; (b) an alignment error of 0.4 ft on each tape length; (c) tape out of level 0.8 ft on each length; or (d) a tension variation of 5 lb from standard on each tape length.

4-40. A line 5 mi long to be laid off must have a standard error less than 2 ft. What standard error per tape length is permissible?

4-41. Determine the acreage of the field for the data of Problem 4-34 by summing the computed areas of the individual triangles.

4-42. Using the data of Problem 4-34, select an appropriate scale and plot the field so that it fits suitably on an $8\frac{1}{2}$ by 11 inch sheet of paper.

4-43. In a nine-sided figure similar to Figure 4-14, the following lengths were measured: $aP = 286.53$, $bP = 258.19$, $cP = 291.07$, $dP = 303.44$, $eP = 260.30$, $fP = 253.16$, $gP = 248.92$, $hP = 347.00$, $iP = 300.22$, $jP = 237.54$, $ab = 320.21$, $bc = 174.08$, $cd = 179.50$, $de = 159.73$, $ef = 198.84$, $fg = 121.66$, $gh = 150.29$, $hi = 167.13$, $ij = 148.98$ and $ja = 147.93$ ft. Compute the angles at each exterior corner of the field.

4-44. Determine the acreage of the field for the data of Problem 4-43 by summing the computed areas of the individual triangles.

4-45. Using the data of Problem 4-43, select an appropriate scale and plot the field so that it fits suitably on an $8\frac{1}{2}$-by 11-inch sheet of paper.

BIBLIOGRAPHY

Colcord, J. E., and F. H. Chick. 1968. "Slope Taping." *ASCE, Journal of the Surveying and Mapping Division* 94(no. SU2):137.

Smirnoff, M. V. 1952. "The Use of the Subtense Bar." *Surveying and Mapping* 12(no. 4): 390.

Wagner-Smith, R. W. 1961. "Errors in Measuring Distances by Offsets." *Surveying and Mapping* 21(no. 1):73.

5 ELECTRONIC DISTANCE MEASUREMENT

5-1. INTRODUCTION. A major advance in surveying in recent years has been the development of electronic distance-measuring instruments (EDMIs). These devices determine lengths based on phase changes that occur as electromagnetic energy of known wavelength travels from one end of a line to the other and returns.

The first EDM instrument was introduced in 1948 by Swedish physicist Erik Bergstrand. His device, called the *geodimeter* (an acronym for geodetic distance meter), resulted from attempts to improve methods for measuring the velocity of light. The instrument transmitted visible light and was capable of accurately measuring distances up to about 25 mi (40 km) at night. In 1957 a second EDM apparatus, the *tellurometer*, designed by Dr. T. L. Wadley and introduced in South Africa, transmitted invisible microwaves and was capable of measuring distances up to 50 mi (80 km) or more, day or night.

The potential value of these early EDM models to the surveying profession was immediately recognized; however, they were expensive and not readily portable for field operations. Furthermore, measuring procedures were lengthy and mathematical reductions to obtain distances from observed values were difficult and time-consuming. In addition, the range of operation of the first geodimeter was limited in daytime use. Continued research and development have overcome all these deficiencies.

The chief advantages of electronic surveying are the speed and accuracy with which distances can be measured. If a line of sight is available, long or short lengths can be measured over bodies of water or terrain that is inaccessible for taping. With modern EDM equipment, distances are automatically displayed in digital form in feet or meters, and many have built-in microcomputers that give results internally reduced to horizontal and vertical components. Their many significant advantages have revolutionized surveying procedures and gained worldwide acceptance. The long-distance measurements possible with EDM equipment make use of radios for communication, which is an absolute necessity in modern practice.

5-2. CLASSIFICATION OF EDM INSTRUMENTS. One system for classifying EDMIs is by wavelength of transmitted electromagnetic energy; the following categories exist:

1. *Electro-optical* instruments, which transmit either modulated laser or infrared light having wavelengths within or slightly beyond the visible region of the spectrum.
2. *Microwave* equipment, which transmits microwaves with frequencies in the range of 3 to 35 GHz corresponding to wavelengths of about 1.0 to 8.6 mm.

Another classification system for EDMIs is by operational range. It is rather subjective, but in general two divisions fit into this system: *short* and *medium* range. The short-range group includes those devices whose maximum measuring capability does not exceed about 5 km. Most equipment in this division is the electro-optical type and uses infrared light. These instruments are small, portable, easy to operate, suitable for a wide variety of field surveying work, and used by many practitioners.

Instruments in the medium-range group have measuring capabilities extending to about 100 km and are either the electro-optical (using laser light) or microwave type. Although frequently used in precise geodetic work, they are also suitable for land and engineering surveys. Longer-range devices also available can measure lines longer than 100 km, but they are not generally used in ordinary surveying work. Most operate by transmitting long radio waves, but some employ microwaves. They are used primarily in oceanographic and hydrographic surveying and navigation.

5-3. PRINCIPLES OF EDM INSTRUMENT OPERATION. In general, EDM equipment measures distances by comparing a line of unknown length to the known wavelength of modulated electromagnetic energy. This is similar to relating a needed distance to the calibrated length of a steel tape.

Electromagnetic energy propagates through the atmosphere in accordance with the following equation:

$$V = f\lambda \tag{5-1}$$

where V is the velocity of electromagnetic energy, in meters per second;[1] f the modulated frequency of the energy, in hertz;[2] and λ the wavelength, in meters.

[1] The velocity of electromagnetic energy in a vacuum is 299,792.5 km/sec. Its speed is slowed somewhat in the atmosphere according to the equation $V = c/n$, where c is the velocity in a vacuum and n the atmospheric *index of refraction*, which varies but is approximately equal to 1.0003.

[2] The *hertz* (Hz) is a unit of frequency equal to 1 cycle/sec.

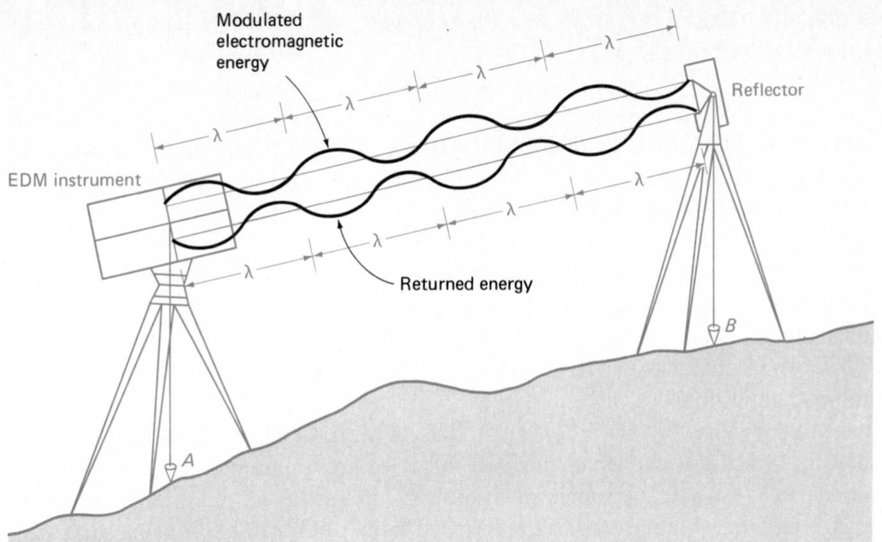

Figure 5-1. Generalized EDM procedure.

With EDMIs, frequency can be precisely controlled but velocity varies with atmospheric temperature, pressure, and humidity. Thus wavelength and frequency must vary in conformance with Eq. (5-1). For accurate electronic distance measurement, therefore, the atmosphere must be sampled and corrections made accordingly.

The generalized procedure of measuring distance electronically is depicted in Figure 5-1. An EDM device, centered by means of a plumb bob or optical plummet over station A, transmits a *carrier signal* of electromagnetic energy upon which a reference frequency has been superimposed or *modulated*. The signal is returned from station B to the receiver, so its travel path is double the slope distance AB. In Figure 5-1, the modulated electromagnetic energy is represented by a series of sine waves having wavelength λ. Any position along a given wave can be specified by its *phase angle*, which is 0° at its beginning, 180° at the midpoint, and 360° at its end.

EDM devices used in surveying operate by measuring *phase shift*. In this procedure, the returned energy undergoes a complete 360° phase change for each even multiple of exactly one-half the wavelength separating the line's endpoints. If, therefore, the distance is precisely equal to a full multiple of the half-wavelength, the indicated phase change will be zero. In Figure 5-1, for example, stations A and B are exactly eight half-wavelengths apart; hence, the phase change is zero. When a line is not exactly an even multiple of the half-wavelength (the usual case), the fractional part is measured by the instrument as a nonzero phase angle or phase change. If the precise length of a wave is known, the fractional part can be converted to distance.

EDMIs directly resolve the fractional wavelength but do not count the full cycles undergone by the returned energy in traveling its double path. This ambiguity is resolved, however, by transmitting additional signals of lower frequency and longer wavelength.

5-4. ELECTRO-OPTICAL INSTRUMENTS. As previously noted, electro-optical EDMIs of today transmit either laser or infrared light as their carrier signals. Earlier models used tungsten or mercury lamps. Their operating range was relatively short, especially during the day, primarily because of excessive atmospheric scatter of this incoherent light. They were also bulky and required a large power source. Coherent light produced by gas lasers now permits measurements of long distances in the daylight with small portable instruments. The K&E Rangemaster III, shown in Figure 5-2(c), for example, has a range of more than 60 km.

Approximately 40 different models of short-range electro-optical EDM devices that use a carrier of infrared light are now manufactured. Their range is restricted to a few kilometers by power limitations of the gallium arsenide (GaAs) diode, which produces the infrared light, but for much routine surveying this is adequate. Perhaps the greatest advantage of infrared light as a carrier is that its intensity can be directly modulated, thus considerably simplifying equipment using this source of radiation. These instruments have become so small (some weigh less than 2 lb) that they can be mounted directly on theodolites. This enables distance and angle measurements to be made with a single setup. Figure 5-2(a) and (b) show two infrared electro-optical EDMIs with different theodolite mounts; the former attached to the telescope, the latter fastened to the standards. The instrument of Figure 5-2(c) has an independent tripod setup. Many currently manufactured instruments can be adapted for all three mounts.

Specific operating principles of different electro-optical instruments vary, which makes it impractical to analyze all of them here in detail. For descriptive purposes two Hewlett-Packard instruments widely used in the United States will be described: the first-generation Model 3800 and the second-generation Model 3805A (shown in Figure 5-3). These are representative of most electro-optical EDM devices currently employed. Discussion will be highly simplified without resorting to descriptions of specific electronic components.

Figure 5-4 is a generalized schematic diagram of the operating characteristics of a Hewlett-Packard 3800. The transmitter uses a GaAs diode which emits *amplitude-modulated* (AM) infrared light. Frequency of modulation is precisely controlled by a crystal oscillator. The modulation process may be thought of as similar to passing light through a stove pipe in which a damper plate is spinning at a precisely controlled rate or frequency. When the damper is closed, no light passes. As it begins to open, light intensity increases to a maximum at a phase angle of 90° with the plate completely open. Intensity reduces to zero again with the damper closed at a phase angle of 180°, and so on. This intensity variation or amplitude modulation is properly represented by sine waves.

Local atmospheric ground pressure and temperature are determined by an operator at the time of the measurement, and an environmental correction factor based on them is taken from a chart. A correction factor is dialed into

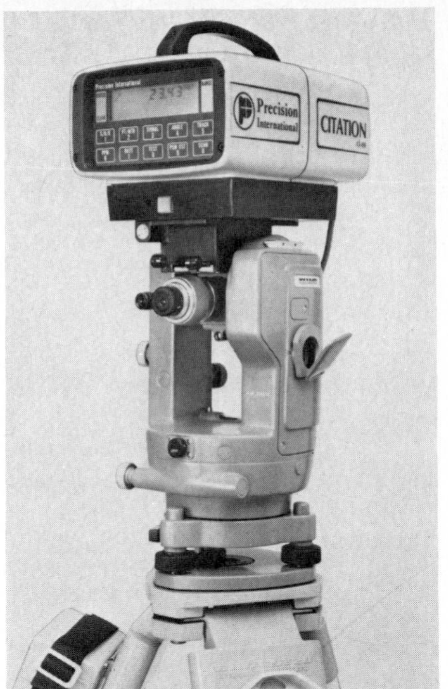

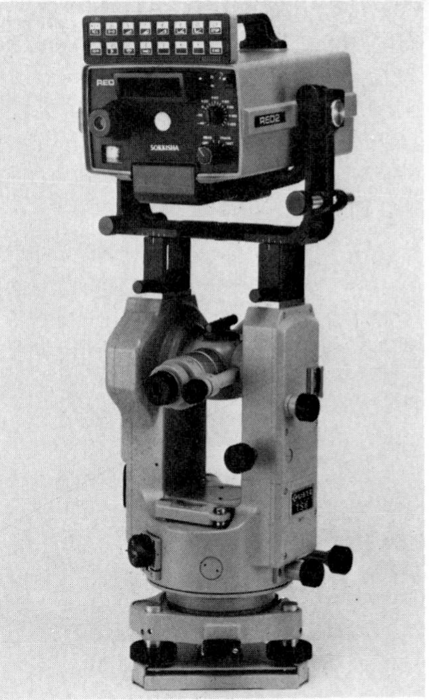

Figure 5-2. Electro-optical EDM instruments. (a) Citation CI 450 infrared instrument with theodolite telescope mount. (Courtesy Wild Heerbrugg Instruments, Inc.) (b) RED 2 infrared system attached to theodolite standards. (Courtesy the Lietz Company.) (c) Rangemaster III laser instrument with independent tripod mount. (Courtesy Keuffel & Esser.)

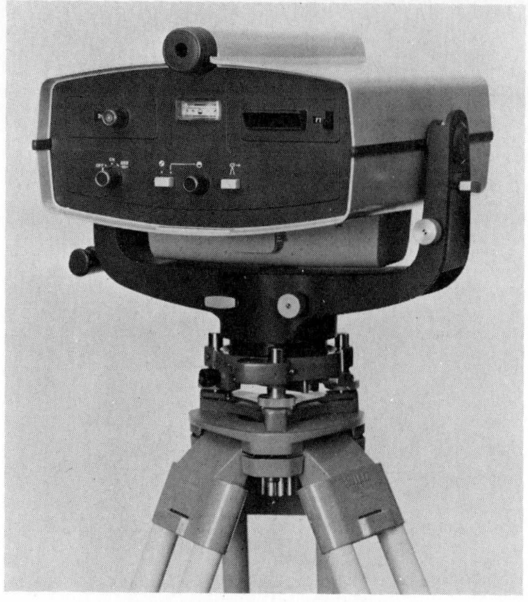

Figure 5-3. Hewlett-Packard 3805A distance meter. (Courtesy Hewlett-Packard Co.)

Figure 5-4. Generalized block diagram of operation of Hewlett-Packard 3800.

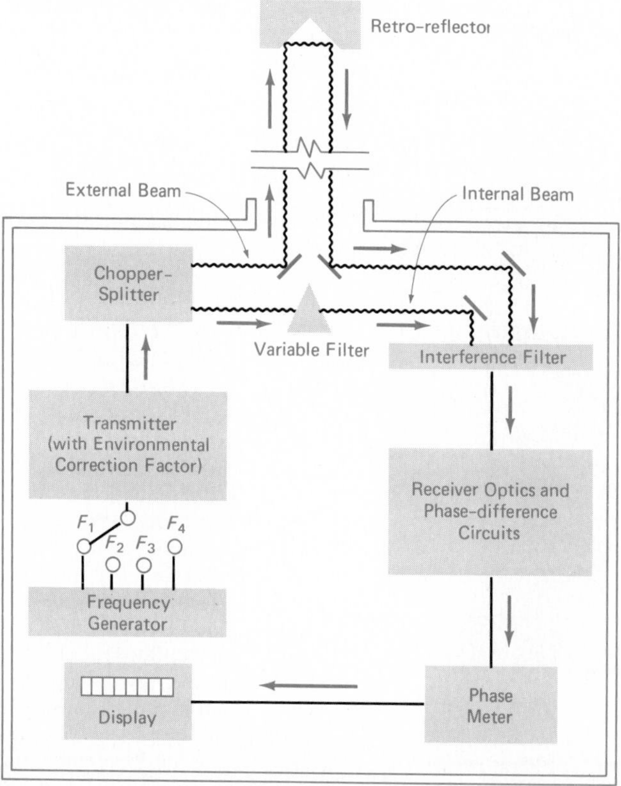

Figure 5-5. Triple retro-reflector. (Courtesy Hewlett-Packard Co.)

the transmitter to slightly vary the frequency so a constant wavelength is maintained despite atmospheric variations, thereby eliminating the need to mathematically adjust measured distances later. Note that humidity has a negligible effect on propagation of infrared light and hence is not measured.

A chopper-splitter divides the light emitted from the diode into two separate beams: an *external* measurement beam and an *internal* reference beam. The external one is carefully aimed at a retro-reflector which has been centered over a point at the other end of the line (Figure 5-5). The reflector returns this beam to the receiver. The internal beam passes through a variable density filter and is reduced in intensity to a level equal to that of the returned external signal, enabling a more accurate measurement to be made. Both internal and external signals go through an interference filter, which eliminates all undesirable energy such as sunlight. The internal and external beams then pass through components to convert them into electrical energy while preserving the phase-shift relationship resulting from their different travel path lengths. A phase meter converts this phase difference into direct current having a magnitude proportional to the differential phase. This current is connected to a *null meter*, which automatically adjusts itself to null the current. The fractional wavelength is converted to distance during the nulling process and displayed on instrument dials.

To resolve the ambiguity of an unknown number of full cycles a wave has undergone, the Hewlett-Packard 3800 transmits four different modulation frequencies, F_1, F_2, F_3, and F_4, as indicated on the schematic diagram of Figure 5-4. A model measuring in feet uses modulation frequencies of 24.5 MHz, 2.45 MHz, 245 kHz, and 24.5 kHz, respectively. Wavelengths associated with these frequencies are 10, 100, 1000, and 10,000 ft long, respectively.[3]

Assume that an HP 3800 shows a digital display of 3417.14 ft from measuring a line. The three rightmost digits, 7.14, are obtained when the phase meter is nulled while transmitting frequency F_1. Distance 7.14 is equivalent to a phase shift of (7.14/10) × 360°, or 257°. Frequency F_2 is then sent and nulled, yielding a fraction of 100 ft for the second digit to the left of the decimal, or number 1. Frequencies F_3 and F_4 are transmitted in turn to get the digits 4 and 3, respectively. It should be evident that the high resolution of a measurement is secured using the 10-ft wavelength; the longer ones resolve only the unknown number of shorter wavelengths.

Some important technological advancements have been incorporated to update the HP 3800 to the current-generation Model 3805A. The addition of a microprocessor made possible a totally new method to determine phase shifts of the measured signal and display the results. The null meter was replaced by an electronic *phase detector*, which converts the phase difference into a number or count having a magnitude proportional to the phase angle. This count is sent to the internal microprocessor where the distance is computed and displayed by means of a light-emitting diode (LED). Also important, the HP 3805A uses only two frequencies of modulation; 15 MHz and 75 KHz. The 15-MHz frequency, having a wavelength of 20 m (360° of phase shift equals a distance of 10 m), provides high-resolution distance data. The 75-KHz frequency, having a wavelength of 4000 m (360° of phase shift equals 2000 m), provides the low-resolution information.

In the HP 3805A, atmospheric pressure and temperature are not compensated for by varying transmitter frequency. Instead, a value representative of the combined correction factor is input directly into the microprocessor from a control on the instrument's front panel. This value is used to mathematically adjust the distance following a measurement. The HP 3805A has a maximum range of 1 mi. After manual initiation of the measurement process, switching modulation frequencies and phase detection is controlled automatically by the built-in computer, and slope distance is displayed within 6 sec. Its accuracy is purported to be ±(7 mm + 7 ppm).[4]

The method of measurement used by the three instruments of Figure 5-2 is basically the same as that of the HP 3805A. The Citation CI 450, Figure 5-2(a), and the RED 2, Figure 5-2(b), have maximum ranges of approximately 4 and 3 km, respectively, and are useful for a variety of general survey

[3] The true modulation wavelength of the 24.5 MHz frequency is 40 ft, but doubler circuitry multiplies the modulation frequency by 2. Furthermore, the wave travels double the distance being measured. A 360° phase change therefore has an "effective" wavelength of 10 ft. The wavelength of 40 ft is calculated as follows:

$$\lambda = \frac{299{,}792.5 \text{ km/sec} \times 1000 \text{ m/km}}{24{,}500{,}000 \text{ cycles/sec (Hz)} \times 0.3048 \text{ m/ft}} = 40 \text{ ft/cycle}$$

[4] ppm = parts per million. In a distance of 3417 ft, this portion of the error would be ±0.024 ft.

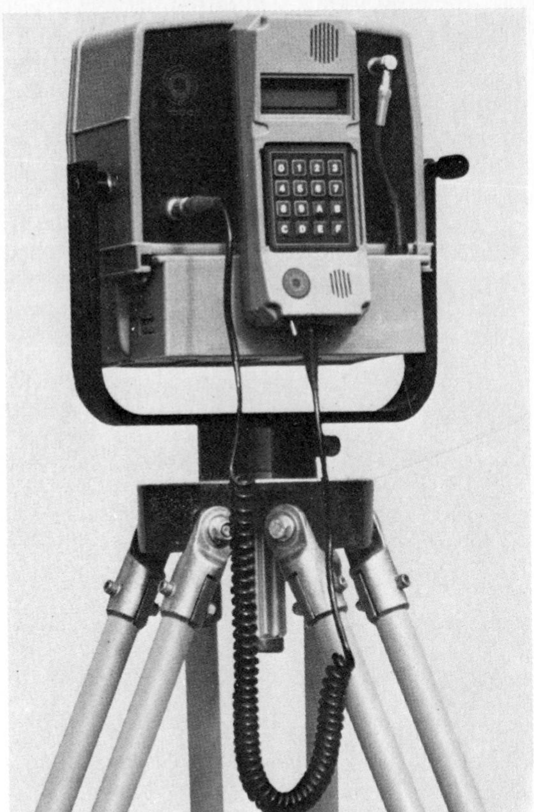

Figure 5-6. Microfix 100C microwave EDM instrument. (Courtesy Telludist, Inc.)

projects. Both have purported accuracies of ±(5 mm ± 5 ppm), will automatically compute and display horizontal and vertical components of a slope distance after manual entry of the vertical angle via the keyboard, and can be operated in a "tracking" mode. In tracking, a required distance (horizontal, vertical, or slope) is entered by means of the keyboard, and the difference between the desired distance and that to the reflector is rapidly updated and displayed. This feature, extremely useful in construction stakeout, is described in Section 24-2. The Rangemaster III [Figure 5-2(c)] has a longer range and accuracy of ±(5 mm + 1 ppm). It is particularly well suited for control surveys.

5-5. MICROWAVE INSTRUMENTS. The measurement signal used by microwave devices consists of frequency modulation (FM) superimposed on the carrier wave. Like electro-optical instruments, microwave equipment operates on the phase-shift principle and uses varying frequencies to resolve ambiguities in the unknown number of full wavelengths in a distance. The range of microwave devices is comparatively long, and they can operate in fog or light rain. Measurements made in such adverse weather conditions are somewhat limited in range, however.

A complete microwave EDM system consists of two portable identical units. Each includes all components necessary to make measurements: transmitter, receiver, antenna, circuitry, and built-in communication arrangement.

Figure 5-7. Geodimeter Model 140 total-station instrument. (Courtesy AGA Geodimeter, Inc.)

Units are centered by plumb bobs or optical plummets over the terminal points of a course, with one instrument functioning in the "master" mode, the other in the "remote" mode. Either may be operated as master or remote by simply changing a switch position.

Measurement with microwave devices requires an operator at each end of the line to take a set of readings while using the instrument in its master mode. Since both units contain temperature-stabilized wavelength calibration, this procedure gives two independent measurements of the distance and a valuable check. The operators, who may not be in sight of each other, coordinate their work procedures by communicating on the built-in radio telephone.

The Microfix 100C shown in Figure 5-6 is a lightweight and compact microwave EDMI. It can be adapted to fit standard theodolites and measures distances up to 60 km to an accuracy of $\pm$(15 mm + 3 ppm). The measurement system is fully automatic and after pointing has been accomplished, takes only 5 sec. This device can also operate in a "tracking" mode.

5-6. TOTAL-STATION INSTRUMENTS. *Total-station instruments* (also called electronic tacheometers) combine an EDM instrument, electronic digital theodolite, and computer in one unit. The electronic digital theodolite, described in more detail in Section 10-14, automatically measures and displays horizontal and vertical angles. Total-station instruments simultaneously measure distance, as well as direction, and transmit the results automatically to a computer. The horizontal and vertical angles and slope distance can be displayed; then upon keyboard commands, horizontal and vertical distance components are instantaneously computed and displayed. If coordinates of the occupied station and reference azimuth are input to the system, coordinates of the sighted point are immediately obtained. This information can all be stored on magnetic tape or in solid-state memory devices, thereby eliminating the need to manually record data. These instruments are of tremendous value in all types of surveying.

The Geodimeter Model 140 total-station instrument shown in Figure 5-7 has a distance range of approximately 6 km with an accuracy of ±(5 mm + 5 ppm) and measures angles to ±2 sec. Figure 10-14 shows a similar instrument, the Hewlett-Packard Model 3820, which has a range of approximately 5 km with an accuracy of ±(5 mm + 5 ppm) and angle accuracy of ±3 sec.

5-7. ERRORS IN ELECTRONIC DISTANCE MEASUREMENT. Sources of error in EDM work may be personal, instrumental, or natural. Personal errors include misreading, improperly setting over stations, and incorrectly measuring meteorological factors and instrument heights.

If EDM equipment is carefully adjusted and precisely calibrated, instrumental errors should be extremely small. As noted earlier, manufacturers specify accuracies of their products in two parts: a constant error and a value proportional to the distance measured. Listed errors vary for different instruments, but the constant portion is usually about ±5 mm, while the proportion is generally near 5 parts per million (ppm). The constant error is most significant on short distances; for example, with an instrument having a constant error of ±5 mm, a measurement of 50 m is good to only $\frac{5}{50{,}000}$, $\frac{1}{10{,}000}$, or 100 ppm. For very long distances the constant error becomes negligible, and the proportional part more important.

EDM equipment should be checked against a first-order base line at regular time intervals to assure its accuracy and reliability. By comparing the base-line length and distance obtained electronically, a *measurement constant* is ascertained. A correction for this systematic error can then be applied to all subsequent measurements. The constant thus determined combines the amount by which the "electrical center" of the instrument is offset forward or back of the plummet, and for electro-optical equipment, the *reflector constant* (any offset of the "optical center" of the reflector).

Although calibration using a first-order base line is preferred, if one is not available, the constant can still be obtained. In this procedure, three stations, *A*, *B*, and *C*, are established on a straight line, with distance *AC* roughly 1 mi and *B* approximately midway between *A* and *C*. Measure the total length *AC* and the two parts *AB* and *BC*. For these measurements, the following equation can be written:

$$AC + K = (AB + K) + (BC + K)$$

from which

$$K = AC - AB - BC \tag{5-2}$$

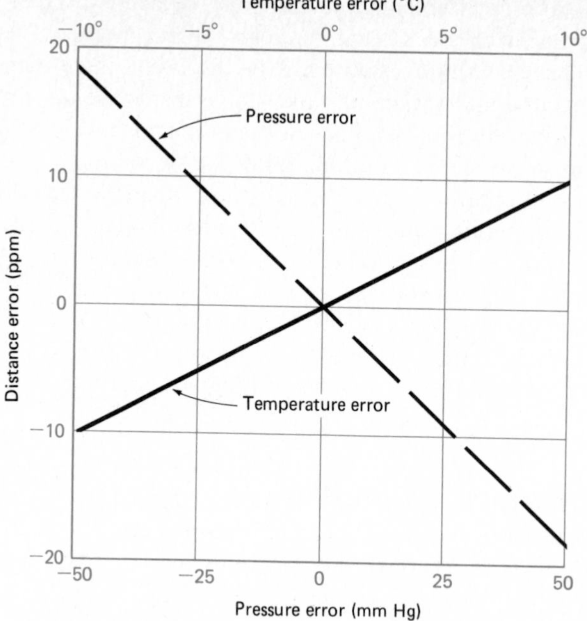

Figure 5-8. Errors in EDM produced by temperature and pressure errors (based on atmospheric temperature and pressure of 15°C and 760 mm of mercury).

where K is the measurement constant to be added to measured distances. The procedure, including centering of the EDM instrument and reflector, should be repeated several times and the average value of K adopted. Since different reflectors have varying offsets, the test should be run with each one used and the results marked on them to avoid confusion.

Multiple reflections of microwaves from a ground or water surface can set up a condition designated as *ground swing*, which affects the accuracy of readings taken with microwave instruments. Errors from this source can be reduced by elevating the master and remote units as high above ground as possible and taking four measurements, two from each end of the line, and averaging.

Natural errors in EDM operations stem primarily from atmospheric variations in temperature, pressure, and humidity which affect the index of refraction and modify the wavelength of electromagnetic energy. These three variables are measured and considered in accurate microwave distance determination, but humidity can be neglected when using electro-optical instruments.

Most EDMIs handle atmospheric variables directly during the measuring process, as described for the Hewlett-Packard 3805A. For older instruments, corrections must be made mathematically later. Equipment manufacturers provide tables and charts that give the necessary correction factors and explain the reduction process.

The magnitude of error in electronic distance measurement due to atmospheric temperature and pressure effects is indicated in Figure 5-8. Note that a 10°C change, and a pressure difference of 25 mm of mercury, each produces a distance error of about 10 ppm. Humidity is determined with a psychrometer which gives wet and dry bulb temperatures. An error of 1.5°C in the difference of the two bulbs is equivalent to approximately 10 ppm in distances established using microwave instruments.

5-8. COMPUTING HORIZONTAL DISTANCES FROM SLOPE DISTANCES. All EDM equipment measures slope distance between stations. Some instruments can reduce these distances to their horizontal components automatically if the vertical angle is input. With many models this cannot be done, so reduction must be carried out manually. A slope length should, of course, first be corrected for instrumental and atmospheric conditions.

Reduction of slope distances to horizontal can be based on difference in elevation or on vertical angle. If difference in elevation is used, during field operations heights h_e of the EDM instrument and h_r of the reflector above their stations are measured and recorded (see Figure 5-9). If elevations of stations A and B in Figure 5-9 are known, any Eq. (4-4) through (4-6) will reduce the slope distance to horizontal with the value of d (the difference in elevation between EDMI and reflector) computed as follows:

$$d = (\text{elev}_A + h_e) - (\text{elev}_B + h_r) \tag{5-3}$$

EXAMPLE 5-1

A slope distance of 165.360 m (corrected for meteorological conditions) was measured from A to B, whose elevations were 447.401 and 445.389 m above datum, respectively. Find the horizontal length of line AB if heights of the EDMI and reflector were 1.417 and 1.615 m above their respective stations.

SOLUTION

By Eq. (5-3):

$$d = (447.401 + 1.417) - (445.389 + 1.615) = 1.813 \text{ m}$$

By Eq. (4-5):

$$H = 165.360 - \frac{(1.813)^2}{2 \times 165.360} = 165.350 \text{ m}$$

If vertical angle α between the transmitted energy's inclined path and horizontal (see Figure 5-9) is obtained at the time of measuring slope distance L, then Eq. (4-3) is applicable to reduce the length to its horizontal component. For most precise work, especially for longer lines, the vertical angle should be measured direct, reversed, and averaged. Also, the mean obtained from both ends of the line will compensate for earth curvature and refraction.

In some cases, rather than measuring the required angle α, the true vertical angle α_t of Figure 5-9 between points A and B may be measured in a separate operation using a theodolite and graduated rod. In that situation, should a significant height difference exist between the EDMI and reflector as shown in

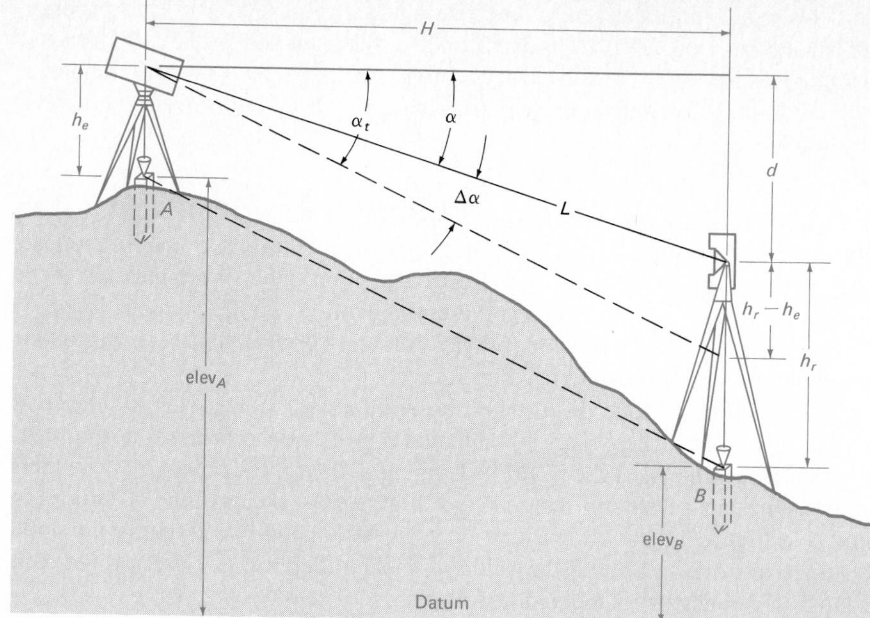

Figure 5-9. Reduction of EDM slope distance to horizontal.

Figure 5-9, angle α for use in Eq. (4-3) can be obtained by adding $\Delta\alpha$ algebraically to α_t where $\Delta\alpha$ in seconds is calculated by

$$\Delta\alpha'' = \frac{(h_r - h_e)\cos\alpha_t}{L} \times 206{,}265 \text{ sec/rad} \tag{5-4}$$

Because algebraic addition of $\Delta\alpha$ to α_t is required, the sign obtained from Eq. (5-4) is retained and α_t is considered plus if point B is above A, and minus if below.

Angle α can be measured directly, thus avoiding computation of $\Delta\alpha$ (even if a height difference exists between the EDMI and reflector), by sighting on the graduated rod to a reading that compensates for the height difference. Assume, for example, that h_e and h_r at the time of distance measurement were 5.4 and 6.2 ft, respectively. For vertical angle measurement, if the theodolite has an h.i. (height of instrument above station) of 5.5 ft, a reading of 6.3 ft sighted on the rod will produce the required angle α. If the EDMI and theodolite fit the same tribrachs, the theodolite can be placed on the tripod when distance measurement is completed, but before the tripod is moved. Then, by sighting the reflector, the required vertical angle α is obtained in spite of height differences. Computations or special field procedures to account for height differences are best avoided, and this can usually be done by setting the EDM instrument and reflector in the same vertical position—for example, at eye level.

If the EDMI is mounted on a theodolite as shown in Figure 5-2(a) and (b), the energy will be transmitted from a point offset vertically above the theodolite's horizontal axis. Vertical angle α_m, measured to the center of the reflector,

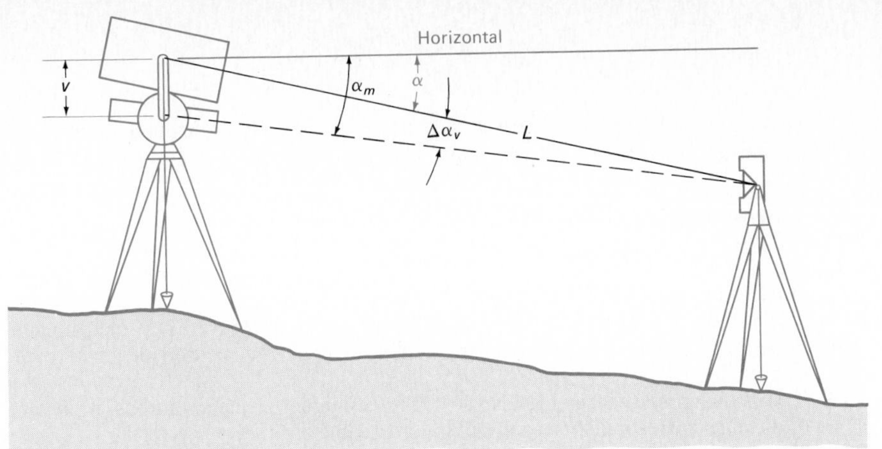

Figure 5-10. Correction for vertical offset between theodolite and EDMI mounted on standards.

will normally be obtained at the time of distance measurement. If the EDM instrument is mounted on the telescope so it tilts with the theodolite's line of sight, correction for the vertical offset is not needed, and the horizontal distance is obtained using L and α_m in Eq. (4-3). If the mount is on the standards, however, as shown in Figure 5-10, a correction $\Delta\alpha_v$ will be needed to account for the vertical offset v. This angle (in seconds) is

$$\Delta\alpha_v'' = \frac{v \cos \alpha_m}{L} \times 206{,}265 \text{ sec/rad} \tag{5-5}$$

The correction must be subtracted from α_m (which increases vertical angles for downhill sights and decreases them for uphill ones) to get the correct angle for use in Eq. (4-3). Again, this correction can be avoided by sighting below the reflector a distance v when measuring the vertical angle.

For long lines, errors caused by ignoring height differences between EDM instrument and reflector, or vertical offsets of the transmission optics above the theodolite, are insignificant. They become important on short lengths, however.

EXAMPLE 5-2

A vertical angle of $-7°25'$ (downhill) was measured to the center of a reflector, and a slope distance of 153.72 ft was obtained. If the EDM instrument was mounted on the standards and offset vertically 0.66 ft, what is the corrected horizontal distance? (Assume the theodolite and reflector heights are equal.)

By Eq. (5-5):

$$\Delta\alpha_v'' = \frac{0.66 \cos 7°25'}{153.72} \times 206{,}265 = 878 \text{ sec} = 14.6 \text{ min}$$

$$\alpha = -7°25' - 0°14.6' = -7°39.6'$$

By Eq. (4-3):

$$L = 153.72 \cos(-7°39.6') = 152.35 \text{ ft}$$

Note that if the vertical offset had been ignored, the reduced horizontal distance would have been 152.43 ft, in error by 0.08 ft.

PROBLEMS

5-1. State some advantages of electronic distance measurement.

5-2. Analyze the differences between electro-optical and microwave EDM instruments.

5-3. Explain briefly how a distance can be measured by the method of phase comparison.

5-4. If electromagnetic energy travels 186,000 mi/sec under given conditions, what unit of distance corresponds to each millimicrosecond of time?

5-5. The speed of electromagnetic energy through the atmosphere at standard barometric pressure of 29.92 in of mercury is accepted as 299,792.5 km/sec for measurements with an EDM instrument. What time lag in the equipment will produce an error of 50 ft in the distance to a target 50 mi away?

5-6. How are variations in the propagation of electromagnetic energy due to atmospheric conditions accounted for in measuring distances with the Hewlett-Packard 3805A?

5-7. List the modulation frequencies of the electromagnetic energy transmitted by a Hewlett-Packard 3805A. Which frequency is used to obtain the precision in measurements? Explain why.

5-8. If an EDM instrument has a purported accuracy capability of $\pm$(5 mm + 5 ppm), what error can be expected in a measured distance of (a) 2000 ft? (b) 800 m? (c) 2 mi?

5-9. If a certain EDM instrument has an accuracy capability of $\pm$(7 mm + 7 ppm), what is the precision of measurements, in terms of $1/x$, for line lengths of (a) 100 ft? b) 500 ft? (c) 3000 m?

5-10. To calibrate an EDM instrument, distances AC, AB, and BC along a straight line were measured as 2438.29 m, 1206.48 m, and 1231.84 m, respectively. What is the instrument constant for this instrument? Compute the length of each segment corrected for the instrument constant?

5-11. Discuss the various systematic errors that affect electronic distance measurements.

5-12. Which causes a greater error in a line measured with an EDMI?
(a) A disregarded 5°C temperature variation from standard or
(b) a neglected atmospheric pressure difference from standard of 5 mm of mercury?

5-13. Assuming the actual temperature and pressure at measurement time are 15°C and 760 mm Hg, what temperature difference, in degrees Fahrenheit, will produce an error of 0.01 ft on a line determined with an EDM instrument to be (a) 1500 ft? (b) 800 m?

5-14. Assuming the temperature and pressure at the time of measurement are 15°C and 760 mm Hg, what pressure difference, in inches of mercury, will produce an error of 0.01 ft on a line determined with an EDM instrument to be (a) 1500 ft? (b) 800 m?

5-15. If the temperature and pressure at the time of measurement are 15°C and 760 mm Hg, what will be the error in the electronic measurement of a line 10 mi long if the temperature at the time of observing is recorded 5°C too low?

5-16. For actual temperature and pressure of 15°C and 760 mm Hg, in EDM work, what neglected temperature difference from standard causes a measurement error of (a) 1 part in 100,000? (b) 1 part in 250,000?

5-17. What angle of phase change has the 15-MHz frequency of an HP 3805A undergone to produce a distance reading of (a) 1968.27 ft? (b) 742.957 m?

5-18. In Figure 5-9, h_e, h_r, elev$_A$, elev$_B$, and the measured slope length L were 5.10, 4.35, 825.75, 987.35, and 1284.29 ft, respectively. Calculate the horizontal length between A and B.

5-19. Similar to Problem 5-18, except that the values are, respectively, 1.205, 1.804, 643.21, 568.29, and 940.07 m.

5-20. In Figure 5-9, h_e, h_r, α_t, and the measured slope length L were 5.25 ft, 4.50 ft, $-13°27'30''$, and 875.29 ft, respectively. Calculate the horizontal length between A and B.

5-21. Similar to Problem 5-20, except h_e, h_r, L, and α_t were 1.52 m, 1.84 m, 245.06 m, and $9°24'15''$, respectively.

5-22. In Figure 5-10, a vertical angle of $+10°45'50''$ was recorded. The EDM instrument was standard-mounted and offset a distance of 0.75 ft vertically above the theodolite axis. If the theodolite and reflector heights were equal, what is the corrected horizontal distance for a recorded slope distance of 179.48 ft?

5-23. Similar to Problem 5-22, except that the recorded vertical angle, vertical EDM instrument offset, and slope distance were $-8°06'20''$, 0.22 m, and 77.54 m, respectively.

BIBLIOGRAPHY

Bell, T. P. 1978. "A Practical Approach to Electronic Distance Measurement." *Surveying and Mapping* 38(no. 4):335.

Crisp, R. 1979. "Electronic Surveying and On-Site Recording of Geodetic Data." *Bulletin, American Congress on Surveying and Mapping,* no. 67, p. 15.

Greene, J. R. 1977. "Accuracy Evaluation in Electro-Optical Distance Measuring Instruments." *Surveying and Mapping* 37(no. 3):247.

Kesler, J. M. 1973. "EDM Slope Reduction and Trigonometric Leveling." *Surveying and Mapping* 33(no. 1):61.

Kivioja, L. A. 1978. "The EDM Corner Reflector Constant Is Not Constant." *Surveying and Mapping* 38(no. 2):143.

McDonnell, P. W., Jr. 1982. "1982 EDMI Survey." *Point of Beginning* 7(no. 3):24.

———. 1983. "1983 EDMI Survey." *Point of Beginning* 8(no. 3):11.

Romaniello, C. G. 1977. "EDM 1976." *Surveying and Mapping* 37(no. 1):25.

Saxena, N. K. 1975. "Electro-Optical Short Range Surveying Instruments." *ASCE Journal of the Surveying and Mapping Division* 101(no. SU1):137.

Witte, B. U., and W. Schwarz. 1982. "Calibration of Electro-Optical Rangefinders—Experience Gained and General Remarks Relative to Calibration." *Surveying and Mapping* 42(no. 2):151.

6 LEVELING—THEORY, METHODS, EQUIPMENT

PART I THEORY

6-1. INTRODUCTION. *Leveling* is the general term applied to any of the various processes by which elevations of points or differences in elevation are determined. It is a vital operation in producing necessary data for mapping, engineering design, and construction. Leveling results are used to (a) design highways, railroads, and canals having grade lines that best conform to existing topography; (b) lay out construction projects according to planned elevations; (c) calculate volumes of earthwork; (d) investigate drainage characteristics of an area; and (e) develop maps showing general ground configurations.

6-2. DEFINITIONS. Basic terms in leveling are defined in this section, some of which are illustrated in Figure 6-1.

Vertical Line. A line that follows the direction of gravity as indicated by a plumb line.

Level Surface. A curved surface that at every point is perpendicular to the plumb line (the direction in which gravity acts). Level surfaces are approximately spheroidal in shape. A body of still water is the best example. For small areas, a level surface is sometimes treated as a plane surface.

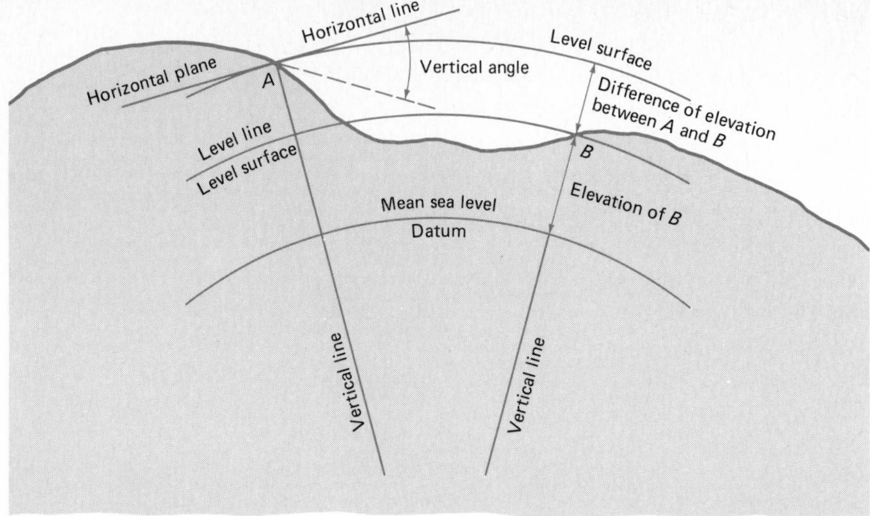

Figure 6-1. Leveling terms.

Level Line. A line in a level surface—therefore, a curved line.

Horizontal Plane A plane perpendicular to the direction of gravity. In plane surveying, a plane perpendicular to the plumb line.

Horizontal Line. A line in a horizontal plane perpendicular to the vertical.

Datum. Any level surface to which elevations are referred (for example, mean sea level). Also called *datum plane*, though not actually a plane.

Mean Sea Level (MSL). The average height of the sea's surface for all stages of the tide over a 19-year period. It was arrived at from readings, usually taken at hourly intervals, at 26 stations along the Atlantic and Pacific oceans and the Gulf of Mexico. The elevation of the sea differs from station to station depending on local influences of the tide; for example, at two points $\frac{1}{2}$ mi apart on opposite sides of an island in the Florida Keys, it varies by 0.3 ft. Therefore, to provide a common reference for elevations throughout North America, it was necessary to adopt a mean sea level. Scientists at the National Oceanic and Atmospheric Administration report that shrinking polar icecaps may be causing the earth's sea level to rise at a rate slightly more than 0.1 in a year since 1940—triple the rate of the preceding 50 years. If this phenomenon continues, low-lying coastal areas will be affected in future years.

Tidal Datum. The average height of all high waters observed from 1960 to 1978. Revised definitions of tidal datums—mean high water (MHW), mean higher high water (MHHW), mean low water (MLW), and mean lower low water (MLLW)—became effective in 1980. Those for high water (HW) and low water (LW) were not changed. The tidal datums are important in surveys to locate property and marine boundaries, fishing rights in tidal waters, and the limits of swamp and overflowed lands.

National Geodetic Vertical Datum (NGVD). The nationwide reference surface for elevations throughout the United States was made available to local surveyors by the National Geodetic Survey with the establishment of thousands of bench marks around the country.[1] It was obtained by a least squares adjustment in 1929 of all first-order leveling in the United States and Canada. Adjustments held all 26 tidal stations (21 in the United States, 5 in Canada) fixed and referenced the NGVD to MSL. Since the 1929 adjustment, more than 625,000 km of level lines have been added to the National Vertical Control Network. Another readjustment program to be completed in 1987 will include observing 110,000 km of new level lines, converting 900,000 km of leveling into a data base format, and determining improved elevations for 480,000 bench marks. Another project will replace half the destroyed marks.

Convergence of Level Surfaces. A phenomenon due to flattening of the earth in the polar direction so that level surfaces at different elevations are not parallel. This condition requires an *orthometric correction* for long north-south level circuits in precise work. Its value, which is relatively small, is a function of the latitude and elevation of the level circuit. On a line of levels run from Seattle to Los Angeles, a correction of approximately 2 ft would be required.

Elevation. The vertical distance from a datum, usually the NGVD, to a point or object. If the elevation of point A is 802.46 ft, A is 802.46 ft above some datum.

Bench Mark (BM). A relatively permanent object, natural or artificial, having a marked point whose elevation above or below an adopted datum is known or assumed. Common examples are metal disks set in concrete, large rocks, nonmovable parts of fire hydrants, and curbs.

Leveling. The process of finding elevations of points or their difference in elevation.

Vertical Control. A series of bench marks or other points of known elevation established throughout a project, also termed *basic control* or *level control*. The basic vertical control for topographic mapping of the United States was derived from first- and second-order leveling. Less precise third-order

[1] Locations and elevations of bench marks can be obtained from the National Geodetic Information Center, National Ocean Survey, NOAA, Rockville, MD 20852. Telephone: (301) 443-8631.

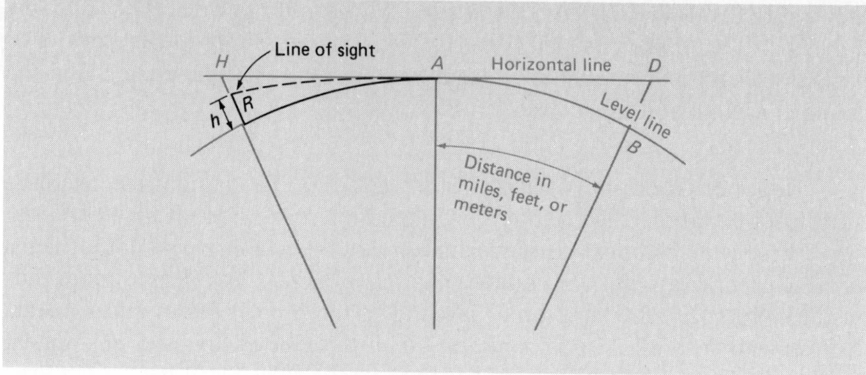

Figure 6-2. Curvature and refraction.

leveling is satisfactory for filling gaps between second-order bench marks as well as for many other projects (see Chapter 20).

6-3. CURVATURE AND REFRACTION. From the definitions of a level surface and a horizontal line, it is evident the latter departs from a level surface because of *curvature* of the earth. In Figure 6-2 the deviation DB from a horizontal line through point A is expressed approximately by the formula

$$C_f = 0.667M^2 = 0.0239F^2 \tag{6-1a}$$

or

$$C_m = 0.0785K^2 \tag{6-1b}$$

where the departure of a level surface from a horizontal line is C_f in feet or C_m in meters, M is the distance in miles, F the distance in thousands of feet, and K the distance in kilometers.

Since points A and B are on a level line, they have the same elevation. If the line of sight is horizontal, the earth's curvature causes a rod held at B to be read too high by length BD.

Light rays passing through the earth's atmosphere are bent or refracted toward the earth's surface as shown in Figure 6-3. Thus, a theoretically horizontal line of sight, like AH in Figure 6-2, is bent to the curved form AR. Hence, an object at R appears to be at H, and the reading on a rod held at R is diminished by length RH.

The effect of refraction in making objects appear higher than they really are (and therefore rod readings too small) can be remembered by noting what happens when the sun is on the horizon, as in Figure 6-3. At the moment when the sun has just passed *below* the horizon, it is seen just *above* the horizon. The sun's diameter of approximately 32 min is roughly equal to the average refraction on a horizontal sight.

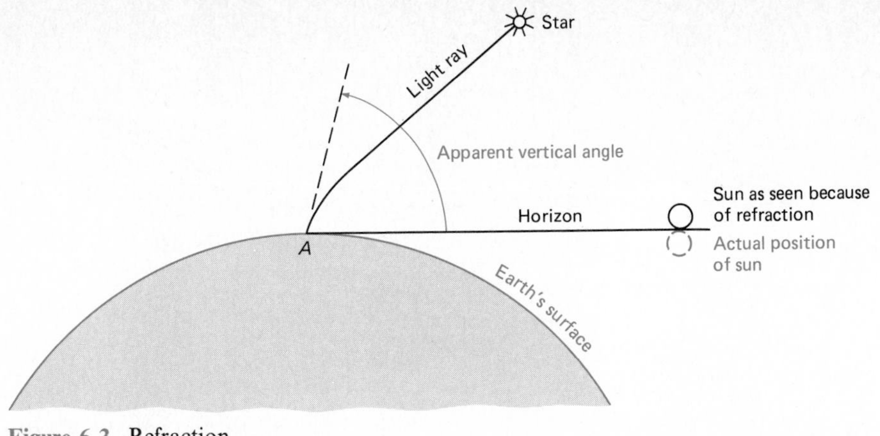

Figure 6-3. Refraction.

Displacement resulting from refraction is variable. It depends on atmospheric conditions, the length of line, and the angle a sight line makes with the vertical. For a horizontal sight, refraction R_f in feet or R_m in meters is expressed approximately by the formulas

$$R_f = 0.093M^2 = 0.0033F^2 \quad (6\text{-}2a)$$

or

$$R_m = 0.011K^2 \quad (6\text{-}2b)$$

This is about one-seventh the effect of curvature of the earth, but in the opposite direction.

The combined effect of curvature and refraction, h_f in feet or h_m in meters, is approximately

$$h_f = 0.574M^2 = 0.0206F^2 \quad (6\text{-}3a)$$

or

$$h_m = 0.0675K^2 \quad (6\text{-}3b)$$

For sights of 100, 200, and 300 ft, $h_f = 0.00021$ ft, 0.00082 ft, and 0.0019 ft, respectively, or 0.00067 m for a 100-m length. It will be explained in Section 7-4 that although the combined effects of curvature and refraction produce rod readings that are slightly too large, proper field procedures can generally completely eliminate the error due to these causes.

6-4. METHODS TO DETERMINE DIFFERENCES IN ELEVATION. Differences in elevation have traditionally been determined by taping, differential leveling, barometric leveling, and indirectly by trigonometric leveling. Brief descriptions of

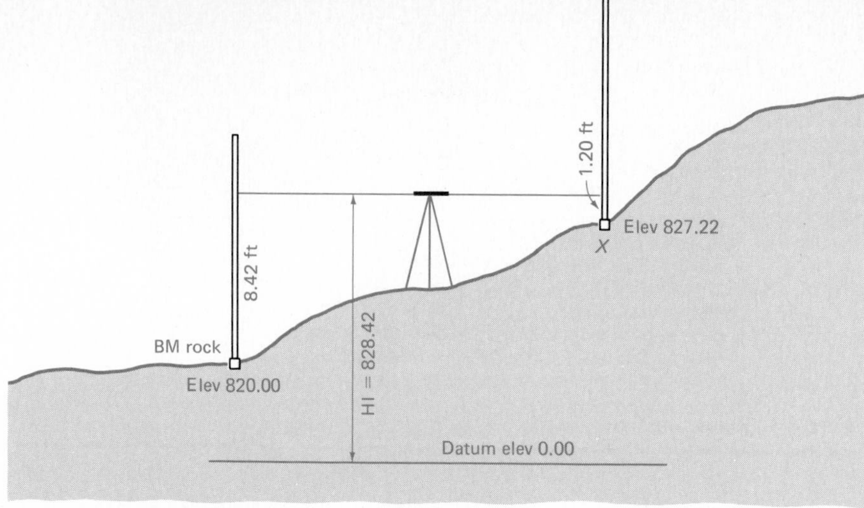

Figure 6-4. Differential leveling.

these methods follow. Newer techniques described in Sections 20-14 and 20-15 utilize inertial and satellite doppler systems.

6-4.1. TAPING METHOD. Application of a tape to a vertical line between two points is sometimes possible. This method is used to measure depths of mine shafts and in the layout and construction of multistory buildings. When a water or sewer line is being laid, a graduated pole or rod may replace the tape.

6-4.2. DIFFERENTIAL LEVELING. In this most commonly employed method, a horizontal line of sight is established by means of a level vial or automatic compensator. A telescope with suitable magnification is used to read vertical distances from fixed points on graduated rods.

The basic procedure is illustrated in Figure 6-4. An instrument is set up approximately halfway between BM Rock and point X. Assume the elevation of BM Rock is known to be 820.00 ft. After leveling the instrument, a plus sight taken on a rod held on the BM gives a reading of 8.42 ft. A *plus sight* (+S), also termed *backsight* (BS), is the reading on a rod held on a point of known or assumed elevation. This reading is used to compute the *height of instrument* (HI), defined as the vertical distance from datum to the instrument line of sight. Direction of the sight—whether forward, backward, or sideways—is not important. The term *plus sight* is preferable to *backsight*, but both are used. Adding the plus sight 8.42 ft to the elevation of BM Rock, 820.00, gives an HI of 828.42 ft.

Turning the telescope to bring into view a rod held on point X, a *minus sight* (−S), also called *foresight* (FS), of 1.20 is obtained. A minus sight is defined as the rod reading on a point whose elevation is desired. The term *minus sight* is preferable to *foresight*. Subtracting the minus sight 1.20 ft from the HI, 828.42, gives the elevation of point X as 827.22 ft.

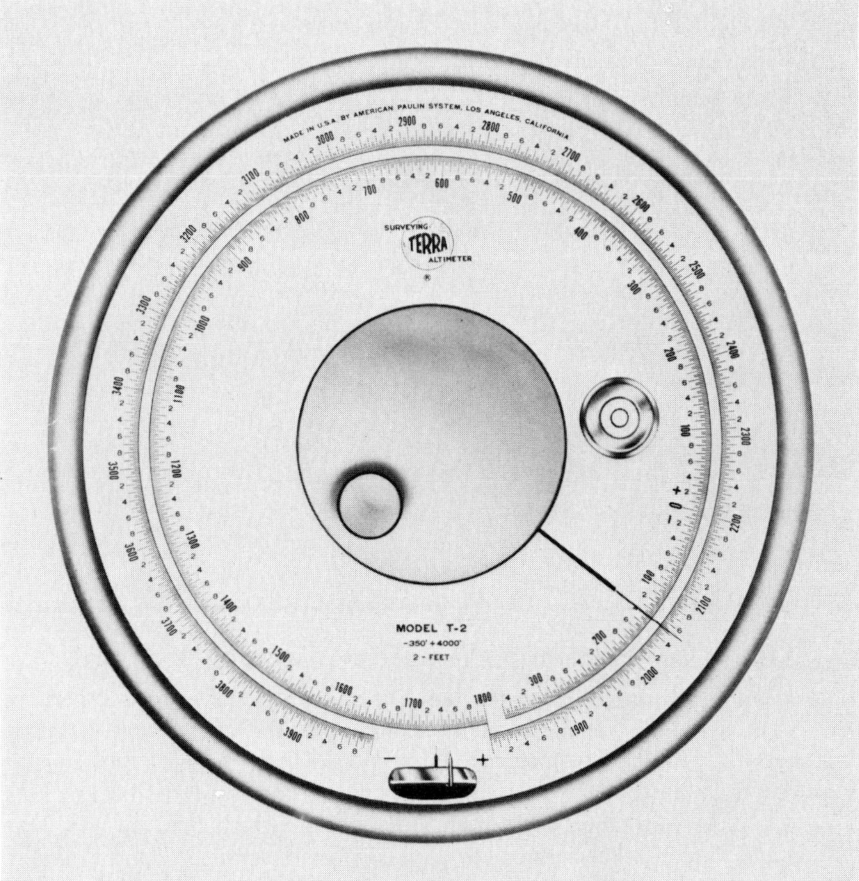

Figure 6-5. Surveying altimeter. (Courtesy American Paulin System.)

All leveling theory and applications can thus be expressed by two equations, which are repeated over and over:

$$\text{Elev} + \text{BS} = \text{HI} \tag{6-4}$$

$$\text{HI} - \text{FS} = \text{Elev} \tag{6-5}$$

6-4.3. BAROMETRIC LEVELING. The barometer, an instrument that measures air pressure, can be used to find relative elevations of points on the earth's surface. Figure 6-5 shows a surveying altimeter. Calibration of the scale on different models is in multiples of 1 or 2 ft, $\frac{1}{2}$ or 1 m.

Air pressures are affected by circumstances other than difference in elevation—for example, sudden shifts in temperature and changing weather conditions due to storms. Also, during each day, a normal variation in barometric pressure amounting to perhaps 100 ft difference in elevation occurs. This variation is known as the *diurnal range.*

In barometric leveling, one or more control barometers remain on a bench mark (base) while the *roving* instrument is taken to points whose elevations are

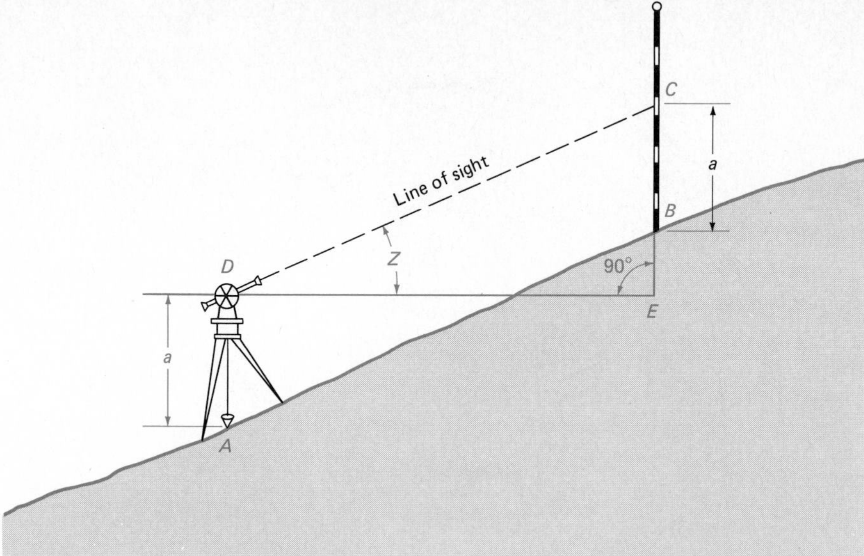

Figure 6-6. Trigonometric leveling.

desired. Readings are made on the bases at stated intervals of time, perhaps every 10 min, and the elevations recorded along with the temperature and time. Readings of the roving barometer are taken at critical points and adjusted later in accordance with changes observed at the control points. Methods of making field surveys by barometer have been developed in which one, two, or three bases may be used. Other methods employ leapfrog or semileapfrog techniques.

The barometric method is particularly suited for work in rough country where extensive areas must be covered but a high order of accuracy is not required. In stable weather conditions, and using several barometers, elevations to within ± 2 to 3 ft are possible.

6-4.4. TRIGONOMETRIC LEVELING. The difference in elevation between two points can be determined by measuring the (a) inclined or horizontal distance between them and (b) vertical angle to one point from a horizontal plane through the other. Thus, in Figure 6-6 if slope distance AB or DC and vertical angle EDC are measured, then the difference in elevation between A and B is $EC = DC \sin EDC$. When a horizontal distance is used, the equation involves the tangent function.

Trigonometric leveling is commonly used in topographic work and over very rugged terrain. New total-station instruments (see Section 5-6) have made trigonometric leveling easy and rapidly accomplished.

The Suunto clinometer is a hand-held optical reading instrument useful for a variety of rough surveying tasks, including trigonometric leveling. It has three graduated scales in (a) degrees from 0 to $\pm 90°$ for reading vertical angles, (b) grade percentages from 0 to 150% for measuring slopes, and (c) 0 to ± 200 ft on the topo scale for determining distances. It can be read to 1° or 1% and,

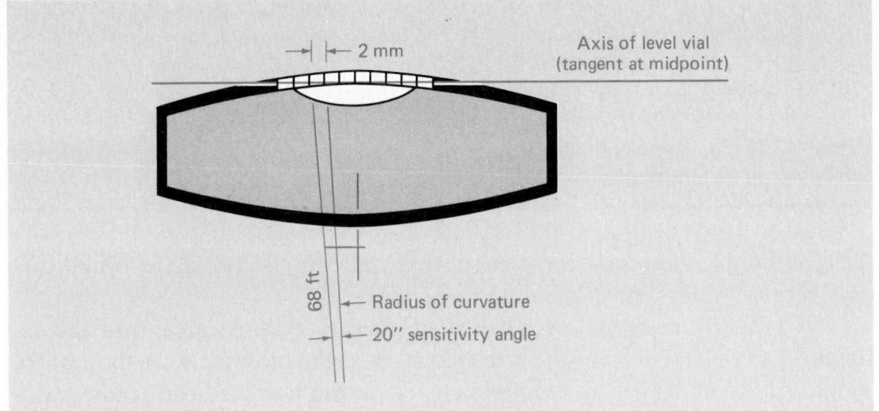

Figure 6-7. Level vial.

near the zero level, estimated to 10 min or 0.2%. The device weighs only 4.2 oz and is used by surveyors, engineers, builders, foresters, and others to measure heights, slopes, and vertical angles.

PART II
EQUIPMENT FOR DIFFERENTIAL LEVELING

6-5. TYPES OF LEVELS. The types of instruments used in differential leveling are the *wye* level, *dumpy* level, *tilting* dumpy level, *self-leveling* (*automatic*) level, *laser tracking* level, and *precise geodetic* level. For less accurate work, a hand level is often employed. All are described in this chapter except the precise geodetic level, which is discussed in Chapter 20, and the laser tracking level discussed in Chapter 24. Transits and theodolites can also be used for leveling. They are described in Chapter 10 and their use as levels is related in Section 11-17.

Although each instrument differs somewhat in design, all have a telescope and means for orienting the line of sight in a horizontal plane. Except for automatic levels, orientation is accomplished by leveling screws and vials.

6-6. LEVEL VIALS. A level vial is a glass tube, sealed at both ends, that contains a sensitive liquid and small air bubble. The liquid must be nonfreezing, quick-acting, and relatively stable in length for normal temperature variations. Purified synthetic alcohol has generally replaced the mixture of alcohol and ether formerly used. The uniformly spaced graduations etched on the tube's exterior surface locate the bubble's relative position. On present-day vials, the divisions are generally 2 mm long, but 0.01-ft and 0.1-in spacings have also been utilized.

The *axis* of a level vial is an imaginary longitudinal line tangent to the upper inside surface at its midpoint. When the bubble is centered in its run, the axis should be a horizontal line, as in Figure 6-7.

Sensitivity of a level vial is determined by the radius of curvature established in grinding. The larger the radius, the more sensitive a bubble. A highly

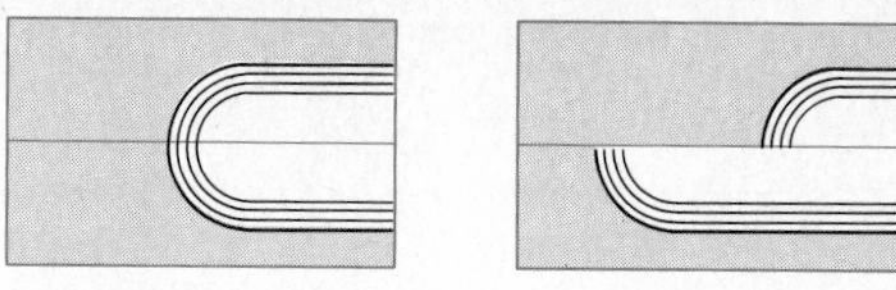

Figure 6-8. Coincidence type of level vial correctly set in left view; twice the deviation of the bubble shown in right view. (Courtesy of Kern Instruments, Inc.)

sensitive bubble, necessary for precise work, may be a handicap in rough surveys because more time is required to center it.

A properly designed level has a vial sensitivity correlated with the resolving power (resolution) of its telescope. A slight movement of the bubble should be accompanied by a minute change in the observed rod reading at a distance of about 200 ft. Sensitivity of a level vial is expressed in two ways: by the (a) angle, in seconds, subtended by one division on the scale and (b) radius of the tube's curvature.

If one division subtends an angle of 20 sec at the center, it is called a 20-sec bubble. Because variable division lengths have been used, this is not always a fair comparison. A 20-sec bubble on a vial with 2-mm divisions has a 68-ft radius. The sensitivity of level vials on wye and dumpy levels ranges from approximately 20 to 90 sec, the usual value being 20 sec for 2-mm divisions.

Theodolites, EDMIs, rod levels, tribrachs, and other equipment have circular bubble vials with sensitivity ranges from 5–15 min per 2-mm division. Other methods of stating their sensitivity have been proposed but are not yet standardized.

Figure 6-8 illustrates the *coincidence*-type level bubble used on precise equipment. The bubble is centered by bringing the two ends together to form a smooth curve. A prism splits the bubble and makes the two ends visible simultaneously.

A noncoincident-type bubble can be centered with an accuracy of about one-tenth of its sensitivity. A coincident bubble with opposite motions of the two ends and the magnification provided in an optical reading instrument can be centered to an accuracy of perhaps one-fortieth its sensitivity. Thus, the precision possible in centering, along with the bubble sensitivity, affects the results.

6-7. TELESCOPES.

The telescope on a dumpy level (Figure 6-9) is a metal tube containing four main parts: objective lens, negative lens, reticle, and eyepiece.

Objective Lens. This compound lens, securely mounted in the main tube's object end, has its optical axis reasonably concentric with the tube axis. Some light that strikes a lens is lost (approximately 4 to 5%) by reflection and absorption, even though the rays are perpendicular to its surface. For a compound lens system, the light reduction is multiplied and may become critical. Loss by reflection is practically eliminated by a thin ($\frac{1}{4}$ wavelength of light) uniform coating evaporated onto the lens surfaces that are in contact with air. The coating has an index of refraction smaller than that for glass. Loss of light by absorption usually is not serious unless the lenses are very thick.

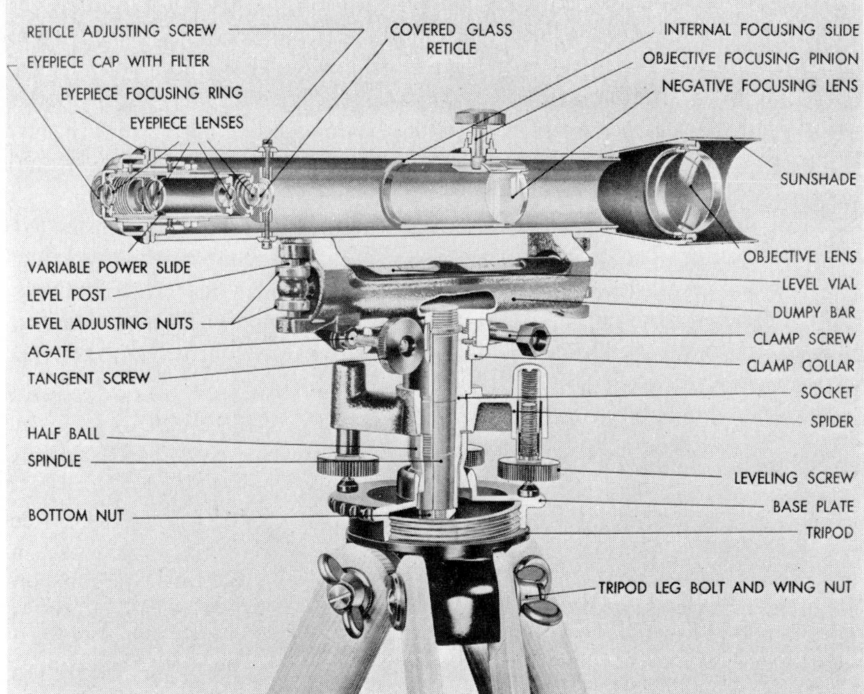

Figure 6-9. Dumpy level.

Negative Lens. The negative lens should be mounted in a sliding tube so its optical axis coincides with that of the objective lens. This arrangement focuses rays of light through the objective lens onto the reticle plane. It is important that the slide tube, and cylinder into which it is received, be exactly fitted so there will be no deviation of either lens axis—objective or negative—during the focusing process from maximum to minimum distances. If a faraway object is sighted—for example, the rod on a long sight in reciprocal leveling (see Section 7-5)—the distance from a lens to the image it forms is called *focal length.*

Reticle. A reticle is a pair of lines mounted near the main tube eyepiece and located at the principal focus of the objective optical system. The point of intersection of these lines, together with the objective system optical center, forms the directing axis of the telescope commonly called the *line of collimation* or *line of sight*. The reticle is mounted by two pairs of opposing capstan screws placed at right angles to each other, one pair horizontal and the other vertical, to facilitate adjusting the line of sight (see Appendix A).

Cross hairs formerly were made of spider web or filaments of platinum or glass stretched across an annular ring (doughnut). In many newer instruments, a thin glass plate with lines ruled and etched, and filaments of dark metal deposited in them, serves as the reticle. Additional lines parallel to and equidistant from the cross lines are added when desired. If a glass reticle is used, the extra lines are shortened to avoid confusing them with the primary ones. The reticle is mounted to place the lines in a horizontal-vertical position.

Eyepiece. The eyepiece is a microscope with magnification of about 35 diameters (diam) for viewing the image focused by the objective lens system at the plane of the reticle. It may consist of two lenses (giving an inverted image to the eye) or four lenses (producing an erected image). The former gives a slightly better optical acuity but may temporarily confuse beginners. Eyepieces are furnished with a focusing movement to accommodate the difference in vision of individual observers.

The focusing process is the most important function to be performed in using a telescope. Today's telescopes generally are internal-focusing (an interior auxiliary lens moves on a rack), although some older ones in service were external-focusing [the objective lens moves in accordance with Eq. (6.6)]. Dust and wear affect the slide and may disturb the optical axis of the external type. An internal-focusing telescope is more dust-resistant.

The fundamental principle of lenses is given by the formula

$$\frac{1}{f_1} + \frac{1}{f_2} = \frac{1}{f} \tag{6-6}$$

where f_1 is the distance from lens to image at the reticle plane, f_2 the distance from lens to object (see Figure 15-1), and f the lens focal length. The focal length is a constant for any particular set of lenses. Therefore, as distance f_1 changes, f_2 must also change.

Since the reticle remains fixed in the telescope tube, the distance between it and the eyepiece must be adjusted to suit the eye of an individual observer. This is done by bringing the cross hairs to a clear focus—that is, making them appear as black as possible when sighting at the sky or a distant, light-colored object. Once this has been accomplished, the adjustment need not be changed for the same observer, regardless of sight length, unless the eyes tire from long observation time or high telescopic magnification.

After the eyepiece has been adjusted, objects are brought to sharp focus at the plane of the cross hairs by moving the objective lens. If the cross hairs appear to travel over the object sighted when the eye is shifted slightly in any direction, *parallax* exists. Either the objective lens, the eyepiece, or both must be adjusted to eliminate this effect if accurate work is to be done.

The "visible" rule *near-far, far-near* may be helpful to beginners using an old external-focusing telescope. If the object is near the observer, the objective lens is run far out. When the object is far away, the objective lens is brought nearer to the observer's eye.

A level vial attached to the telescope tube, when in adjustment, has its axis exactly parallel to the line of sight. Centering the level bubble, therefore, makes the line of sight horizontal.

6-8. OPTICS.[2] A brief discussion of surveying-instrument optics is desirable before leaving the subject of telescopes. The purpose of a telescope is to create

[2] This section on optics includes some material from *Optical Alignment Equipment* published by Keuffel & Esser Company and reprinted here by permission.

for an observer a picture of the cross wire's position on the target with the greatest possible clarity and precision. This end is attained by skillful design and perfection in manufacture to secure the combination of optical qualities best suited to a particular application. Optical factors include resolving power, magnification, definition, eye distance, size of pupil, and field of view.

Specifications alone can seldom indicate the true qualities of one telescope compared with those of another. The most important test for a telescope, and in fact the only true test, is a simultaneous comparison with another one under the same conditions.

Explanations of some important optical terms follow.

Resolving Power. The ability of a lens to show detail is termed *resolving power*. It is measured by the smallest angular distance, expressed in seconds of arc, between two points just far enough apart to be distinguished as separate objects rather than a single blurred one. Its value is usually stated as the maximum number of lines per millimeter that can be seen as separate lines in the image.

The maximum resolving power that theoretically can be attained with a telescope when the optical parts are perfectly designed and exactly placed depends entirely on the diameter of that part of the objective lens actually used (the effective aperture). The resolving power of an objective lens is independent of magnification. It can be computed by the empirical formula

$$R = \frac{5.5 \text{ sec}}{D} \tag{6-7}$$

where R is the angle that can be resolved, in seconds, and D the diameter of the lens aperture, in inches. For example, if the objective lens of a certain telescope has an aperture 1.18 in in diameter, its resolving power is 4.7 sec.

The accepted theoretical standard for resolving power of a human eye is 60 sec, although a value between 80 and 90 sec would probably be more practical. Hence, the resolving power of an objective lens has to be brought at least to this limit by magnification. If the angular distance resolved by the telescope is 4.7 sec, this resolving power must be magnified 13 times to obtain 60 sec. Since the eyesight of different observers varies, more magnification is always used.

Magnification. The value of magnification (power) is the ratio of the apparent size of an object viewed through a telescope to its size as seen by the unaided eye from the same distance. Magnification varies slightly when the telescope focus is changed; therefore, it is affected somewhat by the object distance.

Although telescopic magnification must be greater than $60/R$, there is a limiting point beyond which it is impossible to increase magnification without sacrificing definition. This point is reached when magnification becomes greater than 2 or 3 times $60/R$. For larger values, quality of the image seen is impaired and accuracy with which the line of sight can be made to coincide with a target is reduced.

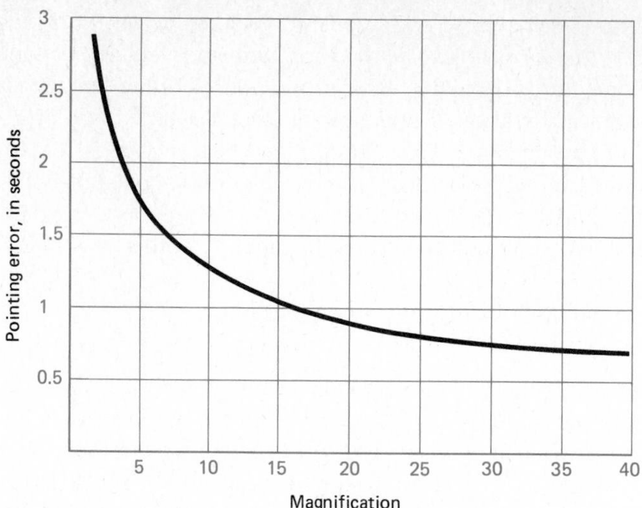

Figure 6-10. Relation between magnification and pointing error. (Courtesy Keuffel & Esser Company.)

Certain disadvantages result from the use of too high a magnification, even when the objective lens is large enough to give the necessary resolution. With high magnification the field of view is reduced, and any heat waves, turbulence, or vibration causes the image of an object to move over the cross hairs too fast for accurate observation. The magnifying power of telescopes on modern levels ranges from 26 to 42 diam, and averages perhaps 32 diam.

Definition. *Definition* is a term used to define the overall results produced by a telescope. Better definition permits objects to be seen more clearly through the telescope. It depends on a number of optical features and is the quality that provides the greatest pointing accuracy.

Since definition is a relative term, it can be determined best by comparing the appearance of the same object when viewed through the telescope to be tested and through a telescope with which the observer is familiar.

Pointing Accuracy. The exactness with which the line of sight can be directed toward a target, or a rod aligned, is called *pointing accuracy.* It depends on magnification, definition, arrangement of the cross hairs, and design of the target or scale sighted. The general relationship between magnification and pointing accuracy for telescopes having the same definition is shown in Figure 6-10.

6-9. LEVEL BARS AND SUPPORTS FOR WYE AND DUMPY LEVELS. Telescopes of wye and dumpy levels rest upon vertical supports at each end of a horizontal member called the *level bar.* The bar is centered on an accurately machined vertical spindle that sits in a conical socket of the leveling head. The spindle makes the level bar revolve in a horizontal plane when the instrument is properly adjusted.

6-10. LEVELING HEAD. For wye and dumpy levels, the conical socket into which the vertical axis of the level bar fits is carried by four large leveling screws. They rest upon a plate fastened to the tripod top and are in two pairs at right angles to each other. The telescope is placed alternately over each pair of opposite screws, which are turned until the bubble remains in the vial center for a complete revolution. The line of sight then generates a horizontal plane.

Most modern instruments have three rather than four leveling screws. The three-screw arrangement is faster and not subject to the rocking that takes place in the four-screw type when two opposite screws are turned up or down slightly more than the other two. The disadvantage of the three-screw type is that slight differences in elevation of the line of sight result if all three screws are turned up or all three turned down. Manipulation of the four-screw head does not change the telescope elevation. Also, after the threads become worn on a three-screw leveling head, there is some loss of rigidity and the screw must be replaced. Tightening one screw of each pair in the four-screw arrangement results in a clamping action and produces a stable setup.

6-11. WYE LEVEL. The wye level, now almost obsolete, has a nonfixed telescope that rests in supports called *wyes* because of their shape. Curved clips, hinged at one end and pinned at the other, fasten the telescope in place.

A wye level is simpler to adjust than a dumpy level because the telescope can be lifted from the wyes and turned end for end. This feature permits one person to complete all adjustments. More adjustments are required for the wye level, but they are easier to make. This advantage is lost if the telescope collars or the wyes' bearings upon which they rest become worn. The instrument then must be adjusted the same way as a dumpy level.

6-12. DUMPY LEVEL. A dumpy level (see Figure 6-9) has the telescope rigidly attached to, and parallel with, the level bar. The level vial is set in the level bar and thereby protected somewhat. The vial always remains in the same vertical plane as the telescope, but screws at each end permit vertical adjustment or replacement of the vial.

Advantages of a dumpy level over the wye level are (a) simpler construction with fewer movable parts, (b) fewer adjustments to be made, and (c) probably longer life for the adjustments. A disadvantage is that one adjustment requires two persons and is more time-consuming. This problem can be eliminated if two points of known elevation are established several hundred feet apart and fixed targets set up on them. (Adjustments of levels are described in detail in Appendix A.)

6-13. TILTING DUMPY LEVEL. Tilting dumpy levels are used for the most precise work and also widely employed for general purposes. A bull's-eye (circular) spirit level is available for quick approximate leveling by means of the leveling screws, or a ball-and-socket arrangement (with no leveling screws) permits the head to be tilted and locked nearly level. Exact level is obtained by tilting or rotating the telescope slightly in a vertical plane about a fulcrum at the vertical axis of the instrument without changing its height. A micrometer screw under the eyepiece controls this movement.

Figure 6-11. Tilting level with micrometer. (Courtesy Keuffel & Esser Company.)

Figure 6-12. GK 23-C tilting level. (Courtesy Kern Instruments, Inc.)

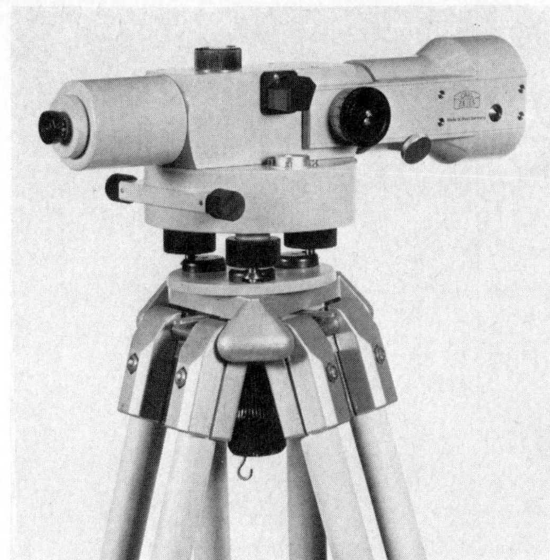

Figure 6-13. Zeiss automatic Ni 2 level with micrometer. (Courtesy of Carl Zeiss, Inc.)

The tilting feature saves time and increases accuracy, since only one screw need be manipulated to keep the line of sight horizontal as the telescope is turned about a vertical axis. The telescope bubble is viewed through a system of prisms from the observer's normal position behind the eyepiece. A prism arrangement splits the bubble image into two parts. Centering the bubble is accomplished by making the images of the two ends coincide.

The tilting dumpy level shown in Figure 6-11 has a four-screw leveling head, 30X magnification, resolving power of 4 sec, minimum focusing distance of 6 ft, glass reticle, and sensitivity of the level vial equal to 20 sec/2 mm.

Figure 6-12 shows a tilting level with a ball-and-socket leveling head. It is characterized by a short telescope, streamlined construction, small size, and light weight. The instrument has a telescope level vial with a sensitivity of 18 sec/2 mm, a centering accuracy of ± 4 sec, telescope magnification of 30X, and weighs only 3.3 lb. A 2.44-in-diameter horizontal glass circle that can be read with its microscope by estimation to 1 min is incorporated.

6-14. AUTOMATIC LEVELS. Automatic levels of the type pictured in Figure 6-13 incorporate a self-leveling feature. With most of these instruments, rough leveling using a three-screw leveling head approximately centers a bull's-eye bubble, although some models have a ball-and-socket arrangement. After the bull's-eye bubble is manually centered, a compensator takes over, automatically levels the line of sight, and keeps it level. The operating principle of one type of compensator is shown schematically in Figure 6-14. Note that the optical pendulum compensator is an entirely different system for establishing a level line of sight than that used to make the line of sight parallel with a level-vial axis.

Automatic levels have become popular for general use because of the ease and rapidity of their operation. Some are precise enough for second-order and even first-order work if a parallel-plate micrometer is attached to the telescope front as an accessory. When the micrometer plate is tilted, the line of sight is

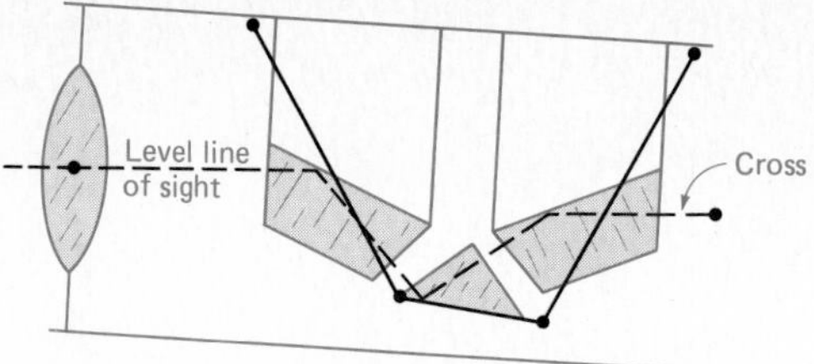

When telescope tilts up, compensator swings backward.

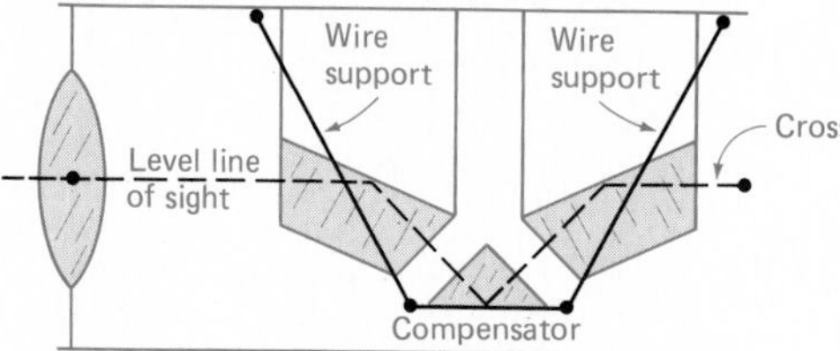

Telescope horizontal

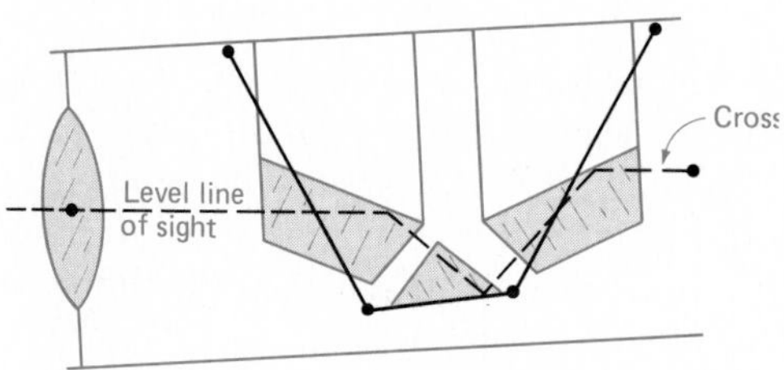

When telescope tilts down, compensator swings forward.

Figure 6-14. Compensator of a self-leveling level. (Courtesy Keuffel & Esser Company.)

displaced parallel to itself and decimal parts of rod graduations can be measured.

Under certain internal and environmental conditions, an air-damped type of automatic level compensator can stick. To check, with the instrument leveled and focused, read the target, tap the tripod, and after it vibrates, determine whether the same reading is obtained. Also, some unique compensator problems, such as residual stresses in the flexible links, can introduce systematic errors if not corrected by an appropriate observational routine on first-order work.

A new approach to leveling instrument design employs a free surface of mercury to make the sight line horizontal or nearly so. Thus, spirit level bubbles or self-leveling compensator pendulums are not required.

6-15. TRIPODS. Several types of tripods are available. The legs may be fixed or adjustable in length, and solid or split. All models are shod with metallic conical points and hinged at the top where they connect to a metal head. An adjustable-leg tripod is advantageous for setups in rough terrain or in a shop, but the type with a fixed-length leg may be slightly more rigid. The split-leg model is lighter than the solid type but less rugged. A wide-framed tripod, first used on European

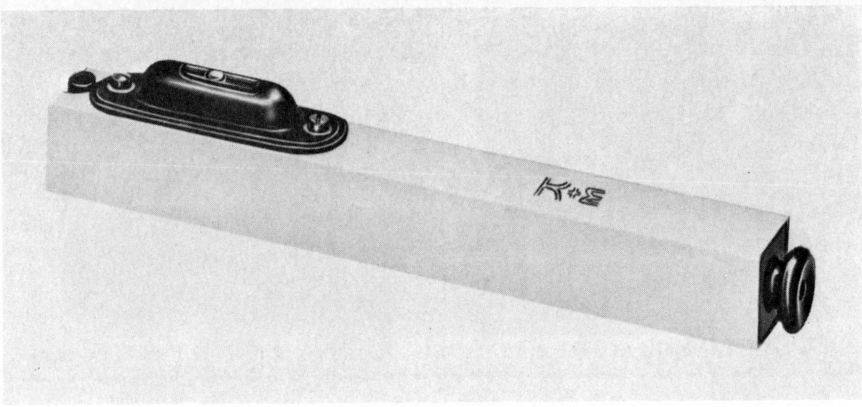

Figure 6-15. Hand level. (Courtesy Keuffel & Esser Company.)

instruments, is now available from American manufacturers. A sturdy tripod in good condition is necessary to obtain the best results from a fine instrument. An 8-ft extension-leg tripod is available and very useful when high setups are necessary to sight above cornfields, brush, and low obstructions.

In the past, many different thread types were used on tripods. The standard now adopted by all American manufacturers is eight threads per inch on a $3\frac{1}{2}$-in-diameter cap. Most European and Japanese models have a $\frac{5}{8}$-in-diameter bolt.

6-16. HAND LEVEL. The hand level (Figure 6-15) is a hand-held instrument used on low-precision work and for checking purposes. It consists of a brass tube 6 in long having a plain glass objective and peep-sight eyepiece. A small level vial mounted above a slot in the tube is viewed through the eyepiece by means of a prism or 45°-angle mirror. A horizontal line extends across the tube.

A prism or mirror occupies only half the inside of the tube, and the other part is open to provide a clear sight through the objective. Thus, the rod being sighted and the reflected image of the bubble are visible beside each other with the cross line superimposed.

The instrument is held in one hand and leveled by raising or lowering the objective end until the cross line bisects the bubble. The tube can be steadied by making a tripod with a thumb on the cheekbone, first finger on the forehead, and eyepiece against the brow. Holding the level beside a staff or, better still, resting it in a Y-shaped stick increases the accuracy.

Stadia lines reading 1:10 may be included (see Chapter 15). Magnification of 4X is provided for observing the bubble and cross wire. The rod is seen through plain glass, but 2X magnification is available on some models. The length of sight is limited, therefore, to the distance at which a rod can be read with natural vision or very small magnification.

The Abney hand level and clinometer, shown in Figure 6-16, has limited application in measuring vertical angles and slopes and for direct leveling. It

Figure 6-16. Abney hand level and clinometer. (Courtesy Keuffel & Esser Company.)

includes an arc graduated in degrees up to 90°, a vernier (see Section 6-19) reading to 10 min, and several scales for slopes ranging from a ratio of 1:1 to 1:10.

6-17. LEVEL RODS. Level rods are made of wood, fiberglass, or metal and have graduations in feet and decimals, or meters and decimals.

Two main classes of rods are:

1. Self-reading rods, which can be read by the instrument operator while sighting through the telescope and noting the apparent intersection of the cross wire on the rod. This is the most common type.
2. Target rods having a movable target that is set by a rodperson at the position indicated by signals from the level operator.

A wide choice of patterns, colors, and graduations on single-piece, two-piece, and three-section leveling rods is available. The various types, usually named for cities or states, include the Philadelphia, New York, Boston, Troy, Chicago, San Francisco, and Florida rods.

Rods for general leveling and for special purposes such as slope staking can be made by fastening a flexible ribbon of treated fabric to a wooden frame. Such strips divided in various ways can be purchased from manufacturers. The direct-reading Lenker level rod [Figure 6-17(d)] has numbers in reverse order on an endless graduated steel-band face strip that can be revolved on the rod's end rollers. Figures run down the rod and can be brought to a desired reading—for example, the elevation of a bench mark. Rod readings are preset for the backsight and then, due to the reverse order of numbers, foresight readings give elevations directly without calculating HIs and subtracting FSs.

A self-reading rod consisting of a wooden frame and an Invar strip (graduated in decimals of meters) to eliminate the effects of humidity and temperature changes is used on precise work. The Invar strip, attached at its ends only, is free to slide in grooves on each side of the wooden frame.

The Philadelphia rod, a combination self-reading and target rod, is the most common type in college surveying instrument rooms. A 13-ft model is described in detail in Section 6-18. Other lengths are also made, with the 12-ft rod being popular.

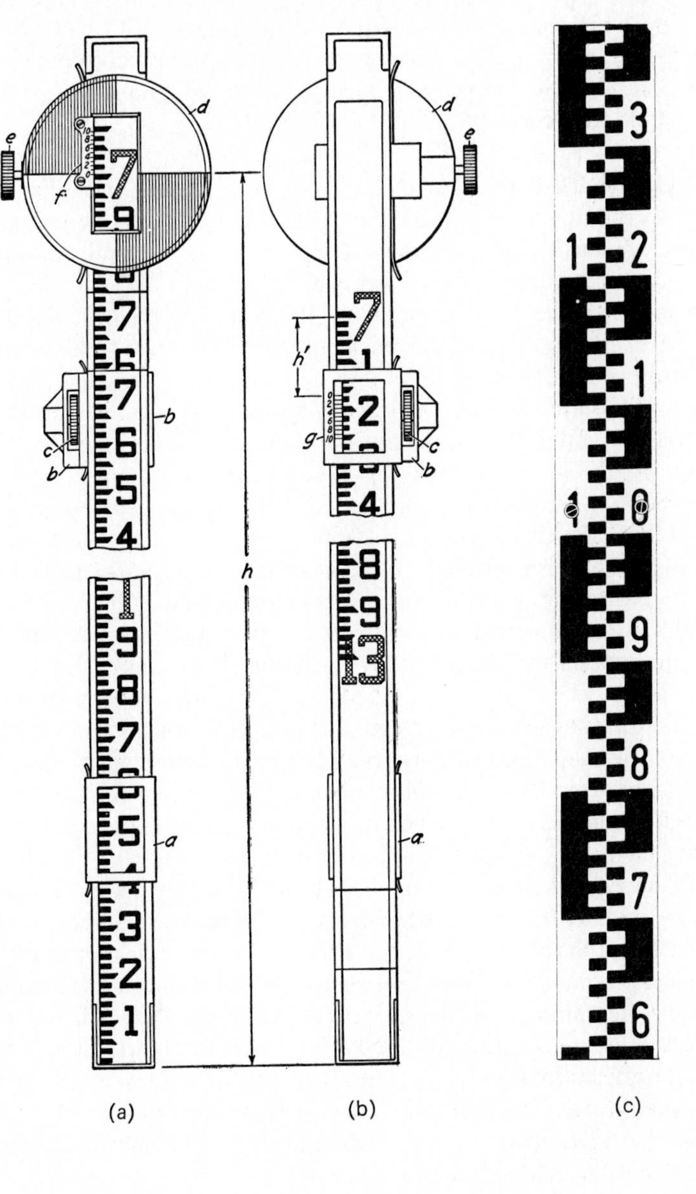

Figure 6-17. (a) Philadelphia rod (front). (b) Philadelphia rod (rear). (c) Leveling rod with metric graduations. (Courtesy Wild Heerbrugg Instruments, Inc.) (d) Lenker direct-reading rod. (Courtesy Lenker Manufacturing Company.)

The Chicago rod, consisting of independent sections (usually three) which fit together but can be disassembled, is widely used on construction surveys; the San Francisco model has separate sections that slide past each other to extend or compress its length, and is generally employed on control, land, and other surveys. Both are conveniently transported in vehicles.

Safety in traffic and near heavy equipment is an important consideration. The Quadpod, an adjustable stand that clamps to any leveling rod, reduces traffic hazards and labor costs.

6-18. PHILADELPHIA ROD. The Philadelphia rod shown in Figure 6-17(a) and (b) consists of two sliding sections graduated in hundredths of a foot and joined by brass sleeves *a* and *b*. The rear section can be locked in position by a clamp screw *c* to provide any length from a *short rod* for readings of 7 ft or less, to a *long rod* (*high rod*) for readings up to 13 ft. *When the high rod is needed, it must be fully extended.* Graduations on the front faces of the two sections read continuously from zero at the base to 13 ft at the top for the high-rod setting.

Rod graduations are accurately painted, alternate black and white spaces 0.01 ft wide. The 0.1- and 0.05-ft marks are emphasized by spurs extending the black painting. Tenths are designated by black figures, foot marks by red numbers, all straddling the proper graduation. A Philadelphia rod can be read accurately with a level at distances up to 250 ft.

On long sights, or when readings to the nearest 0.001 ft are desired, a target *d* may be used. Circular, oval, and angular targets are made. All are approximately 5 in high and painted red and white in alternate quadrants. A clamp *e* and vernier scale *f* are part of the target. For readings of less than 7 ft, the target is set at the proper elevation in accordance with directions given by the observer. When the rod is extended, the target is clamped at 7.000 ft and the rear section raised to bring the target to the correct height. Divisions on the back of the rod are marked from 7 to 13 ft in a downward direction. As the rod is extended, a fixed vernier scale *g* attached to sleeve *b* enables the rodperson to read the target height.

Philadelphia rods are made of carefully selected, kiln-dried, well-seasoned hardwood and graduated in accordance with rigid specifications. They *should not* be used as seats or for pole vaulting, nor should they be left leaning against a tree or building, or laid on any surface with the graduated face down. Hands must be kept off the painted markings, particularly in the 3- to 5-ft section, where a worn face will make the rod unfit for use. Letting the rod down "on the run" batters both sections and may change the vernier reading to less than 7.000 ft—for example, to 6.998 ft. If this happens, the target must be set to the same reading—that is, 6.998 ft for high rods. Figure 6-17(c) shows the front face of a typical leveling rod with metric graduations.

6-19. VERNIERS. A vernier is a short auxiliary scale set parallel to and beside a primary scale. It provides fractional parts of the smallest main-scale divisions without interpolation. Figure 6-18 depicts a simple type of *direct* vernier used on leveling rods. Somewhat more complicated verniers for transits are described and shown in Sections 10-7 and 10-8.

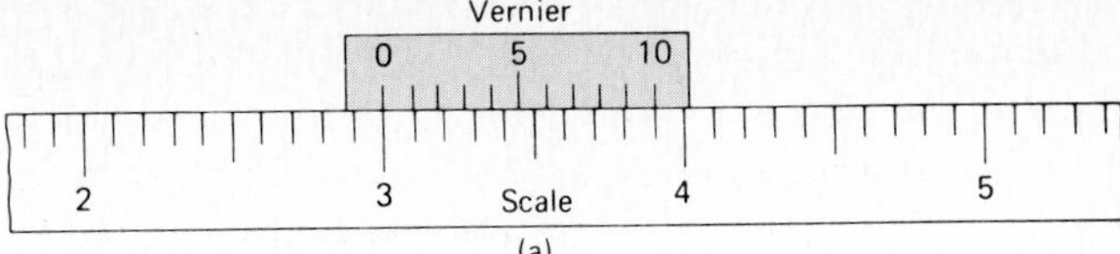

(a)

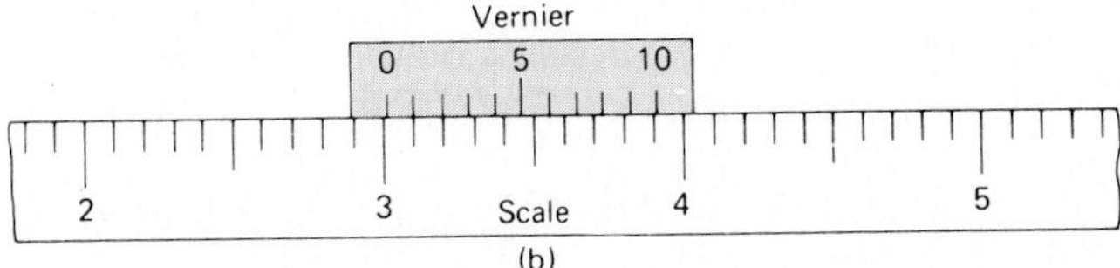

(b)

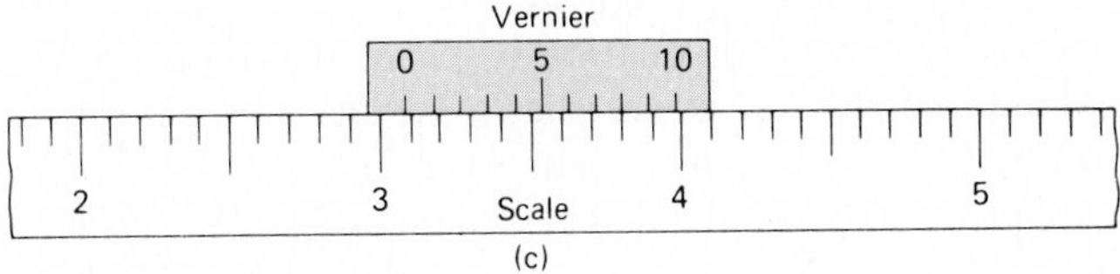

(c)

Figure 6-18. Vernier.

As illustrated in Figure 6-18(a), the vernier has n divisions in a space covered by $(n - 1)$ of the smallest scale divisions. Then

$$(n - 1)d = nv \tag{6-8}$$

where d is the length of a scale division and v the length of a vernier division. This is the fundamental basis for all vernier construction. For most level-rod verniers, $n = 10$, $d = 0.01$ ft, and $v = 0.09/10 = 0.009$ ft.

In Figure 6-18(a), the reading is 0.300. If the vernier is moved so that its first graduation from zero coincides with the first-scale graduation beyond 0.300, as in Figure 6-18(b), the vernier index has moved a distance equal to

$$d - v = 0.010 - 0.009 = 0.001 \text{ ft}$$

The reading is therefore 0.301 ft.

If the vernier is moved so the second vernier graduation is coincident with the graduation representing 0.32 on the scale, movement from the position in Figure 6-18(a) has been $2(d - v) = 0.002$ ft. Thus, the fractional part of a scale division from the preceding scale graduation to the vernier index is read by determining the number of the vernier line that is coincident with *any* scale graduation. The reading in Figure 6-18(c) is therefore 0.308 ft.

Expanding Eq. (6-8),

$$nv = nd - d$$

and

$$d - v = \frac{d}{n} \tag{6-9}$$

For a vernier, $d - v$ is the smallest reading obtainable without interpolating. It is termed the *least count* of the vernier and from Eq. (6-9) is given by an expression d/n; that is,

$$\text{least count} = \frac{\text{value of the smallest division on the scale}}{\text{number of divisions on the vernier}}$$

An observer *cannot* be certain that readings of the scale and vernier are correct until the least count has been determined.

In selecting the vernier line that is coincident with a scale division, an observer must assume a position directly behind the lines, or over them, to avoid parallax. The first and/or second graduation on each side of the apparently coincident lines should be checked to see that a symmetrical pattern is formed about them. In Figure 6-18(c), vernier graduations 6 and 10 fall inside (toward division 8) the scale lines by equal distances; therefore, 8 is the correct reading.

PROBLEMS

6-1. Compute and tabulate the combined effect of curvature and refraction on level sights of 50, 75, 100, 150, and 200 m.

6-2. Similar to Problem 6-1, except for sights of 50, 200, 500, and 1000 ft.

6-3. On a large lake without waves, how far from shore is a boat with a 30-ft mast as it disappears from the view of a person lying at the water's edge?

6-4. Similar to Problem 6-3, except a person whose eye height is 1.8 m above the water's edge.

6-5. Readings on a line of differential levels are taken to the nearest 0.01 ft. For what maximum distance can the earth's curvature and refraction be neglected?

6-6. Similar to Problem 6-5, except readings are to the nearest millimeter.

Successive plus and minus sights taken on a downhill line of levels are listed in Problems 6-7 and 6-8. What error results from curvature and refraction?

6-7. 100, 175; 160, 210; 175, 200; 220, 290 ft.

6-8. 40, 55; 35, 50; 20, 25; 60, 70 m.

6-9. Explain how surveyors and engineers can often ignore the error caused by curvature and refraction in leveling work.

6-10. With the bubble centered, a 250-ft sight gives a reading of 5.26 ft. After moving the bubble three divisions off-center, the reading is 5.47 ft. For 0.1-in vial divisions, what is the (a) vial radius of curvature in feet and (b) angle in seconds subtended by one division?

6-11. Similar to Problem 6-10, except each vial division is 0.01 ft.

6-12. Similar to Problem 6-10, except each vial division is 2 mm.

6-13. Why should a long base be used in trigonometric leveling?

6-14. Compute the sensitivity of a level vial with 0.01-ft graduations and a 103-ft radius of curvature.

6-15. An observer fails to check the bubble and it is off $1\frac{1}{2}$ divisions on a 200-ft sight. What error results for a 15-sec bubble?

6-16. Similar to Problem 6-15, except the 15-sec bubble is off two divisions on an 8-m sight.

6-17. An observer using a new instrument with a 30-sec vial does not know the tube division length—2 mm, 0.1 in, or 0.01 ft. What is the error for each type if the bubble is off one division on a 250-ft sight?

6-18. Sunshine on the forward end of a 20-sec/2-mm level vial bubble draws it off one division, giving a 7.14-ft plus sight on a 220-ft shot. Compute the correct reading.

6-19. Describe two ways to determine the approximate magnifying power of a telescope.

6-20. Why should sunshades and dustcaps on external-focusing telescopes always be removed or replaced by turning them clockwise?

6-21. What errors may be introduced in using a telescope's focusing screws?

6-22. Define optical axis and nodal point of a lens.

6-23. List in tabular form, for comparison, the advantages and disadvantages of a tilting level versus an automatic level.

6-24. A slope distance of 150 ft and 20% slope are read with an Abney level. Compute the horizontal distance and elevation difference.

6-25. If a BS of 4.32 ft is taken on BMA, elevation 858.27, and a FS of 9.03 ft read on point x, calculate the HI and elevation of point x.

6-26. State two causes of parallax when using surveying equipment.

6-27. List four conditions which can cause properly set BM monuments to settle and/or move.

6-28. Define a retrograde vernier.

Sketch scales and verniers for the conditions noted in Problems 6-29 through 6-31.

6-29. A rod is divided into 0.05-ft spaces. Required: a vernier to permit readings to (a) 0.01 ft, and (b) 0.0025 ft.

6-30. A builder's scale has $\frac{1}{2}$-in divisions. Required: readings to $\frac{1}{32}$ in; scale has 1° graduations, readings to 10 min.

6-31. A vernier has 12 divisions and readings are required to 5 sec. Another scale has 5 spaces per inch, the vernier 10 divisions.

6-32. State one advantage and one disadvantage of an inverting telescope.

6-33. What is the advantage of a reversion vial?

6-34. Compare the advantages of a three-screw leveling head, and a four-screw arrangement.

BIBLIOGRAPHY

Berry, R. M. 1977. "Observational Techniques for Use with Compensator Leveling Instruments for First-order Levels." *Surveying and Mapping* 37(no. 1):15.

Hou, C. Y., S. A. Veress, and J. E. Colcord. 1972. "Refraction in Leveling." *Surveying and Mapping* 32(no. 2):231.

Kivioja, L. A. 1979. "New Mercury Leveling Instruments." *Surveying and Mapping* 39(no. 1):61.

Kulp, E. F. 1970. "High Precision Levels with Automatic Instruments." *ASCE Journal of the Surveying and Mapping Division* 96(no. SU1):121.

Kunitomi, D. S. 1972. "Elevation Changes Due to Tides, Long Beach, California." *ASCE Journal of the Surveying and Mapping Division* 98(no. SU2): 137.

Lippold, H. R., Jr. 1980. "Readjustment of the National Geodetic Vertical Datum." *Surveying and Mapping* 40(no. 2):155.

Maddux, W. S. 1982. "Datum Extrapolation by Simultaneous Comparison of Partial Tidal Cycles." *Surveying and Mapping* 42(no. 2):139.

Quinn, F. H. 1976. "Pressure Effects on Great Lakes Vertical Control." *ASCE Journal of the Surveying and Mapping Division* 102(no. SU1):31.

Straub, H. W. 1973. "Computation of Magnitude of Vertical Refraction in Leveling Operations." *Canadian Surveyor* 27:279.

The Surveyor and the Law. 1980. "Changes in Tidal Datums by 1980 National Datum Convention" [from *Federal Register*, 45:207]. *Surveying and Mapping* 40 (no. 1):88.

Weidner, J. P. 1979. "Re: Nearly Flat to the Transit = MHW vs Vegetation, by Greulich, G." *Surveying and Mapping* 39(no. 1):68.

7 LEVELING—FIELD PROCEDURES AND COMPUTATIONS

7-1. INTRODUCTION. Chapter 6 covered the basic theory of leveling, briefly sketched simple procedures, and showed examples of most types of leveling equipment.

This chapter is devoted to handling equipment, running and adjusting simple leveling loops, and performing some project surveys to obtain data for field and office. Construction and other surveys, along with those of higher order to establish the nationwide vertical control networks, will be covered in later chapters.

7-2. CARRYING AND SETTING UP A LEVEL. The safest way to transport a leveling instrument in a car is to leave it in the container. The case closes properly only when the instrument is set perfectly in the padded supports.

A level should be removed from its container by lifting the level bar or base, *not* by grasping the telescope. The head must be screwed snugly on the tripod. If the head is too loose, the instrument is unstable; if too tight, it may "freeze" on a U.S.-type thread tripod. A grain of sand, roughness in the threads, or a change in temperature makes the head stick. Spreading the tripod legs until the instrument head almost touches the ground often helps to release the head if it freezes.

The legs of a tripod must be tightened correctly. If each leg falls slowly of its own weight after being placed in a horizontal position, it is properly adjusted. Clamping them too tightly strains the plate and screws. If the legs are loose, a wobbly setup results.

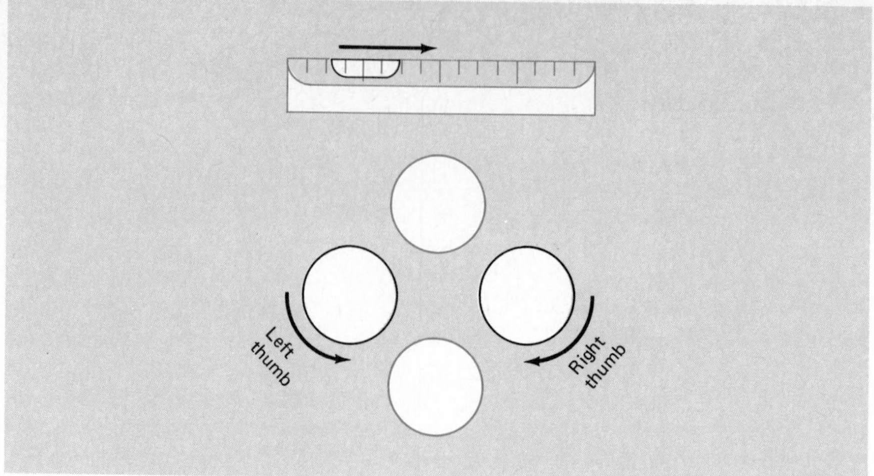

Figure 7-1. Use of leveling screws.

It is generally unnecessary to set a level over any particular point. Therefore, it is inexcusable to have the base plate badly out of level before using the leveling screws. On side-hill setups, placing one leg on the uphill side and two on the downhill slope eases the problem. On very steep slopes, some instrument operators prefer two legs uphill and one downhill for stability. The most convenient height of setup is one that enables the observer to sight through the telescope without stooping or stretching.

In leveling a four-screw head, the telescope is rotated until it is over two opposite screws. The bubble is approximately centered by using the thumb and first finger of each hand to adjust the opposite screws. This procedure is repeated with the telescope over the other two leveling screws. Time is wasted by centering the bubble exactly on the first try, since it will be thrown off during the cross leveling. Working with each pair of screws about three times should complete the job.

Leveling screws are turned in opposite directions at the same speed by both hands, unless the intention is to tighten or loosen the leveling head. A simple rule, illustrated in Figure 7-1, is: *A bubble follows the left thumb.*

If one hand turns faster than the other, the screws loosen, the head rocks on two screws, or they bind. Final precise adjustment can be made with one screw only. Leveling screws should be snug, not wrench-tight, to save time and avoid damage to threads and base plate. A good observer senses the proper setting of all leveling screws to permit ready movement without jamming the threads. Instruments should be leveled on the base plate before being replaced in the case.

For automatic and tilting levels that have a three-screw head and bull's-eye bubble, the telescope is aligned over one screw and thus made perpendicular to a line through the other two. The bull's-eye bubble is centered by alternately turning one screw and then the other two. The telescope need not be rotated during the process.

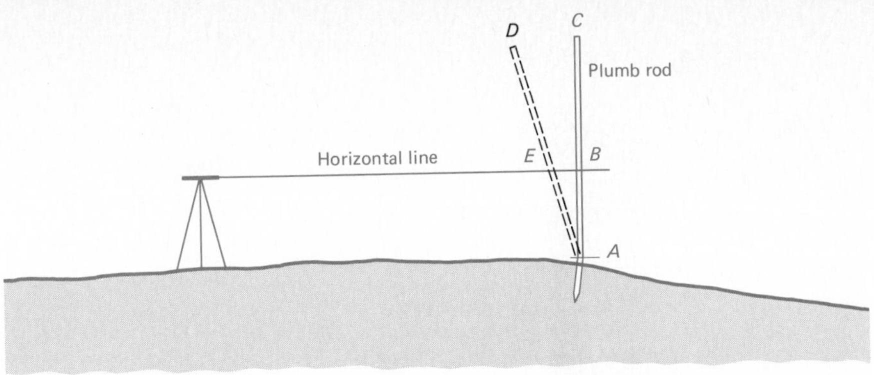

Figure 7-2. Plumbing rod.

Inexperienced instrument operators working on a steep hillside are likely to find, after completing the leveling process, that the telescope is too low for sighting the upper turning point or bench mark. To avoid this, set the instrument without attempting to level it carefully and have the bubble somewhat back of center. Sight the rod, and if it is visible for this placement, it obviously will also be seen when the instrument is leveled. As an alternative, a hand level can be used to check for proper height of the setup before precisely leveling the instrument.

7-3. DUTIES OF A RODPERSON. The duties of a rodperson are relatively simple. Like a tape handler, however, a rodperson can nullify the best efforts of an observer by failing to follow a few basic rules.

A level rod must be held plumb to give the correct reading. In Figure 7-2, point *A* is below the line of sight by distance *AB*. If the rod is tilted to position *AD*, an erroneous reading *AE* is obtained. It can be seen that the smallest reading possible, *AB*, is the correct one and secured only when the rod is plumb.

Waving the rod is a procedure used to ensure that the rod is plumb when a reading is taken. The process consists of *slowly* tilting the rod top, first toward the instrument and then away from it. The observer watches the readings alternately increase and decrease, and then selects the minimum value, the correct one. Beginners tend to swing the rod too fast and through too long an arc. Small errors can be introduced depending on the kind of mark used. A rounded-top monument, steel spike, or thin edge make an excellent mark.

On still days the rod can be plumbed by letting it balance of its own weight while lightly supported by the fingertips. An observer makes certain the rod is plumb in the lateral direction by checking its coincidence with the vertical wire and signals for any adjustment necessary. The rodperson can save time by sighting along the side of the rod to line it up with a telephone pole, tree, or side of a building. Plumbing along the line toward the instrument is more difficult, but holding the rod against the toes, stomach, and nose will bring it close to a plumb position.

A rod level of the type shown in Figure 7-3 ensures *fast* and *correct* rod plumbing. Its L shape is designed to fit the rear and side faces of a rod while the bull's-eye bubble is centered to plumb the rod in *both directions*.

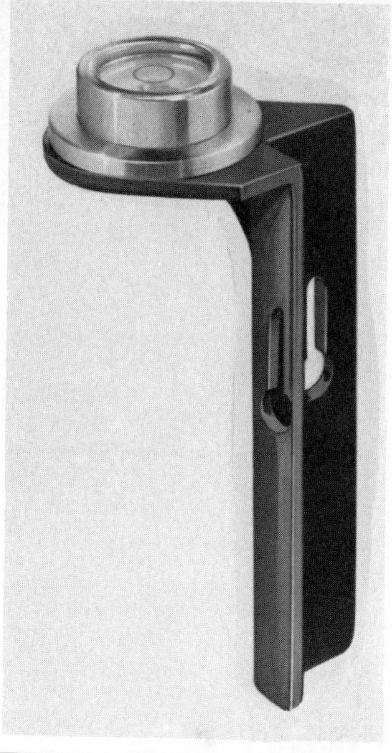

Figure 7-3. Rod level. (Courtesy Keuffel & Esser Company.)

Example 7-1

In Figure 7-2, $AB = 10$ ft and $EB = 6$ in. What error results?

Using the rightmost term of Eq. (4-5) in Section 4-13, the error e is

$$e = \frac{d^2}{2L} = \frac{0.5^2}{2 \times 10} = 0.012 \text{ ft or } 0.01 \text{ ft}$$

Errors of this magnitude are serious, whether the results are carried out to hundredths or thousandths. They make careful plumbing necessary, particularly for high-rod readings.

7-4. DIFFERENTIAL LEVELING. Figure 7-4 illustrates the procedure followed in differential leveling. Several instrument setups were required to complete the runs "out" and "back." Field notes are given in Plate D-3.

The positions on which a rod is held to extend a line from one setup to the next are called *turning points* (TPs). A turning point is defined as a solid point on which both a plus and a minus sight are taken on a line of levels. Horizontal distances for the plus and minus sights should be made approximately equal by pacing, stadia measurements, or counting rail lengths if working along a track, noting pavement joints if beside a concrete-surfaced highway, or by some other easy method. This will eliminate errors due to instrument maladjustment (most important) and the combined effects of the earth's curvature and refraction as shown in Figure 7-5, where e_1 and e_2 are the combined curvature and refraction errors for the plus and minus sights, respectively. Because D_1 and D_2 are equal, e_1 and e_2 are also equal. In calculations the former is added and the later subtracted; thus, they cancel each other.

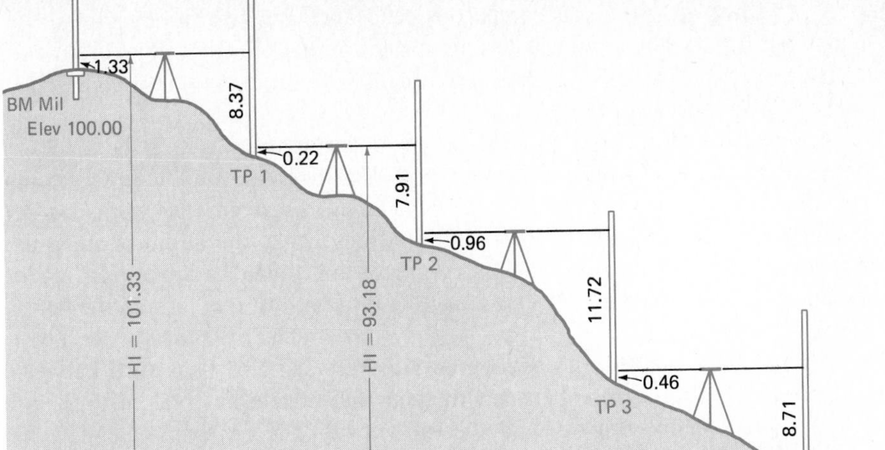

Figure 7-4. Differential leveling.

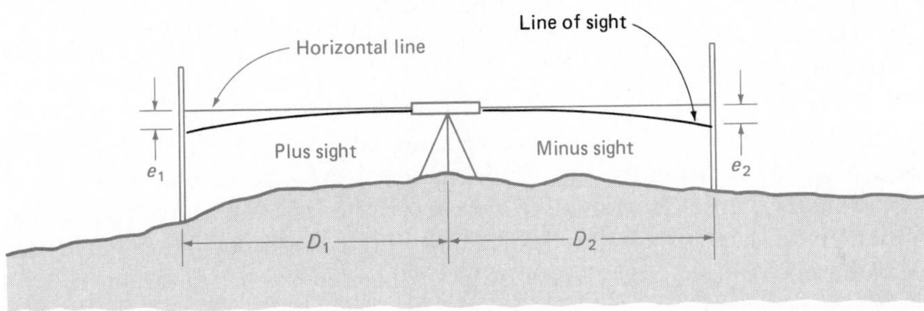

Figure 7-5. Balancing plus and minus sight distances to cancel errors due to curvature and refraction.

On slopes it may be somewhat difficult to balance lengths of plus and minus sights, but usually it can be done by following a zigzag path.

A bench mark is described in a field book the first time used and thereafter referred to by noting the page number on which it was recorded. Descriptions begin with the general location and must include enough details to enable a person unfamiliar with the area to find the mark readily (see Appendix D, Plates D-3 and D-5). A bench mark is usually named for some prominent object it is on, or near, to aid in describing its location; one word is preferable. Examples are BM River, BM Tower, BM Corner, and BM Bridge. On extensive surveys, bench marks are numbered consecutively. This is an advantage in identifying relative positions along a line but more subject to mistakes in field marking or recording.

Turning points are also numbered consecutively but not described in detail, since they are merely a means to an end and usually will not have to be relocated. If possible, however, it is advisable to select turning points that can be relocated, so if reruns are necessary because of blunders on long lines, field work can be reduced.

Before a party leaves the field, all possible note checks must be made to detect any mistakes in arithmetic and verify achievement of an acceptable closure. *The algebraic sum of the plus and minus sights applied to the first elevation should give the last elevation.* This computation checks the addition and subtraction for all HIs and TPs unless compensating mistakes have been made. When carried out for each left-hand page of tabulations, it is termed the "page check."

Important work is checked by leveling forward and backward between end points. The difference between the *rod sum* (algebraic total of plus and minus sights) on the run out and the rod sum on the run back is called the "*loop misclosure.*" Specifications, or the purpose of the survey, fix the permissible loop misclosure (see Section 7-13). If the allowable misclosure is exceeded, one or more additional runs must be made. Note that *a new instrument setup has to be made before starting the return run to get a complete check.* In Plate D-3, a minus sight of 8.71 was read on BM Rutgers to finish the run out, and a plus sight of 11.95 was recorded to start back, showing that a new setup had been made. Otherwise an error in reading the final minus sight would be accepted for the first plus sight on the run back. An even better check is secured by tying the run to more than one bench mark.

The difference in elevation between end points is considered to be the average of rod sums on the runs out and back. Where a number of interlocking "loops" are used in a network of levels, an approximate "loop adjustment" method (see Section 7-15) or a more rigorous least squares adjustment can be used to distribute the misclosure. True elevations are secured by starting from a bench mark whose elevation above mean sea level is known, and checked by closing on the same or another bench mark. If this is not possible, an assumed elevation may be used and all later values converted to true elevations by applying a constant.

A lake or pond undisturbed by wind, inflow, or outflow or even a slow-moving stream can serve as an extended turning point. Stakes driven flush with the lake or stream surface, or rocks whose high points are at this level would be used.

Double-rodded lines of levels are sometimes used on important work. Plus and minus sights are taken on two TPs using two rods from each setup, and the readings carried in separate noteform columns. A check on each instrument setup is obtained if the HI agrees for both lines.

Flying levels can be run at the end of a workday to check the results of a long line of levels run in one direction only. Longer sights and fewer setups are employed; the purpose is only to detect any large mistakes:

Three-wire leveling, formerly used mainly in precise work, is now common on projects that require only ordinary precision. Readings of the upper, middle, and lower cross wires are averaged to obtain a better value. A check is secured by noting the difference between the middle and upper wire and between the

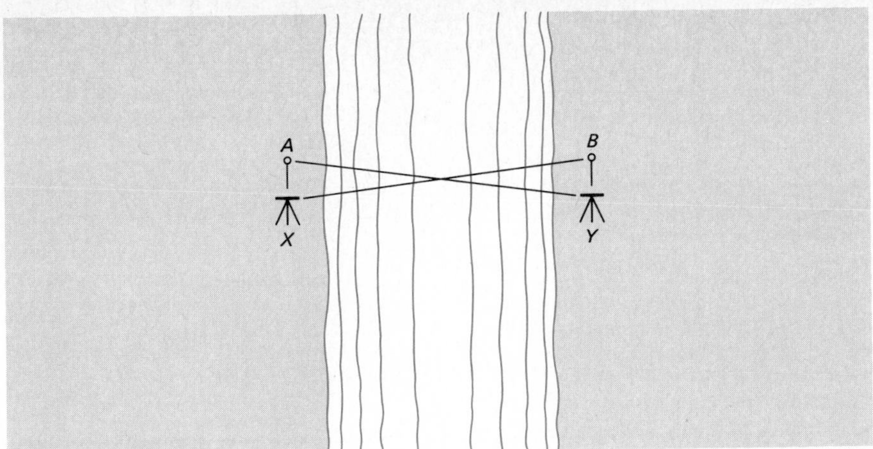

Figure 7-6. Reciprocal leveling.

middle and lower wire. If these fail to agree within one or two of the smallest units being read, the readings are repeated. The intercept between upper and lower wires provides the sight distance for checking the lengths of plus and minus sights. This procedure is described in more detail in Section 20-16.

7-5. RECIPROCAL LEVELING. Topographic features such as rivers, lakes, and canyons make it difficult or impossible to keep plus and minus sights short and equal. Reciprocal leveling is utilized at such locations.

As shown in Figure 7-6, a level is set up on one side of a stream at *X*, near *A*, and rod readings are taken on points *A* and *B*. Since *XB* is very long, several readings are taken for averaging. This is done by reading, turning the leveling screws to throw the instrument out of level, releveling, and reading again. The process is repeated two, three, four, or more times. Then the instrument is moved close to *Y* and the same procedure followed.

The two differences in elevation between *A* and *B*, determined with an instrument at *X* and *Y*, normally will not agree because of curvature, refraction, and personal and instrumental errors. Refraction changes can occur if there is a long delay before making the observations at *Y*. An average of the two elevation differences is accepted as the correct value if their precision appears satisfactory. This procedure, a method of reversion, is also used in adjusting levels and transits. Plate D-4 is a sample set of reciprocal leveling notes. Refinements of this technique have been developed for crossing wider obstacles and obtaining the highest precision.

7-6. PROFILE LEVELING. On route surveys for highways or pipelines, for example, elevations are required at every 100-ft (or 30-m) station; at angle points (points marking changes in direction); at breaks in the ground surface slope; and at critical points such as roads, bridges, and culverts. When plotted, these elevations show a *profile*—a line depicting ground elevations at a vertical section along a survey line. For most engineering projects, profiles are taken along

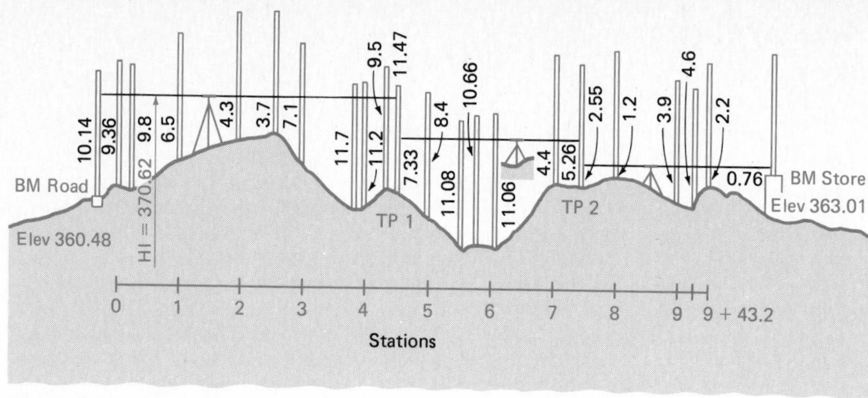

Figure 7-7. Profile leveling.

the center line, which is staked out in 100-ft stations or, if necessary because of rough terrain, in 50- or 25-ft (15- or 10-m) increments.

Profile leveling, like differential leveling, requires establishing turning points on which both plus and minus sights are read. In addition, any number of *intermediate foresights* (minus sights) are taken on points along the line from each instrument setup as shown in Figure 7-7. Plate D-5 is a sample set of notes for that profile.

As presented in the notes, a plus sight is taken upon a bench mark and intermediate sights read on the stations; at breaks in the ground surface; and at critical points, until the limit of accurate sighting distance is reached. A turning point is then selected, the instrument moved ahead, and the process repeated. The level itself is usually set up *off* the center line so that sights of more uniform length can be procured. Bench marks placed out of the way of future construction are established along the route on a long line.

It is evident that when the "page check" is made on arithmetic computations, only the minus sights taken on turning points can be used. For this reason, and to isolate the points to be plotted, a separate column is preferred for intermediate sights.

Readings on paved surfaces, such as concrete roadways, curbs, and sidewalks, can be taken to 0.01 ft. Readings on the ground smaller than 0.1 ft are not practical.

An *elevation meter* used on roads is a mechanical or electromechanical device on wheels, towed by a car or truck, that measures slope and distance and then automatically and continuously integrates and records their product in differences of elevation. A fourth-order accuracy profile can be secured at speeds of 30 mi/hr.

7-7. AIRBORNE PROFILE RECORDER. Development of an airborne profile recorder was begun in the early 1940s to obtain a profile of the terrain beneath the path of an aircraft carrying radar equipment. The ground profile with respect to an *isobaric* reference surface (having equal barometric pressures) was obtained by measuring the elapsed time required for radar signals to travel from

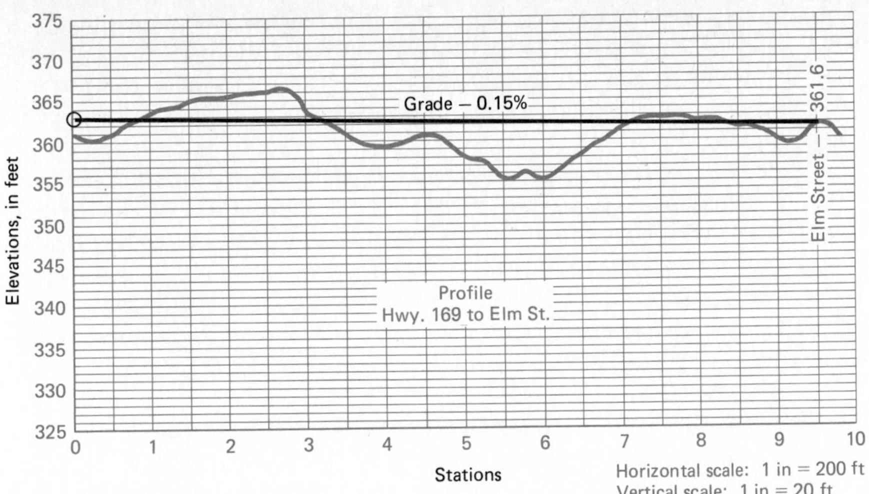

Figure 7-8. Plot of Profile.

an aircraft to the ground, be reflected, and return to the aircraft. By the late 1950s, equipment was capable of 10-ft accuracies in flat terrain and 20 ft over mountainous areas.

Accuracy of airborne profile recording systems has been improved through the use of laser energy. The system operates similarly to electronic distance-measuring devices in that the frequency of laser energy is varied, the output signal is modulated onto a carrier wave, and height is determined by utilizing the phase-shift principle.

Flying heights are normally kept low during laser profiling to get better accuracy. Altitudes of the aircraft above the terrain are usually recorded in digital form. The accuracy of laser airborne profile recorders is quite phenomenal, with tests showing elevations to be correct within 1 ft at flying heights of 1000 ft. Resolution of the system actually approximates 0.1 ft (a precision of $\frac{1}{10{,}000}$) but the accuracy of the isobaric reference datum limits the terrain profile correctness to the higher figure.

7-8. DRAWING AND USE OF A PROFILE. Profiles are plotted on special paper of the type shown in Figure 7-8. The vertical lines are spaced $\frac{1}{4}$ in apart, with every 10th line heavier. Horizontal lines are $\frac{1}{20}$ in apart, with each 5th line thicker and every 50th line still heavier.

The vertical scale of a profile is generally exaggerated with respect to the horizontal scale to make differences in elevation more pronounced. A ratio of 10:1 is frequently used, but flatness or roughness of the terrain determines the desirable proportions. Thus, for a horizontal scale of 1 in = 100 ft, the vertical scale might be 1 in = 10 ft. The heaviest lines of Plate A paper make blocks $2\frac{1}{2} \times 2\frac{1}{2}$ in, which are best suited to a scale of 1 in = 40 ft (or 400 ft) horizontally and 1 in = 4 ft (or 40 ft) vertically (used in many old railroad surveys). The scale actually employed should be plainly marked.

Plotted profiles are used for many purposes, such as (a) determining depth of cut or fill on proposed highways, railroads and airports; (b) studying grade-crossing problems; and (c) investigating and selecting the most economical grade, location, and depth for sewers, pipelines, tunnels, irrigation ditches, and other projects.

Rate of grade (or *gradient* or *percent grade*) is the rise or fall in feet per 100 ft, or meters per 100 m. Thus, a grade of 2.5% means a $2\frac{1}{2}$-ft difference in elevation per 100 ft horizontally. Ascending grades are plus; descending grades, minus. A grade line chosen to give somewhat equal cuts and fills is shown in Figure 7-8. The process of staking grades is described in Chapter 24.

The term *grade* is also used to denote the elevation of the finished surface on an engineering project.

7-9. GRID, CROSS-SECTION, OR BORROW-PIT LEVELING. Grid leveling is a method for locating contours (and topographic features) by staking an area in squares of 10, 20, 50, 100, or more feet (or comparable meter lengths) and determining the corner elevations. Rectangular blocks, say 50 × 100 ft, that have the longer sides roughly parallel with the direction of most contour lines may be preferable on steep slopes. The grid size chosen depends on the project extent and accuracy required.

The same process, termed "borrow-pit leveling," is employed on construction jobs to ascertain quantities of earth, gravel, rock, or other material to be excavated or filled. The procedure is covered in Section 27-10 and Plate D-6.

7-10. USE OF THE HAND LEVEL. A hand level can be used for some types of surveying when a low order of accuracy is sufficient. The instrument operator takes a plus and minus sight while standing in one position and then moves ahead to repeat the process. A hand level is useful, for example, in cross-sectioning to obtain a few additional rod readings on sloping terrain where a turning point would otherwise be required.

7-11. SIZE OF FIELD PARTY. Ordinary differential leveling can be done by a two-person party if the observer keeps notes. On precise leveling, an observer, umbrella holder (unless an automatic level that reacts slowly to temperature changes is used), notekeeper, and two rodpersons are needed.

A self-reading rod is frequently utilized on borrow-pit and profile leveling; hence, a party of two is sufficient. Having a third person to keep notes relieves the observer of this task and enables the party to move faster.

7-12. SIGNALS. The distance between personnel and noise from traffic or other sources make it necessary to communicate by hand signals on many surveys if radios are not available. Special signals to fit unusual situations can be invented for a particular need. They should simulate as closely as possible the action to be taken. Observers must remember they have the advantage of telescopic magnification and give clear signals that cannot be misunderstood by a rod-person using only natural vision. Equipping the rod holder with a small telescope is helpful. Some typical gestures employed in leveling are in the following list.

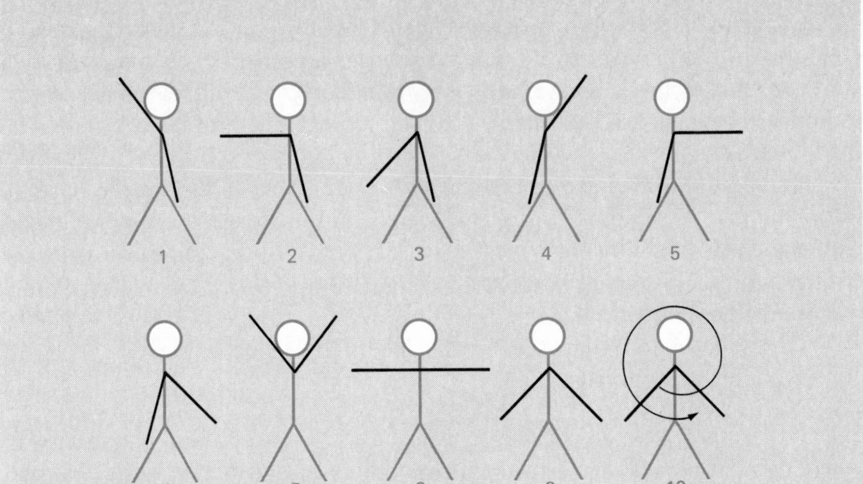

Figure 7-9. Signals for numbers.

Plumb rod. If the rod is to the right of a plumb position, the observer extends the right arm full length upward and inclined to the vertical. This position is maintained until the rod is plumb.

Establish a TP. Either the observer or rodholder may give this signal by holding one arm straight up and moving it in a horizontal circle (indicating "turn").

High rod. Observer extends both arms horizontally and sideways, palms up, and brings them together overhead.

Raise for red. On very close sights, the red foot-mark numbers may be out of the telescope's field of view. If a signal is needed, the observer holds one arm straight forward, palm up, and raises the arm slowly to a position about 45° above the horizontal.

Raise target. Observer raises an extended arm above shoulder height, holding it high if considerable movement is required. The arm is moved toward a horizontal position as the target approaches the desired setting.

Lower target. Same as "raise target," but the extended arm is held below the shoulder and moved up.

Clamp target. Observer waves one hand in a vertical circle with the arm in a horizontal position.

TP or BM. Rodperson holds the rod horizontally above head height and then places it on the TP or BM. This signal is used in differential leveling to distinguish between intermediate sights on TP's or BM's for the benefit of observer and notekeeper.

All right. Arms are extended sideways, palms forward, and waved up and down several times. This is employed by any member of a party in all types of surveying.

Signals for number. One system used is sketched in Figure 7-9.

7-13. PRECISION. Precision in leveling, as in taping, is determined by repeating measurements or tying-in to control points. The elevation of a bench mark may be found by leveling over two different routes, from other bench marks, or by a closed circuit of levels returning to the point of beginning. If an accurately established bench mark is available near the end of a line run, a check should be made upon it.

Loop misclosures are compared with permissible values on the basis of either number of setups or distance covered. Various organizations set precision standards based on their project requirements. For example, on a simple construction survey, an allowable loop closure of $C = 0.03 \text{ ft} \sqrt{n}$ might be used, where n is the number of setups.

The type of formula used by the National Geodetic Survey (NGS) to compute allowable misclosures is

$$C = m\sqrt{K} \tag{7-1}$$

where C is the allowable misclosure, in millimeters; m a constant, in millimeters; and K the length of section or loop leveled, in kilometers.

The NGS specifies constants of 3, 4, 6, 8, and 12 mm for its five classes of leveling now designated respectively as (1) first order—class I, (2) first order—class II, (3) second order—class I, (4) second order—class II, and (5) third order. A single run is allowed for third order and second order class II. The orders of accuracy recommended for different types of projects are given in Section 20-3.

As an example, if differential levels are run from an established BM A to BM B, $\frac{3}{4}$ km away and back, with elevation differences of 3.0556 m and 3.0620 m, respectively, the misclosure is 0.0064 m. Then by Eq. (7-1),

$$m = \frac{C}{\sqrt{K}} = \frac{6.4}{\sqrt{\frac{3}{4} + \frac{3}{4}}} = 5.2 \text{ mm} \tag{7-2}$$

and the leveling meets the allowable 6-mm tolerance for second order—class I work.

If sights of 300 ft are taken, thereby spacing instrument setups at 600 ft, approximately $5\frac{1}{2}$ setups/km would be made. For second order—class I leveling, the allowable misclosure would then be

$$E_c = \frac{6}{\sqrt{5.5}}\sqrt{n} = 2.6\sqrt{n} \tag{7-3}$$

where E_c is the allowable closure in millimeters and n the number of times the instrument is set up.

Other values of m can be specified to meet the precision required and the average length of sight. On first-order leveling the length of sight is varied during the day to conform to atmospheric conditions, with a maximum of 50 m for class I and 60 m for class II.

7-14. ADJUSTMENTS OF SIMPLE LEVEL CIRCUITS. Since permissible misclosures are based on lengths of lines or numbers of setups, it is logical to adjust

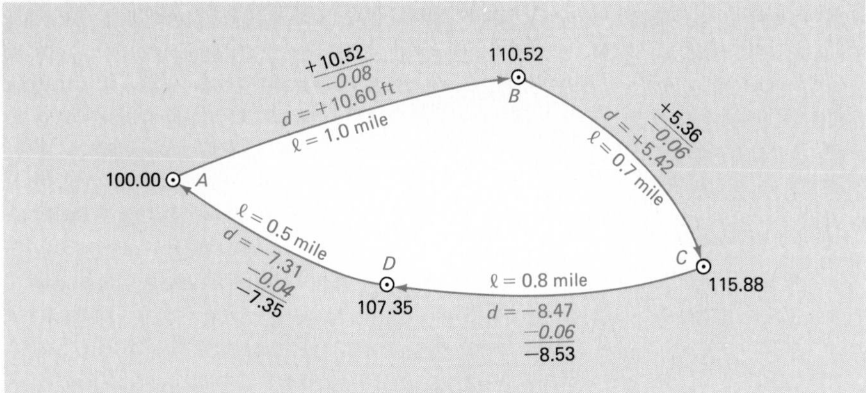

Figure 7-10. Adjustment of level circuit based on lengths of lines.

elevations on these bases. Elevation differences and lengths of lines are shown for a circuit in Figure 7-10. The misclosure found by algebraic summation of the elevations differences is +0.24 ft. Adding lengths of lines yields a total circuit length of 3.0 mi. Elevation adjustments are then (0.24 ft/3.0) times the corresponding lengths, giving −0.08, −0.06, −0.06, and −0.04 ft (shown in the figure). The adjusted elevation differences are used to get the final elevations (also shown in the figure) of the bench marks, based on elevation of BM A = 100.00.

Level circuits with different lengths and routes are sometimes run from scattered reference points to obtain the elevation of a given bench mark. The most probable value for a bench mark can then be computed from a weighted mean of the observations, the weights varying inversely with line lengths.

7-15. ADJUSTMENT OF MULTILOOP CIRCUITS. In running level circuits, especially long ones, it is recommended that some turning points or bench marks used in the first part of the circuit be included again on the return run. This creates a multiloop circuit, and if a blunder or large error exists, its location can be isolated to one of the smaller loops. This saves time because only the smaller loop containing the blunder or error need be rerun. An example is shown in Figure 7-11, where BM B was used a second time in the circuit, thus creating two loops.

Although least squares is the best method for adjusting circuits that contain two or more loops, several approximate procedures are available. The two-loop circuit with corresponding field notes shown in Figure 7-11 will be adjusted to illustrate one approximate method. In the figure, the order in which instrument setups were made is given in parentheses. The asterisk in the field notes identifies the second elevation determination of BM B.

Loop 2, the outer circuit, is adjusted first. From the notes, the elevation of BM B was first computed as 100.62, and after running levels around loop 2, its elevation was 100.70, giving a misclosure of +0.08 ft. Distributing it based on the

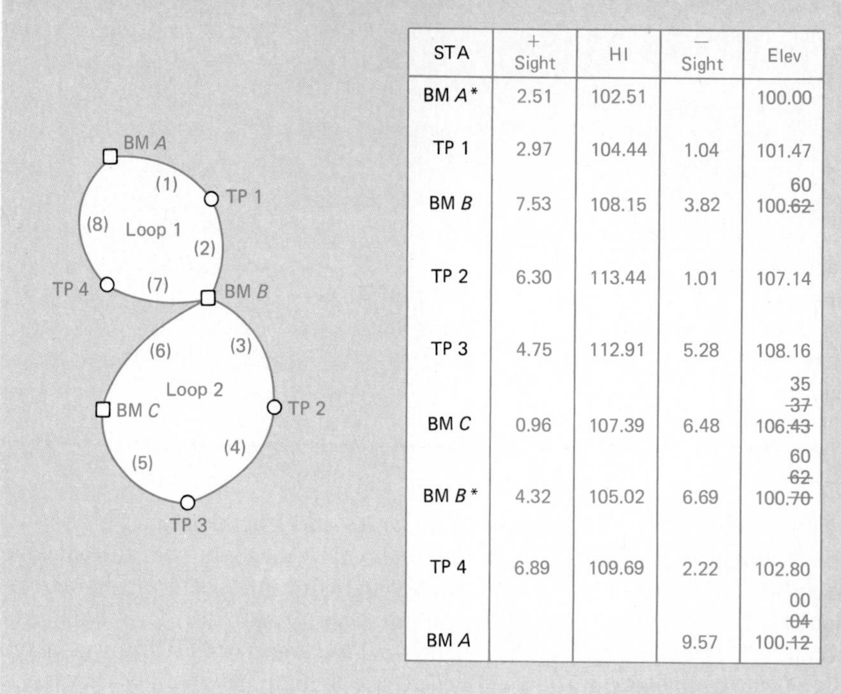

STA	+ Sight	HI	− Sight	Elev
BM *A**	2.51	102.51		100.00
TP 1	2.97	104.44	1.04	101.47
BM *B*	7.53	108.15	3.82	60 100.~~62~~
TP 2	6.30	113.44	1.01	107.14
TP 3	4.75	112.91	5.28	108.16
BM *C*	0.96	107.39	6.48	35 ~~37~~ 106.~~43~~
BM *B**	4.32	105.02	6.69	60 ~~62~~ 100.~~70~~
TP 4	6.89	109.69	2.22	102.80
BM *A*			9.57	00 ~~04~~ 100.~~12~~

Figure 7-11. Two-loop level circuit and corresponding field notes showing adjustment.

number of instrument setups in the loop gives a correction of $-0.08/4 = -0.02$ ft per setup. Thus, BM *C* and BM *B** receive corrections of $3 \times (-0.02) = -0.06$ ft and $4 \times (-0.02) = -0.08$ ft, respectively. Their initially corrected elevations of 106.37 and 102.62 are shown in the notes directly above their crossed-out original unadjusted values.

Because the elevation of BM *B** has been corrected by -0.08 ft and setups 7 and 8 depend on that point as a reference, the elevation of BM *A* must also be corrected by -0.08 to 100.04 ft. The final elevation of BM *A* must be 100.00, however, so a $+0.04$-ft misclosure error exists in loop 1. This is also distributed on the basis of the number of instrument setups in the loop, as $-0.04/4 = -0.01$ ft per setup. BM *B* and BM *A* are therefore corrected by $2 \times 0.01 = -0.02$ ft and $4 \times (-0.01) = -0.04$ ft, respectively, giving adjusted elevations of 100.60 and 100.00 ft.

Instrument setups 3 through 6 depend on BM *B* as a reference elevation. Whereas BM *B* has just been corrected by -0.02 ft, elevations of BM *C* and BM *B** are also adjusted by -0.02 ft. This completes the adjustment. Final elevations for all bench marks are shown above the crossed-out original and initial adjusted values. The result is not the same as would be obtained by distributing the 0.12 misclosure equally to all points in a single-loop adjustment.

In circuits with three or more loops, the same general procedure is followed. Adjustment always begins with the outer loop.

7-16. SOURCES OF ERROR IN LEVELING. All leveling measurements are subject to three sources of error: (1) instrumental, (2) natural, and (3) personal.

Instrumental Errors

Instrument Not in Adjustment. The most important adjustment of a dumpy or wye level makes the line of sight parallel to the level-vial axis (except for automatic levels), so that a horizontal plane, rather than a conical surface, is generated as the telescope is revolved. Serious errors in rod readings result if the instrument is not so adjusted but are canceled if the horizontal lengths of plus and minus sights are kept equal to utilize the principle of reversion (see Appendix A). The error is systematic and may be serious in going up or down a steep hill where all plus sights are longer or shorter than all minus sights, unless care is taken to run a zigzag line.

Cross Hair Not Exactly Horizontal. Reading the rod near the center of the horizontal cross hair will eliminate or minimize this potential error.

Rod Not Correct Length. Inaccurate divisions on a rod cause errors in measured elevation differences similar to those resulting from incorrect markings on a tape. Uniform wearing of the rod bottom makes HI values too large, but the effect is canceled when included in both plus and minus sights. Rod graduations should be checked by comparing them with those on a standardized tape.

Tripod Legs Loose. Tripod leg bolts that are too loose or too tight allow movement or strain that affects the instrument head. Loose metal tripod shoes cause unstable setups.

Natural Errors

Curvature of the Earth. As noted in Section 6-3, a level surface curves away from a horizontal plane at the rate of 0.667 M^2, or about 8 in per 1 mi. The effect of curvature of the earth is to increase the rod reading. Equalizing lengths of plus and minus sights cancels the error due to this cause.

Refraction. Light rays coming from an object to the telescope are bent, making the line of sight a curve concave to the earth's surface, and thereby decrease rod readings. Balancing the lengths of plus and minus sights usually eliminates errors due to refraction. Large and sudden changes in atmospheric refraction may be important in precise work, however. Errors due to refraction tend to be random over a long period of time but could be systematic on one day's run.

Temperature Variations. Heat causes leveling rods to expand, but the effect is not important in ordinary leveling.

If the level vial of a dumpy or tilting level is heated, the liquid expands and the bubble shortens. This does not produce an error (although it may be inconvenient), unless one end of the tube is warmed more than the other and the bubble therefore moves toward the heated end. Other parts of the instrument warp because of uneven heating, and this distortion affects the adjustments. Shading the level by means of a cover when carrying it, and by an umbrella when it is set up, will reduce or eliminate heat effects. These precautions are followed in precise leveling.

Air boiling or heat waves near the ground surface or adjacent to heated objects make the rod appear to wave and prevent accurate sighting. Raising the line of sight by high tripod setups, taking shorter sights and avoiding any that pass close to heat sources (such as buildings and stacks), and using the lower magnification of a variable-power eyepiece reduce the effect.

Wind. Strong wind causes the instrument to vibrate and makes the rod unsteady. Precise leveling is not attempted on windy days.

Settlement of the Instrument. Settlement of the instrument after a plus sight has been taken makes the minus sight too small and therefore the recorded elevation of the next point too large. The error is cumulative in a series of setups on soft material. Unusual care is required in setting up a level on spongy ground, blacktop, or ice. Readings must be taken in quick order, perhaps using two rods and two observers to preclude walking around the instrument. Alternating the order of taking plus and minus sights helps somewhat.

Settlement of a TP. This condition causes an error similar to that resulting from settlement of the instrument. It can be avoided by selecting firm, solid turning points or, if none are available, using a steel *turning pin.*

Personal Errors

Bubble Not Centered. In working with dumpy or tilting levels, errors caused by the bubble not being exactly centered at the time of sighting are the most important of any, particularly on long sights. If the bubble goes off between the plus and minus sights, *it must be recentered before the minus sight is taken.* Experienced observers develop the habit of checking the bubble before and after each sight, a procedure simplified by having a mirror-prism arrangement permitting simultaneous view of the level vial and rod.

Parallax. Parallax caused by improper focusing of the objective and/or eyepiece lens results in incorrect rod readings. Careful focusing eliminates this problem.

Faulty Rod Readings. Incorrect rod readings result from parallax, poor weather conditions, long sights, improper target setting and rodding, and other causes, including mistakes such as those due to careless interpolation and transposition of figures. Short sights selected to fit weather and instrument conditions reduce the number of reading errors. If a target is used, the rodperson should read the rod for the plus sight and have the observer check it independently as the rodperson passes on the way to the next TP. The observer stops to read the minus-sight setting before moving ahead for the next setup.

Rod Handling. Serious errors caused by improper rod handling (plumbing) are eliminated by using a rod level that is in adjustment. Banging the rod on a TP for the second (plus) sight may change the elevation of a point.

Target Setting. The target may not be clamped at the exact place signaled by the observer because of slippage. A check sight should always be taken after the target is clamped.

7-17. MISTAKES. A few common mistakes in leveling are listed here.

Use of Long Rod. If the vernier reading on the back of a damaged rod is not exactly 6.500 or 7.000 for the short rod, the target must be set to read the same value before extending the rod.

Holding the Rod in Different Places for the Plus and Minus Sights on a TP. The rodperson can avoid such mistakes by using a well-defined point or by outlining the rod base with keel.

Reading a Foot Too High. This mistake occurs because the incorrect foot mark is in sight near the cross line; for example, an observer may read 5.98 instead of 4.98. Noting the foot marks above and below the horizontal cross line will prevent this mistake.

Waving the Ordinary Flat-bottom Rod While Holding It on a Flat Surface. This action produces an error because the rotation is about the rod edges instead of the center or front face. In precise work, plumbing with a rod level or other means is preferable to waving.

Recording Notes. Mistakes in recording, such as transposing figures, entering values in the wrong column, and making arithmetic errors, can be minimized by having the notekeeper mentally estimate the reading, repeat the value called out by an observer, and make the standard field-book checks on rod sums and elevations.

Touching Tripod During Reading Process. Beginners may center the bubble, put one hand on the tripod while reading a rod, and then remove the hand while checking the bubble, which has now returned to center but was off during the observation.

7-18. REDUCING ERRORS AND ELIMINATING MISTAKES. Errors in leveling are reduced (but never eliminated) by careful adjustment and manipulation of both instrument and rod (see Appendix A for procedures) and establishing standard field methods and routines. The following routines prevent most large errors or quickly disclose mistakes: (a) checking the bubble before and after each reading (if an automatic level is not being used), (b) using a rod level, (c) keeping the horizontal lengths of plus and minus sights equal, (d) running lines forward and backward, and (e) making the usual field-book arithmetic checks.

PROBLEMS

7-1. Why is it advisable to set up a level with all three tripod legs on, or in, the same material (concrete, blacktop, soil), if possible?

7-2. Does spreading the tripod legs until the instrument head almost touches the ground help release a frozen head?

7-3. Compute the distance a rod extended for a 13-ft reading must be out of plumb to introduce an error of 0.01 ft.

7-4. Similar to Problem 7-3 except for a 4-m reading and an error of 1 mm.

7-5. What error results on a 200-ft sight with a level if the rod reading is 9.00 but the top of the 12-ft rod is 4 in out of plumb?

7-6. Prepare a set of level notes for the data listed. Elevation of BM 7 is 652.54 ft. Total loop distance is 1800 ft. What order of leveling is represented?

POINT	+S (BS)	−S (FS)
BM 7	9.43	
TP 1	6.78	8.36
BM 8		9.82
BM 8	7.26	
TP 2	3.91	9.40
TP 3	7.22	5.53
BM 7		1.47

7-7. Similar to Problem 7-6 except elevation of BM 7 is 521.13 and the loop distance is 1200 ft.

7-8. A differential leveling loop started and closed on BM Rivet, elevation 496.20 ft. The BS and FS distances were kept approximately equal. Readings taken in order are 3.76, 8.34; 4.62, 9.51; 6.17, 7.22; 9.04, 6.93; and 10.16, 1.79. Prepare and check notes.

7-9. A level set up midway between X and Y reads 8.53 on X and 6.27 on Y. When moved within a few feet of X, readings of 7.48 on X and 5.26 on Y are recorded. What is the true elevation difference and the rod reading required on Y to put the instrument in adjustment?

7-10. A level is set up near C (elev 221.618 m) and then near D. Rod readings taken in order are: $C = 1.884$ mm, $D = 1.417$ m, $D = 1.292$ m, and $C = 1.765$ m. Compute the elevation of D and the reading on C to adjust the instrument. (See Section A-4.3 in Appendix A.)

7-11. The peg adjustment test shows that a level's line of sight is inclined downward 0.006 ft/100. What is the allowable difference between BS and FS distances at each setup (neglecting curvature and refraction) to keep elevations correct within 0.001 ft?

7-12. Prepare a set of profile leveling notes for the data listed and show the page check. Elevation of BM A is 287.52 ft. Rod readings are: BS on BM A 2.86; IFS on 1 + 00 5.3; FS on TP 1, 10.56; BS on TP 1, 11.02; IFS on 2 + 00 12.09, on 3 + 00 6.32, FS on TP 2, 9.15; BS on TP 2, 4.28; IFS on 3 + 64 2.0, on 4 + 00 2.6, on 5 + 00 5.7; FS on TP 3, 8.77; BS on TP 3, 4.16; FS on BM B 9.08.

7-13. Same as Problem 7-12 except elevation of BM $A = 413.25$ ft and BS on BM A is 6.21 ft.

7-14. Plot the short profile in Problem 7-12 and select a grade line between stations 0 + 00 and 5 + 00 to balance cut and fill areas.

7-15. What is the grade of the hypotenuse of a 20°–70° right triangle if one side of the 20° angle is horizontal?

7-16. Reciprocal leveling gives the following readings in ft from a setup near A: on A, 2.437; on B, 8.254, 8.259, 8.257. At the setup near B: on B, 11.334; on A, 5.148, 5.152, 5.149. Elevation of B is 1462.793. Compute the misclosure and elevation of A.

7-17. Reciprocal leveling across a canyon provides the data listed. The correct elevation of *Y* is 348.216 ft. Elevation of *X* required. Instrument at *X*: +S = 3.254; −S = 6.817, 6.813, 6.815. Instrument at *Y*: +S = 8.362; −S = 4.798, 4.799, 4.799.

7-18. Differential leveling between BMs *A*, *B*, *C*, *D*, and *A* gives elevation differences in feet of −15.632, +32.458, +38.214, and −55.025, and distances in miles of 0.4, 0.5, 0.6, and 0.5, respectively. If the elevation of *A* is 653.214, compute the elevations of BMs *B*, *C*, and *D* and the order of leveling.

7-19. Leveling from BM *X* to *W*, BM *Y* to *W*, and BM *Z* to *W* gives differences in elevation of −30.24, +26.20, and +10.18 ft, respectively. Distances in feet are $XW = 2000$, $YW = 4000$, and $ZW = 3000$. True elevations of the BMs in feet are $X = 460.82$, $Y = 404.36$, and $Z = 420.47$. What is the adjusted elevation of BM *W*?

7-20. A level rod with a $1\frac{1}{2}$-in square base is held on a flat concrete TP. A level line of sight hits just above the rod base. In waving the rod, it tips alternately on the front and rear plate edges. A minimum reading of 0.14 ft results as the rod is tipped on the rear edge. What is the true rod reading?

7-21. After running a line of levels between BM Sign and BM Road, examination showed that the level rod had a repaired base plate on the bottom, thus making the rod too long. Is the elevation determined for BM Road correct? Explain.

7-22. A line of levels with 16 setups (32 rod readings) was run from BM Point to BM Pond with readings taken to the nearest 0.01 ft; hence, any one could have an error of ±0.005 ft. For reading errors only, what total error might be expected in the BM Pond elevation?

7-23. Same as Problem 7-22 except for 24 setups and readings to the nearest 2 mm with possible error of ±1 mm.

7-24. Compute the permissible error of closure for the following lines of levels: (a) a 12-mi circuit of third-order levels, (b) a 25-km-long line of second order—class I levels, and (c) a 40-km loop of first order—class I levels.

7-25. From your study of the theory of errors, taping, and leveling, why is it usually easier to do good leveling than good taping?

7-26. Explain how errors due to lack of instrument adjustment can be practically eliminated in running a line of differential levels.

7-27. How can errors due to settlement of the instrument and rod be reduced or eliminated?

7-28. What errors in leveling are eliminated by keeping the lengths of plus and minus sights equal?

7-29. What are the primary differences in running "ordinary" and "precise" three-wire leveling?

7-30. Compare the open (expanded) style of differential level notes in Appendix D, Plate D-3, with the closed (compact) arrangement sometimes used. Which type is simpler to follow in checking?

7-31. List four considerations that govern a rodperson's selection of TPs and BMs.

7-32. When would a double-rodded line of levels be preferred to runs in and out?

7-33. Sketch a simple arm signal by which an observer or rodperson can inform the other that a two-way radio has failed.

BIBLIOGRAPHY

Berry, R. M. 1977. "Observational Techniques for Use with Compensator Leveling Instruments for First-order Levels." *Surveying and Mapping* 37(no. 1):15.

Bouen, H. L. 1962. "Combination Leveling Rod and Leveling Method." *Surveying and Mapping* 22(no. 4):561.

Federal Geodetic Control Committee. 1974. *Classification, Standards of Accuracy, and General Specifications of Geodetic Control Surveys.* U.S. Department of Commerce, National Ocean Survey, Rockville, MD.

Federal Geodetic Control Committee. 1975. *Specifications to Support Classification, Standards of Accuracy, and General Specifications of Geodetic Control Surveys.* U.S. Department of Commerce, National Ocean Survey, Rockville, MD.

Geisler, M., and H. Papo, 1967. "Evaluation of Accuracy of Precise Leveling." *ASCE Journal of the Surveying and Mapping Division* 93(no. SU2):103.

Kulp, E. F. 1970. "High Precision Levels with Automatic Instruments." *ASCE Journal of the Surveying and Mapping Division* 96(no. SU2):121.

Selley, A. P. 1977. "A Trigonometric Level Crossing of the Strait of Belle Isle." *Canadian Surveyor* 31:249.

Whalen, C. T., and E. I. Balacz. 1977. "Test Results of First-order Class III Leveling." *Surveying and Mapping* 37(no. 1):45.

8 ANGLES, BEARINGS, AND AZIMUTHS

8-1. INTRODUCTION. The location of points and orientation of lines frequently depend on the measurement of angles and directions. In surveying, directions are given by bearings and azimuths.

As described in Section 2-2, angles measured in surveying are classified as *horizontal* or *vertical*, depending on the plane in which they are measured. Horizontal angles are the basic measurements needed for determining bearings and azimuths. Use of vertical angles is explained elsewhere in this text.

Angles are measured *directly* in the field by a compass, transit, theodolite, or sextant. They can be constructed without measurement on a planetable sheet (see Chapter 18). The compass, transit, and theodolite are discussed in succeeding chapters.

An angle can be measured *indirectly* by the tape method described in Sections 4-18.2 and 4-18.3, and its value computed from the relationship of known quantities in a triangle or other simple geometric figure.

Three basic requirements determine an angle. As shown in Figure 8-1, they are (1) *reference* or *starting line*, (2) *direction of turning*, and (3) *angular distance* (value of the angle). Methods of computing bearings and azimuths described in this chapter are based on those three elements.

8-2. UNITS OF ANGLE MEASUREMENT. A purely arbitrary unit defines the value of an angle. The *sexagesimal* system used in the United States and many other countries is based on degrees, minutes, and seconds, with the last unit further divided decimally. In Europe the *grad* is a standard unit (see Section 2-3).

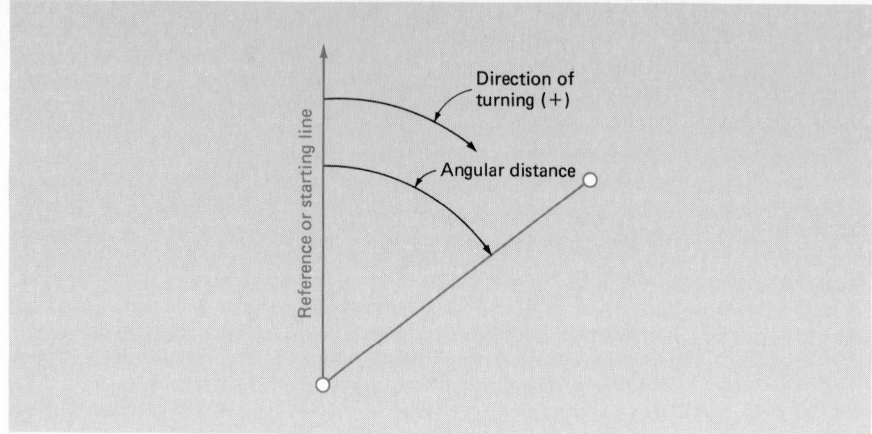

Figure 8-1. Basic requirements in determining an angle.

Radians may be more suitable in computations, and in fact are employed extensively in electronic computers, but the sexagesimal system will continue to be used in most U.S. surveys rather than decimal degrees, radians, or grads.

8-3. KINDS OF HORIZONTAL ANGLES. The kinds of horizontal angles most commonly measured in surveying are (1) *interior angles*, (2) *angles to the right*, and (3) *deflection angles*. Because they differ considerably, the kind used must be clearly indicated in field notes.

Figure 8-2. Closed polygon. (a) Clockwise interior angles (angles to the right). (b) Counterclockwise interior angles (angles to the left).

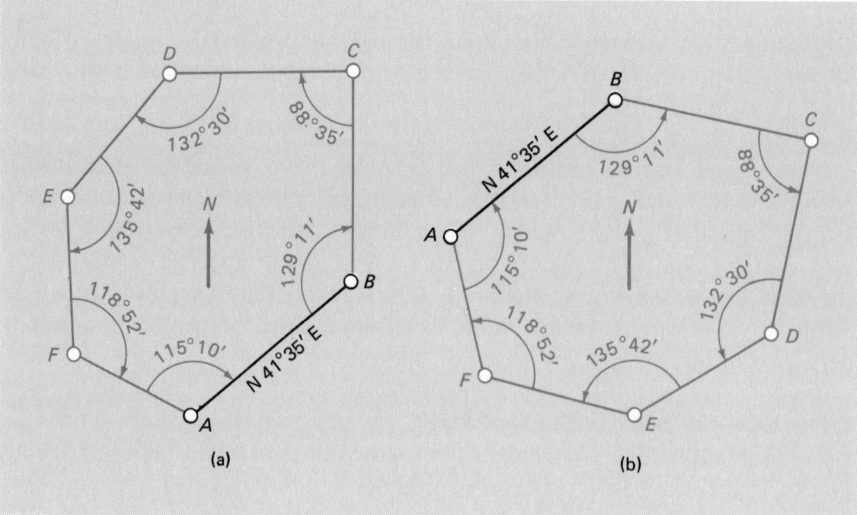

Interior angles, shown in Figure 8-2, are on the inside of a closed polygon. *Exterior angles*, located outside a closed polygon, are explements of interior angles. The advantage to be gained by measuring them is their use as a check, since the sum of the interior and exterior angles at any station must equal 360°.

As illustrated in Figure 8-2, interior angles can be turned clockwise (right) or counterclockwise (left). By definition, *angles to the right* are measured clockwise from the *rear* to the *forward* station. Note: As a survey progresses, stations are commonly identified by consecutive alphabetical letters (as in Figure 8-2) or increasing numbers. Thus, the interior angles of Figure 8-2(a) are also angles to the right. *Angles to the left*, turned counterclockwise from the rear station, are illustrated in Figure 8-2(b). Note that the polygons of Figure 8-2 are "right" and "left"—that is, similar in shape but turned over like the right and left hands. Figure 8-2(b) is shown only to emphasize a serious mistake that occurs if clockwise and counterclockwise angles are mixed. Therefore, a uniform procedure should be adopted, such as *always measuring angles clockwise if possible*, and the direction of turning noted in the field book with a sketch.

Deflection angles (Figure 8-3) are measured right (clockwise, considered plus) or left (counterclockwise, minus) from an extension of the back line to the forward station. Deflection angles are always smaller than 180°, and the direction of turning is defined by appending an R or L to the numerical value. Thus, the angle at *B* in Figure 8-3 is right (R) and that at *C* is left (L).

Figure 8-3. Deflection angles.

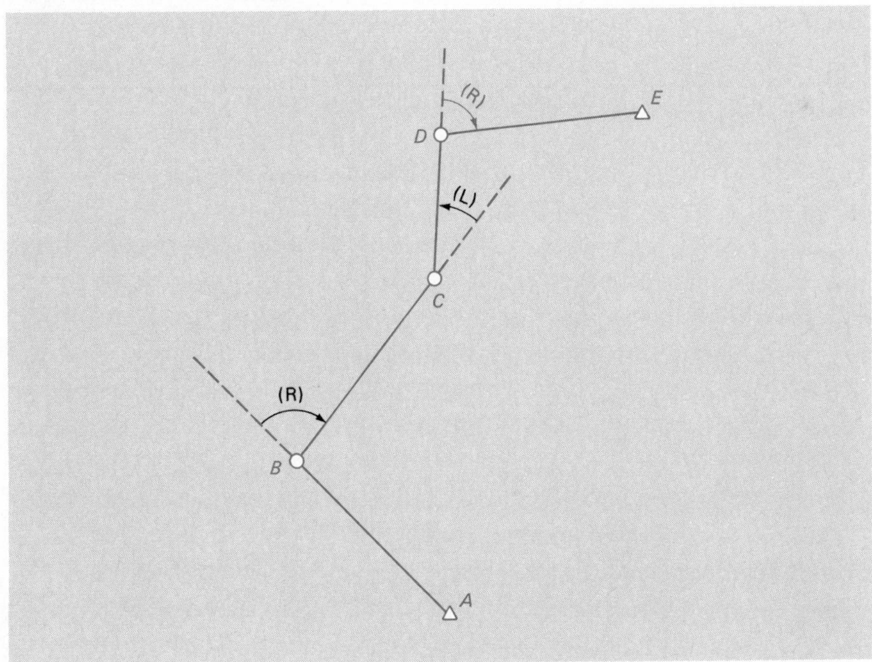

8-4. DIRECTION OF A LINE. The direction of a line is the horizontal angle between it and an arbitrarily chosen reference line called a *meridian.* Different meridians are used. An *astronomic* (sometimes called "true") meridian is the north-south reference line through the earth's geographic poles.

A *magnetic* meridian is defined by a freely suspended magnetic needle that is influenced by the earth's magnetic field only. A magnetic pole is the center of convergence of magnetic meridians.

An *assumed* meridian can be established by merely assigning any arbitrary direction—for example, taking a certain street line to be true north. The directions of all other lines are then found in relation to it. Disadvantages of using an assumed meridian are the difficulty, or perhaps impossibility, of reestablishing it if the original points are lost, and its nonconformance with other surveys and maps.

Surveys based on a state or other plane coordinate system employ a *grid* meridian for reference. Grid north is the direction of true north for a selected central meridian and held parallel to it over the entire area covered by a plane coordinate system (see Chapter 21).

Obviously the terms *true* or *due* north, if used in a survey, must be explained, since they may not specify a unique line.

Other types of meridians discussed in later chapters include *guide, central, prime,* and *local.*

8-5. BEARINGS. Bearings represent one system of designating the directions of lines by means of an angle and quadrant letters. The bearing angle of a line is the acute horizontal angle between a reference meridian and the line. The angle is measured from either the north or south toward the east or west to give a reading smaller than 90°. The proper quadrant is shown by a letter N or S preceding the angle and E or W following it. An example is N80°E.

Figure 8-4. Bearing angles.

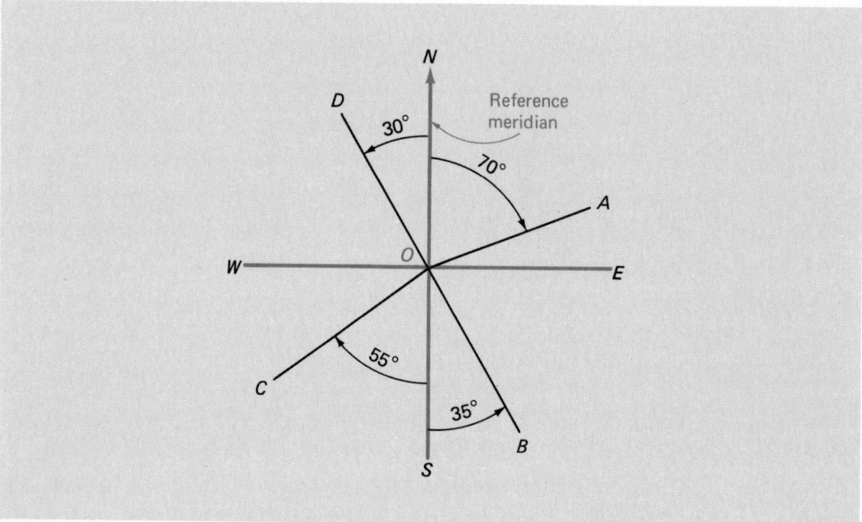

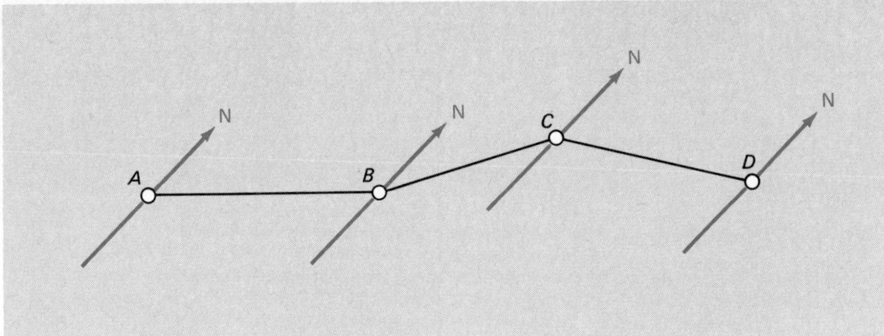

Figure 8-5. Forward and back bearings.

In Figure 8-4, all bearings in quadrant *NOE* are measured clockwise from the meridian. Thus, the bearing of line *OA* is N70°E. All bearings in quadrant *SOE* are counterclockwise from the meridian, so *OB* is S35°E. Similarly, the bearing of *OC* is S55°W and that of *OD*, N30°W.

True bearings are measured from the local astronomic or true meridian, *magnetic* bearings from the local magnetic meridian, *assumed* bearings from any adopted meridian, and *grid* bearings from the appropriate grid meridian. Magnetic bearings can be obtained in the field by observing a magnetic needle in a compass box, and used along with measured angles to get *computed bearings*.

In Figure 8-5 assume a compass is set up successively at points *A*, *B*, *C*, and *D*, and bearings read on lines *AB*, *BA*, *BC*, *CB*, *CD*, and *DC*. Bearings *AB*, *BC*, and *CD* are called *forward bearings*; those of *BA*, *CB*, and *DC*, *back bearings*. Forward bearings have the same numerical value as back bearings but opposite letters. If bearing *AB* is N44°E, bearing *BA* is S44°W.

In land surveys, the term *record bearing* refers to that quoted in a previous survey, *deed bearing* to one used in a property deed description.

8-6. AZIMUTHS. Azimuths are angles measured *clockwise* from any reference meridian. In plane surveying, azimuths are generally measured from north, but astronomers, the military, and the National Geodetic Survey use south as the reference direction.

As shown in Figure 8-6, azimuths range from 0 to 360° and do not require letters to identify the quadrant. Thus, the azimuth of *OA* is 70°; of *OB*, 145°; of *OC*, 235°; and of *OD*, 330°. It is necessary to state in the field notes, at the beginning of work, whether azimuths are measured from north or south.

Azimuths may be *true*, *magnetic*, *grid*, or *assumed*, depending on the meridian used. They may also be *forward* or *back* azimuths. Forward azimuths are converted to back azimuths, and vice versa, by adding or subtracting 180°. For example, if the azimuth of *OA* is 70°, the azimuth of *AO* is 70° + 180° = 250°. If the azimuth of *OC* is 235°, the azimuth of *CO* is 235° − 180° = 55°.

Azimuths can be read on the graduated circle of a transit or repeating theodolite after the instrument has been oriented properly. This can be done

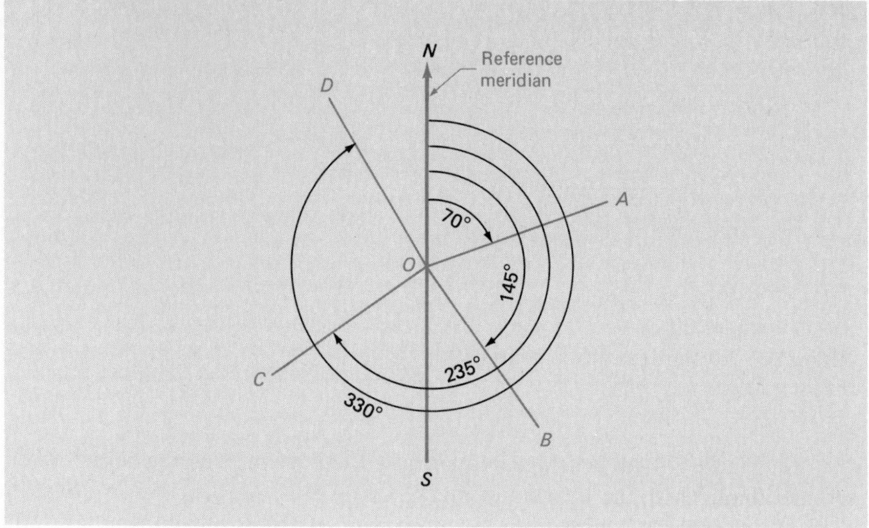

Figure 8-6. Azimuths.

by sighting along a line with its known azimuth on the circle and then turning to the desired course. Azimuths (*directions*) are used advantageously in some topographic control and other surveys as well as in computations.

8-7. COMPARISON OF BEARINGS AND AZIMUTHS. Because bearings and azimuths are encountered in so many surveying operations, a comparative summary of their properties given in Table 8-1 may be helpful. Bearings are readily computed from azimuths by noting the quadrant in which the azimuth falls, then converting as shown in the table.

TABLE 8-1. **COMPARISON OF BEARINGS AND AZIMUTHS**

BEARINGS	AZIMUTHS
Vary from 0 to 90°	Vary from 0 to 360°
Require two letters and a numerical value	Require only a numerical value
May be true, magnetic, grid, assumed, forward, or back	Same
Measured clockwise and counterclockwise	Measured clockwise only
Measured from north and south	Measured from north only in any one survey, or from south only

Example directions for lines in the four quadrants (azimuths from north):

Bearings	Azimuths	
N54°E	54°	
S68°E	112°	(180° − 68°)
S51°W	231°	(180° + 51°)
N15°W	345°	(360° − 15°)

8-8. CALCULATION OF BEARINGS. Many types of surveys, especially those for *traverses*, require computation of bearings (or azimuths). A traverse is a series of distances and angles, or distances and bearings, or distances and azimuths, connecting successive points. The boundary lines of a piece of property form a "closed-polygon" type of traverse. A highway survey from one city to another normally is an "open" traverse, but if possible it should be closed by tying-in on points of known coordinates near the starting and finishing points. Traverses are described in detail in Chapter 12.

Computation of the bearing of a line is simplified by drawing a sketch similar to those in Figures 8-7 and 8-8, showing all data. In Figure 8-7, assume the bearing of line *AB* in Figure 8-2(a) is N41°35′E, and the angle at *B* turned clockwise (to the right) from known line *BA* is 129°11′. Then the bearing angle of line *BC* is 180° − (41°35′ + 129°11′) = 9°14′, and from the sketch, the bearing of *BC* is N9°14′W.

In Figure 8-8, the clockwise angle at *C* from *B* to *D* was measured as 88°35′. The bearing of *CD* is 88°35′ − 9°14′ = S79°21′W. Continuing this technique, the bearings in Table 8-2 have been determined for all lines in Figure 8-2(a).

The bearing of any starting course *must* be recomputed as a check using the last angle. Any discrepancy shows that (1) an arithmetic error was made

Figure 8-7. Computation of bearing *BC* of Figure 8-2(a).

Figure 8-8. Computation of bearing *CD* of Figure 8-2(a).

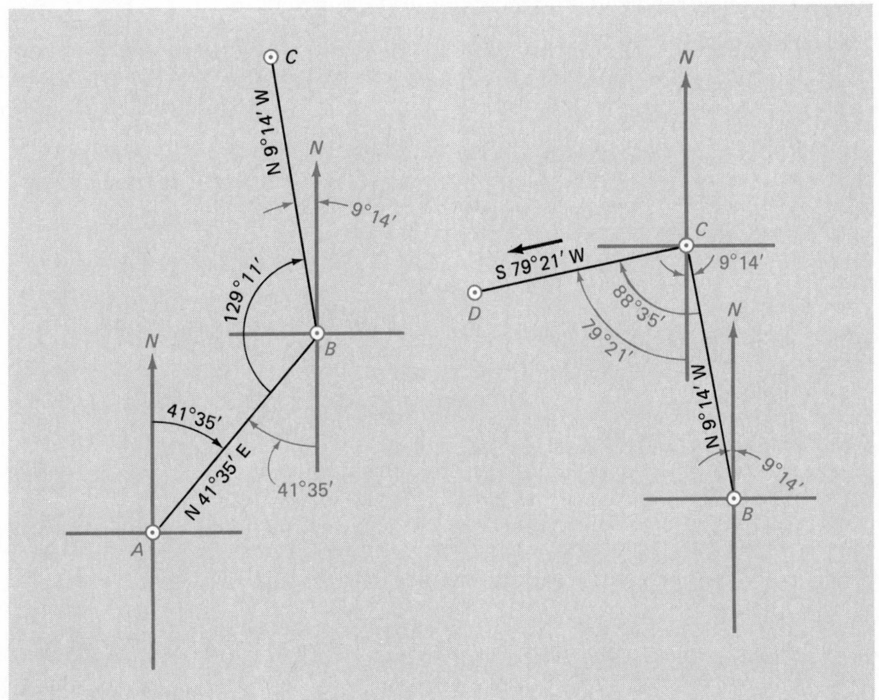

TABLE 8-2. BEARINGS OF LINES IN FIGURE 8-2(a)

COURSE	FIGURE 8-2(a)
AB	N41°35′E
BC	N9°14′W
CD	S79°21′W
DE	S31°51′W
EF	S12°27′E
FA	S73°35′E
AB	N41°35′E ✓

or (2) the angles were not properly adjusted prior to computing bearings. In Table 8-2, note that the bearing of *AB* in Figure 8-2(a), obtained by employing the 115°10′ angle measured at *A*, yields a bearing of N41°35′E, which agrees with the starting bearing.

Traverse angles must be adjusted to the proper geometric total before bearings are computed. In a closed-polygon traverse, the sum of interior angles equals $(n - 2)180°$, where n is the number of sides (courses). If the traverse angles failed to close by, say, 2 min and were not adjusted prior to computing bearings, the original and computed check bearing of *AB* will differ by the same 2 min, assuming there are no other calculating errors.

Figure 8-9. Computation of azimuth *BC* of Figure 8-2(a).

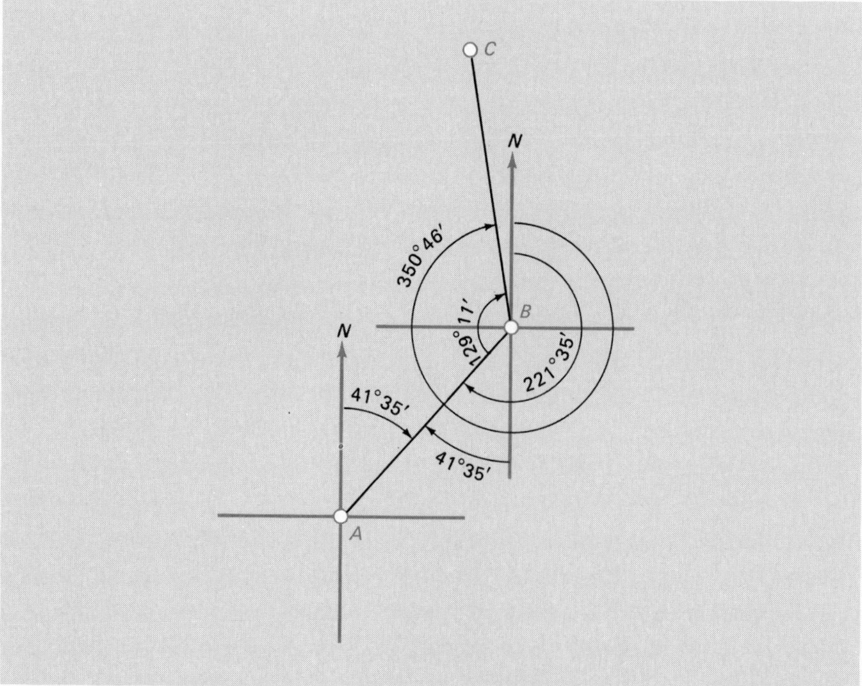

TABLE 8-3. **COMPUTATION OF AZIMUTHS (AZIMUTHS FROM NORTH)**

ANGLES TO THE RIGHT [FIGURE 8-2(a)]	
41°35′ = *AB*	31°51′ = *ED*
+ 180°00′	+ 135°42′
221°35′ = *BA*	167°33′ = *EF*
+ 129°11′	+ 180°00′
350°46′ = *BC*	347°33′ = *FE*
− 180°00′	+ 118°52′
170°46′ = *CB*	466°25′
+ 88°35′	− *360°00′
259°21′ = *CD*	106°25′ = *FA*
− 180°00′	+ 180°00′
79°21′ = *DC*	286°25′ = *AF*
+ 132°30′	+ 115°10′
211°51′ = *DE*	401°35′
− 180°00′	− *360°00′
31°51′ = *ED*	41°35′ = *AB*

* When a computed azimuth exceeds 360°, the correct azimuth is obtained by merely subtracting 360°.

8-9. COMPUTING AZIMUTHS. Many surveyors prefer azimuths over bearings for directions of lines because they are easier to work with, especially when computing traverses with electronic computers. Sines and cosines of azimuth angles provide correct algebraic signs for latitudes and departures as discussed in Chapter 13.

Azimuth calculations, like those for bearings, are best made with the aid of a sketch. Figure 8-9 illustrates the computations for azimuth *BC* in Figure 8-2(a). Azimuth *BA* is found by adding 180° to azimuth *AB*: 180° + 41°35′ = 221°35′. Then clockwise angle *B*, 129°11′, is added to azimuth *BA* to get azimuth *BC* = 221°35′ + 129°11′ = 350°46′. The calculations are conveniently handled in tabular form.

Table 8-3 lists the calculations for all azimuths of Figure 8-2(a). Note again that a check is secured by recomputing the starting course azimuth using the last angle.

8-10. MISTAKES. Some mistakes made in using bearings and azimuths are:

1. Confusing magnetic and true bearings.
2. Mixing clockwise and counterclockwise angles.
3. Jumbling bearings and azimuths.
4. Failing to change bearing letters when using the back bearing of a line.

5. Using an angle at the wrong end of a line in computing bearings—that is, using angle A instead of angle B when starting with line AB as a reference.
6. Not including the last angle to recompute the starting bearing or azimuth as a check—for example, angle A in traverse $ABCDEA$.
7. Subtracting 360°00′ as though it were 359°100′ instead of 359°60′, or using 90° instead of 180° in bearing computations.
8. Adopting an assumed reference line that is difficult to reproduce.
9. Failing to adjust traverse angles before computing bearings or azimuths if there is a misclosure.
10. Orienting an instrument a second time by resighting on magnetic north.
11. Reading degrees and decimals from a calculator as though they are degrees, minutes, and seconds.

PROBLEMS

Make the conversions noted in Problems 8-1 and 8-2.

8-1. Convert 82°15′ to radians; 92.65 grads to degrees, minutes, and seconds; and 4200 mils to degrees, minutes, and seconds.

8-2. Convert 198°30′20″ to radians; 52°29′42″ to grads; and 216.76 grads to degrees, minutes, and seconds.

In Problems 8-3 and 8-4, convert the north azimuths to bearings.

8-3. 54°16′, 154°18′, 261°10′, 312°38′.

8-4. 71°42′, 134°27′, 195°00′, 285°26′.

8-5. Why were angles to the left measured in some old traverses?

8-6. Deflection angles are more dependent on good instrument adjustment than direct angles, so why are they used in route surveys?

8-7. Why have azimuths been referred to south rather than north by astronomers and the National Geodetic Survey?

8-8. How are back bearings and back azimuths used?

Convert the bearings in Problems 8-9 and 8-10 to azimuths from north and compute the angle smaller than 180° between successive bearings.

8-9. N15°12′E, S37°52′E, S49°37′W, N81°26′W.

8-10. N35°08′E, S73°51′E, S90°00′W, N15°23′W.

Compute the azimuth from south of line CD in Problems 8-11 through 8-13. (Azimuths of AB are also from south.)

8-11. Azimuth AB = 150°39′; clockwise angles ABC = 174°28′, BCD = 62°47′.

8-12. Bearing AB = S14°22′W; clockwise angles ABC = 83°17′, BCD = 95°05′.

8-13. Azimuth AB = 191°04′; clockwise angles ABC = 125°10′, BCD = 207°16′.

8-14. Compute the interior angles of Lot 16 in Figure 22-2.

8-15. Calculate the interior angles of Lot 50 in Figure 22-2.

8-16. For a bearing DE = S47°13′E and angles to the right, compute the bearing of FG if angle DEF = 147°19′ and EFG = 201°52′.

Course AB of a five-sided field runs due north. Station C is westerly from B. Compute and tabulate the bearings and azimuths from north for each side of the clockwise interior-angle traverse in Problems 8-17 and 8-18.

8-17. $A = 75°$, $B = 135°$, $C = 60°$, $D = 70°$, $E = 200°$.
8-18. $A = 73°18'$, $B = 125°08'$, $C = 119°06'$, $D = 91°13'$, $E = 131°15'$.

In Problems 8-19 and 8-20, compute and tabulate the bearings of a regular hexagon, given the starting bearing of side *AB*.
8-19. Bearing of *AB* = S50°10′E. (Station *C* is easterly from *B*.)
8-20. Bearing of *AB* = N12°24′E. (Station *C* is westerly from *B*.)

8-21. Similar to Problem 8-19 except for a regular pentagon.
8-22. Similar to Problem 8-20 except for a regular octagon.
8-23. Calculate the course bearings for the following deflection-angle open (route survey) traverse. Bearing 0 + 00 to 7 + 53.2 = S43°12′E; then 6°15′R to 11 + 77.8; 9°01′L to 14 + 29.3; 4°53′R to 20 + 06.6; and 2°34′L to 25 + 48.0. Sketch the traverse.
8-24. Similar to Problem 8-23, except the first course bearing is N39°23′W.

Compute all bearings for a closed traverse *ABCDEFGHIJA* that has clockwise interior angles, using the fixed bearings listed in Problems 8-25 and 8-26. $A = 183°18'$, $B = 119°53'$, $C = 177°36'$, $D = 94°25'$, $E = 152°29'$, $F = 140°35'$, $G = 162°19'$, $H = 126°01'$, $I = 98°16'$, $J = 185°08'$. Similar to Problems 8-25 and 8-26, except calculate azimuths from north for the fixed azimuths noted in Problems 8-27 and 8-28.
8-25. Bearing *AB* = S52°47′E.
8-26. Bearing *DE* = S52°47′E.
8-27. Azimuth *AB* = 210°48′.
8-28. Azimuth *CD* = 346°42′.

8-29. The true (geodetic) azimuth of a long line *XY* is 72°16′35″. The true azimuth of *YX* is 252°16′37″. Explain.
8-30. Balance the deflection angles of the following closed traverse and compute the bearings. The bearing of *AB* is N48°10′W. The angles are *B* = 95°23′L, *C* = 84°37′L, *D* = 62°10′L, *E* = 33°22′L, *F* = 43°27′R, and *A* = 127°50′L.
8-31. How can you quickly verify the formula $(n - 2)180°$ for the sum of interior angles of a polygon so it need not be memorized?
8-32. Explain three methods to determine the azimuth of a line.
8-33. An angle *APB* is measured at different times using various instruments and procedures. The results, which are assigned certain weights, are as follows: 33°09′27″, wt 4; 33°09′24″, wt 3; and 33°09′25″, wt 1. What is the most probable value of the angle?
8-34. Similar to Problem 8-33, but with an additional measurement of 33°09′28″, wt 2.
8-35. What is a disadvantage in running a transit traverse by azimuths compared with direct angles?

9 THE COMPASS

9-1. INTRODUCTION. The compass has been used by navigators and others for many centuries to determine directions. Prior to the invention of the transit and sextant, a compass furnished surveyors with the only practical way to measure directions and horizontal angles.

The surveyor's compass, like the Gunter's chain, has now become little more than a museum piece. Nevertheless, an understanding of them and their vagaries is necessary to check and retrace original land lines on which they were used. Also, the compass is still employed for rough engineering surveys and remains a valuable tool for geologists, foresters, and others.

Engineers' transits (see Chapter 10) are equipped with a compass. In fact, the early design of American transits was based on having a long compass needle over the instrument center and an erecting telescope. The smaller size of transits and theodolites today is due to the shorter internal-focusing telescope and omission of the compass (which is available for mounting as an accessory).

9-2. THEORY OF THE COMPASS. A compass consists of a magnetized steel needle mounted on a pivot at the center of a graduated circle. Unless disturbed by a local attraction (see Section 9-5), the needle points toward magnetic north (in the northern hemisphere). The north and south magnetic poles were located at approximately latitude 76.8°N, longitude 101.5°W, and latitude 65.4°S, longitude 139.4°E, respectively, in 1980 (USGS data). They move daily, perhaps as far as 30 mi.[1]

The earth's magnetic forces align the needle and pull or dip one end of it below the horizontal position. The *angle of dip* varies from 0° near the equator to 90° at the magnetic poles. In the northern hemisphere, the south end of the needle is weighted with a very small coil of wire to balance the dip effect and keep it horizontal. Position of the coil can be adjusted to conform to the latitude in which the compass is used. Weights on transit compasses are set for an average latitude of 40°N and usually do not have to be changed for any location in the United States.

As the compass box is turned, the needle continues to point toward magnetic north and gives a reading that is dependent on the graduated circle position.

9-3. MAGNETIC DECLINATION. *Declination* is the horizontal angle from a true geographic meridian to a magnetic meridian. Navigators call this angle *variation* of the compass; the armed forces use the term *deviation.*

An east declination occurs if the magnetic meridian is east of true north; a west declination if it is west of true north. The declination at any location can be obtained (if there is no local attraction) by establishing a true meridian from astronomical observations and then reading the compass while sighting along the true meridian.

A line on a map or chart connecting points that have the same declination is called an *isogonic line.* The line made up of points that have zero declination is termed the *agonic line.* On it the magnetic needle defines true north as well as magnetic north.

Figure 9-1 is an isogonic map covering the United States for the year 1980. The agonic line (heavy, full) cuts diagonally across the country through Wisconsin, Illinois, Indiana, Kentucky, Tennessee, Alabama, and Florida, but is gradually moving westward. Points to the west of the agonic line have east declinations; points to the east have west declinations. As a memory aid, the needle can be thought of as pointing *toward* the agonic line. Note there is a 42° difference in declination between the states of Maine and Washington. This is a huge change if a pilot flies by compass between the two states!

The *annual change* in declination shown by dashed lines on larger and more detailed isogonic maps, and on Figure 9-1, aids in estimating the declination a few years before and after the chart date.

Secular change (see Section 9-4) for longer intervals should be computed from available tables which extend back to the earliest times likely to be significant in such problems. The best way to determine the declination at a given location on any date is to make an astronomical observation or use existing

[1] Studies of magnetism in some rocks containing iron particles show the small metal pieces solidified and aligned in the direction of the magnetic poles like a compass needle. North-seeking poles of the metal fragments point to the north magnetic pole location. Researchers have found that in past eons, the north and south magnetic poles have switched positions many times! (*Nature and Science*, 2 March 1970, p. 4.)

control lines. If this is not possible, an approximate declination can be obtained from the National Geodetic Survey or an isogonic map.

9-4. VARIATIONS IN MAGNETIC DECLINATION.

Magnetic declinations at any point vary over time. Variations can be categorized as secular, daily, annual, and irregular.

Secular Variation. Because of its magnitude, this is the most important of the variations. Unfortunately, no general law or mathematical formula has been found to predict secular variation, and its past behavior can be described only by means of detailed tables and charts derived from observations. Records which have been kept at London for four centuries show a range in magnetic declination from 11°E in 1580, to 24°W in 1820, back to 8°W in 1960, 6°58′W in 1975, and 6°10′W in 1980. Secular variation changed the magnetic declination at Baltimore, Maryland, from 5°11′W in 1640 to 5°41′W in 1700, 0°35′W in 1800, 5°19′W in 1900, 7°25′W in 1950, 7°43′W in 1960, 8°43′W in 1975, and 9°30′W in 1980 (with an annual change of 6.8′W).

In retracing old property lines run by compass or based on the magnetic meridian, it is necessary to allow for the difference in magnetic declination at the time of the orignal survey and at the present date. The difference is generally due mostly to secular variation.

Daily Variation. Daily variation of the magnetic needle's declination causes it to swing through an arc averaging approximately 8 min for the United States. The needle reaches its extreme easterly position at about 8 A.M. and its most westerly position at about 1:30 P.M. Mean declination occurs at around 10:30 A.M. and 8 P.M. These hours and the daily swing range change with latitude and season of the year, but neglect of the daily variation is normally well within the range of error expected in compass readings.

Annual Variation. This periodic swing amounts to less than 1 min of arc and can be neglected. It must not be confused with the annual change (the amount of the secular-variation change in one year) shown on some isogonic maps.

Irregular Variations. Unpredictable magnetic disturbances and storms can cause short-term irregular variations of a degree or more.

9-5. LOCAL ATTRACTION.

The magnetic field is affected by metallic objects and direct-current electricity, both of which cause a local attraction. For example, when set up beside an old-time streetcar with overhead power lines, the compass needle would swing toward the car as it approached, then follow it until it was out of effective range.

If the source of an artificial disturbance is fixed, all bearings from a given station will be in error by the same amount. Angles calculated from bearings taken at the station will be correct, however.

Local attraction is present if the forward and back bearings of a line differ by more than the normal observation errors. Consider the following compass

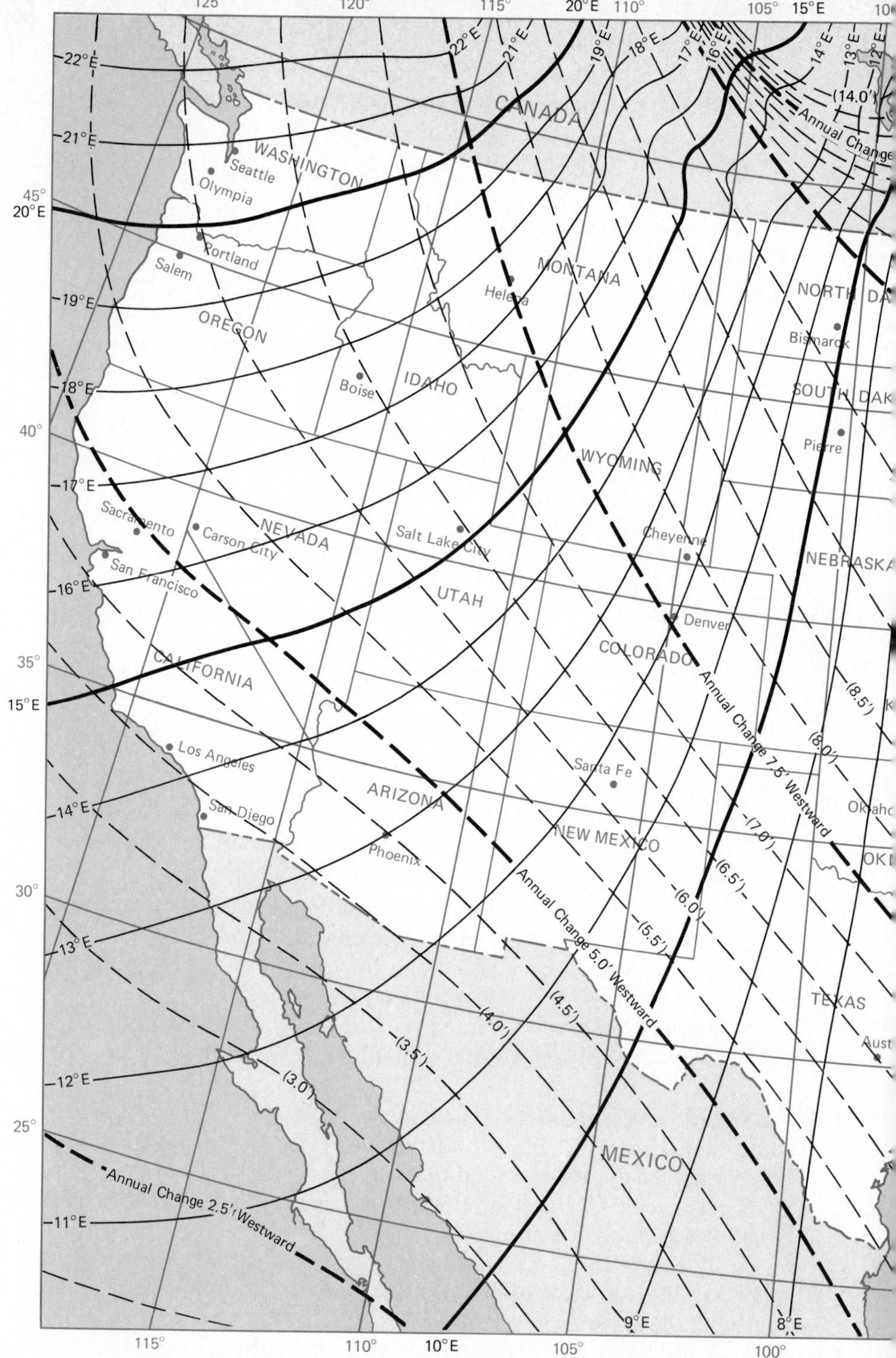

Figure 9-1. Distribution of magnetic declination in the United States for 1980. (Courtesy

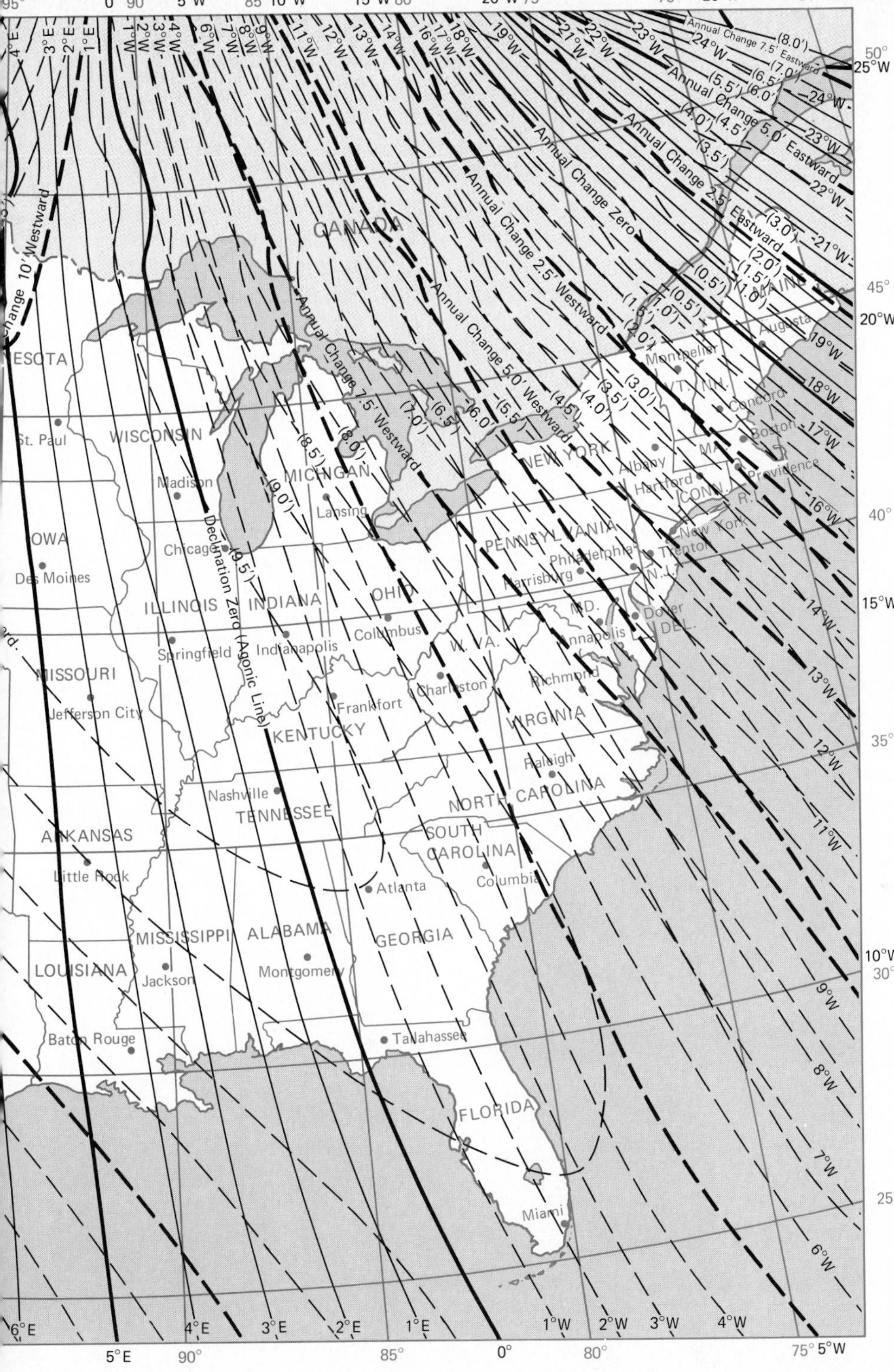

U.S. Geological Survey.)

bearings read on a series of lines:

AB	N24°15′W	*CD*	N60°00′E
BA	S24°10′E	*DC*	S61°15′W
BC	N76°40′W	*DE*	N88°35′E
CB	S76°40′E	*ED*	S87°25′W

Forward bearing *AB* and back bearing *BA* agree reasonably well, indicating that little or no local attraction exists at *A* or *B*. The same is true for point *C*. However, the bearings at *D* differ from corresponding bearings taken at *C* and *E* by roughly 1°15′ to the west of north. Local attraction therefore exists at point *D* and deflects the compass needle 1°15′ to the west of north.

It is evident that to detect local attraction, successive stations on a compass traverse have to be occupied and forward and back bearings read, even though the directions of all lines could be determined by setting up an instrument only on alternate stations.

9-6. THE SURVEYOR'S COMPASS. A surveyor's compass is shown in Figures 9-2 and 9-3. George Washington and thousands of surveyors who followed him used this type of instrument to run land lines which still determine property holdings and therefore must be retraced. So it is important to understand its construction and properties.

Figure 9-2. Surveyor's compass. (Courtesy W. & L. E. Gurley.)

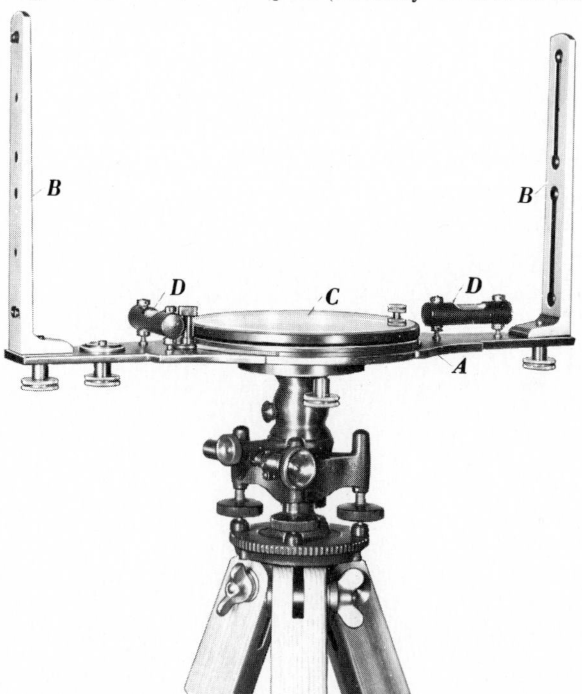

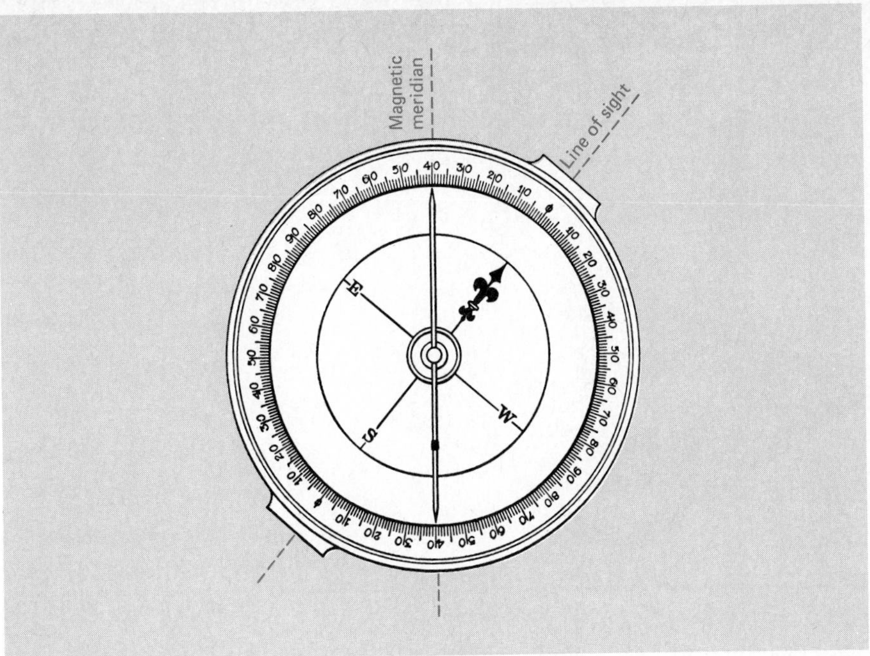

Figure 9-3. Compass box.

The compass circle is graduated in degrees or half-degrees but can be read to perhaps 5 or 10 min by estimation. The instrument consists of a metal base plate *A* (see Figure 9-2) with two vertical sight vanes *B* at the ends. The sight vanes are strips of metal with vertical slits to define the line of sight. The compass box is at *C*, and two small level vials *D* are mounted on the plate perpendicular to the box and each other. When the bubbles in these vials are centered, the plate and compass are horizontal and ready for use.

The compass box (see Figure 9-3) has a conical point at its center to support the needle and a glass cover to protect it. A circular scale at the outer rim of the box is graduated in degrees and half-degrees. Zero marks are at the north and south points in line with the sight-vane slits. Graduations are numbered in multiples of 10°, clockwise and counterclockwise from 0° at the north and south, to 90° at the east and west. As the sight vanes and compass box are revolved, the needle establishes the bearing of the line observed, which can be read by estimation to the nearest 10 or 15 min.

Note that letters E and W on the compass box are reversed from their normal positions to give *direct readings* of bearings. Thus, in Figure 9-3, the sight-line bearing through the vanes is N40°E.

Accuracy of a compass depends on the sensitivity of its needle. A sensitive needle is readily attracted toward a small piece of iron held nearby but settles back in the original position each time the stimulus is removed. Sensitivity itself results from the needle having (a) proper shape and balance, (b) strong magnetism, (c) a sharp conical pivot point, and (d) a smooth cup that bears on the

Figure 9-4. Compass. (Courtesy Keuffel & Esser Company.)

pivot. Tapping the glass cover releases a needle that does not swing freely. Touching the cover with a moistened finger removes static electricity that may otherwise affect the needle.

Remagnetizing the needle is relatively easy, but reshaping the pivot is difficult. To retain a conical shape of the pivot point and prevent blunting to a spherical or flat form that produces sluggishness, the needle should be lifted from the pivot when not in use. A lever arm is provided to raise the needle from the pivot and press it against the glass cover when the instrument is being moved or boxed.

Early compasses were supported by a single leg called a *Jacob staff.* A ball-and-socket joint and clamp were used to level the instrument and set the plate in a horizontal position. Later-version compasses are mounted on a base with a four-screw leveling head, as shown in Figure 9-2.

The compass box of a transit is similar in construction to the surveyor's compass. Zero marks at the north and south points are on a line parallel with and beneath the telescope's line of sight. A special adjustment on some compasses swings the graduated circle through an arc to compensate for a given declination, thus permitting "true" bearings to be read from the needle. Because of different local attractions at succeeding stations, its use may not be very practical.

9-7. THE FORESTER'S AND GEOLOGIST'S COMPASS. Figure 9-4 shows a type of compass employed by geologists and the U.S. Forest Service. It can be used as a hand-held instrument or supported on a staff or tripod.

The instrument is made of aluminum and has brass sights and a declination adjustment for the raised (upper) graduated compass ring. The beveled (lower) ring is used to turn right angles or to measure vertical angles by placing an edge of the base on a level surface.

Figure 9-5. Brunton pocket transit. (Courtesy Keuffel & Esser Company.)

9-8. BRUNTON COMPASS. Figure 9-5 shows a Brunton pocket transit, which combines the main features of a sighting compass, prismatic compass, hand level, and clinometer. The instrument is convenient and sufficiently accurate for topographic and preliminary surveys of many kinds. It can be hand-held or mounted on a Jacob staff or tripod. The device is widely used by geologists.

A Brunton compass consists of a brass case hinged on two sides. The cover at the left has a fine mirror and center line on the inside face. A hinged sighting point at the extreme left and the sighting point at the far right are folded outward when the instrument is in use. The bearing of a line is read from the compass needle position, while the point observed is reflected through the sight vane on a mirror. A declination adjustment can be made by revolving the raised compass ring.

The clinometer (vertical-angle) arc, inside the compass ring, is graduated to degrees and read to the nearest 5 min by a vernier on the clinometer arm. To read vertical angles or grade percentages, the compass is held vertically instead of horizontally. Another arc gives grade percentages for both elevation and depression. A Brunton pocket transit measures $2\frac{3}{4} \times 2\frac{3}{4} \times 1$ in and weighs approximately 8 oz.

Another small and convenient instrument is the liquid-filled Suunto compass. Graduated in degrees, readings can be estimated to 10 min. The compass measures $3 \times 2 \times \frac{9}{16}$ in.

9-9. TYPICAL PROBLEMS. Typical problems in compass surveys require the conversion of true bearings to magnetic bearings, magnetic bearings to true bearings, and magnetic bearings to magnetic bearings for the declinations existing at different dates.

EXAMPLE 9-1

Assume the magnetic bearing of a property line was recorded as S43°30′E in 1862. The magnetic declination at the survey location was 3°15′W. The true bearing is needed for a subdivision property plan.

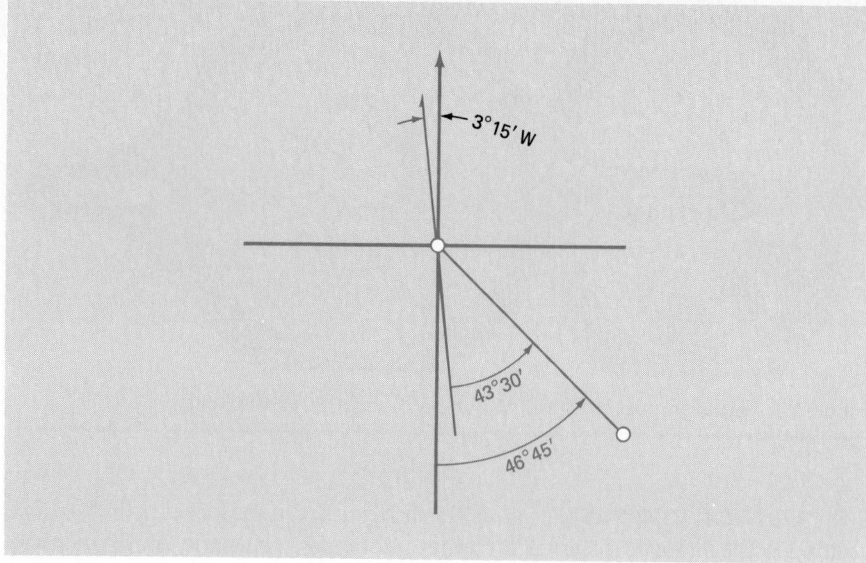

Figure 9-6. Computing true bearings from magnetic bearings and declinations.

SOLUTION

A sketch similar to Figure 9-6 makes the relationship clear and should be used by beginners to avoid mistakes. True north is designated by a full-headed long arrow and magnetic north by a half-headed shorter arrow. The true bearing is seen to be S43°30′E + 3°15′ = S46°45′E. Using different colored pencils to show the direction of true north, magnetic north, and lines on the ground helps clarify the sketch.

Figure 9-7. Computing magnetic bearing changes due to magnetic declinations.

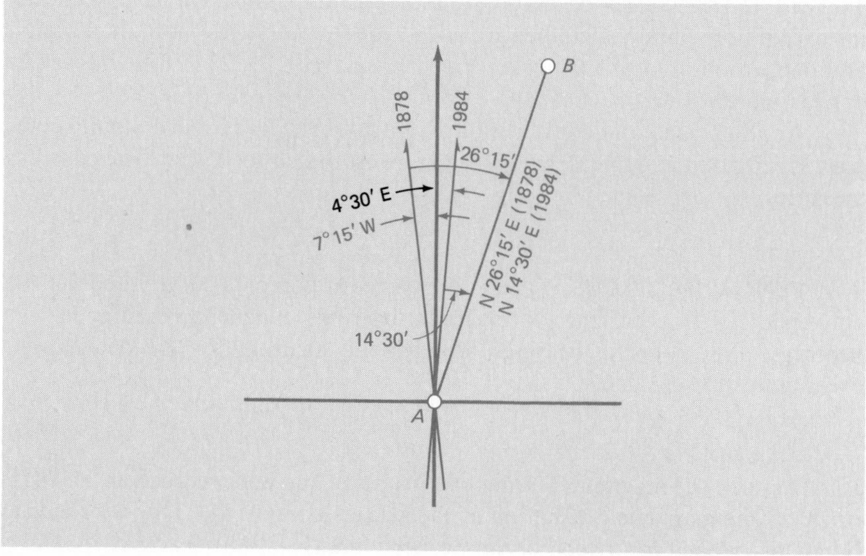

EXAMPLE 9-2

Assume the magnetic bearing of line *AB* read in 1878 was N26°15′E. The declination at that time and place was 7°15′W. In 1984 the declination is 4°30′E. The magnetic bearing in 1984 is needed.

SOLUTION

The declination angles are shown in Figure 9-7. The magnetic bearing of line *AB* is equal to the earlier-date bearing minus the sum of the declination angles, or

$$N26°15'E - (7°15' + 4°30') = N14°30'E$$

9-10. SOURCES OF ERROR IN COMPASS WORK. Some sources of errors in using the compass are:

1. Compass out of level.
2. Pivot not sharp or off the graduated circle center.
3. Needle or sight vanes bent.
4. Magnetism of needle weak.
5. Magnetic variations.
6. Local attraction caused by overhead power lines, underground ore deposits, chaining pins, metal range poles, loose-leaf field books, a penknife, parked car nearby, and so on.

9-11. MISTAKES. Some fairly typical mistakes in compass work are:

1. Reading the wrong end of the needle.
2. Setting off the declination on the wrong side of north.
3. Declination set off when reading magnetic bearings.
4. Parallax (reading while looking from the side of the needle instead of over and along it).
5. Failing to check the forward and back bearings when possible.
6. Not making a sketch showing known and desired data.

PROBLEMS

9-1. At the present rate of declination change (see Figure 9-1), approximately how fast in miles per year, and in what direction, is the agonic line moving across the United States?

9-2. Determine from Figure 9-1 the approximate declinations in 1980 and 1990 at Boston, Miami, Los Angeles, and Seattle.

9-3. What is the total difference in magnetic declination between New York City and San Francisco?

9-4. Assuming a constant declination annual rate of change in Figure 9-1 (it has not been), compute the approximate declination at Atlanta in 1900. Check your answer in an old surveying book.

9-5. List three possible causes of short-term irregular variations of more than a degree. Omit storms and items listed in Section 9-10.

9-6. Would a vernier be useful on a surveyor's compass? Explain.

9-7. Magnetic declination at a monument is 3°45′E. What is the magnetic bearing of true north? Of true south? Of true west?

9-8. Explain why the letters E and W on a compass (see Figure 9-3) are reversed from their normal positions.

9-9. The magnetic bearing of an old survey line recorded as N7°15′E is now N2°00′W. What are the magnetic declination change and direction?

9-10. The magnetic bearing of line *XY* in 1929 was N46°28′E and the declination 2°30′E. The declination now is 1°15′W. What magnetic bearing should be used to retrace line *XY*?

9-11. Where on the earth's surface are the magnetic lines of force horizontal? Where vertical?

9-12. How would you determine magnetic declination in the field?

9-13. Can local attraction be defined by setting up at one station?

9-14. Classify the type of error resulting from local attraction.

9-15. A person travels due north on a meridian for 1 mi, due east on a parallel of latitude 1 mi, due south on a meridian 1 mi, and is back at the starting point. Where was the beginning point?

9-16. Does local attraction at a point affect the size of an angle computed from the magnetic bearings read there? Explain.

9-17. The observed bearing of a line is S88°22′W. Its correct bearing is N89°12′W. Find the value and direction of local attraction.

9-18. After finding no local attraction at hub *A*, bearings read were AB = N36°W, BA = S37$\frac{1}{4}$°E, and BC = S40$\frac{1}{2}$°W. What is the acute angle at *B* and the correct bearing of *BC*?

9-19. Readings of a surveyor's compass to check local declination were N12°45′W, N12°45′W, N12°40′W, N12°45′W, N12°50′W, and N12°45′W. Compute the declination's most probable value and its standard error.

Convert magnetic bearings to true bearings in Problems 9-20 and 9-21.

9-20. N62°25′W, declination 10°E.

9-21. S11°45′E, declination 12°30′W.

What magnetic bearing is needed to retrace line *CD* for the conditions stated in Problems 9-22 through 9-25.

	1875 MAGNETIC BEARING	1875 DECLINATION	PRESENT DECLINATION
9-22.	N64°15′E	2°20′W	3°10′E
9-23.	S83°15′W	1°22′E	0°52′W
9-24.	N89°45′W	1°56′E	3°18′W
9-25.	S54°30′E	5°46′W	8°33′E

Calculate the magnetic declination of line *EF* in 1870 based on the following data from an old survey record.

	1870 MAGNETIC BEARING	PRESENT MAGNETIC BEARING	PRESENT MAGNETIC DECLINATION
9-26.	S00°15′E	S02°30′E	5°20′E
9-27.	S61°15′W	S50°30′W	15°37′E
9-28.	N02°30′W	N01°15′W	3°00′E
9-29.	N22°00′E	N24°15′E	3°30′E

Compute any local attractions and list the corrected bearings of *BC* and *CD* for Problems 9-30 and 9-31. The accepted bearing of *AB* is N34°E.

9-30. BA = S33°W, BC = S68°E, CB = N70°W, CD = S0°.

9-31. BA = S35°W, BC = N82°E, CB = S86°W, CD = N75°W.

The observed compass bearings of a five-sided traverse *ABCDEA* are given in Problems 9-32 and 9-33. Calculate the interior angles and explain the closure.

	AB	*BC*	*CD*	*DE*	*EA*
9-32.	N20°00′E	N64°45′E	N85°15′W	S39°30′W	N88°30′E
9-33.	S64°30′E	S12°45′E	S85°00′W	N29°45′E	N43°15′E

BIBLIOGRAPHY

Barker, N. 1978. "The Cardinal Points." *Surveying and Mapping* 38 (no. 3):203.

Boyum, B. H. 1982. "The Compass That Changed Surveying." *Professional Surveyor* 2:28.

Huey, S. E. 1952. "Surveying with Magnetic Compass." *Surveying and Mapping* 12 (no. 4):293.

Sipe, F. H. 1980. *Compass Land Surveying.* Landmark Enterprises, Rancho Cordova, CA 95670.

10 THE TRANSIT AND THEODOLITE

10-1. INTRODUCTION. Transits and theodolites are perhaps the most universal surveying instruments. Although their primary use is for accurate measurement or layout of horizontal and vertical angles, they are also commonly employed for a wide variety of other tasks such as determining horizontal and vertical distances by stadia (see Chapter 15), prolonging straight lines, and low-order differential leveling.

The main components of a transit or theodolite are a sighting telescope, two graduated circles mounted in mutually perpendicular planes, and level vials. Prior to measuring angles, the "horizontal" circle is oriented in a horizontal plane by means of the level vials, which automatically puts the other circle in a vertical plane. Horizontal and vertical angles can then be measured directly in their respective planes of reference.

There is no internationally accepted understanding among surveyors on the exact difference denoted by the terms *transit* and *theodolite.* In Europe the term *transiting theodolite* was originally applied to this type of angle-measuring instrument. The word *transiting* meant the telescope could be *plunged, reversed,* or *inverted.* Europeans eventually dropped the adjective and retained the name *theodolite,* while Americans shortened the term to *transit.*

Through the years, in addition to adopting different names for American and European angle-measuring instruments, divergent basic design characteristics emerged, which now are the generally accepted criteria for distinguishing a transit from a theodolite. American transits, shown in Figures 10-1 and 10-2,

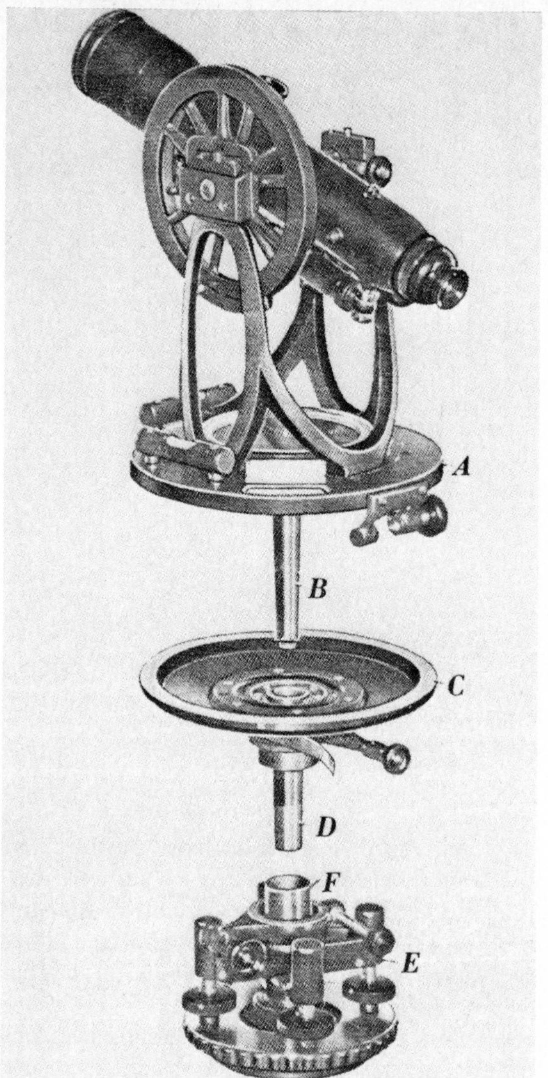

Figure 10-1. Transit parts. *A*, upper plate; *B*, inner spindle; *C*, lower plate; *D*, outer spindle; *E*, leveling head; *F*, socket. (Courtesy W & L. E. Gurley.)

have an "open" design with metal circles read by means of verniers. European theodolites feature an "enclosed" design (see Figures 10-9 through 10-12) and, except for electronic digital theodolites (see Section 10-14), employ glass circles. Readings are taken from either finely graduated glass scales or micrometers, which are viewed through internal microscopic optical systems. Other distinctions are described subsequently in this chapter. A few instruments, such as that in Figure 10-3, combine some design features of both the transit and theodolite, are called *optical reading transits.* They have glass circles and are read from glass verniers viewed through magnifying eyepieces.

In general, theodolites are capable of greater precision and accuracy in angle measurements than transits. Because of this and other advantages (lighter

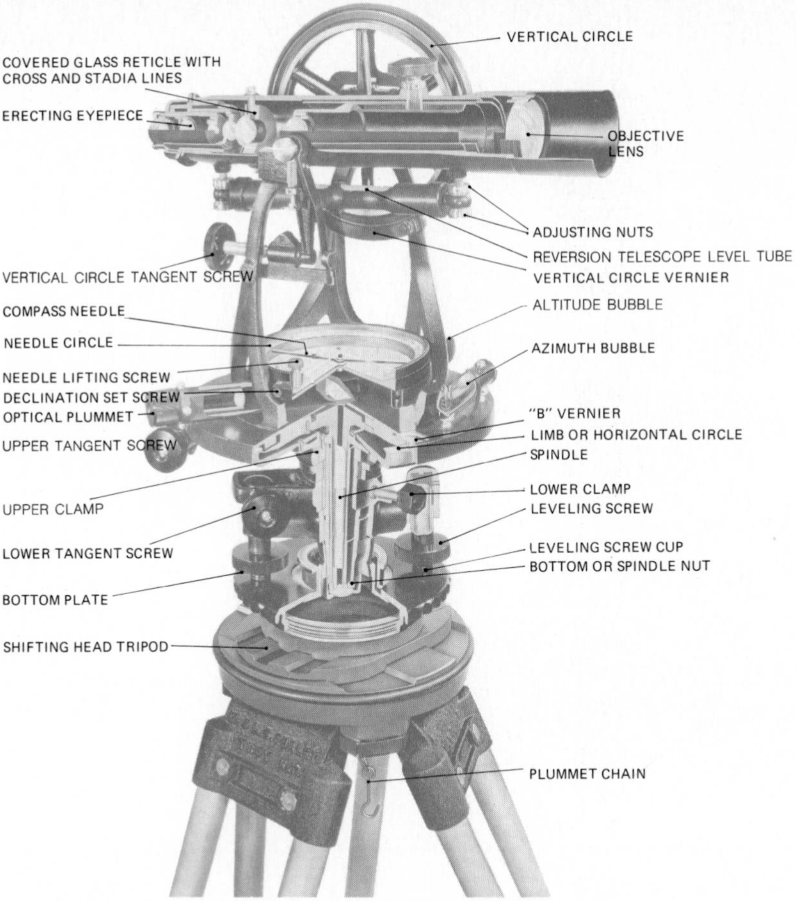

Figure 10-2. American transit. (Courtesy W. & L. E. Gurley.)

weight, easier reading, etc.), theodolites are rapidly replacing transits in the United States. In spite of the differences between the two types of instruments, both operate on the same basic principles, and the parts and relationships described for transits are directly applicable to theodolites.

PART I
THE TRANSIT

10-2. PARTS OF A TRANSIT. Transits are manufactured for general and special uses but all have three main parts: (1) alidade and upper plate, (2) lower plate, and (3) leveling head. These are shown in their relative positions in Figure 10-1 and assembled in Figure 10-2. Reference to these figures will lead to a better understanding of the descriptions given in the following sections.

The various parts of a transit and its operation can best be learned by actually examining and handling an instrument. Once a transit has been taken apart and assembled, even though it be an old or damaged one retired from

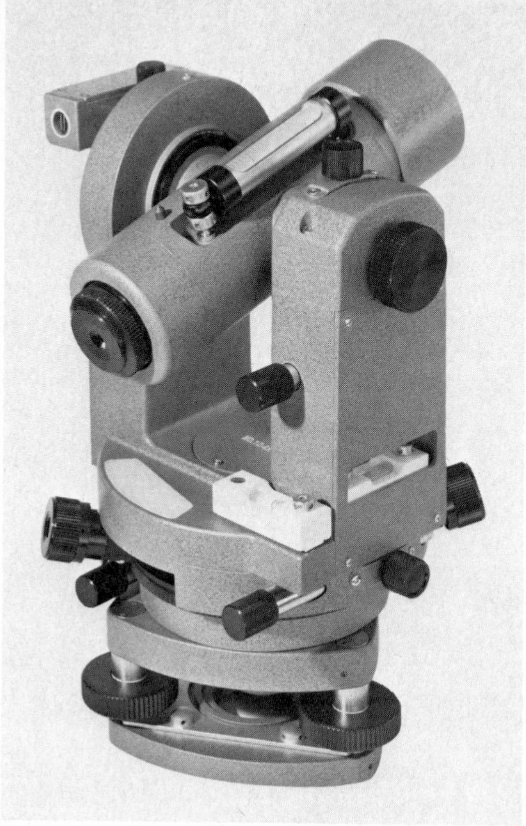

Figure 10-3. Optical reading transit. (Courtesy Dietzgen Corp.)

service, the precise machining and construction are certain to increase respect for such fine equipment.

10-3. UPPER PLATE. The alidade containing an upper plate (*A* in Figure 10-1) is a horizontal circular plate combined with a vertical *spindle B*, which enables the plate to revolve about a vertical axis. The tapered design of American transit spindles assures that despite wear, unless damaged by dirt or an accident, they will still seat and center properly. Attached to the plate are two level vials, one parallel with the telescope (*altitude bubble*) and the other at right angles to it (*azimuth bubble*) (see Figure 10-2), and two *verniers*, referred to as *A* and *B*, set 180° apart. Provisions are made for adjusting verniers and level vials.

Two vertical *standards*, either the A or U type, are cast as an integral part of the upper plate to support the horizontal *cross arms* of the telescope in bearings. The telescope revolves in a vertical plane about the center line through arms called the *horizontal* (*or transverse*) *axis.*

The telescope, similar to that of a dumpy level (Section 6-7), contains an eyepiece, a reticle with one vertical and three horizontal lines, and an objective lens system. A sensitive vial is attached to the telescope tube so the transit can be used as a leveling instrument on work where lower magnification and lesser sensitivity of the telescope vial are satisfactory.

The telescope is said to be in the *normal* or *direct* position when the level vial is below it. Turning the telescope on its horizontal axis puts the level vial above, and the instrument is said to be in a *plunged*, *inverted*, or *reversed* mode. To permit use of the telescope for leveling in either the normal or inverted position, a *reversion vial* (curved and graduated on both its top and bottom so it is usable in both positions) is desirable.

A *vertical circle* or *arc* supported by a cross arm turns with the telescope as it is revolved. The arc normally is divided into $\frac{1}{2}°$ spaces with readings to the nearest minute obtained from a vernier having 30 divisions. The vernier is mounted on one standard with provisions for adjustment. If set properly, it should read zero when the telescope bubble is centered. If out of adjustment, a constant *index error* is read from the arc with the bubble centered and must be applied to all vertical angles, with appropriate sign, to get correct values.

The upper plate also contains the *compass box* and holds the *upper tangent screw*.

A *vertical circle clamp* (for the horizontal axis) is tightened to hold the telescope horizontal or at any desired inclination. After the clamp is set, a limiting range of vertical movement can be obtained by manipulating the *vertical-circle tangent screw* (also called the *slow-motion screw*.)

10-4. LOWER PLATE. The lower plate (*C* in Figure 10-1) is a horizontal circular plate graduated on its upper face. Its underside is attached to a vertical, hollow, tapered spindle *D* into which the upper plate fits precisely. The upper plate completely covers the lower plate, except for two openings where the verniers exactly meet the graduated circle.

The *upper clamp* (see Figure 10-2) fastens the upper and lower plates together. A small range of movement is possible after clamping by using the *upper tangent screw* located on the upper plate.

10-5. LEVELING HEAD. The leveling head (*E* in Figure 10-1) consists of a bottom horizontal plate and a "spider," with four *leveling screws*. The leveling screws, set in cups to prevent scoring the bottom plate, are partly or completely enclosed for protection against dirt and damage. The bottom plate has a collar threaded to fit on the tripod head.

A socket (*F* in Figure 10-1) of the leveling head includes a *lower clamp* (see Figure 10-2) to fasten the lower plate. The *lower tangent screw* is used to make precise settings after the lower clamp is tightened. The base of the socket is fitted into a ball-and-socket joint resting on the bottom plate of the leveling head, on which it slides horizontally. A *plummet chain* attached to the center of the spindle holds a plumb-bob string. An *optical plummet*, which is a telescope through the vertical center (spindle), is available on some transits. It points vertically when the plates are level and is viewed at right angles (horizontally by means of a prism for ease of observation.

A recapitulation of the use of the various clamps and tangent screws may be helpful to a beginner. The vertical-circle clamp and tangent screw on one standard control movement of the telescope in the vertical plane. The upper clamp fastens the upper and lower plates together, and an upper tangent screw permits a small differential movement between them. A lower clamp fastens

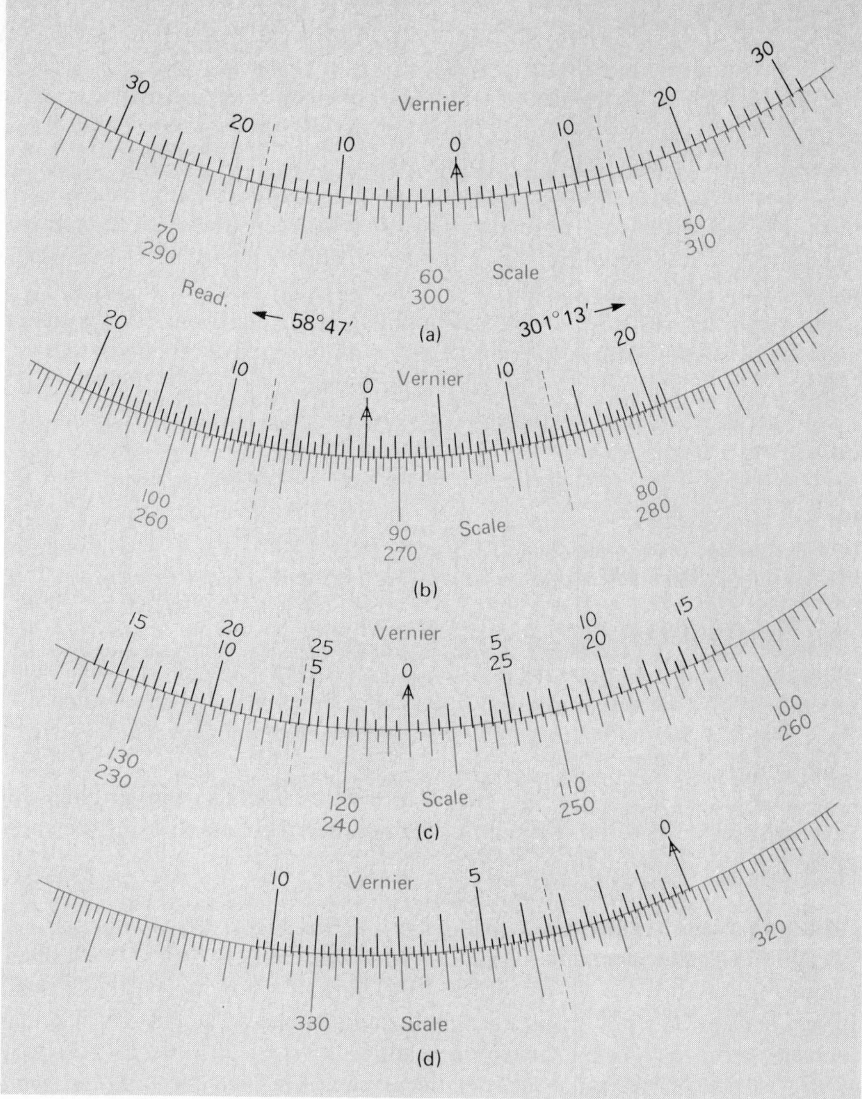

Figure 10-4. Transit verniers.

the lower plate to the socket, after which a lower tangent screw turns the plate through a small angle. If the upper and lower plates are clamped together, they will, of course, move freely as a unit until the lower clamp is tightened.

Tripods for transits, either fixed- or adjustable-leg types, are used interchangeably for levels.

10-6. SCALES. The horizontal circle of the lower plate may be divided in various ways, but generally the circle is graduated into 30- or 20-min spaces. For convenience in measuring angles to the right or left, graduations are numbered from 0 to 360° both clockwise and counterclockwise. Figure 10-4 shows these arrangements. On newer transits, the numbers are slanted to show the direction in which the circle should be read. Different-length lines mark the 10, 5, 1°, and other major graduations.

The horizontal circles of more precise instruments are graduated in divisions of 10 or 15 min. The outer set of numbers on some old transits runs from 0 to 180° and back to 0°. Obsolete instruments had a circle divided into quadrants like a compass box. Graduations from 0 to 360° facilitate reading azimuths and direct angles and therefore have replaced the quadrant system of numbers, which was used for reading bearings.

Vertical circles of most transits are graduated into 30-min spaces. They are usually numbered from zero at the bottom (for a horizontal sight) to 90° in both directions (for vertical sights), and then back to zero at the top. This facilitates reading both elevation and depression angles (see Section 11-15) with the telescope in either a direct or an inverted position.

Transit circles are graduated automatically by means of a precise dividing machine. After each line is cut by a sharp tool, a precision gear moves the tool ahead for the next cut. Any small error in the gears is adjusted by a compensating cam. Under a microscope, the division slashes look somewhat rough, but to the naked eye they are smooth. The marks are painted to make them stand out clearly. Graduations on transit scales are correct to within about 2 sec.

10-7. TRANSIT VERNIERS. Vernier principles were demonstrated in Section 6-19 and the least count given by the following relationship:

$$\text{least count} = \frac{\text{value of the smallest division on the scale}}{\text{number of divisions on the vernier}}$$

The combinations of scale graduations and vernier divisions generally used on transits are shown in Table 10-1. Three types of verniers used on transits are shown in Figure 10-4.

Direct, or Single, Vernier [Figure 10-4(d)]. This is read in only one direction and must therefore be set with the graduations *ahead of the zero* (*index*) *mark in the direction to be turned.*

TABLE 10-1. **TRANSIT SCALES AND VERNIERS**

SCALE GRADUATIONS	VERNIER DIVISIONS	LEAST COUNT	FIG. NO.
30′	30	1′	10-4(a)
20′	40	30″	10-4(b)
30′	60	30″	10-4(c)
15′	45	20″	
10′	60	10″	10-4(d)

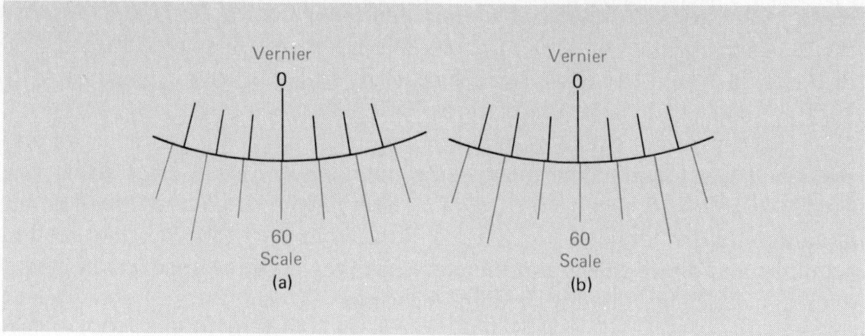

Figure 10-5. Vernier reading.

Double Vernier [Figures 10-4(a), (b), and (c)]. A double vernier can be read either clockwise or counterclockwise, with only one-half being used at a time. Once the index mark is set coincident with 0°00′ on the circle, or any known value, an observer is not limited to turning angles in one direction only.

Folded Vernier [Figure 10-4(c)]. This type avoids the long vernier plate required by the normal double vernier. Its length is that of a direct vernier with half the graduations placed on each side of the index mark. Except possibly for vertical circles, the use of folded verniers is not justified by space or cost savings and is likely to cause reading mistakes.

10-8. READING TRANSIT VERNIERS. A vernier is read by finding a graduation on it that coincides with *any* division on the circle scale. There will be two such matching lines on a double vernier, one for a clockwise angle and the other for a counterclockwise angle. A vernier index shows the number of degrees (and sometimes the multiple of 10, 15, 20, or 30 min) passed over on the scale. The coincident vernier graduation gives directly the additional part of a degree. The second divisions on each side of the apparently matching lines should be checked for symmetry of pattern.

In Figure 10-5(a), the vernier index (zero) mark is set exactly opposite a graduation on the scale, since the distances between the second vernier and the second scale division on both sides of zero are equal. If two sets of lines appear to be almost coincident and a symmetrical pattern is formed, as in Figure 10-5(b) by 0 and the first division on the left, a reading halfway between them can be interpolated.

Figure 10-4(a) shows a double vernier and two sets of numbers on the circle. The reading for the inner set is 58°30′ + 17′ = 58°47′. For the outer circle it is 301°00′ + 13′ = 301°13′. Note that *the vernier is always read in the same direction from zero as the numbering of the circle—that is, on the side of the double vernier in the direction of increasing angle.*

The reading of the inner set of numbers from the double vernier in Figure 10-4(b) is 91°20′ + 07′ = 91°27′; for the outer set it is 268°20′ + 13′ = 268°33′. The folded vernier of Figure 10-4(c) reads 117°05′30″ on the inner row of

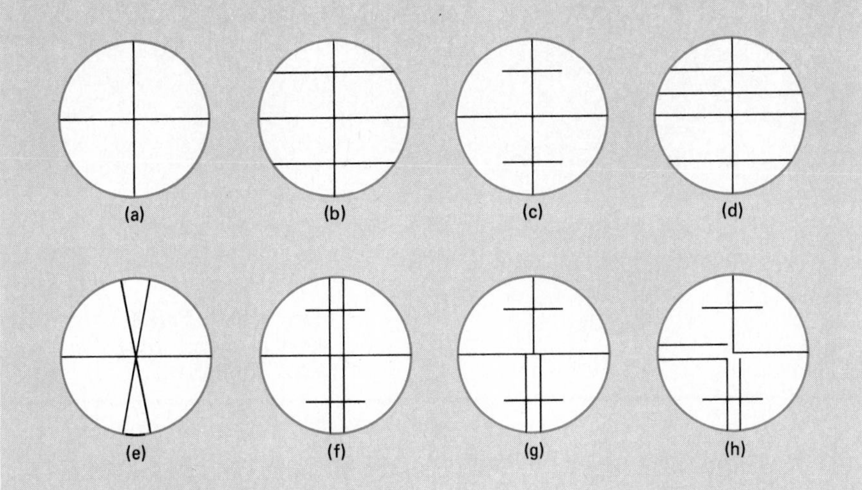

Figure 10-6. Arrangement of cross hairs.

numbers and 242°54′30″ on the outer row. The direct vernier in Figure 10-4(d) reads 321°13′20″ for a clockwise angle.

An understanding of verniers is best obtained by practice in reading various types and by calculating and sketching the least count for different combinations of scale and vernier divisions. Typical mistakes in reading minutes and seconds from verniers result from the following:

1. Not using a magnifying glass.
2. Reading in the wrong direction from zero, or on the wrong side of a double vernier.
3. Failing to determine the least count correctly.
4. Omitting 10, 15, 20, or 30 min when the index is beyond those marks.

10-9. PROPERTIES OF THE TRANSIT. Transits are designed to have a proper balance between magnification and resolution of the telescope, least count of the vernier, and sensitivity of the plate and telescope bubbles. An average length of sight of about 300 ft is assumed in design. Thus a standard 1-min instrument has the following properties:

Magnification, 18 to 28 diam.
Field of view, 1° to 1°30′.
Resolution, 3 to 5 sec.
Minimum focus, about 3 to 7 ft.
Sensitivity of plate levels per 2-mm division, 70 to 100 sec.
Sensitivity of telescope vial per 2-mm division, 30 to 60 sec.
Weight of instrument head without tripod, 11 to $18\frac{1}{2}$ lb.

Cross hairs usually include vertical and horizontal center hairs and two stadia hairs, as shown in Figure 10-6(b) and (c). Short stadia lines, used on glass reticles [Figure 10-6(c)], avoid confusion between the center and stadia hairs.

Figure 10-7. Builder's transit level. (Courtesy Keuffel & Esser Company.)

A quarter hair, located halfway between the upper and middle hairs [Figure 10-6(d)], is sometimes used to increase the range of stadia readings, as noted in Chapter 15.

The X pattern [Figure 10-6(e)] is sometimes incorporated in precise instruments to prevent a rod or object seen at a long distance from being completely hidden behind the vertical hair. It also permits an observer to balance distances between the rod and hairs on both sides of the upper and lower sections to ensure centering, a task the human eye does in a highly efficient manner. The arrangement shown in Figure 10-6(f) or one of the variations in Figure 10-6(g) and (h) likewise avoids covering the object sighted and aids in centering.

A transit is a *repeating instrument* because angles can be measured by repetition any number of times and their total added on the plates. Advantages of the repeating procedure are (1) better accuracy obtained through averaging, and (2) disclosure of errors by comparing values of the single and multiple readings.

A *builder's transit level* is a lower-priced instrument for use on work requiring only short sights and moderate precision. The model shown in Figure 10-7 has a telescope with magnification of 20 diam, resolution of 4.7 sec, telescope level vial with sensitivity of 90 sec/2-mm division, and a horizontal circle and vertical arc reading to 5 min. The cost is approximately one-third that of a standard transit.

10-10. HANDLING AND SETTING UP A TRANSIT. A transit is taken from its box by holding the leveling head, the underside of the lower plate, or the standards (*not* lifting by the telescope). It should be screwed securely on the tripod. A transit carried indoors should be balanced in a horizontal position under one arm, with the instrument head forward. The same method is suitable in areas covered with brush. In open terrain the instrument may be balanced on a shoulder. When a transit is carried, the telescope should be clamped lightly in a position perpendicular to the plates. The plate clamps should be set lightly to prevent swinging, while still permitting ready movement if the instrument is bumped.

The wing nuts on the tripod must be tight to prevent slippage and rotation of the head. They are correctly adjusted if each tripod leg falls slowly of its own weight when placed in a horizontal position. If the wing nuts are overly tight, or if pressure is applied to the legs crosswise (which can break them) instead of lengthwise to fix them in the ground, the tripod is in a strained position. The result may be an unnoticed movement of the instrument head after observations have begun. Tripod legs should be well spread to furnish stability and place the telescope at a height convenient for the observer. Tripod shoes must be tight. Proper field procedures can eliminate most instrument maladjustments, but there is no way to take care of a poor tripod with dried-out wooden legs, except to discard or repair it.

A plumb-bob string hung on a hook at the bottom of the spindle using a slipknot permits raising or lowering the bob without retying and avoids knots. A small slide attachment is also useful in accomplishing this purpose. The plumb bob must be brought directly over a definite point such as a tack in a wooden stake, and the plates leveled. The tripod legs can be moved in, out, or sideways to approximately level the plates before the leveling screws are used. Shifting the legs affects the position of the plumb bob and makes it more difficult to set up a transit than a level.

Two methods are used to bring the plumb bob within about $\frac{1}{4}$ in of the proper point. In the first method, the transit is set over the mark and one or more legs moved to bring the plumb bob into position. One leg may be moved circumferentially to level the plates without greatly disturbing the plummet. Beginners sometimes have difficulty with this method because at the start the transit center is too far off the point, or the plates are badly out of level. Several movements of the tripod legs may then fail to both level the plates and center the plumb bob while maintaining a convenient height of instrument. If an adjustable-leg tripod is used, one or two legs can be lengthened or shortened to bring the bob directly over the point.

In the second method, which is particularly suited to level on uniformly sloping ground, the transit is set up near the point and the plates approximately leveled by moving the tripod legs as necessary. Then, with one tripod leg held in the left hand, another under the left armpit, and the third supported in the right hand, the transit is lifted and placed over the mark. A slight shifting of one leg should bring the plumb bob within perhaps $\frac{1}{4}$ in of the proper position and leave the plates practically level.

The plummet is centered exactly by loosening all four leveling screws and sliding them on the bottom plate using the ball-and-socket shifting-head device, which permits a limited movement. To assure mobility in any direction, the shifting head should be approximately centered on the bottom plate before setting up the instrument, and when boxing it.

A transit is accurately leveled by means of the four leveling screws in somewhat the same manner described for a level. However, each level vial on the upper plate is first set over a pair of opposite screws, and because there are two vials available, the telescope position need not be changed in the leveling process. After the bubbles are carefully centered and the telescope rotated, if they run very far off center, it may be necessary or desirable to adjust the vials as described in Section A-6.1.

If the plumb bob is still over the mark after leveling, the instrument is ready for use. But if the plates were badly out of level, or the leveling screws not uniformly set, the plummet moves off the mark during leveling. The screws must then be loosened and shifted again, and the transit releveled. It is evident that time can be saved by starting with the plates reasonably level to avoid excessive manipulation and possible binding of the screws.

PART II
THE THEODOLITE

10-11. CHARACTERISTICS OF THEODOLITES. Theodolites differ from American transits in general appearance (they are compact, lightweight, and "streamlined") and in design by a number of features, the more important of which are as follows:

1. *Telescopes* are short, have reticles etched on glass, and are equipped with rifle sights or collimators for rough pointing.

2. *Horizontal and vertical circles* are made of glass with graduation lines and numerals etched on the circles' surfaces. The lines are very thin (0.004 mm), short (0.05 to 0.10 mm), and more sharply defined than can be achieved by scribing them on metal. Precisely graduated circles with small diameters can be obtained, and this is one reason the instruments are so compact. Circles are divided into conventional sexagesimal degrees and fractions (360°), or into centesimal "grads" or "grades" (full circle divided into 400^g).

3. The *vertical circle* of most theodolites is precisely indexed with respect to the direction of gravity in one of two ways: (a) by an *automatic compensator* or (b) by a *collimation level* or *index level*, usually the coincidence type connected to the reading system of the vertical circle. Both provide a more accurate plane of reference for measuring vertical angles than the plate levels used on transits.

4. Circle *reading systems* consist basically of a microscope with the optics inside the instrument. A reading eyepiece is generally adjacent to the telescope eyepiece or located on one of the standards. Some instruments have optical micrometers for fractional reading of circle intervals (micrometer scale visible through reading microscope); others are "direct-reading." With most theodolites, a mirror located on one standard can be adjusted to reflect light into the instrument and brighten the circles for daytime use. They can be equipped with a battery-operated internal lighting system for night and underground operation. Some newer theodolites also use the battery-operated system in lieu of mirrors for daytime work.

5. Rotation about the *vertical axis* occurs within a steel cylinder or on precision ball bearings, or a combination of both.

6. The *leveling head* consists of three screws or *cams.*

7. *Bases* or *tribrachs* of theodolites are often designed to permit interchange of instrument and accessories (targets, EDMIs, subtense bar, and so on), without disturbing centering over the survey point. Figure 10-8, for example, shows the placement of a sighting target and an EDM reflector on a theodolite tribrach.

Figure 10-8. Standard tribrachs for most theodolites are designed for interchanging various accessories. Here tribrachs are shown to be compatible with a sighting target (**left**) and an EDM reflector (**right**). (Courtesy Wild Heerbrugg Instruments, Inc.)

8. An *optical plummet*, built into the base or alidade of most theodolites, replaces the plumb bob and permits centering with great accuracy.

9. *Carrying cases* for theodolites are made of steel, lightweight alloy, or heavy plastic. They are generally compact, watertight, and can be locked.

10. *Distance-measuring devices* may be permanent and integral parts of theodolites. Tacheometers, for example, are theodolites that measure slope distance by the stadia principle and automatically convert them to horizontal and vertical components. Some theodolites have built-in EDM devices that permit measuring slope distances and horizontal and vertical angles from a single setup (see Section 5-6).

11. Various *accessories* increase the versatility of theodolites, adapting them for special applications such as astronomical observations. The compass is an accessory rather than an integral part of the theodolite. A gyroscopic attachment is very expensive but valuable for certain applications.

12. *Tripods* are the wide-frame type. Some are all metallic and feature devices for preliminary leveling of the tripod head and mechanical centering ("plumbing") to eliminate the need for a plumb bob or optical plummet.

10-12. REPEATING THEODOLITES. Theodolites are divided into two basic categories: the *repeating* (or *double-center*) type and the *directional* (or *triangulation*) model. Repeating theodolites are equipped with a double vertical axis (similar to that of American transits but usually cylindrical in shape) or a repetition clamp. As in the American transit, this design enables angles to be repeated any number of times and added directly on the instrument circle.

Figures 10-9 and 10-10 show examples of repeating-type theodolites. The optical reading system of each instrument is shown in the small inset figures. Each of these theodolites reads directly to the nearest minute, with estimation possible to 0.1 min. Both instruments have vertical-circle automatic compensators, telescopes with standard eyepieces of 30X magnification, optical plummets, and plate bubble sensitivity of 30 sec/2-mm division.

The reading system of the Lietz TS6 theodolite in Figure 10-9 consists of a graduated glass scale having a span of 1° which appears superimposed on

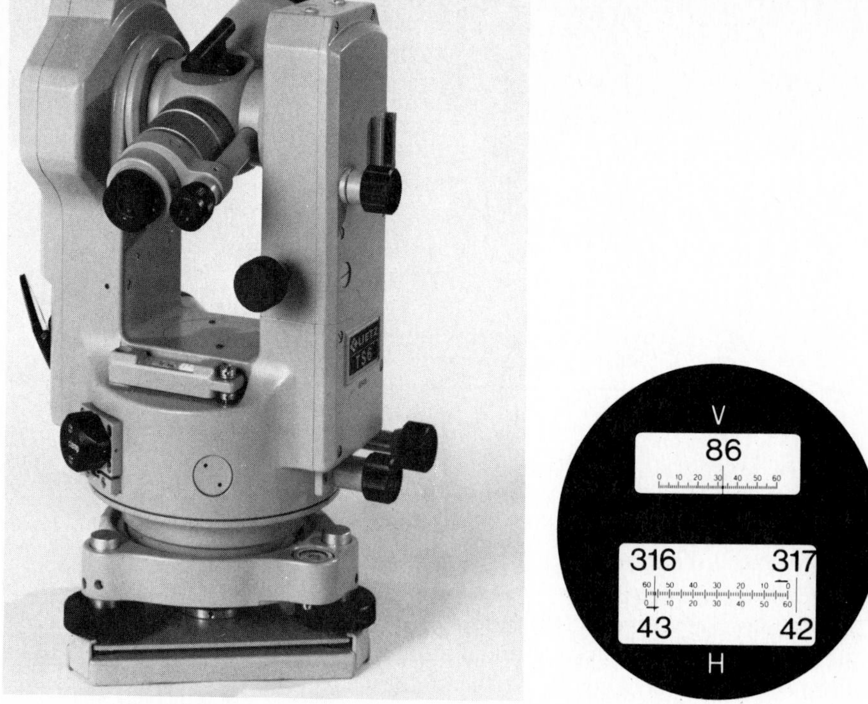

Figure 10-9. Lietz TS6 repeating theodolite. (Courtesy Lietz Company.)

Figure 10-10. T-1 repeating theodolite. (Courtesy Wild Heerbrugg Instruments, Inc.)

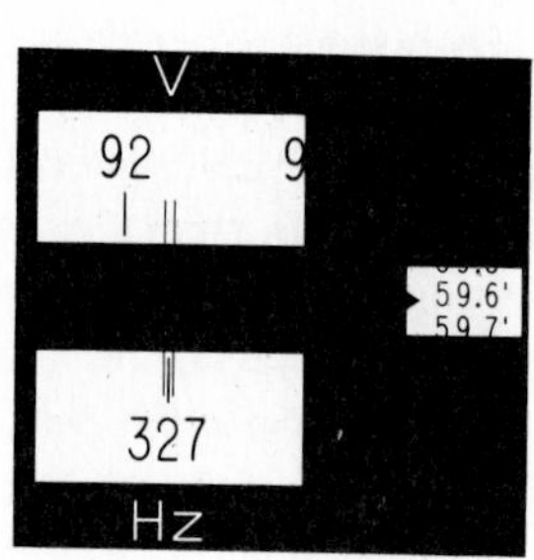

the degree divisions of the main circle. This scale is read directly by means of a microscope whose eyepiece is beside the main telescope. To take a reading, it is simply necessary to observe which degree number lies within the 1° span of the glass scale and select the minute indicated by the index mark. The vertical and clockwise horizontal-circle readings indicated for the TS6 in Figure 10-9 are 86°32.5′ and 316°56.5′, respectively. (The counterclockwise horizontal circle reading is 43°03.5′.) Thus on this instrument the horizontal and vertical circles can be viewed and read simultaneously through the reading microscope.

The reading system of the Wild T-1 in Figure 10-10 consists of an optical micrometer. To make a reading, the operator must first center the reference mark between the two bifilar lines of a degree mark by turning the micrometer knob. The micrometer spans 1° of the main circle and by centering the reference mark, the minutes portion of the angle can be read in the micrometer window on the right side of the reading microscope field of view. The horizontal angle indicated in Figure 10-10 is 327°59.6′. The micrometer must be set again to read the vertical angle.

10-13. DIRECTIONAL THEODOLITES. A directional theodolite is a nonrepeating type of instrument that has no lower motion. "Directions" rather than angles are read. After a sight has been taken on a point, the line's direction is read on the circle. An observation on the next mark gives a new direction, so the angle between the lines can be found by subtracting the first direction from the second.

Directional theodolites have a single vertical axis and therefore cannot measure angles by the repetition method. They do, however, have a *circle-orienting drive* to make a rough setting of the horizontal circle at any desired position.

On all directional theodolites each reading represents the *mean* of two diametrically opposed sides of the circle, made possible because the operator simultaneously views both sides of it through internal optics. This reading procedure, equivalent to averaging readings of the A and B verniers of a transit, automatically compensates for eccentricity errors (see Section 11-18.1).

Typical directional theodolites are shown in Figures 10-11 and 10-12. Each has a micrometer that permits reading the horizontal and vertical circles directly to 1 sec, with estimation possible to the nearest 0.1 sec. Both have automatic compensators for orienting the vertical circle, optical plummets, and plate bubbles with 20 sec/2-mm division sensitivity.

An inset for each figure illustrates the circle reading system of the instrument. The vertical and horizontal circles of the DKM2-A (see Figure 10-11) carry two concentric scales, one with single-line graduations, the other with bifilar divisions. The circle-reading microscope shows a portion of one scale superimposed on the diametrically opposed portion of the other. In reading an angle, the operator turns the optical micrometer knob to shift the two scales until the single-line graduations appear centered within the bifilar lines. As the micrometer is revolved, there is a simultaneous movement of a *field stop*, which frames the number of 10-min divisions in the reading. Centering with the microscope must be done separately for the horizontal and vertical circles.

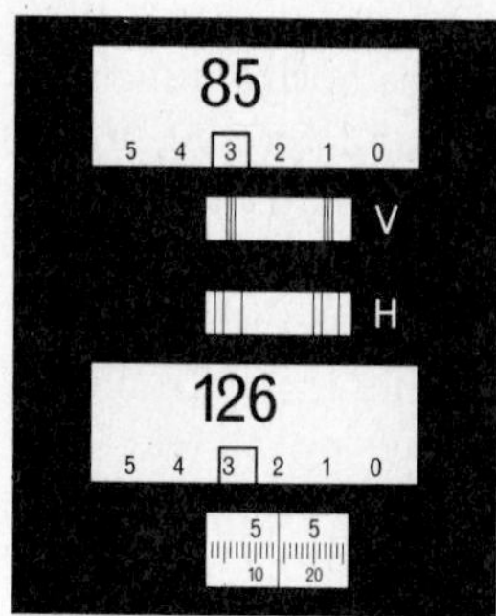

Figure 10-11. DKM2-A directional theodolite. (Courtesy Kern Instruments, Inc.)

Figure 10-12. Th-2 directional theodolite. (Courtesy Carl Zeiss, Oberkochen.)

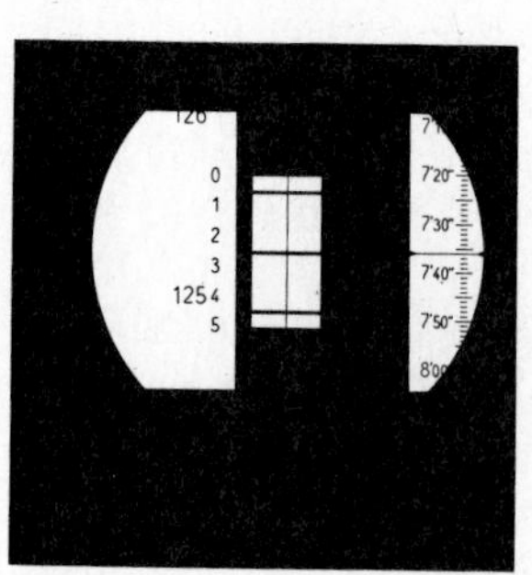

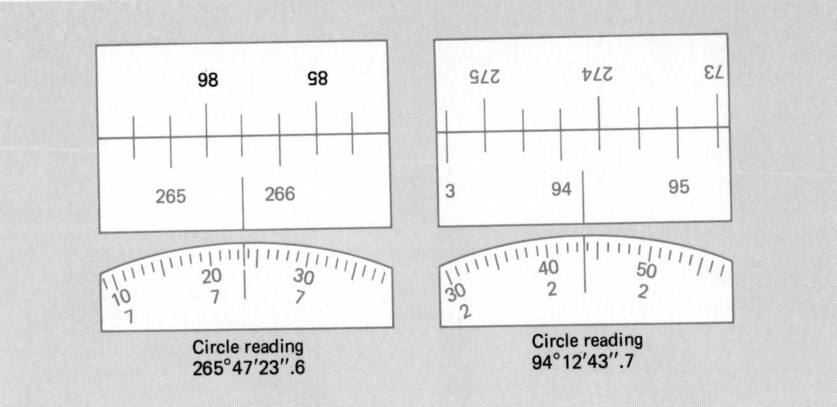

Figure 10-13. Reading system of horizontal circle of older version of T-2 directional theodolite.

In Figure 10-11, the micrometer has been set to read the vertical circle so the single graduations are centered within the bifilar lines of the window labeled V. (Note they are not centered in the H window.) The reading is 85° (seen directly in the upper window) plus 3 × 10 or 30 min (the 3 being taken from within the field stop frame of the same window), plus 5′14.0″ (from the lower window). Thus the vertical circle reading is 85°35′14.0″. The micrometer setting should be repeated before deciding to record 14 sec instead of 13 or 15 sec.

The reading system of the Th-2 shown in Figure 10-12 is similar to that of the DKM2-A. A selector knob permits viewing either the horizontal or vertical circle through the microscope; both circles cannot be seen simultaneously. The center window of Figure 10-12 shows the graduations on diametrically opposed parts of the circle. The micrometer has already been adjusted for reading by making opposite graduations coincide, and in that position the number corresponding to a multiple of 10 min in the reading is indicated in the left window directly beside the number of degrees. Minutes and seconds portions of the reading are taken from the rightmost window. The angle thus indicated in the Th-2 is 125°, plus 4 × 10 min (both from the left window), plus 7′36″ (in the right window), giving a final reading of 125°47′36″.

The reading system of a widely used, older-version T-2 directional theodolite is illustrated in Figure 10-13. In this arrangement, the circles are divided into 20-min spaces and the whole-degree marks numbered.

A reading is obtained by turning the micrometer knob to make the division lines (which move in opposite directions) coincident. Since an average reading of the two sides of the circle is secured, each graduation is counted as 10 min to save a later division by 2. The micrometer range is limited to 10 min. The set of numbers seen upside down is on the circle's opposite side.

In Figure 10-13, a direction of 265°40′ is read by counting the number of divisions (4) between 265° and its diametrically opposite graduation, 85°. A micrometer scale gives the additional minutes and seconds—in this example,

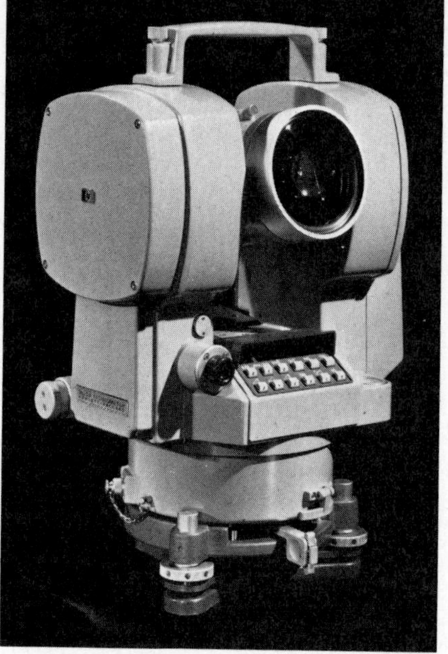

Figure 10-14. Hewlett-Packard 3820 total-station instrument, which combines an electronic digital theodolite and EDMI. (Courtesy Hewlett-Packard Co.)

Figure 10-15. Glass measuring circle with superimposed metal film pattern for automatic resolution of angles in the HP 3820 total-station instrument. (Courtesy Hewlett-Packard Co.)

7′23.6″. The vertical circle, read in the same manner, furnishes the zenith distance (the complement of the vertical angle) to avoid the need for signs.

An additional extremely precise directional instrument used for first-order triangulation, the DKM-3, is shown in Figure 20-8.

10-14. ELECTRONIC DIGITAL THEODOLITES. Modern technological advances have recently stimulated production of electronic digital theodolites that can automatically read and record horizontal and vertical angles. These devices can be used exclusively for angle measurement, but often they are combined with an EDMI and microcomputer to yield a so-called *total-station* instrument such as the HP 3820 shown in Figure 10-14. Total-station units are also sometimes called *electronic tacheometers*, (see Section 5-6).

The design of electronic digital theodolites resembles that of standard theodolites described in preceding sections. The fundamental difference is their ability to automatically resolve angular values and display them externally in digital form, thereby eliminating the need to read circles through a microscope. Either light-emitting diodes (LEDs) or liquid-crystal diodes (LCDs) may be used for the display. The latter draws less power but requires illumination for making night readings.

Methods of automatically achieving angle measurements vary somewhat with these instruments, but the system used in the HP 3820 is representative and described briefly here. The glass measuring circle shown in Figure 10-15 contains a unique metal film pattern. A beam of light is directed through the circle and the intensity that passes varies because of interferences in the pattern. Photodiodes measure these different intensities and convert them to photocurrents, which in turn are resolved by the interfaced computer to yield angular positions of the beam at different locations on the circle. This system is similar to those now used by automatic checkout machines in modern grocery and department stores, which operate by passing a beam of light through unique patterns of black bars with varying thicknesses and spacing.

Both the horizontal and vertical circles of the HP 3820 are equipped with the system just described, and can resolve angles accurately to within ±3 sec. The results are displayed visually in the instrument but can also be recorded automatically in a solid-state memory device. Some other instruments use magnetic-tape storage devices. Once collected, the data can be transferred directly to a computer system for processing.

10-15. HANDLING AND SETTING UP A THEODOLITE. Theodolites should be carefully lifted from their carrying cases by grasping the standards (some newer instruments are equipped with handles for this purpose), and the instrument securely fastened to the tripod by means of a tribrach. The tripod with instrument is placed over the ground point in the manner described for transit setups (see Section 10-10). Beginners can use a plumb bob to approximate the required setup position. Exact centering over the point is done by means of an optical plummet, which provides a line of sight directed downward collinear with the vertical axis of the theodolite. The instrument must be level for the optical plummet to define a vertical line. Most theodolite tribrachs have a relatively insensitive bull's-eye bubble to facilitate rough preliminary leveling before beginning final leveling with the plate bubble. Some tribrachs also contain an optical plummet.

The setup process using an instrument with an optical plummet, tribrach mount with bull's-eye bubble, and adjustable-leg tripod is most easily accomplished in the following steps: (a) adjust the position of the tripod legs by lifting and moving the instrument as a whole until the point is near the optical plummet's line of sight; (b) plant the legs and center the bull's-eye bubble by adjusting the tripod leg lengths (the point will still be nearly on the optical plummet's line of sight); (c) level the instrument using the plate bubble and leveling screws; and (d) loosen the tribrach screw and translate the instrument (do not rotate it) to exactly center the plummet cross hair on the point. Repeat steps (c) and (d) until perfect leveling and centering are accomplished. Before starting, the instrument should be centered on the tripod head to permit maximum translation [step (d)] in any direction.

It was noted earlier that theodolites have a three-screw leveling head and a single plate bubble. To level the instrument, the plate bubble is placed parallel to the line through any two foot screws, centered by turning these two screws, then rotated 90°, and centered again using the third screw only. This process is repeated and carefully checked to ensure that the bubble remains centered. (As with the transit and level, *the bubble moves in the direction of the left thumb when the foot screws are turned.*) A solid tripod setup is essential for theodolites that have very sensitive bubbles, and the instrument must be shaded if set up in bright sunlight; otherwise, the bubble will expand and run toward the warmer end as the instrument is heated.

PROBLEMS

10-1. List the fundamental differences between a transit and a theodolite.
10-2. Explain how and why the spindle and double-centered taper of an American transit seat and center properly unless damaged.
10-3. Why should a transit or theodolite *not* be lifted by holding the telescope?
10-4. What precautions must be taken with an instrument that has its leveling screws set in cups?
10-5. Why are two plate level vials used on a transit instead of just one?
10-6. What is the purpose of loosening the wing nuts of a transit or theodolite tripod before setting up?
10-7. Sketch a transit circle divided into 5-min spaces and a vernier that can be used to read it to the nearest 10 sec.
10-8. Draw a circle and vernier graduated in grads that can be read to $\frac{1}{8}$th grad.
10-9. Determine the least count for the following main scale and vernier combinations:
(a) Fifteen divisions on a transit vernier span 14, 5-min divisions of the main scale.
(b) Thirty divisions on a transit vernier cover 29, 30-min divisions of the main scale.
(c) Ten divisions on a protractor vernier span 9, 2° divisions of the main scale.
10-10. Discuss the basic differences between repeating and directional theodolites.
10-11. Explain the procedure of leveling a theodolite that has three leveling screws and a single plate bubble.
10-12. Describe the various cross-hair configurations available with transits and theodolites. Give the advantage or purpose of each design.
10-13. What are the reasons for having an eyepiece prism on a transit or theodolite?
10-14. List several advantages and disadvantages of an optical plummet compared with a plumb bob.
10-15. Explain "run of the micrometer" on a directional theodolite.
10-16. Can the repetition method be used for measuring vertical angles with a transit or theodolite? Discuss.
10-17. Explain why a directional theodolite cannot be used to measure horizontal angles by repetition.
10-18. Name and describe an instrument used to accurately position a theodolite on top of a tower, exactly above its ground station.
10-19. What is a gyroscopic theodolite, and where can it be used most advantageously?
10-20. Explain how the Hewlett-Packard 3820 total-station instrument is able to automatically measure angles.

BIBLIOGRAPHY

Haller, R. 1963. "Theodolite Axis Systems—Their Design, Manufacture and Precision." *Surveying and Mapping* 23(no. 4):575.
Kissam, P. 1961. "Circle Accuracy, Its Value to the Engineer and the Economy of Its Use." *Surveying and Mapping* 21(no. 2):193.
Kivioja, L. A., and J. E. Pettey. 1973. "Wobbles of the Horizontal Axis of a Theodolite." *Surveying and Mapping* 33(no. 4):481.
McDonnell, P. W., Jr. 1982. "Transit/Theodolite Survey." *Point of Beginning* 7(no. 6):14.
———. 1982. "Transit/Theodolite Survey, Part II." *Point of Beginning* 8(no. 1):18.

11 FIELD OPERATIONS WITH TRANSITS AND THEODOLITES

11-1. INTRODUCTION. As mentioned in Section 10-1, transits and theodolites are employed principally for measuring horizontal and vertical angles. In some cases, unknown angular values must be determined so that positions of points can be calculated; in others, known angles are laid off to establish points at fixed locations given on construction plans. Prolonging a straight line, differential leveling (see Chapter 6), measuring horizontal and vertical distances by stadia (see Chapter 15), and staking alignments are other tasks for which transits and theodolites are commonly employed.

Methods of measuring angles and prolonging straight lines vary depending on the type of instrument used (repeating or direction) and the special conditions and requirements of different surveys. These varying methods are discussed in this chapter.

11-2. RELATIONSHIP OF ANGLES AND DISTANCES. The best-quality surveys result when there is compatibility between the accuracies of measured angles and distances. To select instruments and survey procedures necessary for achieving consistency, and evaluating the effects of errors due to various sources, it is helpful to remember the relationships between angles and distances given here and illustrated in Figure 11-1.

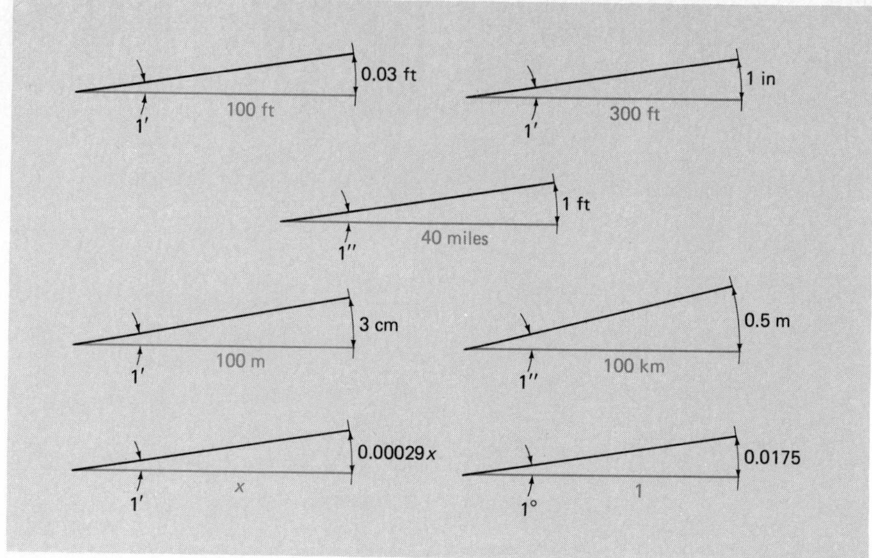

Figure 11-1. Angle and distance relationships.

1′ of arc = 0.03 ft at 100 ft, or 3 cm at 100 m (approx.)
1′ of arc = 1 in at 300 ft (approx.; actually 340 ft)
1″ of arc = 1 ft at 40 mi, or 0.5 m at 100 km (approx.)

$$\sin 1' = \tan 1' = 0.00029 \text{ (approx.)}$$
$$\sin 1^\circ = \tan 1^\circ = 0.0175 \text{ (approx.)}$$

In accordance with the relationships listed, approximately a 1-min error results in a measured angle if the line of sight is misdirected by 1 in over a distance of 300 ft. This illustrates the importance of precisely setting the instrument and targets over their respective points, especially where short sights are involved. If an angle is expected to be accurate to within $\pm\frac{1}{2}$ min for sights of 500 ft, then the distance must be correct to within $500 \times (0.00029/2) = \pm 0.07$ ft for compatibility. To appreciate the precision capabilities of a high-quality theodolite, an instrument reading to the nearest 0.1 sec is theoretically capable of measuring the angle between two points approximately 1 in apart and 40 mi away! (As discussed in Section 11-4, errors in centering the instrument, sighting the point, and reading the circle make it difficult, if not impossible, to actually accomplish this accuracy, however.)

11-3. MEASURING HORIZONTAL ANGLES WITH THE TRANSIT. Horizontal angles are measured with a transit by operating the upper clamp, lower clamp, and tangent screws. The vertical-circle clamp and tangent screw are used to bring the object sighted to the center of the field of view.

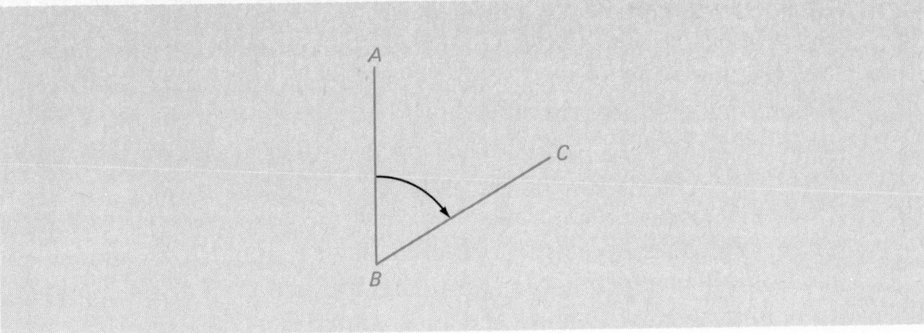

Figure 11-2. Measuring an angle.

Beginners may find it helpful to remember the following rules covering the use of the upper and lower clamps:

1. The lower clamp is used for backsighting only.
2. The upper clamp is used for setting the plates to zero, or any desired angle, and for foresighting.

The upper clamp and tangent screw are used to set 0°00′ (or any desired value) on the plates before sighting along the reference line, and to obtain a differential movement between the plates when foresighting. Expressed in a different way, the lower clamp and tangent screw are used to bring the line of sight along a reference line from which an angle is to be measured. The step-by-step procedure for measuring a *direct* (interior) angle *ABC* in Figure 11-2 is outlined to illustrate operation of the upper and lower motions.

1. Set up the instrument over point *B* and level it. Loosen both motions. Estimate the size of the angle as a check on the value to be obtained.
2. Set the circle to read approximately zero by holding the upper plate while turning the lower by tangential pressure on its underside. Tighten the upper clamp (snug but not wrench-tight). The upper and lower plates are now locked together.
3. Bring the vernier index mark exactly opposite the zero reading by means of the upper tangent screw. Always use a positive (clockwise direction) turn for the final setting for any tangent screw. If the zero is run beyond the point, back off and always finish with a positive motion. This prevents *backlash* (release of the spring tension, which can change the plate position).
4. Sight point *A* through the telescope. Set the vertical cross hair on, or almost on, the center line of the range pole or other object marking *A* by turning the instrument with both hands on the plate edge or on the standards (*not* on the telescope).
5. Tighten the lower clamp. The lower plate is now fastened to the socket.

6. Set the cross hair exactly on the mark by means of the lower tangent screw, finishing with a positive motion. Both motions are now clamped together and thus to the socket, the plates read zero, and the telescope is pointing to A. The transit is therefore *oriented*, since the line of sight is in a known direction with the proper value (0°00′) on the plates. Read the compass bearing for line BA.
7. Loosen the upper clamp and turn the plate until the vertical hair is on, or almost on, point C. The lower plate containing the graduated circle is still clamped to the socket, and the zero graduation continues to point toward A. Tighten the upper clamp.
8. Set the vertical hair exactly on mark C by means of the upper tangent screw.
9. Read the angle on the plates, using the vernier *ahead* of the zero mark (*in the same clockwise direction as the angle was turned.*) Read the compass bearing for line BC. Check the angle by comparing the measured value with the angle computed from compass bearings.

Since instrument setup, line clearing, objects to be sighted, and so on, are ready, little extra time is required to make repeated measurements, which provide checks, eliminate instrumental errors, and yield more reliable results.

11-4. MEASURING ANGLES BY REPETITION WITH A REPEATING INSTRUMENT. If an angle is to be measured by repetition (turned two or more times), the method just described is followed for the first reading. Then, with the reading for the first angle left on the circle, a backsight is taken on A, as before, *using only the lower clamp and tangent screw to retain the angle setting.* The instrument is now oriented in the starting position, but the single-angle value is on the circle, instead of 0°00′.

The upper clamp is loosened, point C sighted again, the upper clamp tightened, and the cross hair brought exactly on the mark with the upper tangent screw. The sum of the first two turnings of the angle is now on the circle. This process can be continued for the number of repetitions desired.

A transit or theodolite should be releveled if necessary after turning the angle, but the *leveling screws must not be used between the backsight and foresight* as required in differential leveling. An even number of repetitions should be taken, half with the telescope *normal* (direct) and half with the telescope *plunged* (reversed). This eliminates by reversion the effects of some possible maladjustments of the instrument described in Appendix A. (Alternative terms for direct and reversed positions of the telescope are *face left* and *face right*, which refer to whether the vertical circle is left or right of the telescope.)

The total angle accumulated on the circle divided by the number of repetitions gives an average value. The total angle may be larger than 360°, making it necessary to add a multiple of 360° to the reading before dividing. It is always desirable, therefore, to record the single angle after the first foresight.

It might be assumed that turning an angle 10, 50, or 100 times would give an increasingly better answer, but this is not true. Experience shows that with a 1-min transit having the usual properties, an average observer can point the instrument (align the vertical wire) within about 2 to 5 sec.

A 1-min vernier can be read to within 30 sec. An angle on the plates of, say, 42°11′29″ would theoretically be called 42°11′ by an experienced observer using a magnifying glass. If the angle on the plates is 42°11′31″, presumably a reading to the nearest minute of 42°12′ would be obtained. In either case, the recorded value would be within 30 sec of the correct angle.

If the transit is in adjustment, leveled, exactly centered, and operated by an experienced observer under suitable conditions, there are only two sources of error in measuring an angle—pointing the telescope and reading the plates. For a 5-sec average pointing error and a maximum discrepancy of 30 sec in setting to zero and in reading a 1-min vernier scale, the number of repetitions needed to strike a balance between readings and pointings is approximately seven. Since an even number should be measured to have equal repetitions of normal and plunged sights, six or eight turnings are usually made.

A general formula for computing the maximum random error resulting from repeated angles can be developed from Eq. (2-9) as

$$E = \frac{1}{N}\sqrt{E_0^2 + 2NE_p^2 + E_R^2} \qquad (11\text{-}1)$$

where E_0 is the error in setting to zero, N the number of repetitions of the angle, E_p the error in pointing, and E_R the error in reading. E_0 and E_R are equal to one-half the least count of the vernier or reading system.

EXAMPLE 11-1

A 1-min transit is used and an angle measured by repetition twice direct and twice reversed (2D, 2R). Assuming pointing errors of ± 3 sec, compute the expected maximum error in the angle. Compare this with the error expected if the angle were measured four times independently and averaged.

SOLUTION

Neglecting small plate graduation errors, there remain two reading errors—initial setting and final reading—but eight pointings. Then, taking E_0 and E_R to be ± 30 sec, by Eq. (11-1) the error in the angle obtained by repetition is

$$\tfrac{1}{4}\sqrt{(30)^2 + 8(3)^2 + (30)^2} = \tfrac{1}{4}(43.3) = 10.8''$$

If measured four times independently and the results averaged, the maximum random error, from Eq. (2-12), would be

$$E = \frac{1}{\sqrt{4}}\sqrt{(30)^2 + 2(3)^2 + (30)^2} = 21.3''$$

The advantage of the repetition method is evident.

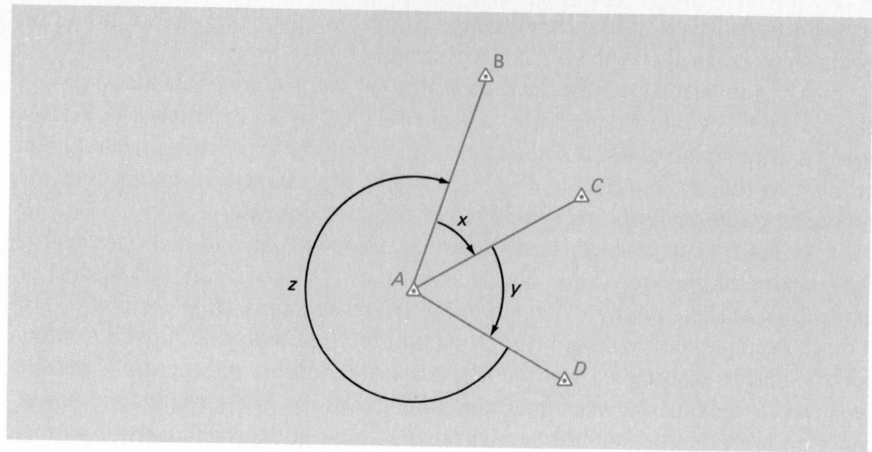

Figure 11-3. Closing the horizon.

Direct angles, measured singly or by repetition, are commonly used in boundary surveys, hydrographic work, and construction.

11-5. CLOSING THE HORIZON. Closing the horizon is the process of measuring the angles around a point to obtain a check on their sum, which should

Figure 11-4. Field notes for closing the horizon.

CLOSING THE HORIZON

Point Sighted	Plate Reading	Angle	Mag Bearing	Angle Comp. from Bear.
	⊼ at point A			
B	3°26′		N22°15′E	
C	45°38′	42°12′		42°15′
C	~~42°08′~~ 47°08′		N64°30′E	
D	107°04′	59°56′		60°00′
D	110°35′		S55°30′E	
B	8°29′	257°54′		257°45′
		360°02′		
Misclosure		0°02′		

equal 360°00′. For example, if in Figure 11-3 only angles x and y are needed, it is desirable also to turn angle z to close the horizon at A. The method provides an easy way for a beginner to test readings and pointings. Figure 11-4 shows the left-hand page of notes covering measurement of the angles in Figure 11-3. The circle readings are changed slightly for each backsight to provide practice in reading the instrument and for checking purposes. Each individual angle is calculated by subtracting the previous reading.

The difference between 360° and the sum of angles x, y, and z is called the *horizon misclosure*. For this example, it was 02 min. Permissible values of misclosure will determine whether the work must be repeated.

Plate D-7 shows a sample set of notes illustrating the measurement of angles with a transit by repetition to close the horizon. In this particular arrangement, the A vernier is set to read zero only at the beginning of the work, and thereafter the final reading for each angle—for example, 253°13′00″—becomes the initial reading for the next angle. Both a *vernier misclosure* (difference between the initial and final vernier readings) and a horizon misclosure are obtained by this rigorous procedure.

11-6. LAYING OFF AN ANGLE WITH A REPEATING INSTRUMENT.

To lay off an angle BAC equal to 25°30′ with an instrument at point A (Figure 11-5), the circle is set to zero and point B is sighted using the lower motion. The upper clamp is loosened, the telescope turned until the circle reads 25°30′, and the upper clamp again tightened. The line of sight establishes AC at the proper angle with AB.

To lay off an angle BAC equal to 25°30′40″ by repetition with a 1-min instrument, an angle BAC' of 25°30′ is laid out as previously described and point C' marked. The angle BAC' is then measured by repetition as many times as the desired precision requires. The difference between angle BAC' and 25°30′40″ can be marked off by measuring distance AC' and locating C by the following relation: Distance $C'C = AC' \tan C'AC$. The angle BAC can then be turned by repetition as a check.

In Figure 11-5, if angle BAC' is found by repetition to be 25°30′20″, then $C'AC = 20''$. If distance AC' is 300 ft, then $C'C = 300 \tan 20'' = 300(0.00029/3) =$ 0.029 ft.

Figure 11-5. Laying out an angle by repetition.

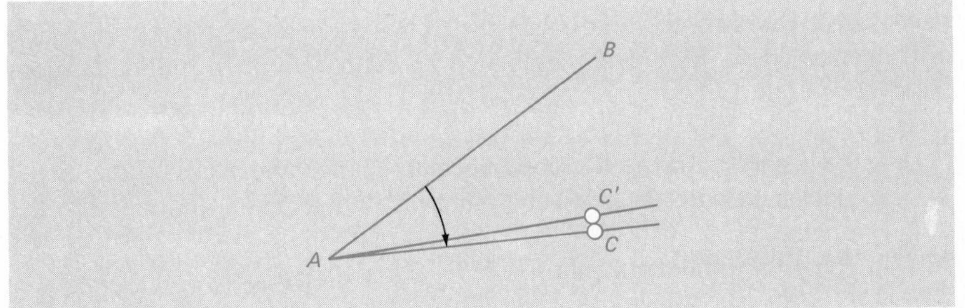

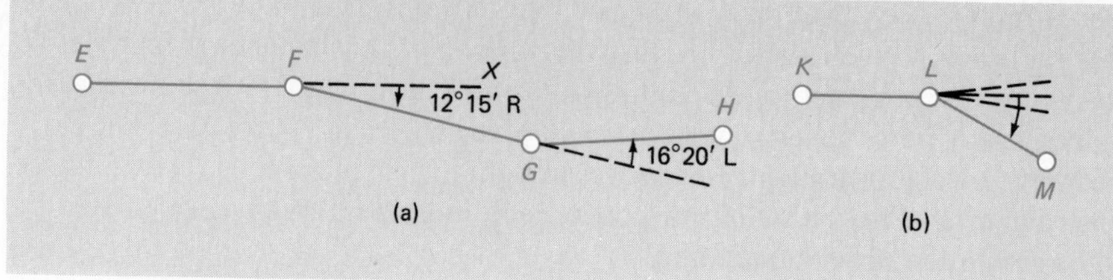

Figure 11-6. Deflection angles.

11-7. DEFLECTION ANGLES. A deflection angle (described in Section 8-3) is a horizontal angle measured from the prolongation of the preceding line, right or left, to the following line. In Figure 11-6(a), the deflection angle at *F* is 12°15′ to the right (12°15′R). At *G* the deflection angle is 16°20′L.

A straight line between terminal points is theoretically the most economical route to build and maintain for highways, railroads, pipelines, canals, and transmission lines. Practically, obstacles and conditions of terrain and land use require bends in the route, but deviations from a straight line are kept as small as possible. Use of deflection angles, rather than large direct angles, is therefore appropriate for easier visualization, sketching, and computation.

If an instrument is in *perfect* adjustment (which is unlikely), the deflection angle at *F* [see Figure 11-6(a)] is measured by setting the circle to zero and backsighting on point *E* with the telescope plunged, then plunging again. The line of sight is now on *EF* extended and directed toward *X*. The upper clamp is loosened, point *G* sighted, the upper clamp tightened, and the vertical wire brought exactly on the mark by means of the upper tangent screw. The vernier will be under the eyepiece end of the telescope, so the observer can read the deflection angle without moving around the transit.

Deflection angles are subject to serious errors if the instrument is not in adjustment, and may be larger or smaller than their correct values depending on whether the line of sight after plunging is to the right or left of the true prolongation [see Figure 11-6(b)].

To eliminate errors from this cause, angles are usually doubled or quadrupled by the following procedure: The first backsight is taken with the circle set at zero and the telescope in the direct position. After plunging, the angle is measured and kept on the circle. A second backsight is taken using the lower motion, retaining the first angle, and keeping the telescope reversed. The telescope is plunged back to the normal position for the foresight, and the angle remeasured. Dividing the total angle by 2 gives an average angle from which instrumental errors have been eliminated by cancellation. In outline fashion, the method is as follows:

1. Backsight with the telescope normal. Plunge and measure the angle.
2. Backsight with the telescope plunged. Plunge again and measure the angle.
3. Read the total angle and divide by 2 for an average.

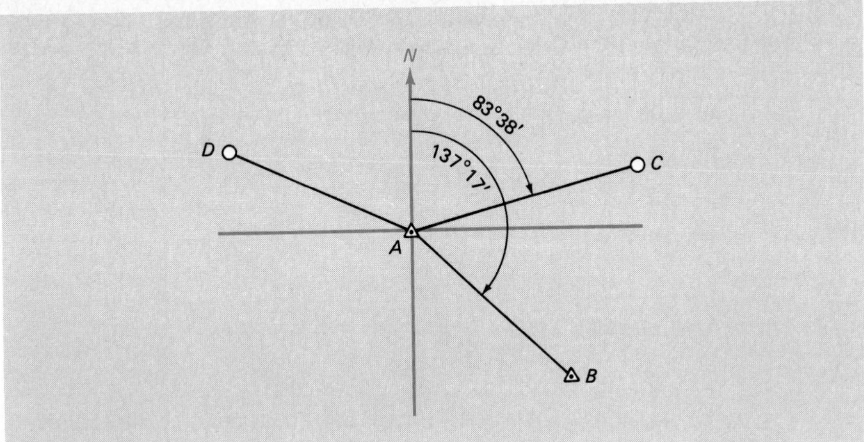

Figure 11-7. Orientation by azimuths.

11-8. AZIMUTHS. Azimuths are measured from a reference direction which itself must be determined from (a) a previous survey, (b) the magnetic needle, (c) a solar or star observation, or (d) assumption. Suppose that in Figure 11-7 the azimuth of line *AB* connecting two triangulation stations is known to be 137°17′ from true north. With a repeating instrument to find the azimuth of any other line from *A*, such as *AC*, first set 137°17′ on the circle numbered in the clockwise direction and backsight on point *B*. The instrument is now oriented, since the line of sight is in a known direction with the appropriate angle on the circle. Loosen the upper motion, turn the telescope clockwise to *C*, and read the clockwise angle. In this case the reading would be 83°38′.

Note that after the lower clamp and tangent screw are used to backsight on point *B*, they are not disturbed regardless of the number of angles read from point *A*. When the circle reads zero, the telescope is pointing true north. As a check, if a transit equipped with a compass is being used, the needle can be lowered and read. If the telescope is pointing north, the needle should read the declination of the place (provided there is no local attraction at point *A*).

In Figure 11-7, if the instrument is set up at point *B* instead of *A*, the azimuth of *BA* (317°17′) or back azimuth of *AB* is put on the circle and point *A* sighted. The upper motion is loosened and sights are taken on points whose azimuths from *B* are desired. Again, if the instrument is turned until the circle reads zero, the telescope points true north.

An alternative and shorter method of orienting the instrument at *B*, *if it plunges correctly*, is to leave the azimuth of *AB* (137°17″) on the circle while backsighting on point *A* with the telescope reversed. The telescope is then plunged to the normal position to bring the line of sight along line *AB* extended with its proper azimuth on the circle.

Figure 11-8 shows the left-hand page of a sample set of notes for a traverse run by azimuths. Note that the survey began with a setup on point *A* and finished by reoccupying *A*. This procedure yields a check on all work, since

AZIMUTH-TAPE TRAVERSE

Point Occupied	Point Sighted	Distance	Azimuth	Mag. Bear.
A	Mag. N		0°00'	Due N
	B	126.24	23°32'	N23°30'E
B	A		203°32'	S23°30'W
	C	82.50	93°51'	S86°15'E
C	B		273°51'	N87°00'W
	D	122.58	137°39'	S42°30'E
D	C		317°39'	N43°00'W
	A	216.35	264°46'	S84°45'W
A	D		84°46'	N84°45'E
	B		23°34'	N23°30'E
	Misclosure		0°02'	

Figure 11-8. Field notes for azimuth traverse (initial orientation on magnetic north).

remeasured azimuth *AB* should equal its starting value. In this example the angular misclosure of 2 min obtained would be adjusted by procedures described in Section 13-2.

11-9. MEASURING WITH A DIRECTIONAL THEODOLITE. As noted in Section 10-13, directional theodolites can be used for determining horizontal angles, but field procedures consist of measuring "directions," which are simply horizontal circle readings taken to successive stations sighted around the horizon. The difference in directions between any two stations is the angle.

Figure 11-9 shows a set of field notes for directions measured at station *A* of Figure 11-3. The notes actually are the results of four "positions," each representing a reading at every station around the horizon with the instrument in both the direct and reversed modes.

Although there is no lower motion on a directional theodolite, the horizontal circle can be roughly indexed to selected values. To distribute readings around the entire circle, and hence minimize possible circle graduation errors, the initial sighting in the direct mode for the first position is set near 0°00′, then advanced approximately 180°/*n* for the first pointing of each successive position, where *n* is the number of positions being measured. For the field notes of Figure 11-9, the initial readings for the four positions in the direct mode were indexed near 0°, 45°, 90°, and 135°, and the beginning readings with the tele-

DIRECTIONS OBSERVED FROM STATION A

Position No.	Station Sighted	Reading Direct	Reading Reversed	Mean	Reduced Direction
(1)	(2)	(3)	(4)	(5)	(6)
		° ′ ″	° ′ ″	″	° ′ ″
1	B	0 00 05	180 00 04	04	0 00 00
	C	42 12 15	222 12 16	16	42 12 12
	D	102 08 33	282 08 28	30	102 08 26
2	B	45 00 03	225 00 06	04	0 00 00
	C	87 12 ~~14~~ 12	267 12 20	16	42 12 12
	D	147 08 35	327 08 28	32	102 08 28
3	B	90 00 08	270 00 05	06	0 00 00
	C	132 12 20	312 12 22	21	42 12 15
	D	192 08 28	12 08 31	30	102 08 24
4	B	135 00 07	315 00 05	06	0 00 00
	C	177 12 15	357 12 19	17	42 12 11
	D	237 08 30	57 08 34	32	102 08 26

Figure 11-9. Field notes for measuring directions.

scope plunged therefore were near 180°, 225°, 270°, and 315°, thus providing a uniform distribution of readings around the circle.

In the field notes of Figure 11-9, the position number is in column (1); the station sighted in column (2); readings taken in the direct and reversed modes in columns (3) and (4), respectively; mean values of the seconds portion for the direct and reversed readings in column (5); and the reduced direction (obtained by subtracting the mean value for station *B* from all other mean directions) in column (6). Note at position number 2, column (6) values of 0°00′00″, 42°12′12″, and 102°08′28″ were obtained by subtracting the initial reading of 45°00′04″ from the other values in column (5) for that position.

The sets of values in column (6) should be compared for agreement and acceptance criteria before leaving the station occupied so additional positions can be measured if necessary. Angles can be calculated from the directions, but in triangulation computations, directions are often preferable.

11-10. SIGHTS AND MARKS. Objects commonly used for sights on plane surveys include range poles, chaining pins, pencils, plumb-bob string, and tripod-mounted targets (see Figure 10-8). For short sights, string is preferred to a range

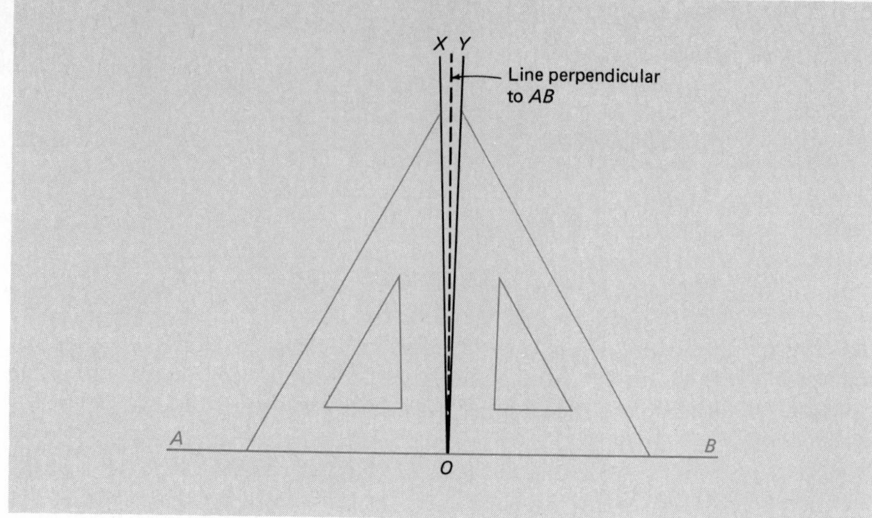

Figure 11-10. Principle of reversion.

pole because the small diameter permits more accurate centering. Small red and white targets of thin-gauge metal or cardboard placed on the string, extend the length of observation possible.

An error is introduced if the range pole sighted is not plumb. The observer must sight as low as possible on the pole when the mark itself is not visible, and the rodperson has to take special precautions in plumbing the pole, perhaps using a rod level or plumb bob.

In layout work on construction and in topographic mapping, *permanent* backsights and foresights may be established. These can be marks on structures such as walls, steeples, water tanks, and bridges, or they can be fixed artificial targets. They provide definite points on which the instrument operator can check orientation without the help of a rodperson.

11-11. PROLONGING A STRAIGHT LINE. On route surveys, straight lines may be continued from one hub through several others. To prolong a straight line from a backsight, the vertical wire is aligned on the back point by means of the lower motion, the telescope plunged, and a point, or points, set ahead on line.

To eliminate the effects of instrument maladjustment, the same procedure used in making a number of adjustments, known as the *principle of reversion*, is employed. The method applied, actually *double reversion*, is termed *double centering*. Figure 11-10 shows a simple use of the principle in drawing a right angle with a defective triangle. Lines OX and OY are drawn with the triangle in "normal" and "reversed" positions. Angle XOY represents twice the error in the triangle at the 90° corner, and its bisector establishes a line perpendicular to AB.

In practice, instruments should always be kept in good adjustment but used as though they might not be.

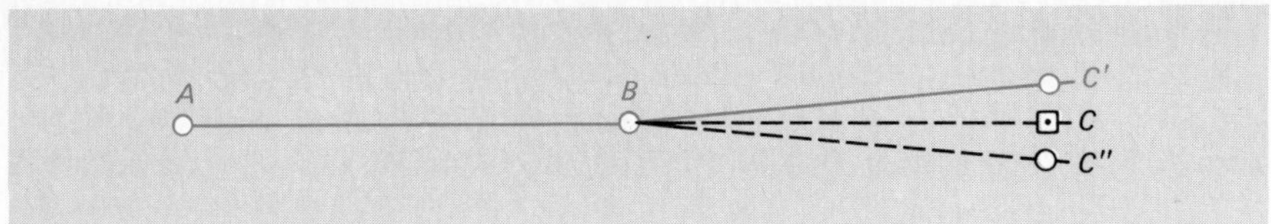

Figure 11-11. Double centering.

In double centering, after the first point C' in Figure 11-11 has been located with the telescope plunged, the lower motion is released and a second backsight taken on point A, this time with the telescope still plunged. The telescope is transited again to its normal position and point C'' marked. Distance $C'C''$ is bisected to get point C, on the line AB prolonged.

In outline form, the procedure is as follows:

1. Backsight on point A with the telescope direct. Plunge to the reversed position and set point C'.
2. Backsight on point A with the telescope reversed. Plunge to a direct position and set point C''.
3. Split the distance $C'C''$ to locate point C. Note that $C'C''$ represents *twice* the plunging error (and, as described in Appendix A, *four times* the adjustment error).

11-12. PROLONGING A LINE PAST AN OBSTACLE. Buildings, trees, telephone poles, and other objects may block survey lines. Four of the various methods used to extend lines past an obstacle are the (1) equilateral-triangle method, (2) right-angle-offset method, (3) measured-offset method, and (4) equal-angle method. Short backsights can introduce and accumulate errors, so procedures using distant points should be followed.

11-12.1. EQUILATERAL-TRIANGLE METHOD. At point B in Figure 11-12(a), a 120° angle is turned off from a backsight on A, and a distance BC of 80.00 ft (or any distance necessary, but preferably not more than one tape length) is measured to locate point C. The instrument is then moved to C, a backsight taken on B, an angle of 60°00′ put on the circle, and a distance $CD = BC = 80.00$ ft laid off to mark point D. The instrument is moved to D, backsighted on C, and an angle of 120°00′ turned. The line of sight DE is now along AB prolonged.

11-12.2. RIGHT-ANGLE-OFFSET METHOD. With instrument setups at points B, F, G, and D in Figure 11-12(b), 90°00′ angles are turned at each hub. Distances FG and $BF = GD$ need be only large enough to clear the obstruction, but longer lengths provide more accurate sights.

The lengths shown in Figure 11-12(a) and (b) permit students to check their taping and instrument manipulation by combining the two methods.

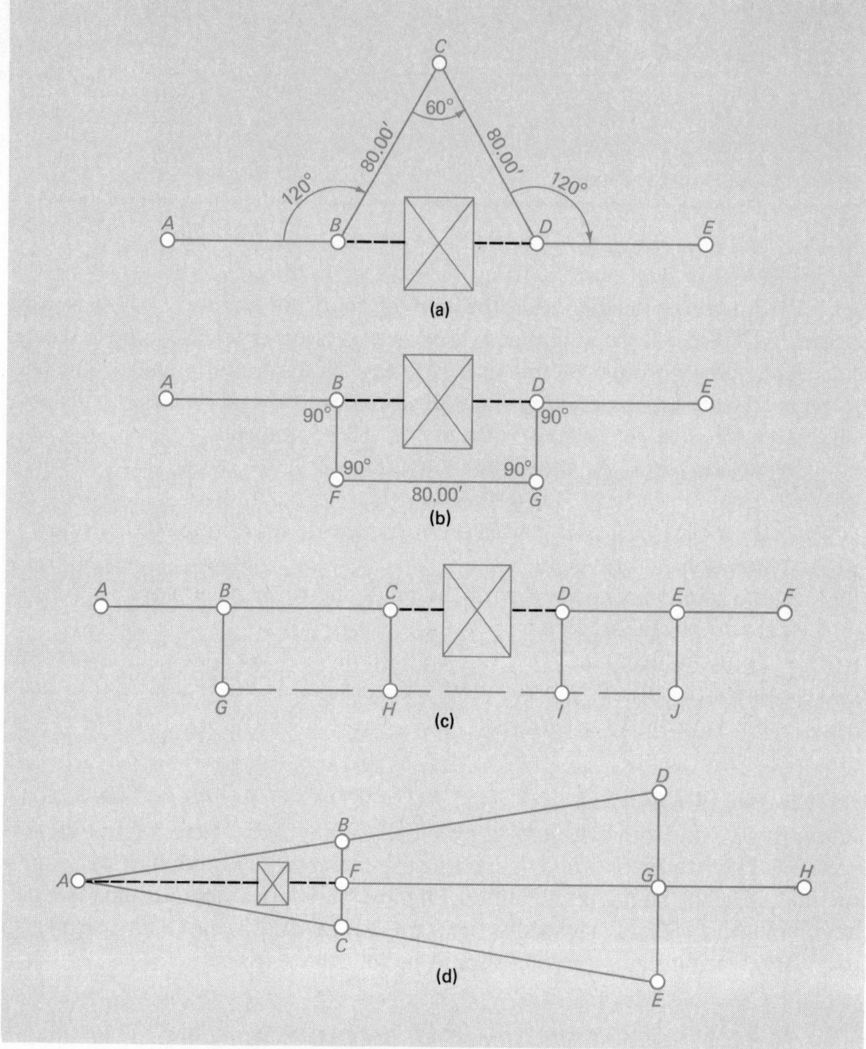

Figure 11-12. Prolonging a line past an obstacle.

11-12.3. MEASURED-OFFSET METHOD. To avoid the four 90° angles with short sights and consequently possible large errors, measured offsets obtained by swinging arcs with the tape can be used [see Figure 11-12(c)]. A long base is established for check points on *GHIJ* if desired. This method is preferred to obtain the accuracy necessary for most survey work.

11-12.4. EQUAL-ANGLE METHOD. This method is excellent when field conditions are suitable. Equal angles just large enough to clear the obstacles are turned from the line at point *A*, and equal distances $AB = AC =$ and $AD = AE$ measured in Figure 11-12(d). The line through points *F* and *G* at the midpoints of *BC* and *DE*, respectively, provides an extension of *AH* through the obstacle. Very little additional clearing is necessary using this method to bypass a large tree on line in wooded or brushy areas.

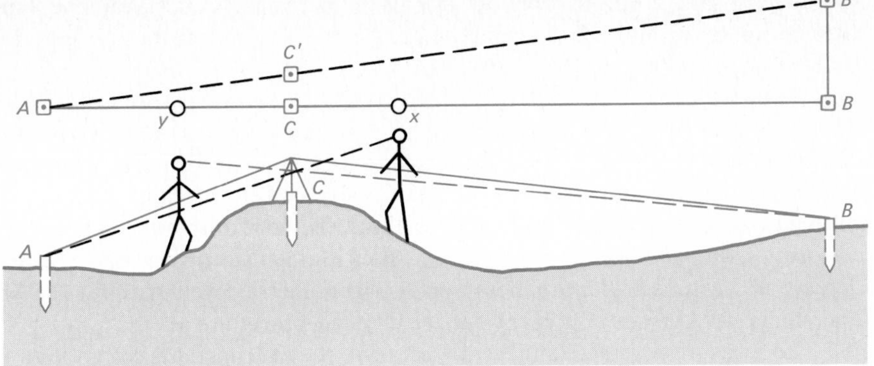

Figure 11-13. Balancing in.

11-13. BALANCING IN. Occasionally it is necessary to set up on a line between two points already established but not intervisible—for example, hubs A and B in Figure 11-13. This process is called *balancing in* or *wiggling in.*

Location of a trial point C' on line is estimated and the instrument set over it. A sight is taken on point A from point C', and the telescope plunged. If the line of sight does not pass through B, the instrument is moved laterally a distance CC' estimated from the proportion $CC' = BB' \times AC/AB$, and the process repeated. Several trials may be required to locate point C exactly, or close enough for the purpose at hand. The shifting head of the instrument is used to make the final small adjustment.

Another method is to have two persons, X able to see hub A and Y having point B visible, as shown in Figure 11-13. Each lines the other in with the visible hub in a series of adjustments, and two range poles are placed at least 20 ft apart on the course established. An instrument set at point C in line with the poles should be within a few tenths of a foot of the required location.

11-14. RANDOM LINE. On many surveys it is necessary to run a random line from a mark X to a nonvisible point Y, which is a known or indeterminate distance away. This problem arises repeatedly in property surveys.

On the basis of compass bearing or information from maps and other sources, a random line, such as XY' in Figure 11-14, is run as close as possible by estimation to the true line XY. Distances XY' and YY' (distance perpendicular to XY'), by which the random line misses point Y, are measured and

Figure 11-14. Random line.

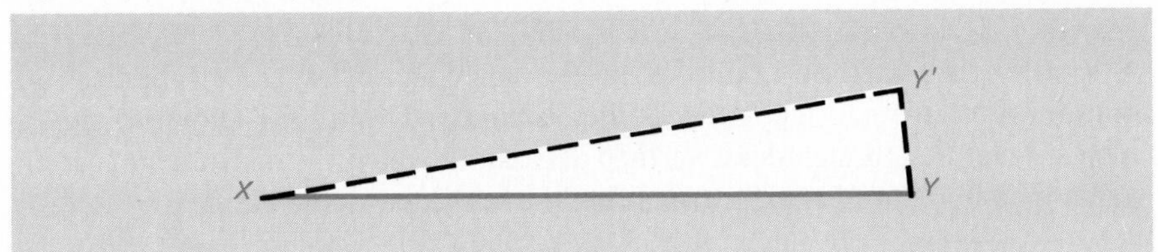

the angle YXY' is found from its calculated tangent. The correct line can then be run by turning off the computed angle $Y'XY$, or points on XY may be set by computed right-angle offsets from XY'.

11-15. MEASURING VERTICAL ANGLES. A vertical angle is the difference in direction between two intersecting lines measured in a vertical plane. As commonly used in surveying, it is the angle above or below a horizontal plane through the point of observation. Angles above the horizontal plane are called *plus angles*, or *angles of elevation*. Those below it are *minus angles*, or *angles of depression*. Vertical angles are measured in trigonometric leveling and in EDM and stadia work as an important part of the field procedure.

To measure a vertical angle with a transit, the instrument is set up over a point and carefully leveled. The bubble in the telescope level vial should remain centered when the telescope is clamped in a horizontal position and rotated 360° about its vertical axis. If the vernier on the vertical arc does not read 0°00′ when the bubble is centered, there is an *index error* that must be added to, or subtracted from, all readings. Confusion of signs is eliminated by placing in the field notes a statement such as "Index error is minus 2 min, to be subtracted from angles of depression and added to angles of elevation."

The horizontal cross line is set approximately on the point to which a vertical angle is being measured, and the telescope clamped. Exact pointing is obtained by using the vertical-circle tangent screw. The vertical circle is read and any index error applied to get the true angle above or below the horizon. The observer calls out the uncorrected angle reading and any required adjustment is made later.

To eliminate the index error resulting from displacement of the vernier on the vertical arc, and lack of parallelism of the line of sight and telescope level vial, an average of two readings should be taken. One is secured with the telescope normal, the second with it inverted. This method requires a transit equipped with a complete vertical circle and reversion level vial.

Measurement of vertical angles with a theodolite follows the same general procedure just described, except that the vertical circle is oriented by either an automatic compensator or an index level vial. If the latter is used, serious errors result if the index level bubble is not centered prior to reading angles. As with the transit, instrumental errors are compensated for by averaging an equal number of direct and reversed observations.

Theodolites are designed so that vertical-circle readings give *zenith angles*. Thus, a reading of 0° corresponds to the telescope pointing vertically (toward the zenith). In face left position, with the telescope horizontal, the reading is 90°, and if the telescope is elevated 30° above horizontal, the reading is 60°. In face right, the horizontal reading is 270°, and with the telescope raised 30° above the horizon, it is 300°.

11-16. MEASURING ANGLES WITH ELECTRONIC DIGITAL THEODOLITES. Except for their method of automatically resolving angles, the mechanical operation of electronic digital theodolites is similar to that for standard instruments. Their design includes a vertical axis about which the instrument turns in azimuth, a horizontal axis for transiting the telescope, and a clamp and tangent screw for pointing. To measure an angle, a backsight is taken using the clamp and tangent screw and an initial value entered into the display. Zero can be set if direct angles are being measured, but any required value may be entered if orienting on a line of known azimuth. The angle is then turned by pointing again, using the clamp and tangent screw, and its value automatically displayed in the instrument. To eliminate instrumental errors and increase precision, angles can be repeated any number of times in both direct and reversed modes and the average taken. Built-in computers will automatically perform the averaging and display the results.

Some special capabilities designed into most electronic digital theodolites enhance their accuracy and expedite operation. The HP 3820, for example, has a built-in computer that orients the vertical circle. Any index error is automatically read and transmitted to the computer, which applies it to measured angles. Another device senses any dislevelment of the horizontal circle, whereupon the computer applies a correction to observed horizontal angles. Thus, precise leveling of the instrument is unnecessary. The automatic reading system averages values of diametrically opposed sides of the circle, thereby correcting for any eccentricities (see Section 11-18.1). Also, a rough drive mechanism permits advancing the circle to initialize readings for different positions and compensate for possible imperfections in marking the circle.

As noted in Section 10-14, the greatest utility of electronic digital theodolites can be realized when they are interfaced with an EDMI, resulting in a combination known as a "total station". A number of total-station instruments are currently available. Figures 5-7 and 10-14 show two examples. Total stations are extremely versatile and useful for almost all types of surveys. Discussions of their application to specific surveys are given in chapters devoted to those specialized tasks.

11-17. TRANSITS AND THEODOLITES AS LEVELS. As noted in Section 6-5, transits and theodolites can be used as levels, although they generally yield results that are somewhat less accurate than those obtained with instruments designed especially for leveling. To use a transit as a level, after centering the plate bubbles of the instrument with the foot screws, the telescope bubble is carefully centered using the vertical-circle, slow-motion screw. If the instrument is in proper adjustment, this will create a horizontal sight line so rod readings can be made, and the transit is operated essentially as a tilting level.

To create a horizontal line of sight with theodolites, after centering the plate bubble using the foot screws, the vertical-circle reading (zenith angle) is set to 90°00.0′ (270°00.0′ if the telescope is inverted), whence the line of sight is horizontal so rod readings can be taken. In the case of micrometer theodolites, this requires setting the micrometer on zero first and then fixing the circle reading with the slow-motion screw.

11-18. SOURCES OF ERROR IN TRANSIT AND THEODOLITE WORK. Errors in transit-theodolite surveys result from instrumental, natural, and personal sources. It is impossible to determine the exact value of an angle and therefore the error in its measured value. Precise results can be obtained, however, by (a) following specified procedures in the field, (b) manipulating the instrument

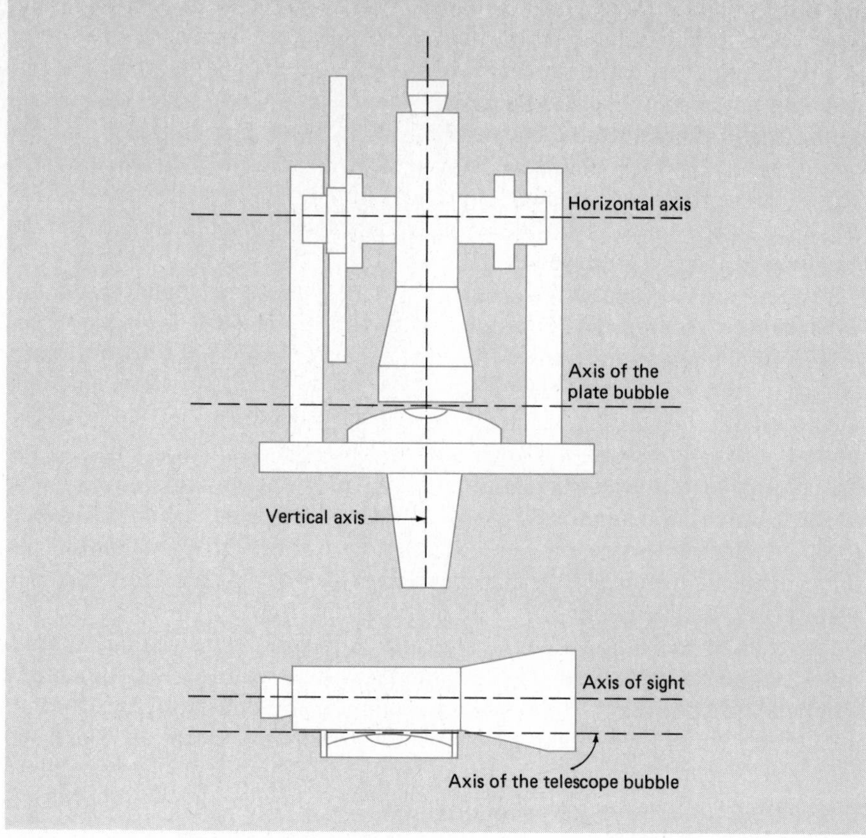

Figure 11-15. Reference lines of a transit.

carefully (for example, eliminating parallax, which can cause serious errors), and (c) checking measurements. Probable values of random errors and the degree of precision secured can be calculated from formulas given in Chapter 2.

11-18.1. INSTRUMENTAL ERRORS. Figure 11-15 shows the reference lines of a transit, which are referred to in the following discussion of instrumental errors. (They also apply to theodolites, with the exception of the telescope bubble axis.) For a properly adjusted transit, these reference lines must bear specific relationships to each other. If they become maladjusted, errors result in measured angles unless proper field procedures are observed. The principal sources of instrumental error follow.

1. *Plate bubbles out of adjustment.* If the axes of the plate bubbles are not perpendicular to the vertical axis, the latter will not be truly vertical when the plate bubbles are centered. This condition causes errors in measured horizontal and vertical angles that *cannot* be eliminated by averaging direct and reversed readings. Plate bubbles are out of adjustment, if, after centering, they run when the instrument is rotated 180° in azimuth. The length of bubble run indicates

double the tilt of the vertical axis, which is therefore made truly vertical by bringing the bubbles back *halfway* using the foot screws. With plate bubbles maladjusted, angles can be measured but it is inconvenient and time-consuming, so the required adjustment should be made. Procedures for making this and other transit and theodolite adjustments are described in Appendix A.

2. *Axis of sight not perpendicular to the horizontal axis.* If this condition exists, as the telescope is plunged, the axis of sight generates a cone whose axis coincides with the horizontal axis of the instrument. The greatest error from this source occurs when plunging the telescope, as in prolonging a straight line or measuring deflection angles. Also, when the angle of inclination of the backsight is not equal to that of the foresight, measured horizontal angles will be incorrect. These errors are eliminated by double centering and by averaging equal numbers of direct and reversed readings.

3. *Horizontal axis not perpendicular to the vertical axis.* This situation causes the axis of sight to define an inclined plane as the telescope is plunged and, therefore, if the backsight and foresight have differing angles of inclination, fallacious horizontal angles will result. Errors from this origin can also be canceled by averaging an equal number of direct and reversed readings.

4. *Axis of the telescope bubble not parallel to the axis of sight.* If this case exists for a transit, the axis of sight is inclined upward or downward when the telescope bubble is centered. It causes an error in vertical angles and rod readings when the transit is used as a level. The effect is eliminated in vertical angles by averaging equal numbers of direct and reversed readings, and in leveling by balancing backsight and foresight distances.

5. *Eccentricity of centers or verniers.* When the A and B transit vernier readings differ by exactly 180° for all positions, the circles are concentric and the verniers correctly set. If the readings disagree by a uniform number other than 180°, the verniers are offset and it is best to use only the A vernier or take the mean of both verniers. If the difference is not constant, eccentricity of centers exists. Readings should be taken at several positions on the circle and the results of the A and B verniers averaged. Theodolites do not have verniers. Eccentricity of centers can be present, however, and errors from this source are minimized by taking readings at several positions on the circle so they are spaced around the entire arc, and the results averaged.

6. *Unsteady tripod.* Tripod leg bolts must be tight so there is neither play nor strain (they can be tapped lightly to relieve any stress before taking the first sight), and the shoes set solidly in the ground. To eliminate strain, some surveyors loosen wing nuts and retighten after planting the legs and before leveling the instrument.

Surveyors may adjust transits in the field by following procedures outlined in Appendix A. Precise theodolites and damaged instruments, however, should be worked on only by experts.

11-18.2. NATURAL ERRORS

1. *Wind.* Wind vibrates a transit and deflects the plumb bob. Shielding the instrument, or even suspending observations on precise work, may be necessary on windy days. An optical plummet is helpful in this situation.

2. *Temperature effects.* Temperature differentials cause unequal expansion of various parts of transits and theodolites. A level bubble is drawn toward the heated end of the vial, as can be checked by blowing on one end of the vial and noting movement of the bubble, then releveling and checking the position of the cross hairs on a target. Temperature effects are reduced by shielding instruments from sources of heat or cold.

3. *Refraction.* Unequal refraction bends the line of sight and may cause an apparent shimmering of the observed object. It is desirable to keep lines of sight well above the ground and avoid sights close to buildings, smokestacks, and even large individual bushes in generally open spaces. In some cases, observations may have to be postponed until atmospheric conditions have improved.

4. *Settling of the tripod.* The weight of an instrument may cause it to settle in soft ground. When a job involves crossing swampy terrain, stakes should be driven to support the tripod legs and work at a given station completed as quickly as possible. Stepping near a tripod leg, or touching one while looking through the telescope, will demonstrate the effect of settlement on the position of the bubble and cross hairs.

11-18.3. PERSONAL ERRORS

1. *Instrument not set up exactly over the point.* The plumb bob or optical plummet position should be checked at intervals during the time a station is occupied, to be certain it remains centered and the instrument is over the point.

2. *Level bubbles not centered perfectly.* The bubbles must be checked frequently, but NEVER releveled between a backsight and foresight—only *before* starting, and *after* finishing, an angle measurement. Note that in leveling, or in measuring vertical angles with a transit, the level vials *parallel* to the telescope are the critical ones. For horizontal angles, the telescope can be elevated or depressed in the vertical plane without affecting measurements if the standards are properly adjusted. Hence the bubble *at right angles* to the telescope is the important one.

3. *Vernier misinterpolated.* Using a magnifying glass and exercising caution will help reduce the size of these errors. Also, the number of minutes on the scale passed over by a vernier index should be estimated to check a reading.

4. *Improper use of clamps and tangent screws.* An observer must form good operational habits and be able to identify the various clamps and tangent screws by their touch without looking at them. Final setting of tangent screws is always made with a positive motion to avoid backlash. Clamps should be tightened just once and not checked again to be certain they are secure.

5. *Poor focusing.* Correct focusing of the eyepiece on the cross hairs, and of the objective lens on the target, is necessary to prevent parallax. Objects sighted should be placed as near the center of the field of view as possible. Focusing affects pointing, which is an important source of error.

6. *Overly careful sights.* Checking and double checking the position of the cross-hair setting on a target wastes time and actually produces poorer results than one fast observation. The cross hair should be aligned quickly and the next operation begun promptly.

7. *Careless plumbing and placement of the rod.* One of the most common errors results from careless plumbing of a rod when only the top can be seen

by the instrument operator because of brush or other obstacles in the way. Another is due to planting a pole off-line behind a point to be sighted.

11-19. MISTAKES. Some common mistakes to guard against are:

1. Sighting on, or setting up over, the wrong point.
2. Calling out or recording an incorrect value.
3. Reading the wrong circle.
4. Turning the wrong tangent screw.
5. Using haphazard field procedures.

PROBLEMS

11-1. Determine the angles subtended for the following conditions.
(a) Diameter of a 2-in pipe sighted by theodolite from 500 ft away
(b) Width of a 1-in stake sighted by transit from $\frac{1}{2}$ mi away
(c) A 6-mm diameter chaining pin seen by theodolite from 50 m away

11-2. What is the error in a measured direction for the situations noted?
(a) Setting a theodolite $\frac{1}{4}$ in to the side of a tack on a 200-ft sight
(b) Lining in the edge (instead of center) of an 8-mm-diameter pencil at 75 m
(c) Sighting the edge (instead of center) of a $1\frac{1}{4}$-in-diameter range pole 100 ft away
(d) Sighting a range pole that is 2 in off-line on an 800-ft sight

11-3. Intervening terrain obstructs the line of sight so only the top of an 8-ft-long range pole can be seen on a 600-ft sight. If the range pole is out of plumb and leaning sidewise $\frac{1}{2}$ in per vertical foot, what maximum angular error results?

11-4. What are the primary sources of error in measuring angles with a transit or theodolite, assuming it is properly leveled and centered?

11-5. Describe a method of reading angles with a transit or theodolite that will eliminate most instrumental errors caused by improper construction and poor adjustment.

11-6. What are the advantages of measuring angles by repetition?

11-7. An angle is to be measured precisely with a transit. (a) What error is eliminated when both A and B verniers are read? (b) How is the error in graduations of the circle minimized? (c) What errors are eliminated by measuring an angle the same number of times with the telescope normal and plunged?

11-8. What are the advantages of closing the horizon around a point where only one angle is actually required?

11-9. In measuring an angle by repetition, the reading after the first turning in a direct position was 260°40.3′. The reading after the fourth turning in the reversed position was 322°42.0′. Determine the angle.

11-10. Similar to Problem 11-9, except the first and fourth readings were 49°36.3′ and 198°26.0′, respectively.

11-11. An interior angle x and its explement y were turned to close the horizon. Each angle was measured once direct and once reversed using the repetition method. Starting with an initial backsight setting of 0°00′ for each angle, the readings after the first and second turnings of angle x were 49°36.4′ and 99°13.0′, and the readings after the first and second turnings of angle y 310°23.2′ and 260°46.0′. Calculate each angle and the horizon misclosure.

11-12. Similar to Problem 11-11, except that the readings for angle x were 136°53.3′ and 273°46.5′ and those for angle y 223°07.0′ and 86°13.8′.

In Figure 11-3, observed directions with a theodolite, normal and plunged, from *A* to points *B*, *C*, and *D* are listed in Problems 11-13 and 11-14. Find the values of the three angles.

11-13. Normal: 25°28′20″, 90°55′42″, 253°06′24″.
Plunged: 205°28′16″, 270°55′44″, and 73°06′22″.

11-14. Normal: 106°52′06″, 186°33′38″, 288°43′28″.
Plunged: 286°52′05″, 06°33′41″, 108°43′29″.

11-15. The angles at point X were measured with a 10-sec theodolite, and based on 12 readings, the standard error of the angle was ± 4.0 sec. If the same procedure is used in measuring the angles of a triangle, what is the standard error of the triangle closure?

11-16. Similar to Problem 11-15, except that eight readings were taken with a standard error of ± 6.0 sec.

11-17. An angle $ABC = 40°16'10''$ must be laid off with a 30-sec transit. After a 400-ft backsight on point A, point C is marked 500 ft away with an angle of 40°16′ set on the plates. Angle ABC, measured six times by repetition, gives a reading of 241°36′36″. What offset at C will give the required angle?

11-18. Similar to Problem 11-17, except that an angle of 27°43′53″ is required to a point 800 ft away and the repeated angle is 166°23′48″.

11-19. Approximately how close to the true value would a horizontal angle be when read by repetition eight times with a 30-sec transit?

11-20. Why is an assumed reference line not desirable for azimuths?

11-21. The line of sight of a transit is out of adjustment by 12 sec. (a) In prolonging a line by plunging the telescope between the backsight and foresight, but not double centering, what angular error is introduced? (b) What linear error results on a foresight of 1000 ft?

11-22. A line PQ is prolonged to point R by double centering. Two foresight points R' and R'' are set. What is the angular error introduced in a single plunging based on the following lengths of QR and $R'R''$, respectively?
(a) 638.95 ft, 0.37 ft
(b) 321.06 m, 4.25 cm
(c) 895.00 ft, $6\frac{3}{8}$ in

11-23. What is the index error of a transit? How is its value obtained and the effect eliminated?

11-24. With the transit telescope normal, a vertical angle to point A is $+5°32'$, and with the telescope plunged it is $+5°36'$. What is the index error? Compute the correct vertical angle to point B if the reading to it is $-6°29'$ with the telescope normal?

11-25. Similar to Problem 11-24, except for readings of $-9°16'$ and $-9°10'$ on point A, and $+16°17'$ on point B.

11-26. Explain the difference between eccentricity and improper graduation of transit plates in their effect on measured angles.

11-27. If pointings can be made with an accuracy of ± 3 sec, what is the maximum error expected in measuring a horizontal angle by repetition for the following situations?
(a) 2 D and 2 R with transit reading and setting error of ± 20 sec each
(b) 4 D and 4 R with theodolite reading and setting error of ± 0.1 min each
(c) 8 D and 8 R with theodolite reading and setting error of ± 2 sec each

11-28. What is the required number of repetitions to measure a horizontal angle to an accuracy of 2 sec with an instrument having a reading and setting error of ± 6 sec? Assume reasonable pointing errors.

11-29. To reduce the effects of circle graduation errors in precise direction measurements with a directional theodolite, if 12 positions are being measured, what should be the approximate circle readings for initial sightings at each position with the telescope direct?

11-30. What error in horizontal angles is consistent with the following linear precisions? Check by Table E-9 in Appendix E.
(a) Linear precision: $\frac{1}{300}$, $\frac{1}{800}$, $\frac{1}{1500}$, $\frac{1}{3000}$, and $\frac{1}{10{,}000}$
(b) Linear precision: $\frac{1}{250}$, $\frac{1}{500}$, $\frac{1}{2000}$, $\frac{1}{5000}$, and $\frac{1}{15{,}000}$
(c) Linear precision: $\frac{1}{400}$, $\frac{1}{1000}$, $\frac{1}{1800}$, $\frac{1}{4000}$, and $\frac{1}{12{,}000}$

11-31. In Figure 11-14, random line XY' of length 2645.50 ft is run and falling distance $Y'Y$ measured as 19.79 ft. What angle $Y'XY$ is necessary to be laid off from line XY' to establish the correct direction of XY?

11-32. List the reference lines or axes of a transit. What relationships should these axes bear to one another for an instrument in perfect adjustment?

BIBLIOGRAPHY

Irish, S. B. 1959. "The Setup Error in Horizontal Angles." *ASCE Journal of the Surveying and Mapping Division* 85 (no. SU1):1.

Madkour, M. F. 1969. "Rounds Versus Positions." *ASCE Journal of the Surveying and Mapping Division* 95 (no. SU1):151.

12 TRAVERSING

12-1. INTRODUCTION. A traverse is a series of consecutive lines whose lengths and directions have been determined from field measurements. *Traversing*, the act of establishing traverse stations and making the necessary measurements, is one of the most basic and widely practiced means of determining the relative locations of points.

There are two basic types of traverses: *closed* and *open*. In a closed traverse, (1) the lines either return to the starting point, thus forming a closed polygon (geometrically and mathematically closed), as shown in Figure 12-1(a), or (2) they finish upon another station that has a positional accuracy equal to or greater than that of the starting point. The second kind (geometrically open, mathematically closed), illustrated in Figure 12-1(b), must have a closing reference direction—for example, line E-$Az\ Mk_2$. Closed traverses provide checks on the measured angles and distances, an extremely important consideration. They are used extensively in control, construction, property, and topographic surveys.

An open traverse (geometrically and mathematically open) (Figure 12-2), consists of a series of lines that are connected but do not return to the starting point or close upon a point of equal or greater-order accuracy. Open traverses are sometimes used on route surveys, but generally should be avoided because they offer no means of checking for errors and mistakes. In open traverses, measurements *must* be repeated to guard against mistakes.

Hubs (wooden stakes with tacks to mark the points), steel stakes, or pipes are set at each traverse station *A*, *B*, *C*, and so on, in Figures 12-1 and 12-2,

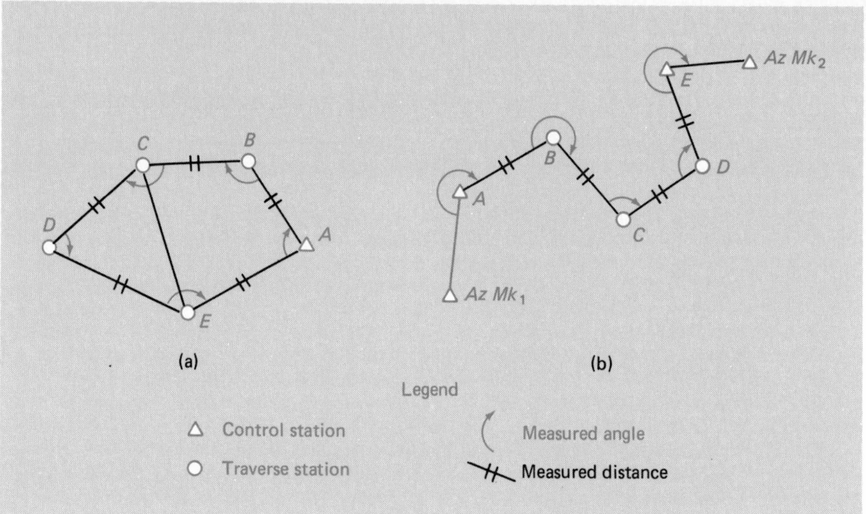

Figure 12-1. Examples of closed traverses.

where a change in direction occurs. Spikes, "P-K" nails, and nails driven through bottle caps are used on blacktop pavement. Chisled or painted marks are made on Portland cement concrete. Traverse stations are sometimes interchangeably called *angle points* because an angle is measured at each one.

12-2. METHODS OF MEASURING TRAVERSE ANGLES OR DIRECTIONS. The methods used in measuring angles or directions of traverse lines vary, and include (a) compass bearings, (b) interior angles, (c) deflection angles, (d) angles to the right, and (e) azimuths.

Figure 12-2. Open traverse.

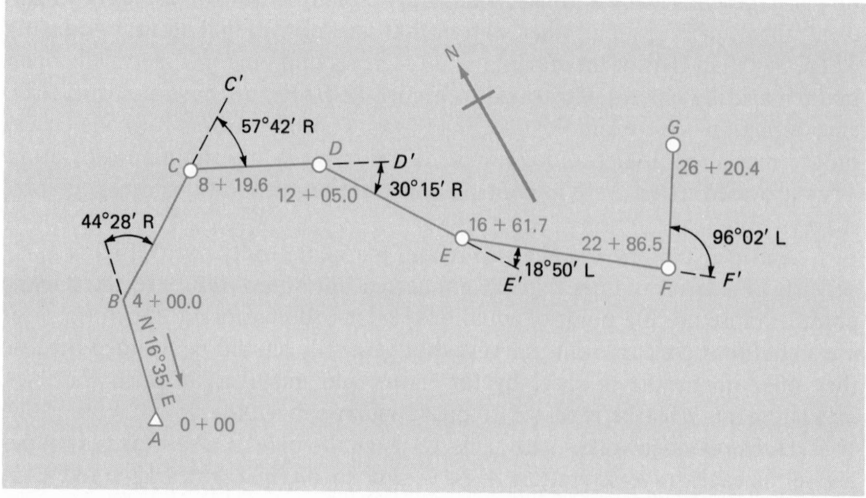

12-2.1. TRAVERSING BY COMPASS BEARINGS. The surveyor's compass was designed for use as a traversing instrument. Bearings are read directly on the compass as sights are taken along the lines (courses) of the traverse. Engineer's transits equipped with compasses can also be employed to read bearings directly. Normally if a bearing traverse is being run with a transit, however, *calculated* bearings based on horizontal circle readings would be used and compass readings utilized only as checks. In this procedure, which is especially appropriate if old surveys are being retraced, the instrument is oriented at each station by backsighting on the previous point with the back bearing set on the plates. The angle from backsight to the next foresight is then computed based on record bearings and applied to the back bearing to cause the telescope to point in the direction of the next course.

12-2.2. TRAVERSING BY INTERIOR ANGLES. Interior angles, such as *ABC*, *BCD*, *CDE*, *DEA*, and *EAB* in Figure 12-1(a), are used almost exclusively on property survey traverses. They may be read either clockwise or counterclockwise, as the survey party progresses around the traverse either to the right or to the left in *ABC* order, as shown. It is good practice, however, to measure all angles clockwise. Consistently following one method reduces mistakes in reading, recording, and plotting. The exterior angles may also be measured to close the horizon (see Section 11-5).

12-2.3. TRAVERSING BY DEFLECTION ANGLES. Route surveys are commonly run by deflection angles measured to the right or left from the lines extended, as indicated in Figure 12-2. A deflection angle is not complete without a designation R or L and, of course, it cannot exceed 180°. Each angle should be doubled or quadrupled to reduce instrument errors, and an average value determined.

12-2.4. TRAVERSING BY ANGLES TO THE RIGHT. Angles measured clockwise from a backsight on the previous line [see Figure 12-1(b)] are called angles to the right, or azimuths from the back line. The procedure used is similar to running an azimuth traverse except that the backsight is taken with the plates set to zero instead of to the back azimuth. The angles can be checked (and improved) by doubling, or roughly tested by means of compass readings. Always turning angles in the clockwise direction eliminates mix-ups in recording and plotting and is suited to the arrangement of circle graduations on all transits and theodolites, including directional instruments.

As noted in Chapter 8, deflection angles can be obtained by subtracting 180° from angles to the right.

12-2.5. TRAVERSING BY AZIMUTHS. Topographic surveys are often run by azimuths, a process that permits reading azimuths of all lines directly, thus eliminating the need to calculate them. In Figure 12-3, azimuths are measured clockwise from the north end of the meridian through the angle points. The transit is oriented at each setup by sighting on the previous station with either the back azimuth on the circle (if angles to the right are turned) or the azimuth of the line on the plates (if deflection angles are turned) as described in Section 11-7.

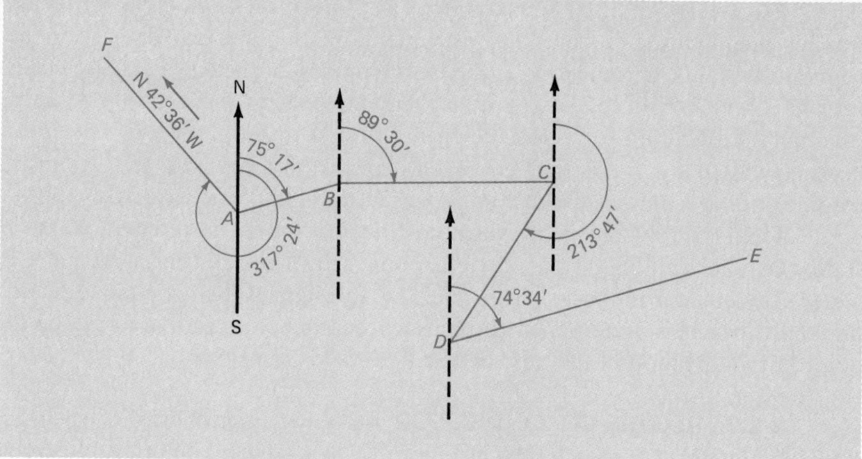

Figure 12-3. Azimuth traverse.

12-3. MEASUREMENT OF LENGTHS. The length of each traverse line is usually obtained by the simplest and most economical method capable of satisfying the required precision on a given project. Electronic devices and taping are used most often and provide the highest order of accuracy. When EDM is employed, the procedure is termed *electronic traversing*. Distances measured in both directions by stadia give control suitable for various types of work, such as low-precision topographic mapping. For some geological and agricultural work, pacing may be accurate enough.

The precision specified for a traverse to locate boundaries is based on land values and survey costs. On construction work, allowable limits of closure depend on the use and extent of the traverse and type of project. Bridge location, for example, demands a high degree of precision.

In closed traverses, each line is measured and recorded as a separate distance. On long, open traverses for highways and railroads, distances are carried along continuously from the starting point. In Figure 12-2, for example, beginning with station 0 + 00 at point *A*, 100-ft stations (1 + 00, 2 + 00, 3 + 00, and 4 + 00) are marked until hub *B* at station 4 + 00.0 is reached. Then stations 5 + 00, 6 + 00, 7 + 00, 8 + 00, and 8 + 19.6 are set along course *BC* to *C*, and so on. The length of an open-traverse line is the difference between stations of its ends; thus, the length of line *BC* is 819.6 − 400.0 = 419.6.

12-4. SELECTION OF TRAVERSE STATIONS. Positions selected to set traverse stations vary with the type of survey. On property surveys, they are placed at each corner if the actual boundary lines are not obstructed and can be occupied. If offset lines are necessary, a stake is located near each corner to simplify the measurements and computations. Long lines and rolling terrain may necessitate extra stations.

On route surveys, stations are set at each angle point and other locations where necessary to obtain topographic data or extend the survey. Usually the center line is run before construction begins, and again after it is completed.

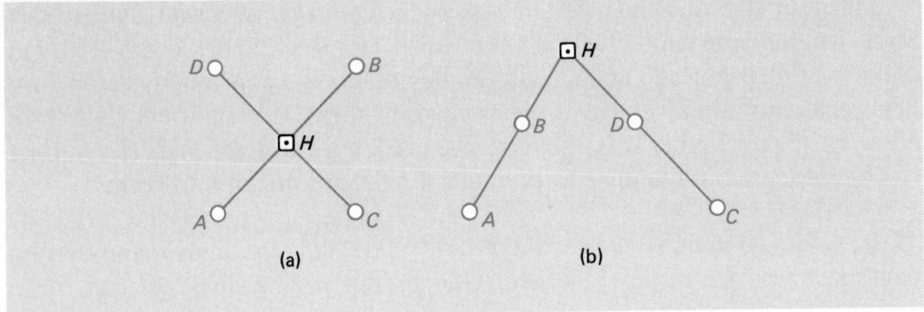

Figure 12-4. Hubs for ties.

An offset traverse is necessary during the earth-moving and roadway-surfacing stages on a highway job.

A traverse run for control of topographic mapping serves as a skeleton to which are referenced such details as roads, buildings, streams, and hills. Location of stations must be selected to permit complete coverage of the area to be mapped. *Spurs* consisting of one or more lines may branch off as open (*stub*) traverses to reach vantage points. Their use should be discouraged, however, because a check on their positions cannot be made.

Traverse stations, like bench marks, may be lost if not properly described and preserved. *Ties* are used to aid in finding a survey point, or to relocate one if it is destroyed. Figure 12-4(a) shows an arrangement of *straddle hubs* well suited to tying-in a point on a highway or elsewhere. The traverse hub *H* may be found by intersecting strings stretched between diagonally opposite ties if the lengths are not too long. Hubs in the position illustrated by Figure 12-4(b) are sometimes used but are not as desirable for stringing. In either configuration, the intersection of the lines of sight of two theodolites set up at *A* and *C*, and simultaneously aimed at *B* and *D*, respectively, will recover the point.

Figure 12-5. Referencing a point.

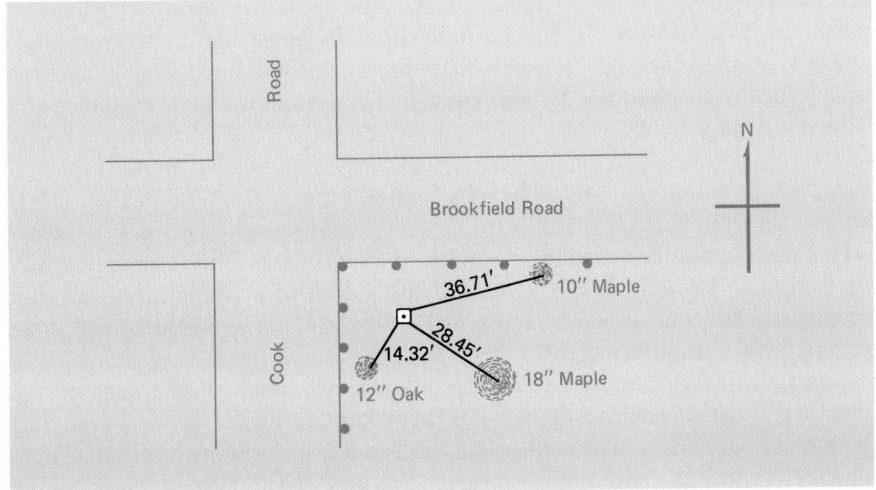

Figure 12-5 and Plate D-2 in Appendix D illustrate typical traverse ties. Short lengths (less than 100 ft) are convenient if a steel tape is being used, but of course the distance to definite and unique points is a controlling factor. Two ties, preferably about at right angles to each other, are sufficient, but three allow for the possibility that one reference mark may be destroyed. Ties to trees can be measured in hundredths of a foot if nails are driven into them.

12-5. ORGANIZATION OF FIELD PARTY. The type of survey and terrain determine the size of party needed. One person pacing distances can run a compass traverse alone; an instrument operator/notekeeper with one rodperson can lay out a stadia traverse; three people—an instrument operator and two tapepersons who also serve as rodpersons—are enough for a transit-tape survey; and a party of two is sufficient for electronic traversing. In most cases, additional personnel for notekeeping and rodding will speed up the work, but increased production must be balanced against the greater cost of operating the crew. A head tapeperson with energy and drive is more valuable than a fast instrument operator in keeping an engineering survey party moving.

On some surveys it is desirable to have a party chief who is free to move around and collect information on lines, stations, reference marks, property owners' names, and other items. The instrument operator or the head tapeperson may serve as party chief, but the chief's range of movement is then limited.

In bush or wooded country, one or two persons with axes and/or brush clippers may be needed to open lines.

12-6. TRAVERSE NOTES. The importance of notekeeping was discussed in Chapter 3. Since a traverse is the end itself on a property survey and the basis for all other data in mapping, a single mistake or omission in recording is one too many. All possible field and office checks must therefore be made. Examples of field notes for interior-angle and azimuth traverses are shown in Plate D-8 and Figure 11-8, respectively.

12-7. ANGLE MISCLOSURE. The misclosure in angle for an interior-angle traverse is the difference between the sum of the measured angles and the geometrically correct total for the polygon. The sum of the interior angles of a closed polygon, $\sum$, is

$$\sum = (n - 2)180^\circ \qquad (12\text{-}1)$$

where n is the number of sides, or angles, in the polygon. This formula is easily derived from known facts. The sum of the angles in a triangle is 180°; in a rectangle, 360°; and in a pentagon, 540°. Thus each side added to the three required for a triangle increases the sum of the angles by 180°.

Figure 12-1(a) shows a five-sided figure in which, if the sum of the measured interior angles equals 540°02′, the angular misclosure is 2 min. The permissible misclosure is based on the occurrence of random errors that may increase or decrease the sum of measured angles. It can be computed by the formula

$$c = K\sqrt{n} \tag{12-2}$$

where n is the number of angles and K a fraction of the (a) least count of a transit vernier or (b) smallest graduation of a theodolite scale, in minutes or seconds. The fraction depends on the number of repetitions used and the angular accuracy desired.

For ordinary transit work, a reasonable value of K is $\frac{1}{2}$ to 1 min, and the permissible misclosure for a pentagon 1 to 2 min.

The algebraic sum of the deflection angles in a closed-polygon traverse equals 360°, clockwise (right) deflections being considered plus and counterclockwise (left) deflections minus. This rule applies if lines do not crisscross, or if they cross an even number of times. When lines in a traverse cross an odd number of times, the sum of right deflections equals the sum of left deflections.

If a transit is used, reading magnetic bearings also aids in locating an angle turned left but mistakenly recorded to the right. Checks are available on bearings computed from the deflection angles of an open traverse. The true direction or bearing of the first line may be determined from two intervisible stations with a known azimuth between them, or from a sun or Polaris observation, as described in Chapter 19. Measured angles are applied to calculate the bearings of all lines including the last, and then comparing it with its known value or the result obtained from another sun or Polaris shot. On long traverses, intermediate lines can be similarly checked. It is important to note the magnetic bearings for rough verification. Although they cannot be read closer than perhaps 15 min, a serious mistake in an angle will be disclosed. For a closed polygon traverse, the bearing of the first line should always be recomputed, using the last angle as a check after progressing around the figure.

A closed polygon azimuth traverse is checked by setting up on the starting point a second time after occupying the successive stations around the traverse and orienting by back azimuths. The azimuth of the first side should be the same as the original value. Any difference is the misclosure. If the first point is not reoccupied, the interior angles computed from the azimuths will automatically check the proper geometric total even though one or more of the azimuths is incorrect.

A *cutoff line*, such as CE in Figure 12-1(a), run between two stations on a traverse, produces smaller closed figures to aid in checking and isolating blunders. The additional measurements increase redundancy and hence precision.

One way to check an open traverse is to run a separate series of lines with the same or a lesser degree of precision to close the traverse. Long sights and stadia distances may be used, for example, to get a rough check.

Another check for an open traverse is to obtain coordinates of the starting and closing points by tying-in to marks of known position, thereby making it a closed traverse, and comparing the computed difference in coordinates with the actual values. State plane coordinate systems have been devised by the National Geodetic Survey for every state (see Chapter 21), and permanent monuments have been set for use by all surveyors. Computation of coordinates for traverse courses is discussed in Chapter 13.

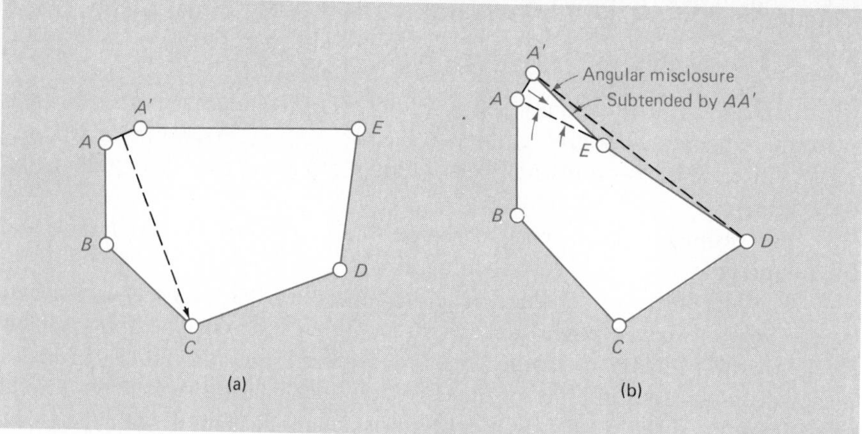

Figure 12-6. Locating angular error.

A numerical or graphical analysis can be employed to determined the location of a mistake and save considerable field time.[1] For example, if the sum of the interior angles of a five-sided traverse gives a bad misclosure—say, 10°03′—it is likely that one mistake of 10° and several small errors of 1 min have been made. A method of graphically locating the station at which the mistake occurred, so only one point need be reoccupied, will be illustrated. The procedure shown for a five-sided traverse can be used for traverses having any number of sides.

In Figure 12-6(a), the traverse has been plotted by using the measured lengths and angles and shows a linear misclosure error AA'. The perpendicular bisector of line AA' points to the angle in error—in this case, C. A correction applied to this angle will swing the traverse through an arc to eliminate the linear error AA'.

If, as in Figure 12-6(b), a second angle point lies near the perpendicular bisector of AA', then the station at which the error has been made is the one for which the angular misclosure, when plotted at that station, is subtended by AA'. Stated differently, if the misclosure angle equals AEA', E is the error station; if it equals ADA', D is the error station.

For a mistake made in one traverse line length, suspect the side parallel with the direction of the closing line.

12-8. SOURCES OF ERROR. Some sources of error in running a traverse are:

1. Errors in measurement of angles and distances.
2. Poor selection of stations, resulting in bad sighting conditions due to (a) alternate sun and shadow, (b) visibility of only the top of the rod, (c) line of sight passing too close to the ground, (d) lines that are too long or too short, and (e) sighting into the sun.
3. Failing to measure angles an equal number of times direct and reversed.

[1] See Dana E. Low, 1954, "Finding Angle-reading Errors in Long Traverses," *Civil Engineering* 24:738.

12-9. MISTAKES. Some mistakes in traversing are:

1. Occupying or sighting on the wrong station.
2. Incorrect orientation.
3. Confusing angles to the right and left.
4. Not taking extra precautions in measuring an angle that has one short side (or two).

PROBLEMS

12-1. Explain the difference between closed and open traverses. Comment on the advisability of using open traverses.

12-2. Prepare a set of typical field notes for the closed interior angle traverse of Figure 12-1(a). (Refer to the example shown in Plate D-8.) Assume the bearing of AB is N37°40′W. Scale the lengths of lines, and measure the angles with a protractor.

12-3. Similar to Problem 12-2, except for angles to the right in the traverse of Figure 12-1(b).

12-4. Similar to Problem 12-2, except for deflection angles in the traverse of Figure 12-1(b).

12-5. Prepare a set of typical field notes for the closed traverse of Figure 8-2(a). Scale the lengths of lines.

12-6. Compute and tabulate bearings for the traverse of Figure 12-2.

12-7. What should be the sum of the interior angles for a closed-polygon traverse that has (a) 7 sides? (b) 10 sides? (c) 13 sides?

12-8. Five interior angles of a six-sided closed-polygon traverse were measured as: $A = 95°10'$, $B = 137°46'$, $C = 69°32'$, $D = 189°15'$, and $E = 120°17'$. The angle at F was not measured. If all measured angles are assumed to be correct, what is the value of angle F?

12-9. What is the correct algebraic sum of the deflection angles 8R and 5L in a closed-polygon traverse, assuming (a) no lines crisscross? (b) lines crisscross once?

12-10. Four deflection angles of a closed five-sided traverse, none of whose lines crisscross, were measured as: $A = 135°28'$R, $B = 82°14'$R, $C = 91°07'$R, and $D = 129°56'$R. The angle at E was not measured. Assuming all measured angles to be correct, what is the value of angle E?

12-11. How can an angular check be obtained on a closed-azimuth traverse? On a closed-azimuth traverse with a cutoff line?

12-12. Why is the method of placing reference hubs shown in Figure 12-4(b) generally not as good as that of Figure 12-4(a)?

12-13. List four pertinent considerations in selecting locations for (a) traverse stations and (b) ties.

12-14. What are the two primary reasons for referencing traverse stations?

12-15. If the standard error for each measurement of a traverse angle is ± 15 sec, what is the expected standard error of the misclosure in the sum of the angles for a six-sided traverse?

12-16. If the angles of a traverse are turned so the 90% error of any angle is $\pm$ 1 min, prove that the 90% error of closure for a 12-sided traverse is equal to $\pm 1'\sqrt{12}$.

12-17. In running an azimuth traverse, if the plate bubbles on a transit run off-center between the backsight and foresight readings, should they be recentered before taking the foresight reading? Explain.

12-18. The true azimuth from north of line AB, obtained from coordinates of monuments at A and B, is 111°28′. After a transit at station A has been oriented on this azimuth, point C is sighted and its azimuth found to be 159°12′. An open deflection angle traverse $ABCDEFGHI$ is then run, and the following angles read: $C = 4°27'$L; $D = 10°48'$L; $E = 7°35'$R; $F = 3°18'$R; $G = 6°52'$R; and H (to I) $= 5°06'$L. The true bearing of *IH* determined by a Polaris observation is N23°20′W. Are the traverse angles acceptable for ordinary work? Explain.

12-19. If station *A* in Problem 12-18 is at station 10 + 00 and hub *I* at station 73 + 50, what precision of linear measurements will be consistent with that of the angles?

The recorded bearings and lengths for a traverse are listed in Problems 12-20 through 12-23. If the lengths are assumed to be correct, which bearing is most likely to be wrong?

12-20. *AB* = N57°11′E, 340 ft; *BC* = S18°55′E, 262 ft; *CD* = S30°05′W, 414 ft; and *DA* = N15°49′W, 378 ft.

12-21. *AB* = N2°58′E, 205 ft; *BC* = N42°58′E, 150 ft; *CD* = S42°02′E, 200 ft; and *DA* = S59°20′W, 297 ft.

12-22. *AB* = N18°03′W, 996 ft; *BC* = S74°27′W, 598 ft; *CD* = S16°43′E, 548 ft; *DE* = S74°47′W, 335 ft; *EF* = S17°13′E, 564 ft; and *FA* = N57°32′E, 956 ft.

12-23. *AB* = N64°30′E, 437 ft; *BC* = S23°30′E, 236 ft; *CD* = S32°30′W, 244 ft; *DE* = N75°30′W, 324 ft, and *EA* = N23°30′W, 163 ft.

BIBLIOGRAPHY

Mattson, D. F. 1973. “Angular Reference Ties.” *Surveying and Mapping* 33(no. 1):33.

O’Neill, J. B. 1952. “Winter Control Traverses in Canada.” *Surveying and Mapping* 12(no. 2):111.

Phillips, J. O. 1967. “Electronic Traverse Versus Triangulation.” *ASCE Journal of the Surveying and Mapping Division* 93(no. SU2):29.

13 TRAVERSE COMPUTATIONS

13-1. INTRODUCTION. Measured angles or directions of a closed traverse are readily investigated before leaving the field. Linear measurements, if taped, even though repeated, are a more likely source of error and must also be checked. Although the calculations are more lengthy than angle checks, with today's pocket calculators they can also be done in the field to determine before leaving whether the traverse meets the required precision. If specifications have been satisfied, the traverse is then adjusted to create perfect "closure" or geometric consistency among angles and lengths; if not, field measurements must be repeated until adequate results are obtained.

Investigation of precision, and acceptance or rejection of the field data, are extremely important in surveying. Adjustment for geometric closure is also crucial; in land surveying, for example, the law generally requires property descriptions to have exact geometric agreement.

The usual procedures followed in traverse computations, all of which are discussed in this chapter, are (1) adjusting angles or directions to fixed geometric conditions, (2) determining bearings or azimuths, (3) calculating latitudes and departures and adjusting them for misclosure, and (4) computing rectangular coordinates of traverse stations.

13-2. BALANCING ANGLES. The first step in traverse calculations is to balance (adjust) the angles to the proper geometric total. This is readily done, since the total error is known (see Section 12-7), although its exact distribution is not.

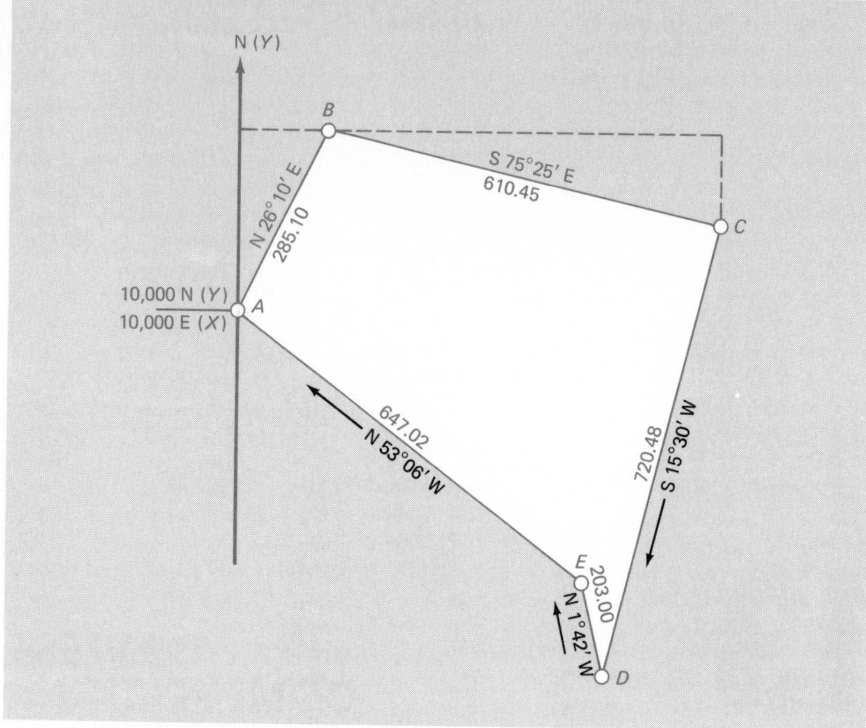

Figure 13-1. Traverse.

Angles of a closed traverse can be simply adjusted to the correct geometric total by applying one of three methods:

1. Arbitrary corrections to one or more angles.
2. Larger corrections to angles where poor observing conditions were present.
3. An average correction found by dividing total angle misclosure by the number of angles.

EXAMPLE 13-1

For the traverse of Figure 13-1, the measured interior angles are given in Table 13-1. Compute the adjusted angles using methods 2 and 3.

SOLUTION

The computations are best arranged as shown in Table 13-1.

The first part of the adjustment consists of summing the interior angles and determining the misclosure according to Eq. (12-1), which in this instance is +1′30″. The remaining calculations are tabulated, and the rationale for the procedures follows.

For work of ordinary precision, it is reasonable to adopt corrections that are even multiples of (a) the least count of the transit vernier, (b) the smallest recorded digit or decimal place for theodolite readings, or (c) the smallest significant digit or decimal place when measuring angles by repetition.

TABLE 13-1. ADJUSTMENT OF ANGLES

	METHOD 2			METHOD 3			
POINT	MEASURED ANGLE	ADJUSTMENT	ADJUSTED ANGLE	MULTIPLES OF AV. CORR.	CORR. ROUNDED TO 30″	SUCCESSIVE DIFF.	ADJUSTED ANGLE
A	100°44′30″	0″	100°44′30″	18″	30″	30″	100°44′
B	101°35′	30″	101°34′30″	36″	30″	0	101°35′
C	89°05′30″	0″	89°05′30″	54″	60″	30″	89°05′
D	17°12′	30″	17°11′30″	72″	60″	0	17°12′
E	231°24′30″	30″	231°24′00″	90″	90″	30″	231°24′
	540°01′30″	90″	540°00′00″			90″	540°00′

In method 2, corrections should be made to angles expected to contain the largest errors. In this case, 30 sec is subtracted from angles *B*, *D*, and *E*, since they have the shortest sights. The adjustment in this manner is shown on the left side of Table 13-1. If only the top of the range pole was visible at *C* from the setup at *B*, the entire correction of 1′30″ could have been subtracted from the angle at *B*.

Method 3 consists of subtracting 1′30″/5 = 18″ from each of the five angles. Since the angles were read in multiples of $\frac{1}{2}$ min, applying corrections of 18 sec gives a false impression of their precision. Therefore it is desirable to establish a pattern of corrections, as shown on the right side of Table 13-1. First a column consisting of multiples of the average correction of 18 sec is tabulated beside the angles. In the next column, each of these multiples is rounded off to the nearest 30 sec. Successive differences (adjustments for each angle) are found by subtracting the preceding value in the rounded-off column of corrections from the one being considered. The adjusted angles obtained by using these adjustments must total exactly the true geometric value. The adjustments fall into a pattern and thus distort the shape of the traverse less than when all of the misclosure is put into one angle.

It should be noted that although the adjusted angles satisfy the geometric condition of a closed figure, they may be no nearer the true values than before adjustment. Unlike corrections for linear measurements, *the adjustments applied to angles are independent of the size of an angle.*

13-3. COMPUTATION OF BEARINGS OR AZIMUTHS. Computation of bearings and azimuths was discussed in Chapter 8. Angles adjusted to the proper geometric total must be used; otherwise the bearing or azimuth of the first line will differ from its computed value by the angular misclosure (found by applying the successive angles around a closed polygon traverse).

13-4. LATITUDES AND DEPARTURES. Closure of a traverse is checked by computing the latitude and departure of each line (course). *The latitude of a course is its orthographic projection on the north-south axis of the survey*, and is equal to the course length multiplied by the cosine of its bearing or azimuth. Latitude is also called *latitude difference*, and *northing* or *southing*.

The departure of a course is its orthographic projection on the east-west axis of the survey, and is equal to the length of the course multiplied by the sine of its bearing or azimuth. Departures are sometimes called *eastings* or *westings*.

In equation form, the latitude and departure of a line are

$$\text{latitude} = L \cos \alpha \tag{13-1}$$

$$\text{departure} = L \sin \alpha \tag{13-2}$$

where L is the length and α the bearing or azimuth of the course.

Departures and latitudes are merely the X and Y components of a line in a rectangular grid system, sometimes referred to as ΔX and ΔY. In traverse calculations, north latitudes and east departures are considered plus; south latitudes and west departures minus. For bearings, angles are always between 0 and 90°; hence their cosines and sines are invariably positive. Proper algebraic signs of latitudes and departures are therefore assigned on the basis of the bearing angle directions, so a NE bearing line has a plus latitude and departure, a SW bearing course gets a minus latitude and departure, and so on.

Azimuths used in computing latitudes and departures range from 0 to 360°, and the algebraic signs of cosines and sines automatically produce the proper algebraic signs of the latitudes and departures. Thus a line with an azimuth of 137°30′ has a negative latitude and positive departure (its cosine is negative and its sine positive); a course of 323°18′ azimuth has a positive latitude and negative departure. Electronic computers and hand calculators with trigonometric functions automatically affix correct algebraic signs to latitudes and departures through those of their cosines and sines, and make azimuths convenient for traverse computations.

13-5. LATITUDE AND DEPARTURE CLOSURE CONDITIONS. In addition to the angular condition of traverse closure discussed in Section 13-2, two other conditions may be enforced in traverse calculations. They are (1) the algebraic sum of all latitudes of closed polygons [see Figure 12-1(a)] must equal zero [or the difference in latitude between the initial and final control points for closed traverses shown in Figure 12-1(b)], and (2) the same applies to departures.

13-6. COMPUTATION OF LATITUDES AND DEPARTURES. Calculation of latitudes and departures and linear misclosure and precision for a closed-polygon traverse will be described by an example. The interior angles of Example 13-1 have been used to compute the bearings shown in Figure 13-1. These are based on a known bearing of N26°10′E for line *AB* and computed by procedures described in Chapter 8. Note on the figure that all bearings (which occupy more space) are lettered outside the traverse, and each length on the interior. This arrangement can be reversed, but a consistent policy should be followed. *Arrows show the correct bearing directions for those courses where the bearings read from left to right but actually run from right to left.*

EXAMPLE 13-2

Based on the bearings and lengths shown in Figure 13-1, calculate the latitudes and departures, linear misclosure, and precision of the traverse.

SOLUTION

Latitudes and departures are computed with the data and results usually inserted in a standard prepared form, arranged as shown in Table 13-2. The forms are printed with column headings and rulings to save time and simplify checking. Sine and cosine columns are normally omitted because it is a waste of time to tabulate them when computing with today's electronic calculators having trigonometric functions.

Latitudes and departures can be computed in a different type of prepared form similar to Table 13-3. This is a combined table, which also includes adjustment of the latitudes and departures. It is often preferred, especially when using electronic calculators.

Summing north (+) and south (−) latitudes gives the misclosure in latitude, −0.70 ft, and summing east (+) and west (−) departures gives the misclosure in departure, +0.53 ft. *Linear misclosure* is the hypotenuse of a small triangle with sides of 0.70 ft and 0.53 ft, and represents the distance from the starting point *A* to the computed point *A′* on the basis of the lengths and adjusted bearings used. In this example the linear misclosure is 0.88 ft (see Table 13-3). A misclosure should not be called an error, since the actual error is never known.

The *precision* for a traverse is expressed by a fraction that has the linear misclosure as a numerator and the traverse perimeter length as the denominator, reduced to reciprocal form. In the example (see Table 13-3) this is 0.88/2466 = 1/2800. The denominator is not carried beyond multiples of 100, or possibly 10, based on the number of significant figures in the values. Note that random and perhaps systematic errors in both angles and distances affect the *computed* precision, and since an angular adjustment was made initially, the actual precision is not obtained.

Various projects require different accuracies, and some surveys must rigidly meet specifications if the work is to be accepted and paid for. Precision requirements for control survey traverses are listed in Chapter 20. The required

TABLE 13-2. COMPUTATION OF LATITUDES AND DEPARTURES

STATION	BEARING	LENGTH	COSINE	SINE	LATITUDE	DEPARTURE
A						
	N26°10′E	285.10	0.897515	0.440984	+255.88	+125.72
B						
	S75°25′E	610.45	0.251788	0.967782	−153.70	+590.78
C						
	S15°30′W	720.48	0.963630	0.267238	−694.28	−192.54
D						
	N1°42′W	203.00	0.999560	0.029666	+202.91	−6.02
E						
	N53°06′W	647.02	0.600420	0.799685	+388.49	−517.41
					Σ = −0.70	Σ = +0.53
A						

TABLE 13-3. BALANCING LATITUDES AND DEPARTURES BY THE (BOWDITCH) COMPASS RULE

			LATITUDE		DEPARTURE		BALANCED		COORDINATES	
STA.	LENGTH	BEARING	NORTH +	SOUTH −	EAST +	WEST −	LATITUDE	DEPARTURE	Y(NORTHING)	X(EASTING)
A									10,000.00	10,000.00
	285.10	N26°10′E	*+0.08* 255.88		*−0.06* 125.72		N255.96	E125.66		
B									10,255.96	10,125.66
	610.45	S75°25′E		*+0.17* 153.70	*−0.13* 590.78		S153.53	E590.65		
C									10,102.43	10,716.31
	720.48	S15°30′W		*+0.21* 694.28		*−0.15* 192.54	S694.07	W192.69		
D									9408.36	10,523.62
	203.00	N1°42′W	*+0.06* 202.91			*−0.05* 6.02	N202.97	W6.07		
E									9611.33	10,517.55
	647.02	N53°06′W	*+0.18* 388.49			*−0.14* 517.41	N388.67	W517.55		
A									10,000.00	10,000.00
	2466.05		+847.28	−847.98	+716.50	−715.97	0.00	0.00		
				+847.28	−715.97					
			Misclosures	−0.70	+0.53					

$$\text{Linear misclosure} = \sqrt{0.70^2 + 0.53^2} = 0.88 \text{ ft}$$

$$\text{Precision} = \frac{0.88}{2466} = \frac{1}{2800}$$

precisions for property surveys may be set by state law (for example, a minimum of $\frac{1}{7500}$ in Minnesota) and by cities and counties. A small city might decree $\frac{1}{5000}$; a large metropolitan area, $\frac{1}{10,000}$.

13-7. METHODS FOR TRAVERSE ADJUSTMENT. For any closed traverse the linear misclosure must be distributed throughout the traverse to close the figure. This is true even though the misclosure is negligible in plotting the traverse at map scale. Five basic methods for traverse adjustment are the (1) arbitrary method, (2) transit rule, (3) compass or Bowditch rule, (4) Crandall method, and (5) least squares method.

13-7.1. ARBITRARY METHOD. The arbitrary method of traverse adjustment does not conform to fixed rules or equations. Rather, the linear misclosure is distributed arbitrarily according to the surveyor's analysis of prevailing field conditions. For example, courses taped over rough terrain necessitating frequent plumbing and breaking tape will likely contain bigger errors than courses on level ground; therefore, they are given larger corrections. The total misclosure is distributed in this discretionary fashion to close the figure mathematically—that is, make the algebraic sum of the latitudes and the algebraic sum of the departures equal zero. This method of traverse adjustment is simple to perform and provides a logical assignment of weights based on the expected accuracy of individual measurements.

13-7.2. TRANSIT RULE. The transit rule theoretically is better for surveys where the angles are measured with greater accuracy than the distances, such as stadia surveys, but it seldom is used in practice because different results are obtained for every possible meridian. Corrections are made by the following rules:

$$\frac{\text{correction in latitude for } AB}{\text{misclosure in latitude}} = \frac{\text{latitude of } AB}{\text{absolute sum of all latitudes}} \tag{13-3}$$

$$\frac{\text{correction in departure for } AB}{\text{misclosure in departure}} = \frac{\text{departure of } AB}{\text{absolute sum of all departures}} \tag{13-4}$$

13-7.3. COMPASS (BOWDITCH) RULE. The compass or Bowditch rule, suitable for surveys where the angles and distances are measured with equal precision, is the rule most commonly used in practice. It is appropriate for a transit-tape survey on which angles are measured to the nearest 1 or $\frac{1}{2}$ min and distances taped to 0.01 ft, and for electronic traverses where distances are measured by EDM instruments and angles by theodolite. Corrections are made by the following rules:

$$\frac{\text{correction in latitude for } AB}{\text{misclosure in latitude}} = \frac{\text{length of } AB}{\text{perimeter of traverse}} \tag{13-5}$$

$$\frac{\text{correction in departure for } AB}{\text{misclosure in departure}} = \frac{\text{length of } AB}{\text{perimeter of traverse}} \tag{13-6}$$

Application of both the transit and compass rules assumes all lengths were measured with equal care and all angles taken with the same precision. Otherwise, suitable weights must be given to individual angles or distances. Small misclosures can be apportioned by inspection.

EXAMPLE 13-3

Balance the latitudes and departures of Example 13-2 using the compass rule.

SOLUTION

In applying corrections by the compass rule, the following modified form of Eq. (13-5) is simpler to use:

$$\frac{\text{correction in latitude}}{\text{(departure) for } AB} = \frac{\text{misclosure in latitude (departure)}}{\text{traverse perimeter}} \times \text{length of } AB$$

The other corrections are likewise found by multiplying a constant—the ratio of misclosure in latitude (departure) to the perimeter—by the successive course lengths.

In Example 13-3 the correction in latitude for AB is

$$\frac{+0.70}{2466} \times 285 = +0.08 \text{ ft}$$

and the correction for the departure of BC is

$$\frac{-0.53}{2466} \times 610 = -0.13 \text{ ft}$$

Each correction is generally lettered in different-colored ink or pencil above the latitude or departure to which it will be applied. In Table 13-3, the corrections are shown in small italic numbers. In this example the adjustments are added algebraically to the north and south latitudes, to bring the totals of the north and south latitudes to the same corrected value. *Note that algebraic signs of corrections are opposite those of their corresponding misclosures.*

The corrections applied to the tabular values should produce a perfect closure. In rounding off, an excess or deficiency of 0.01 ft may result, but this is eliminated by revising one of the corrections.

13-7.4. CRANDALL METHOD. In the Crandall method of traverse adjustment, the angular misclosure is first distributed in equal portions to all the measured angles. The adjusted angles are then held fixed and all remaining corrections placed in the linear measurements through a weighted least squares procedure. The Crandall method is more time-consuming than transit and compass rule procedures but is suitable for adjusting traverses where the linear measurements contain larger random errors than the angular measurements—for example, stadia traverses.

13-7.5. LEAST SQUARES METHOD. The method of least squares, based on the theory of probability, simultaneously adjusts the angular and linear measurements to make the sum of the squares of the residuals a minimum. The method is valid for any type of traverse survey regardless of the relative precision of angle and distance measurements, since each measured quantity can be assigned a relative weight. It provides the best and most rigorous traverse adjustment, but until recently has not been widely used because of the lengthy computations required. Availability of electronic computers has now made these calculations routine and consequently the least squares method has gained popularity.

13-8. RECTANGULAR COORDINATES. Rectangular X and Y coordinates of any point give its position with respect to an arbitrarily selected pair of mutually perpendicular reference axes. The X coordinate is the perpendicular distance, in feet or meters, from the point to the Y axis; the Y coordinate is the perpendicular distance to the X axis. Although the reference axes are discretionary in position, in surveying they are normally oriented so that the Y axis points north-south, with north the positive Y direction. The X axis runs east-west, with positive X being east. Given the rectangular coordinates of a number of points, their relative positions are uniquely defined.

Coordinates are useful in a variety of computations including (1) determining lengths and directions of lines (see Section 13-9); (2) calculating areas of land parcels (see Section 14-6); (3) making certain curve calculations (see Section 25-12); and (4) locating inaccessible points. Coordinates are also advantageous for plotting traverses on base maps (see Section 17-12).

In practice, *state plane coordinate* systems, as described in Chapter 21, are most frequently used as the basis for rectangular coordinates in plane surveys. For calculations, however, any arbitrary system may be used. As an example, coordinates may be arbitrarily assigned to one traverse station. To avoid negative values of X and Y, an origin is assumed south and west of the traverse such that one hub has coordinates $X = 1000$, $Y = 1000$, or any other suitable values. In a closed traverse, assigning $Y = 0$ to the most southerly point and $X = 0$ to the most westerly station saves time in calculations.

Given the Y and X coordinates of any starting point A, the Y coordinate of the next point B is obtained by adding the latitude of line AB to Y_A. Likewise the X coordinate of B is the departure of AB added to X_A. In equation form this is:

$$\begin{aligned} Y_B &= Y_A + \text{latitude } AB \\ X_B &= X_A + \text{departure } AB \end{aligned} \tag{13-7}$$

The process is continued around the traverse, successively adding latitudes and departures until the coordinates of starting point A are recalculated. If these recalculated coordinates agree exactly with the starting ones, a check on the coordinates of all intermediate points is obtained (unless compensating mistakes have been made).

EXAMPLE 13-4

Using the balanced latitudes and departures obtained in Example 13-3 (see Table 13-3) and starting coordinates $X_A = Y_A = 10{,}000.00$, calculate coordinates of the other traverse points.

SOLUTION

The process of successively adding balanced latitudes and departures to obtain coordinates is carried out in the two rightmost columns of Table 13-3. Note that the starting coordinates $X_A = 10{,}000.00$ and $Y_A = 10{,}000.00$ are recomputed at the end to provide a check.

13-9. LENGTHS AND BEARINGS FROM LATITUDES AND DEPARTURES, OR COORDINATES (INVERSION). If the latitude and departure of a line are known, its length and bearing (or azimuth) are readily obtained from the following relationships:

$$\tan \text{bearing (or azimuth)} = \frac{\text{departure}}{\text{latitude}} \tag{13-8}$$

$$\text{length} = \frac{\text{departure}}{\sin \text{bearing (or azimuth)}} \tag{13-9}$$

$$= \frac{\text{latitude}}{\cos \text{bearing (or azimuth)}} \tag{13-10}$$

$$= \sqrt{(\text{departure})^2 + (\text{latitude})^2} \tag{13-11}$$

Equations (13-7) can be rewritten to express latitudes and departures in terms of coordinate differences ΔX and ΔY as follows:

$$\begin{aligned} \text{latitude}_{AB} &= Y_B - Y_A = \Delta Y \\ \text{departure}_{AB} &= X_B - X_A = \Delta X \end{aligned} \tag{13-12}$$

Substituting Eqs. (13-12) into Eqs. (13-8) through (13-11), there results:

$$\tan \text{bearing (or azimuth) } AB = \frac{X_B - X_A}{Y_B - Y_A} = \frac{\Delta X}{\Delta Y} \tag{13-13}$$

$$\text{length } AB = \frac{X_B - X_A (\text{or } \Delta X)}{\sin \text{bearing (or azimuth) } AB} \tag{13-14}$$

$$= \frac{Y_B - Y_A (\text{or } \Delta Y)}{\cos \text{bearing (or azimuth) } AB} \tag{13-15}$$

$$= \sqrt{(X_B - X_A)^2 + (Y_B - Y_A)^2} \tag{13-16a}$$

or

$$= \sqrt{(\Delta X)^2 + (\Delta Y)^2} \tag{13-16b}$$

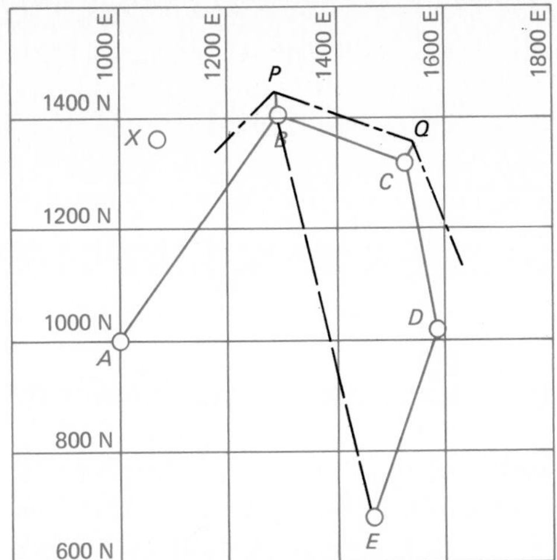

Figure 13-2. Plot of traverse.

The above formulas can be applied to any line whose coordinates are known, whether or not it was actually measured in the survey. Note that X_B and Y_B must be listed first in Eqs. (13-12) and (13-13), so that ΔX and ΔY will have the correct algebraic signs. Computing lengths and directions of lines from latitudes and departures (or coordinates) is called *inversing*.

EXAMPLE 13-5

Data for the traverse from *A* to *E* in Figure 13-2 are given in Table 13-4. Compute the length and bearing of the *cutoff line BE*.

SOLUTION

By Eqs. (13-12):

$$\Delta Y = Y_E - Y_B = 688.3 - 1407.5 = -719.2 \text{ ft}$$

$$\Delta X = X_E - X_B = 1463.9 - 1290.6 = 173.3 \text{ ft}$$

TABLE 13-4. COMPUTATIONS FOR A CLOSING LINE

			LATITUDE		DEPARTURE		COORDINATES	
POINT	LENGTH	BEARING	NORTH	SOUTH	EAST	WEST	Y	X
A							1000.0	1000.0
	500.5	N35°30′E	407.5		290.6			
B							1407.5	1290.6
	251.6	S70°10′E		85.4	236.7			
C							1322.1	1527.3
	310.4	S10°50′E		304.9	58.3			
D							1017.2	1585.6
	350.7	S20°18′W		328.9		121.7		
E							688.3	1463.9

By Eq. (13-13):

$$\tan \text{bearing} = \frac{173.3}{-719.2} \quad \text{and} \quad \text{bearing } BE = S13°33'\text{E}$$

Also by Eqs. (13-15) and (13-16b):

$$\text{length } BE = \frac{-719.2}{\cos 13°33'} = \sqrt{173.3^2 + 719.2^2} = 739.8 \text{ ft}$$

13-10. COORDINATE COMPUTATIONS IN BOUNDARY MEASUREMENTS. Computation of a bearing from the known coordinates of two points on a line is a common problem in boundary measurements. If the lengths and directions of lines from traverse points to the corners of a field are known, the coordinates of the corners can be determined and the lengths and bearings of all sides calculated.

EXAMPLE 13-6

In Figure 13-2, *BC* is a traverse line and *PQ* the property line, which cannot be run directly because of obstructions. The measured lengths and azimuths are: for *BP*, 42.5 ft and 354°50′; for *CQ*, 34.6 ft and 26°40′. Compute the length and bearing of property line *PQ*.

SOLUTION

From the latitudes and departures of these lines, the coordinates of *P* and *Q* are found as follows:

	Y	X		Y	X
B	1407.5	1290.6	C	1322.1	1527.3
BP	+42.3	− 3.8	CQ	+30.9	+15.5
P	1449.8	1286.8	Q	1353.0	1542.8

From the coordinates of *P* and *Q*, the length and bearing of line *PQ* are found in the following manner:

	Y	X
Q	1353.0	1542.8
P	−1449.8	−1286.8
PQ	$\Delta Y = -96.8$	$\Delta X = 256.0$

By Eq. (13-13):

$$\tan \text{bearing } PQ = \frac{256.0}{-96.8} \quad \text{and} \quad PQ = \text{S}69°17'\text{E}$$

By Eq. (13-16b):

$$\text{length } PQ = \sqrt{(96.8)^2 + (256.0)^2} = 273.7 \text{ ft}$$

By continuing this method around the field, coordinates of all corners and lengths and bearings of all the lines can be determined.

13-11. TRAVERSE ORIENTATION BY COORDINATES.

If the coordinates of one traverse station, like A in Figure 13-2, and a visible point X are known, the direction of line AX can be computed and used to orient the transit or theodolite at A. In this way, azimuths and bearings of traverse lines are obtained without the necessity of making astronomical observations. This procedure is followed in various cities which have control monuments and coordinate systems.

Federal, state, and local agencies will ultimately provide closely spaced permanent monuments whose coordinates are based on precise control surveys. Such marks will permit accurate location of the corners of any piece of property, either by coordinates or by lengths and true bearings.

13-12. STATE PLANE COORDINATE SYSTEMS.

Under ordinary circumstances, rectangular coordinate systems for plane surveys would be limited in size due to earth curvature. However, the National Geodetic Survey (NGS) developed statewide coordinate systems for each state in the United States which retain an accuracy of 1 part in 10,000 or better while fitting curved earth distances to plane grid lengths.

State plane coordinates are related to latitude and longitude, so control survey stations set by the NGS and others can be tied to the systems. As additional stations are set and their coordinates determined, they too become usable reference points. Ultimately, local surveys, and accurate restoration of obliterated or destroyed marks having known coordinates, will be simplified.

Some cities and counties have their own coordinate systems for use in locating street, sewer, property, and other lines. Because of their limited extent and the resultant discontinuity at city or county lines, such local systems are less desirable than a statewide grid.

Military grids are used to pinpoint the locations of objects by coordinates for fire control and other purposes.

A more extended discussion of state plane coordinates is given in Chapter 21.

13-13. TRAVERSE COMPUTATION USING ELECTRONIC DEVICES.

Electronic calculators and computers have become commonplace in surveying computations and are especially useful in computing traverses. Pocket calculators with trigonometric functions are now almost indispensable to a modern surveyor. They have the advantage of portability, thus permitting traverse computations to be made and field data verified for closure before returning to the office.

Figure 13-3. Programmable desk-top computer for survey calculations interfaced with automatic plotter. (Courtesy Hewlett-Packard Co.)

Various programmable desk-top calculators of the type shown in Figure 13-3 have been developed specifically for surveying work. Some manufacturers supply traverse computation and other standard programs for use with their machines. These devices are capable of handling large traverses and performing least squares adjustment. Some, such as that of Figure 13-3, are interfaced with an accessory to plot traverses automatically at any selected scale.

Programs to guide the computer can be written in a standard computer language. FORTRAN and BASIC are two common ones used to program surveying and engineering problems. A simple traverse computation program written in the BASIC language is given in Appendix C. It performs the steps of Table 13-3, including computations of latitudes and departures, calculation of linear misclosure and precision, and adjustment by the compass (Bowditch) rule. In addition, the program computes traverse point coordinates and area within polygon traverses using the coordinate method discussed in Section 14-6.

13-14. SOURCES OF ERROR IN TRAVERSE COMPUTATIONS. Some sources of error in traverse computations are:

1. Adjustment of angles, latitudes, and departures that do not duplicate the actual occurrence of errors.

2. Carrying out corrections beyond the number of decimal places in the original measurements.

13-15. MISTAKES. Some more common mistakes in traverse computations are:

1. Failing to adjust the angles before computing bearings.
2. Applying angle adjustments in the wrong direction and failing to check the angle sum for proper geometric total.
3. Interchanging latitudes and departures, or their signs.
4. Confusing the signs of coordinates.

PROBLEMS

13-1. The sum of nine interior angles of a closed-polygon traverse, each read to the nearest minute, is 1259°54′. Balance the angles by methods 1, 2, and 3, and state any assumptions made.

13-2. The misclosure of a 10-sided deflection-angle traverse *ABCDEFGHIJA* with six left deflections and four right deflections is +5 min. Balance the angles by methods 1, 2, and 3, and state any assumptions made.

13-3. Balance the angles in the following azimuth (from north) traverse by any method. Compute and tabulate the bearings, assuming azimuth *AB* is correct: *AB* = 166°26′; *BC* = 28°35′; *CD* = 58°28′; *DE* = 168°44′; *EF* = 89°39′; *FA* = 280°12′; *AB* = 166°32′. Explain the misclosure.

13-4. Balance the following interior angles of a six-sided closed-polygon traverse using method 3. If the bearing of side *AB* is fixed at N87°20′E, calculate the bearings of the remaining sides. *A* = 89°11′; *B* = 91°30′; *C* = 101°24′; *D* = 156°21′; *E* = 196°36′; *F* = 85°04′. (All angles measured to the right.)

13-5. Compute latitudes and departures, linear misclosure, and precision for the traverse of Problem 13-4 if lengths of the sides (in feet) are as follows: *AB* = 1651.51; *BC* = 547.11; *CD* = 415.91; *DE* = 732.78; *EF* = 534.29; and *FA* = 556.07.

***13-6.** Using the compass (Bowditch) rule, adjust the latitudes and departures of the traverse in Problem 13-5. If the coordinates of station *A* are *X* = 10,000 and *Y* = 10,000, calculate coordinates for the other stations and then the lengths and bearings of lines *AD* and *EB*.

13-7. Balance the following interior angles of a closed-polygon traverse to the nearest 0.1 min using method 3. Compute the bearings assuming a correct bearing of S54°24.2′E for line *HI*. *A* = 121°38.3′; *B* = 91°21.6′; *C* = 183°34.7′; *D* = 56°41.3′; *E* = 315°32.4′; *F* = 49°38.9′; *G* = 255°33.3′; *H* = 5°35.2′; *I* = 180°25.2′. (All angles measured clockwise.)

13-8. Determine latitudes and departures, linear misclosure, and precision for the traverse of Problem 13-7 if lengths of the sides (in feet) are as follows: *AB* = 329.65; *BC* = 263.41; *CD* = 301.20; *DE* = 334.94; *EF* = 996.66; *FG* = 962.54; *GH* = 719.29; *HI* = 1200.16; and *IA* = 622.05.

***13-9.** Using the compass (Bowditch) rule, adjust the latitudes and departures of the traverse in Problem 13-8. If the coordinates of station *A* are *X* = 207,556.88 and *Y* = 95,431.53, calculate coordinates for the other stations and, from them, the lengths and bearings of lines *EA* and *BG*.

***13.10.** Compute and tabulate for the following traverse (a) bearings, (b) latitudes and departures, (c) linear misclosure, and (d) precision. For what type of survey is the accuracy satisfactory?

* The asterisks indicate problems whose solutions are used again in problems for Chapter 14.

LINE	INTERIOR ANGLE (RIGHT)	BEARING	LENGTH (ft)
AB	*A* = 128°57′	N32°21′E	569.10
BC	*B* = 138°03′		818.93
CD	*C* = 110°23′		899.67
DE	*D* = 125°56′		1070.79
EF	*E* = 85°56′		1173.90
FA	*F* = 130°45′		637.14

13-11. In Problem 13-10, if one side and/or angle is responsible for most of the error of closure, which is it likely to be?

13-12. Adjust the traverse of Problem 13-10 using the compass (Bowditch) rule. If the coordinates of point *A* are 5000.00N and 5000.00E, determine the coordinates of all other points. Find the length and bearing of line *AE*.

For the traverses given in Problems 13-13 through 13-15, compute and tabulate the (a) unbalanced latitudes and departures, (b) latitudes and departures adjusted by both the compass and transit rules for comparison, (c) linear misclosure, and (d) precision (lengths in feet).

	COURSE	*AB*	*BC*	*CD*	*DA*
***13-13.**	**BEARING**	N9°28′E	N88°13′E	S15°58′W	N67°32′W
	LENGTH	303.05	466.64	489.92	413.25
***13-14.**	**BEARING**	N32°E	S58°E	S77°W	N58°W
	LENGTH	122.00	366.00	172.14	244.00
***13-15.**	**BEARING**	S66°40′E	N27°05′E	N45°15′W	S27°07′W
	LENGTH	289.49	762.44	302.43	873.38

13-16. Which of the traverse adjustment rules, compass or transit, swings the traverse around through a greater angle? Explain.

13-17. After adjustment of latitudes and departures, does a traverse actually close? Explain.

***13-18.** Compute the linear misclosure, precision, and new bearings for the sides after the latitudes and departures are balanced by the compass (Bowditch) rule in the following traverse.

LINE	LENGTH (m)	LATITUDE (m)	DEPARTURE (m)
AB	261.90	N81.58	W248.78
BC	575.92	S538.33	W204.65
CA	643.90	N456.97	E453.63

***13-19.** Balance by the transit rule the latitudes and departures listed in the following traverse. Calculate the linear misclosure, precision, and new bearings.

LINE	*AB*	*BC*	*CD*	*DA*
Latitude	N190.55	S397.35	S152.20	N358.72
Departure	E64.09	W180.47	W199.33	E315.98

***13-20.** Two measurements which presented unusual difficulty in the field were omitted in the survey of a closed boundary, as shown in the following field notes. Compute the missing data.

LINE	DISTANCE	BEARING	LINE	DISTANCE	BEARING
CD	1058.34	N80°15′E	*EF*	441.06	S17°15′W
DE	937.79	S87°37′E	*FC*	—	—

***13-21.** Similar to Problem 13-20, but ascertain the missing field data indicated in the following field notes. (Note: distance *CD* is approximately 300 feet long).

LINE	DISTANCE	BEARING	LINE	DISTANCE	BEARING
AB	402.64	N88°43′E	*DE*	232.02	—
BC	261.87	S88°25′E	*EA*	466.83	N41°26′W
CD	—	S26°30′W			

The bearings of a closed-polygon traverse are *AB* = N48°15′E; *BC* = S86°10′E; *CD* = S6°10′W; and *DA* = N67°13′W. Which course length is most likely responsible for the closure conditions given in Problems 13-22 through 13-24? Is the course too long or too short?

13-22. Algebraic sum of latitudes = +66.57 ft, of departures = +74.63 ft.
13-23. Algebraic sum of latitudes = −6.68 ft, of departures = −9.97 ft.
13-24. Algebraic sum of latitudes = −9.42 ft, of departures = −1.08 ft.

13-25. Determine the lengths and bearings of the sides of a lot whose corners have the following *X* and *Y* coordinates: *A*(0, 0); *B*(+150.50, +100.00); *C*(−52.78, +158.72); *D*(−94.65, +31.87).

13-26. Compute the lengths and azimuths of the sides of a closed-polygon traverse whose corners have the following *X* and *Y* coordinates: *A*(0, 0); *B*(−231.85, +419.63); *C*(+140.50, +625.28); *D*(+383.91, +215.08).

13-27. Why shouldn't traverse closure precision be carried out to a reciprocal such as 1/9376.85?

13-28. What is the purpose of balancing the angles, and latitudes and departures, of a closed traverse? After balancing latitudes and departures, will the angles close?

13-29. In adjusting measured traverse angles, why aren't adjustments made in proportion to angle sizes?

13-30. What uses are made of the precision for a closed traverse?

13-31. In searching for a record of the length and true bearing of a certain boundary line which is straight between *A* and *B*, the following notes of an old random traverse were found (survey by compass and Gunter's chain, declination 4°45′E). Compute the true bearing and length of *BA*.

LINE	A–1	1–2	2–3	3–B
Magnetic bearing	North	N20°E	East	S46°30′E
Distance (ch)	5.95	17.90	12.07	6.36

13-32. The interior angles of a four-sided closed traverse *ABCD* were all measured by the same observer using the same equipment and procedures, except the angles at *A* and *B* were turned four times (2D, 2R) and those at *C* and *D* eight times (4D, 4R). If the mean values at *A*, *B*, *C*, and *D* were 42°33.3′, 91°15.1′, 77°48.6′, and 148°20.6′, respectively, adjust the angles for misclosure.

13-33. Similar to Problem 13-32, except for a five-sided closed traverse with *A* and *C* turned eight times (4D, 4R) and the others repeated six times (3D, 3R), giving mean values at *A*, *B*, *C*, *D*, and *E* of 19°20.1′, 130°03.6′, 66°59.6′, 106°28.4′, and 217°11.9′, respectively.

BIBLIOGRAPHY

Goussinsky, B. 1952. "Three Notes on Traverses." *Surveying and Mapping* 12(no. 4):397.

Montgomery, C. J. 1965. "Impact of Electronic Computers on Survey Computations." *Surveying and Mapping* 25(no. 1):49.

Navyasky, M. 1974. "Adjusting Observed Directions by Angles Using Equitable Weights." *Surveying and Mapping* 34(no. 4):355.

Root, J. A. 1970. "Computations for Missing Elements of Closed Traverses." *Surveying and Mapping* 30(no. 1):91.

Stoughton, H. W. 1974. "The First Method to Adjust a Traverse Based on Statistical Considerations." *Surveying and Mapping* 34(no. 2):145.

———. 1975. "Computing Missing Elements of a Polygon." *Surveying and Mapping* 35(no. 3):217.

Tarczy-Hornoch, A. 1972. "Remarks on Computations for Missing Elements of Closed Traverses." *Surveying and Mapping* 32(no. 4):523.

Vreeland, R. R. 1969. "Adjustment of Traverses." *ASCE Journal of the Surveying and Mapping Division* 95(no. SU1):25.

14 AREA

14-1. INTRODUCTION. One important reason to compute the area for a tract of land is to include the acreage or square footage in a deed describing the property. Other purposes are to determine the acreages of fields and lakes, and the number of square yards to be surfaced, paved, seeded, or sodded. A special application is determining end areas for earthwork volume calculation (see Chapter 27). In plane surveying, acreage is considered to be the orthogonal projection of the area onto a horizontal surface.

The most common unit of area for lots is the *square foot*; for large tracts it is the acre: 1 acre = 43,560 ft^2 = 10 ch^2 (Gunter's). An acre lot, if square, would thus be 208.71 ft on a side. In the metric system, area is given in square meters or *hectares*: 1 hectare = 10,000 m^2 = 2.471 acres.

14-2. METHODS OF MEASURING AREA. Both field and map measurements are used to determine area. *Field measurement* methods include (1) division of the tract into simple figures (triangles, rectangles, and trapezoids), (2) offsets from a straight line, (3) double meridian distances, and (4) coordinates. The typical system of getting field data for procedure (3) or (4) is traversing, with the property corners serving as hubs of a closed-polygon figure. Before calculating area, the traverse is checked for closure and adjusted by one of the methods described in Section 13-7.

Map measurements are made by (1) dividing the area into triangles, (2) coordinate squares, and (3) running a planimeter over the enclosing lines.

Each process listed is described and illustrated in topics that follow.

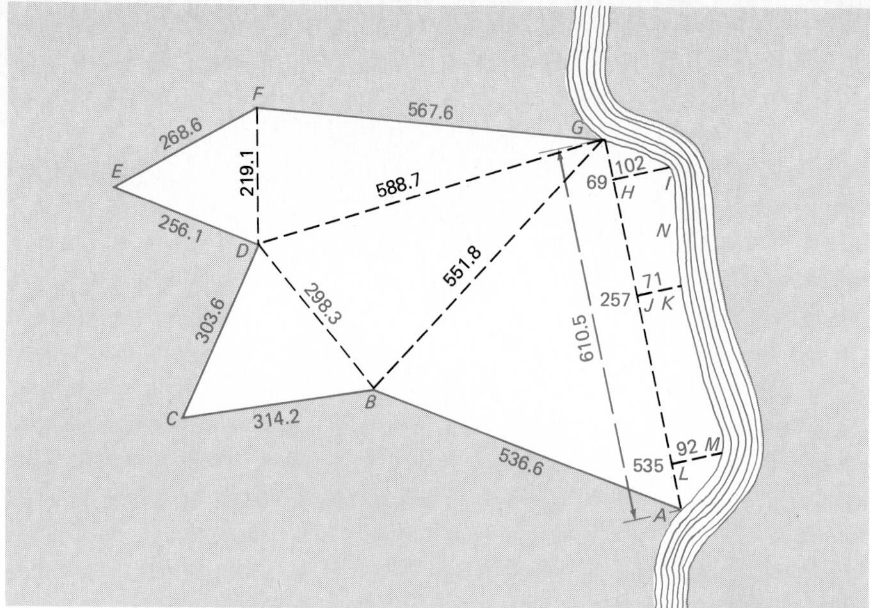

Figure 14-1. Area determination by triangles.

14-3. AREA BY DIVISION INTO TRIANGLES. A tract can be divided into simple geometric figures such as the triangles shown in Figure 14-1. The area of a triangle whose sides are known can be computed by the formula

$$\text{area} = \sqrt{s(s-a)(s-b)(s-c)} \tag{14-1}$$

where a, b, and c are sides of the triangle and

$$s = \tfrac{1}{2}(a+b+c)$$

Another formula for area of a triangle is

$$\text{area} = \tfrac{1}{2}ab \sin C \tag{14-2}$$

where C is the angle included between sides a and b.

The area of a field is the sum of the areas of all triangles. If Eq. (14-1) is used, each side and division line must be measured. This triangle method was used more often prior to invention of the transit for measuring angles. Now, electronic distance-measuring devices and the availability of computers make the method practical again.

14-4. AREA BY OFFSETS FROM STRAIGHT LINES. Irregular tracts can be reduced to a series of trapezoids by right-angle offsets from points at regular intervals along a measured straight line, as indicated in Figure 14-2.

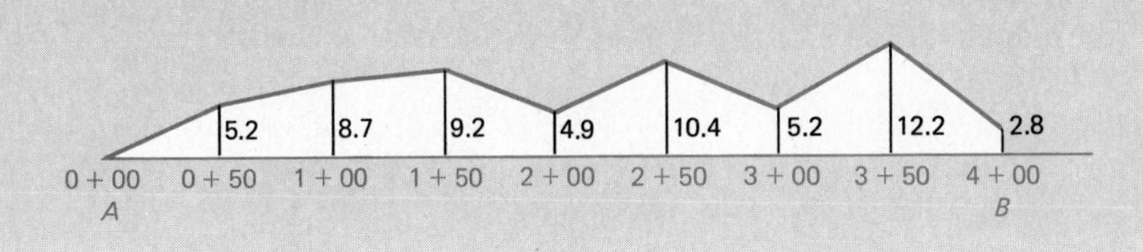

Figure 14-2. Area by offsets.

Area is found by the formula

$$\text{area} = b\left(\frac{h_0}{2} + h_1 + h_2 + \cdots + \frac{h_n}{2}\right) \tag{14-3}$$

where b is the length of a common interval between offsets and $h_0, h_1, \ldots, h_n$ are the offsets.

EXAMPLE 14-1

Compute the area of the tract shown in Figure 14-2.

By Eq. (14-3):

$$50\left(0 + 5.2 + 8.7 + 9.2 + 4.9 + 10.4 + 5.2 + 12.2 + \frac{2.8}{2}\right) = 2860 \text{ ft}^2$$

In the example, a summation of offsets (terms within parentheses) can be secured by the *paper-strip method* in which the area is plotted to scale and the mid-ordinate of each trapezoid is successively added by placing tick marks on a long strip of paper. Area is then obtained by making a single measurement between the first and last tick marks and multiplying by width b.

For irregularly curved boundaries like that in Figure 14-3, the spacing of offsets along the reference line should be adjusted so that connecting straight lines on the curved boundary at offset points accurately define the curves.

Figure 14-3. Area by offsets for a tract with curved boundary.

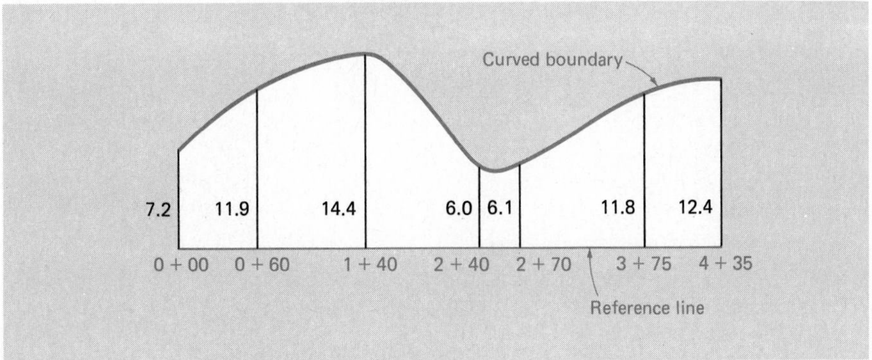

EXAMPLE 14-2

Compute the area of the tract in Figure 14-3.

$$\tfrac{1}{2}[60(7.2 + 11.9) + 80(11.9 + 14.4) + 100(14.4 + 6.0) + 30(6.0 + 6.1) + 105(6.1 + 11.8) + 60(11.8 + 12.4)] = 4492 \text{ ft}^2$$

14-5. AREA BY DOUBLE-MERIDIAN-DISTANCE METHOD. It is convenient to compute the area of a closed figure by the double-meridian-distance method when latitudes and departures of boundary lines are known. *The meridian distance of a traverse course is the perpendicular distance from the center point of the course to the reference meridian.* To ease the problem of signs, a reference meridian usually is placed through the most westerly traverse station.

In Figure 14-4 the meridian distances of courses *AB*, *BC*, *CD*, *DE*, and *EA* are *MM'*, *PP'*, *QQ'*, *RR'*, and *TT'*, respectively.

To express *PP'* in terms of convenient distances, draw *MF* and *BG* perpendicular to *PP'*. Then

$$PP' = P'F + FG + GP$$
$$= \text{meridian distance of } AB + \tfrac{1}{2}\text{ departure of } AB + \tfrac{1}{2}\text{ departure of } BC$$

Figure 14-4. Meridian distances and traverse area computation by DMD method.

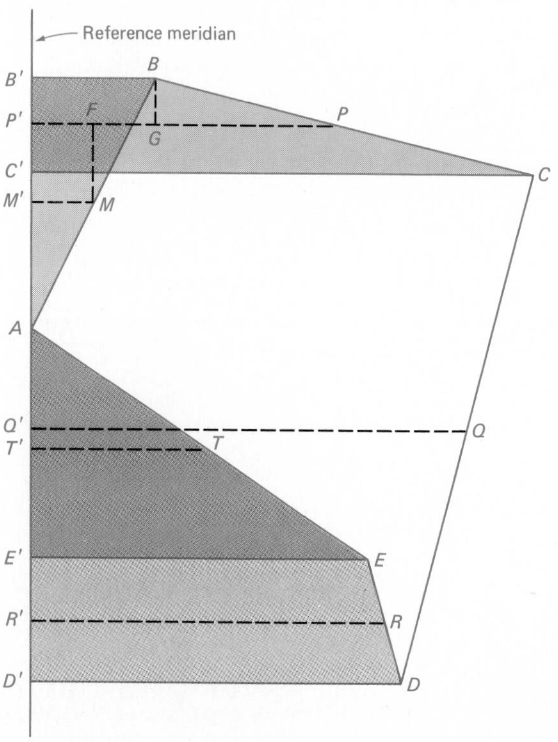

Thus the meridian distance of any course of a traverse equals the meridian distance of the preceding course, plus one-half the departure of the preceding course, plus one-half the departure of the course itself. It is simpler to employ full departures of courses. Therefore *double meridian distances* (DMDs) equal to twice the meridian distances are used, and a single division by 2 is made at the end of the computation.

Based on the considerations described, the following general rule can be applied in calculating DMDs: *The DMD for any traverse course is equal to the DMD of the preceding course, plus the departure of the preceding course, plus the departure of the course itself.* Signs of the departures, east plus and west minus, must be considered. When the reference meridian is taken through the most westerly station of a closed traverse and calculations of the DMDs started with a course through that station, *the DMD of the first course is its departure.* Applying these rules, for the traverse in Figure 14-4,

$$\text{DMD of } AB = \text{departure of } AB$$

$$\text{DMD of } BC = \text{DMD of } AB + \text{departure of } AB + \text{departure of } BC$$

A check on all computations is obtained if the DMD of the last course, after computing around the traverse, is also equal to its departure but has the opposite sign. If there is a difference, the departures were not correctly adjusted before starting, or a mistake was made in the computations.

The area enclosed by traverse $ABCDEA$ in Figure 14-4 may be expressed in terms of the areas of trapezoids (shown with different color shadings) as:

$$B'BCC' + C'CDD' - (AB'B + DD'E'E + AEE') \tag{14-4}$$

The area of each of these figures equals the meridian distance of a course times its balanced latitude. For example, the area of trapezoid $C'CDD' = Q'Q \times C'D'$, where $Q'Q$ and $C'D'$ are the meridian distance and latitude, respectively, of line CD. The DMD of a course multiplied by its latitude equals double the area. *Algebraic summation* of all double areas gives *twice the area* inside the entire traverse.

Signs of the products of DMDs and latitudes must be considered. If the reference line is passed through the most westerly station, all DMDs are positive. The products of DMDs and north latitudes are therefore plus, and of DMDs and south latitudes minus.

EXAMPLE 14-3

Using the balanced latitudes and departures listed in Table 13-3 for the traverse of Figure 14-4, compute the DMDs of all courses.

SOLUTION

The calculations are illustrated in Table 14-1.

EXAMPLE 14-4

Using the DMDs determined in Example 14-3, calculate the area within the traverse.

TABLE 14-1. COMPUTATION OF DMDS

Departure of AB =	+ 125.66	= DMD of AB
Departure of AB =	+ 125.66	
Departure of BC =	+ 590.65	
	+ 841.97	= DMD of BC
Departure of BC =	+ 590.65	
Departure of CD =	− 192.69	
	+1239.93	= DMD of CD
Departure of CD =	− 192.69	
Departure of DE =	− 6.07	
	+1041.17	= DMD of DE
Departure of DE =	− 6.07	
Departure of EA =	− 517.55	
	+ 517.55	= DMD of EA Check

SOLUTION

Computations for area are generally arranged as in Table 14-2, although a combined form may be substituted. Sums of positive and negative double areas are obtained, and the absolute value of the smaller subtracted from that of the larger. The result is divided by 2 to get the area (272,610 ft^2), and by 43,560 to obtain the number of acres (6.258).

If the total of minus double areas is larger than the plus value, it signifies only that DMDs were computed by going around the traverse in a clockwise direction. If the route had been from A through E, D, C, and B and back to A, the total plus double area would have been the greater. Areas carried out beyond the nearest square foot or 0.001 acre cannot be justified for a traverse on which distances were measured to the nearest 0.01 ft and angles read to 1 or $\frac{1}{2}$ min.

As a check, the area can be computed by *double parallel distances* (DPDs). *The DPD for any traverse course is equal to the DPD of the preceding course, plus the latitude of the preceding course, plus the latitude of the course itself.*

TABLE 14-2. COMPUTATION OF AREA BY DMDs AND DPDs

COURSE	BALANCED LATITUDE	BALANCED DEPARTURE	DMD	DOUBLE AREAS +	DOUBLE AREAS −	DPD	DOUBLE AREAS +	DOUBLE AREAS −
AB	N255.96	E125.66	+ 125.66	32,164		+255.96	32,164	
BC	S153.53	E590.65	+ 841.97		129,268	+358.39	211,683	
CD	S694.07	W192.69	+1239.93		860,598	−489.21	94,266	
DE	N202.97	W6.07	+1041.17	211,326		−980.31	5,951	
EA	N388.67	W517.55	+517.55	201,156		−388.67	201,156	
Total	0.00	0.00		444,646	989,866		545,220	00
					444,646			
					2) 545,220			
					272,610 sq ft = 6.258 acres			

The last three columns in Table 14-2 show the area computation by DPDs for the traverse of Figure 14-4. Again, signs of the latitudes, north plus and south minus, must be used in calculating DPDs.

All important surveying computations should be checked by using different methods, or by two persons who use the same system. As an example of good practice, an individual working alone in an office could calculate areas by DMDs and check the results by DPDs. Experienced surveyors and engineers have learned that a half-hour spent checking computations in the field and office can eliminate later lengthy frustrations.

14-6. AREA BY COORDINATES. Computation of area is a simple process for a closed-polygon traverse with known coordinate values for each corner. The procedure can be readily developed by reference to Figure 14-4. Since double the meridian distances $M'M$ and $P'P$ in coordinate terms are $(X_B + X_A)$ and $(X_C + X_B)$, and the latitudes of lines AB and BC are $(Y_B - Y_A)$ and $(Y_C - Y_B)$, respectively, then based on the summation of trapezoidal areas, the following formula for double area can be written:

$$\begin{aligned} 2(\text{area}) = {} & (X_C + X_B)(Y_C - Y_B) + (X_D + X_C)(Y_D - Y_C) \\ & + (X_E + X_D)(Y_E - Y_D) + (X_A + X_E)(Y_A - Y_E) \\ & + (X_B + X_A)(Y_B - Y_A) \end{aligned} \tag{14-5}$$

Equation (14-5) is equivalent to the trapezoid-area formula [Eq. (14-4)], except that the first two products are negative because $(Y_C - Y_B)$ and $(Y_D - Y_C)$ are negative, and the last three products are positive. Thus the double area resulting from Eq. (14-5) is negative but of no consequence because the absolute value is adopted. Expanding and rewriting, Eq. (14-5) is simplified to

$$\begin{aligned} 2(\text{area}) = {} & X_A Y_B + X_B Y_C + X_C Y_D + X_D Y_E + X_E Y_A - X_B Y_A - X_C Y_B \\ & - X_D Y_C - X_E Y_D - X_A Y_E \end{aligned} \tag{14-6}$$

Equation (14-6) can be reduced to an easily remembered form by listing the X and Y coordinates of each point in succession in two columns as shown in Eq. (14-7), *with coordinates of the starting point repeated at the end.* The products noted by diagonal arrows are ascertained, with solid arrows considered minus and dashed ones plus. The algebraic summation of all products is computed and its absolute value divided by 2 to get area.

$$\begin{matrix} Y_A & & X_A \\ Y_B & \times & X_B \\ Y_C & \times & X_C \\ Y_D & \times & X_D \\ Y_E & \times & X_E \\ Y_A & \times & X_A \end{matrix} \tag{14-7}$$

The procedure indicated in Eq. (14-7) is applicable to calculating any size traverse. It is necessary only to consider the algebraic signs of the coordinates,

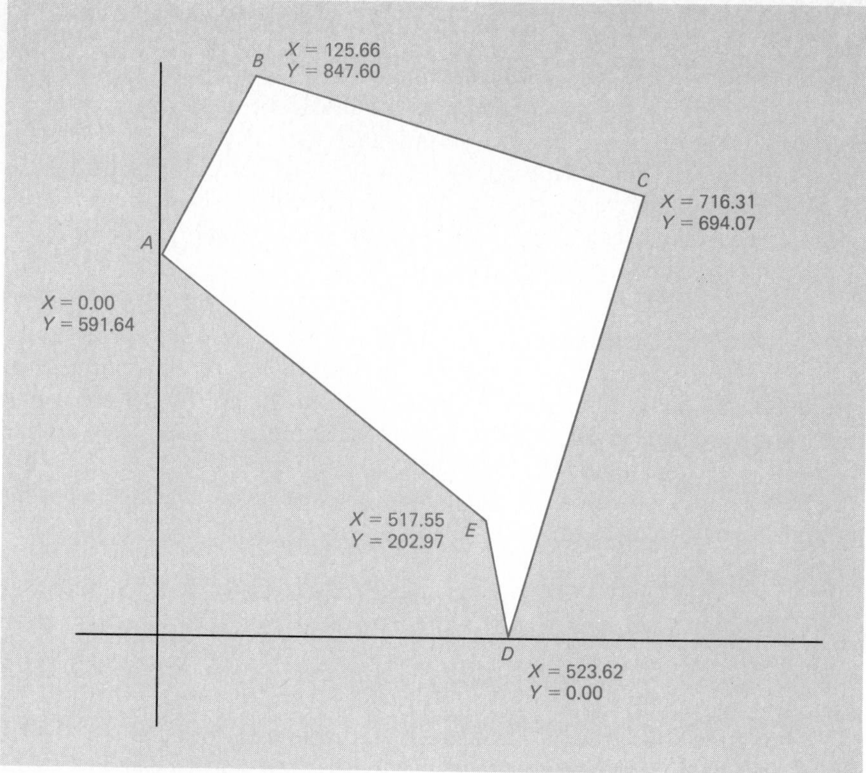

Figure 14-5. Area by coordinates.

and an origin can be selected to make them all positive. Some surveyors assume $X = 0$ for the most westerly point and $Y = 0$ for the most southerly station. Magnitudes of coordinates and products are thereby reduced and the amount of work lessened, since four products will equal zero.

EXAMPLE 14-5

Figure 14-5 illustrates the same traverse used for Examples 14-3 and 14-4, but with coordinates of points reduced so that $X_A = 0.00$ (A is the most westerly station) and $Y_D = 0.00$ (D is the most southerly station); thus, all coordinates are positive. Compute the traverse area by the coordinate method.

SOLUTION

These computations are also best organized for tabular solution. Table 14-3 shows the procedure and results.

Another handy formula, easily derived, for calculating areas within closed-polygon traverses is

$$\text{area} = \tfrac{1}{2}[X_A(Y_E - Y_B) + X_B(Y_A - Y_C) + X_C(Y_B - Y_D) + X_D(Y_C - Y_E) + X_E(Y_D - Y_A)] \quad (14\text{-}8)$$

TABLE 14-3. COMPUTATION OF AREA BY COORDINATES

POINT	*Y*	*X*	DOUBLE AREA MINUS	DOUBLE AREA PLUS
A	591.64	0.00		
B	847.60	125.66	74,345	0
C	694.07	716.31	607,144	87,216
D	0.00	523.62	363,429	0
E	202.97	517.55	0	106,279
A	591.64	0.00	0	306,203
			−1,044,918	+499,698
			499,698	
			2) 545,220	
			272,610 ft²	

Equations (14-5) through (14-8) are all readily programmed for solution by electronic computer. A program written in BASIC for computing traverses, including area by coordinates, is given in Appendix C.

Calculations for purposes of partitioning land—that is, cutting off a portion of a tract for title transfer—can be significantly aided by using coordinates. This subject is discussed in Chapter 22.

14-7. AREA OF PARCELS WITH CIRCULAR BOUNDARIES. The area of a tract that has a circular curve for one boundary, as in Figure 14-6, can be found by dividing the figure into two parts: polygon *ABCDEGFA* and sector *EGF*. The radius $R = EG = FG$ and either central angle *EGF* or length *EF* must be known or computed to permit calculation of sector area *EGF*. If *R* and central angle θ are known, then the area of sector $EGF = \pi R^2 \times (\theta°/360°)$. If chord length *EF*

Figure 14-6. Area with circular curve as part of boundary.

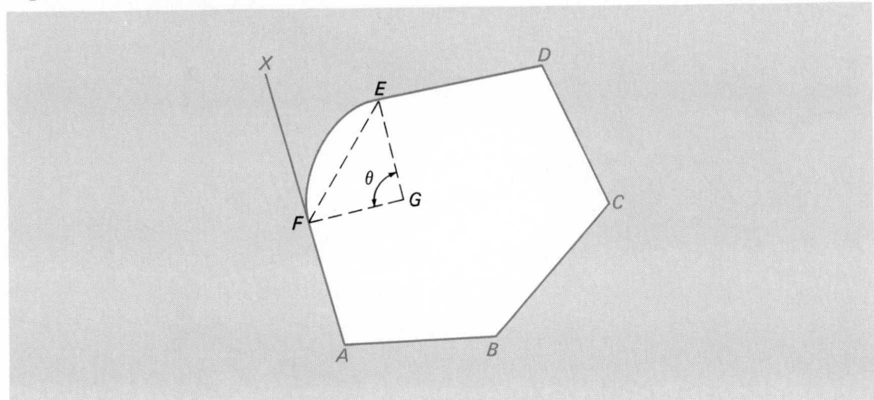

is known, angle $\theta = \sin^{-1}(EF/2R)$ and the preceding equation is used to calculate the sector area. To obtain the tract's total area, the sector area is added to area *ABCDEGFA*, found by the DMD or coordinate method.

As a variation, the length and direction of chord *EF* can be calculated and the area of segment *EF* added so that of *ABCDEFA*. Note that angle *XFE* equals one-half of angle *EGF*.

14-8. AREA FROM A MAP BY TRIANGLES. A traverse can be plotted to scale, divided into triangles, the sides measured, and areas of the triangles found by Eq. (14-1). This method is not as accurate as computations using field measurements, but the procedure is useful for checking purposes.

14-9. AREA BY COORDINATE SQUARES. To find the area within a plotted traverse by coordinate squares, the map is marked off in squares of unit area. The number of complete unit squares included in the traverse is counted and the sum of all partial units estimated. Areas of fractional units can be computed by treating them as trapezoids, but generally this refinement is unnecessary.

A simpler method results from use of transparent paper marked in squares to some scale. The grid is placed over the traverse and the number of squares and partial units counted.

A third method consists of plotting the traverse on coördinate paper and determining the number of units in the manner just described.

14-10. MEASUREMENT OF AREA BY PLANIMETER. A planimeter mechanically integrates area and records the answer on a drum and disk as a tracing point is moved over the outline of the figure to be measured.

Figure 14-7. Mechanical planimeter: *A*, anchor arm; *B*, anchor point; *C*, scale bar; *D*, tracing point; *E*, disk; *F*, drum; *G*, vernier. (Courtesy Keuffel & Esser Company.)

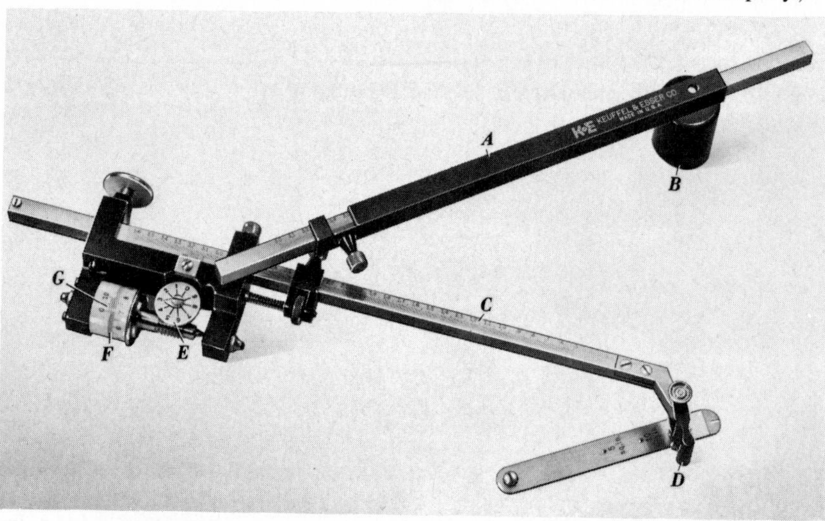

Figure 14-8. Electronic planimeter. (Courtesy Numonics Corporation.)

There are two types of planimeters: mechanical and electronic. The major parts of the mechanical type are a scale bar, graduated drum and disk, vernier, tracing point and guard, and anchor arm, weight, and point. The scale bar may be fixed or adjustable as in Figure 14-7. For the standard fixed-arm planimeter, one revolution of the disk (dial) represents 100 in^2 and one turn of the drum (wheel) represents 10 in^2. The adjustable type can be set to read units of area directly for any particular map scale. The instrument touches the map at only three places: the anchor point, drum, and tracing-point guard.

An electronic planimeter (Figure 14-8) operates similarly to the mechanical type, except that the results are given in digital form on a display console. Areas can be measured in units of square inches or square centimeters, and by setting an appropriate "scale factor" they can be obtained directly in acres or hectares. Some instruments feature multipliers which can automatically compute and display volumes.

As an example of using a mechanical planimeter, suppose the area within the traverse of Figure 14-4 is to be measured. The anchor point beneath the weight is set in a position outside the traverse (if inside, a polar constant must be added), and the tracing point brought over corner *A*. An initial reading—say, 7231—is taken, the 7 coming from the disk, 23 from the drum, and 1 from the vernier. The tracing point is moved along the traverse lines from *A* to *B*, *C*, *D*, and *E* and back to *A*. The point may be guided by a triangle or a straightedge, but normally it is steered freehand. A final reading, perhaps 8596, is made. The difference between the initial and final readings, or 1365, represents the area if the bar was set exactly to the scale of the map. Since the bar setting may not be perfect, it is best to check the planimeter constant by running over the perimeter of a carefully laid out square 5 in on a side, with diagonals of 7.07 in.

Assume the difference between initial and final readings for the 5-in square is 1250. Then

$$5 \text{ in} \times 5 \text{ in} = 25 \text{ in}^2 = 1250 \text{ units}$$

or

$$1 \text{ unit} = \tfrac{25}{1250} = 0.020 \text{ in}^2$$

and

$$\text{Area} = 1365 \text{ units} \times 0.020 = 27.30 \text{ in}^2$$

For a map scale of 1 in = 100 ft, 1 in^2 = 10,000 ft^2 and the area measured is 273,000 ft^2.

As a check on planimeter operation, the outline may be traced in the opposite direction. The initial and final readings at point A should agree within a limit of perhaps 2 to 5 units.

The precision obtained in using a planimeter depends on operator skill, accuracy of the plotted map, type of paper, and other factors. Results correct within $\frac{1}{2}$ to 1% can be obtained by careful work.

A planimeter is most useful for irregular areas, such as that in Figure 14-3, and has applications in many branches of engineering. The planimeter has been widely used in highway offices for determining areas of cross sections, and is helpful in checking computed areas in property surveys.

14-11. SOURCES OF ERROR. Some sources of error in area computations are:

1. Poor selection of intervals and offsets to fit a given irregular boundary properly.
2. Not adjusting latitudes and departures in accordance with true conditions.
3. Using coordinate squares that are too large and therefore make estimation of areas of partial blocks difficult.
4. Incorrect setting of the planimeter scale bar.
5. Running off and on the edge of the map sheet with the planimeter drum.
6. Using different types of paper for the map and planimeter calibration sheet.

14-12. MISTAKES. In computing areas, common mistakes by students include:

1. Forgetting to divide by 2 in the DMD and coordinate methods.
2. Confusing signs of DMDs, coordinates, latitudes, departures, or areas.
3. Failing to check an area computation by a different method.

4. Poor selection of the origin, resulting in minus values for DMDs and coordinates.
5. Not drawing a sketch to scale or general proportion for visual check.
6. Not verifying the planimeter scale constant by tracing a known area.

PROBLEMS

14-1. Compute the area enclosed within polygon *BDEFGB* of Figure 14-1.
14-2. Calculate the area enclosed within polygon *BCDGAB* of Figure 14-1.
14-3. Compute the area enclosed between line *GHJLA* and the shoreline of Figure 14-1 using the offsets method.
14-4. Determine the area between the traverse line from station 5 + 42 to station 6 + 50 and Black Creek of Figure 16-3 by the offsets method.
14-5. Ascertain the area between a lake and a straight line from which offsets are taken at irregular intervals as follows (all distances in chains):

OFFSET POINT	*A*	*B*	*C*	*D*	*E*	*F*	*G*
Distance from *A*	0	1.28	3.61	4.52	5.76	7.95	8.80
Offset	0.29	0.75	1.21	2.03	2.92	1.64	0.30

14-6. Calculate by DMDs the area within the traverse of Problem 13-13.
14-7. Compute by DMDs the area enclosed by the traverse of Problem 13-14.
14-8. Calculate by DMDs the area within the traverse of Problem 13-15.
14-9. Compute the area enclosed in the traverse of Problem 13-13 using coordinates.
14-10. Determine the area within the traverse of Problem 13-14 employing coordinates.
14-11. By the coordinate method, find the area enclosed by the traverse of Problem 13-15.
14-12. Compute the area within the traverse of Problem 13-20 using the DMD method. Check by DPDs.
14-13. Calculate the area inside the traverse of Problem 13-21 by DMDs, and check by DPDs.
14-14. Calculate the area within the traverse of Problem 13-6 utilizing the coordinate method.
14-15. Determine the area inside the traverse of Problem 13-9 by the coordinate method.
14-16. Compute the area enclosed by the traverse of Problem 13-10 using the DMD method. Check by coordinates.
14-17. Find the area within the traverse of Problem 13-18 employing the DMD method. Check by coordinates.
14-18. Ascertain the area of the lot in Problem 13-25.
14-19. Calculate the area inside the traverse of Problem 13-26.
14-20. Determine the area within the closed traverse of Problem 13-19 by the DMD method. Check by DPDs.
14-21. Plot the traverse of Problem 13-13 to a scale of 1 in = 100 ft. Determine its surrounded area using a planimeter.
14-22. Similar to Problem 14-21, except for the traverse of Problem 13-14.
14-23. Plot the traverse of Problem 13-15 to a scale of 1 in = 200 ft and find its enclosed area using a planimeter.
14-24. The latitudes and departures (in 100-ft tape lengths) for a closed-polygon traverse *ABCDEFGA* follow. *AB*: N lat. = 1, E dep. = 6; *BC*: S lat. = 9, W dep. = 6; *CD*: S lat. = 3, E dep. = 6; *DE*: S lat. = 7, W dep. = 8; *EF*: N lat. = 11, W dep. = 1; *FG*: N lat. = 4, W dep. = 6. Calculate the (a) latitude and departure of line *GA*, (b) length of line *GA*, and (c) area within the traverse in acres by both DMDs and coordinates.

14-25. A closed-polygon traverse *ABCDEFGHA* has courses with latitudes and departures (in hundreds of feet) as specified. *AB*: N lat. = 5, E dep. = 2; *BC*: N lat. = 6, W dep. = 4; *CD*: S lat. = 3, W dep. = 5; *DE*: S lat. = 4, W dep. = 6; *EF*: S lat. = 2, E dep. = 8; *FG*: S lat. = 5, W dep. = 6; *GH*: S lat. = 2, E dep. = 7; *HA*: N lat. = 5, E dep. = 4. (a) Which is the most westerly station? (b) The most southerly? (c) What are the length and bearing of the line connecting stations *H* and *E*? (d) Compute by DMDs the area enclosed in acres. (e) Check by DPDs and coordinates.

14-26. Compute by DMDs the area in acres within a closed-polygon traverse *ABCDEFA* by placing the axis through the most westerly station. Latitudes and departures (in engineer's chains) follow. *AB*: N lat. = 5, E dep. = 4; *BC*: N lat. = 0, E dep. = 6; *CD*: S lat. = 4, E dep. = 4; *DE*: S lat. = 4, W dep. = 6; *EF*: S lat. = 2, W dep. = 10; *FA*: N lat. = 5; E dep. = 2.

Compute the unknown elements and area enclosed by the traverses of Problems 14-27 and 14-28, selecting an axis that eliminates all negative DMDs.

14-27.

COURSE	*AB*	*BC*	*CD*	*DA*
Bearing	Due east	S28°00′W	Due west	Unknown
Length (ft)	310.5	165.8	110.0	Unknown

14-28.

COURSE	*AB*	*BC*	*CD*	*DA*
Bearing	N55°00′W	S35°00′W	N65°00′E	Unknown
Length (ft)	460.00	402.50	201.25	Unknown

14-29. Calculate the area of a piece of property bounded by a traverse and circular arc described as follows: *AB*, S40°00′W, 400.0 ft; *BC*, S80°E, 400.0 ft; *CD*, N35°00′W, 200.0 ft; *DA*, a circular arc tangent to *CD* at point D.

14-30. Similar to Problem 14-29, except that *CD* is 185.0 ft.

14-31. Divide the area of the lot in Problem 14-29 into two equal parts by a line through point *B*. List in order the lengths and bearings of all sides for each parcel.

14-32. Partition the lot in Problem 14-30 into two equal areas by means of a line parallel to *BC*. Tabulate in clockwise consecutive order the lengths and bearings of all sides.

14-33. Lot *ABCD* between two parallel street lines is 150.00 ft deep and has a 100-ft frontage (*AB*) on one street and a 120-ft frontage (*CD*) on the other. Interior angles at *A* and *B* are equal, as are those at *C* and *D*. What distances *AE* and *BF* should be laid off by a surveyor to divide the lot into two equal areas by means of a line *EF* parallel to *AB*?

14-34. The area of a field plotted on a map to a scale of 1 in = 200 ft is measured with a fixed-arm planimeter having a constant of 10, and 5.238 revolutions of the wheel are recorded. What is the area of the field in square feet and in acres?

14-35. A planimeter tracing point, initial reading 1162, is run clockwise around the sides of a 5-in square and gives a reading of 1287 when returned to the starting point. If the beginning reading on hub *A* of a plotted traverse is 1322, and 1976 when guided around the closed traverse back to *A*, compute the enclosed area in square feet if the plot is at a scale of 100 ft/in.

14-36. In Problem 14-24, if a tape assumed to be 100.00 ft long is found after standardization to be 100.07 ft long, how large an error in area is produced?

14-37. Write a computer program for calculating areas within closed polygons by the coordinate method.

14-38. Write a computer program for calculating areas within closed polygons by the DMD method.

14-39. Calculate the interior angles in polygon *ABCDEFG* of Figure 14-1. If the bearing of line *AB* is N73°45′W, compute bearings of the other courses.

BIBLIOGRAPHY

Ahmed, F. A. 1983. "Area Computation Using Salient Boundary Points." *ASCE Journal of Surveying Engineering* 109(no. 1):54.

Griggs, F. E. 1967. "Positive Calculation of Area by Coordinates." *ASCE Journal of the Surveying and Mapping Division* 93(no. SU1):1.

Hickerson, T. F. 1967. *Route Location and Design*, 5th ed. New York: McGraw-Hill.

Meyer, C. F., and D. W. Gibson. 1980. *Route Surveying and Design*, 5th ed. New York: Harper & Row.

PART TWO

15 STADIA

15-1. INTRODUCTION. The "stadia method," referred to as "tacheometry" in Europe, is a rapid and efficient way to measure distances accurately enough for trigonometric leveling, some traverses, and the location of topographic details. Furthermore, a two- or three-person party can replace the three or four people usually required in transit-tape surveys.

Stadia comes from the Greek word for a unit of length originally applied in measuring distances for athletic contests—hence, our modern "stadiums." The word denoted 600 Greek units (equivalent to our "feet"), or 606 ft 9 in by present-day American standards.

The term *stadia* is now applied to the cross wires and rod used in making measurements, as well as to the method itself. Stadia readings can be taken with transits, theodolites, alidades, and levels.

New total-station equipment combines a theodolite, EDMI, and recording-computing capabilities to automatically reduce slope distances and vertical angles. Readouts of horizontal lengths and elevation differences—even coordinates—are produced. Thus, the new equipment can reduce the size of field parties and take over many stadia projects. Nevertheless, stadia principles and methods provide basic concepts and will likely continue in use indefinitely.

15-2. MEASUREMENT BY STADIA FOR HORIZONTAL SIGHTS. Besides the center horizontal cross wire, a transit or theodolite reticle for stadia work has two additional horizontal wires spaced equidistant from the center one, as illustrated in Figure 15-6. The interval between stadia wires in most surveying instruments

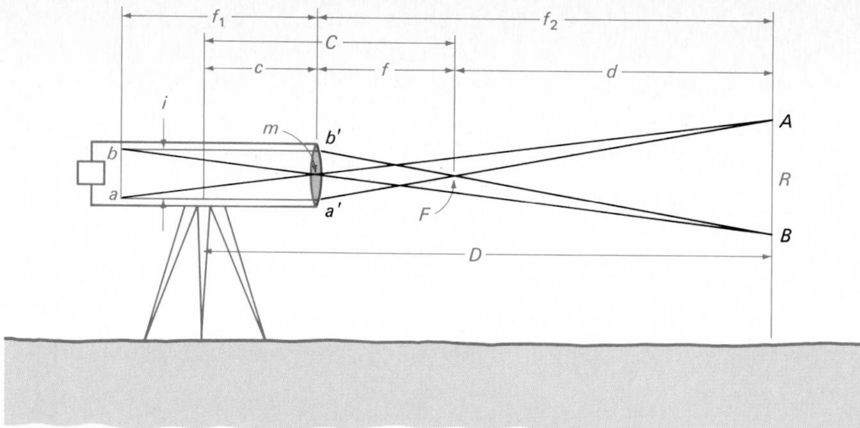

Figure 15-1. Principle of stadia; external-focusing telescope.

gives a vertical intercept of 1 ft on a rod held 100 ft away (or 1 m for a 100-m length). Thus the distance to a rod decimally divided in feet, tenths, and hundredths can be read directly to the nearest foot. This is sufficiently precise for locating topographic details such as rivers, bridges, and roads which are to be plotted on a map having a scale smaller than 1 in = 100 ft, and sometimes for scales even larger such as 1 in = 50 ft.

The stadia method is based on the principle that in similar triangles corresponding sides are proportional. In Figure 15-1 depicting an *external-focusing telescope*, light rays from points A and B passing through the lens center form a pair of similar triangles AmB and amb. Here $AB = R$ is the rod intercept (*stadia interval*) and ab is the spacing between the stadia wires.

Standard symbols used in stadia measurements and their definitions are as follows:

f = *focal length* of the lens (a constant for any particular compound objective lens). It can be determined by focusing on a distant object and measuring the distance between the center (actually the nodal point) of the objective lens and the reticle.[1]

f_1 = image distance or length from the center (actually the nodal point) of the objective lens to the plane of the cross wires when the telescope is focused on some definite point.

f_2 = object distance or length from the center (actually the nodal point) of the objective lens to a definite point when the telescope is focused on that point. When f_2 is infinite, or very large, $f_1 = f$. [See Eq. (6-6).]

i = spacing between the stadia wires (ab in Figure 15-1).

f/i = *stadia interval factor*, usually 100.

[1] In a lens system, any ray in the object space directed toward the first or front nodal point will emerge in the image space from the second (rear) nodal point parallel to its former direction. Thus, the nodal points establish the optical axis of the system.

c = distance from the instrument center (vertical axis) to the objective lens center. It varies slightly as the objective lens moves in or out for different sight lengths but is generally considered to be a constant.

$C = c + f$. C is called the *stadia constant*, although it varies slightly with c.

d = distance from the focal point in front of the telescope to the face of the rod.

$D = C + d$ = distance from the instrument center to the rod face.

From similar triangles of Figure 15-1,

$$\frac{d}{f} = \frac{R}{i} \quad \text{or} \quad d = R\frac{f}{i}$$

and

$$D = R\left(\frac{f}{i}\right) + C \qquad (15\text{-}1)$$

Fixed stadia wires in transits, theodolites, levels, and alidades are carefully spaced by instrument manufacturers to make the stadia interval factor f/i equal to 100. The stadia constant C ranges from about 0.75 to 1.25 ft for different external-focusing telescopes, but is usually assumed to equal 1 ft. The only variable on the right side of the equation, then, is R, the rod intercept between stadia wires. In Figure 15-1, if the intercept R is 4.27 ft, the distance from instrument to rod is $427 + 1 = 428$ ft.

An older type of external-focusing telescope has been described, since a simple drawing correctly shows the relationships. The objective lens of an *internal-focusing telescope* (the type now used in surveying instruments) remains fixed in position while a movable negative-focusing lens between the objective lens and plane of the cross wires changes direction of the light rays. As a result, the stadia constant is so small that it can be assumed equal to zero.

Disappearing stadia hairs were used in some older instruments to prevent confusion with the center horizontal hair. Modern glass reticles with short stadia lines and a full-length center line [see Figure 10-6(c)] accomplish the same result more effectively.

The stadia interval factor should be determined the first time an instrument is used, although the manufacturer's exact value posted inside the carrying case will not change unless the cross hairs, reticle, or lenses are replaced or adjusted in older models.

To determine the interval factor, rod intercept R for a horizontal sight of known distance D is read. Then, in an alternate form of Eq. (15-1), the stadia interval factor is $f/i = (D - C)/R$. As an example, at a known distance of 300.0 ft, a rod interval of 3.01 was read. Values for f and c were measured as 0.65 and 0.45 ft, respectively; hence, $C = 1.1$ ft. Then $f/i = (300.0 - 1.1)/3.01 = 99.3$. Accuracy in determining f/i is increased by averaging values from several lines whose measured lengths vary from about 100 to 500 ft by 100-ft increments.

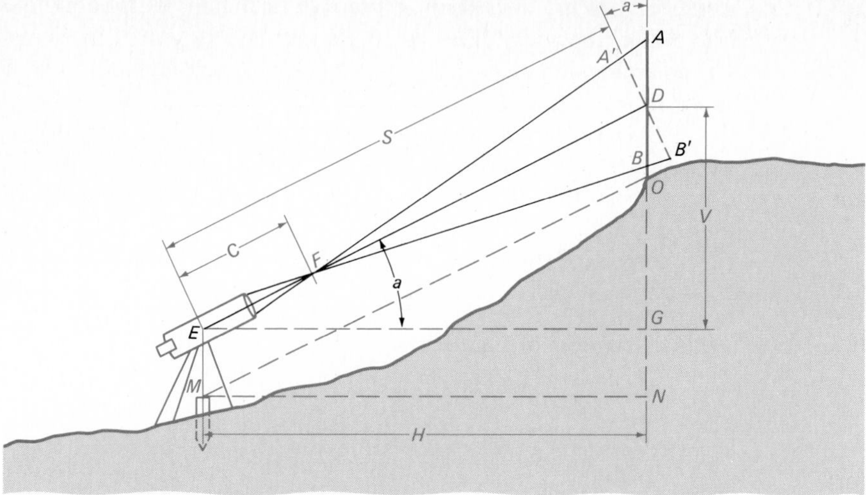

Figure 15-2. Inclined stadia measurement.

15-3. MEASUREMENT BY STADIA FOR INCLINED SIGHTS. Most stadia shots are inclined because of varying topography, but the intercept is read on a *plumbed rod* and the slope length "reduced" to horizontal and vertical distances.

In Figure 15-2, a transit is set over point M and the rod held at O. With the middle cross hair set on point D to make DO equal to the height of instrument EM, the vertical angle (angle of inclination) is a. Note that in stadia work the height of instrument (h.i.) is defined as the height of the line of sight above the point occupied (*not* HI, height above the datum as in leveling).

Let S represent the slope distance ED; H, the horizontal distance $EG = MN$; and V, the vertical distance $DG = ON$. Then

$$H = S \cos a$$
$$V = S \sin a$$

If the rod could be held normal to the line of sight at point O, a reading $A'B'$, or R', would be obtained, making

$$S = R' \frac{f}{i} + C$$

Since it is not practical to hold the rod at an inclination angle a, it is plumbed and reading AB, or R, taken. For the small angle at D on most sights, it is sufficiently accurate to consider angle $AA'D$ as a right angle. Therefore

$$R' = R \cos a$$

and

$$S = R \frac{f}{i} \cos a + C \tag{15-1}$$

or

$$H = R\frac{f}{i}\cos^2 a + C\cos a \tag{15-2}$$

For small angles and external-focusing telescopes, $C = 1$ ft approximately, and

$$H = R\frac{f}{i}\cos^2 a + 1 \tag{15-3}$$

If $f/i = K$, then

$$H = KR\cos^2 a + 1 \tag{15-4}$$

To avoid multiplying R by $\cos^2 a$, which is a large decimal number, the formula for H can be rewritten for use in computation as

$$H = KR - KR\sin^2 a + C \tag{15-5}$$

The vertical distance is found by the formula

$$V = S\sin a = \left(R\frac{f}{i}\cos a + C\right)\sin a$$

or

$$V = R\frac{f}{i}\sin a\cos a + C\sin a \tag{15-6}$$

For small angles, $\sin a$ is very small and the quantity $C\sin a$ can be neglected. Substituting $\frac{1}{2}\sin 2a$ for $\sin a\cos a$, the formula becomes

$$V = KR(\tfrac{1}{2}\sin 2a) \tag{15-7}$$

In the final form generally used, K is taken as 100 and the formulas for reduction of inclined sights to horizontal and vertical distances are

$$H = 100R - 100R\sin^2 a + 1 \qquad \text{(external focus)} \tag{15-8}$$

or

$$H = 100R - 100R\sin^2 a \qquad \text{(internal focus)} \tag{15-8a}$$

and

$$V = 50R\sin 2a \tag{15-9}$$

Tables, diagrams, special slide rules, and electronic calculators have been used by surveyors to obtain quick solutions of these formulas. Table E-1 in Appendix E lists horizontal and vertical distances for a rod intercept of 1 ft and vertical angles from 0 to 16° (74 to 90° and 90 to 106° for readings from the zenith). Using the tables to check the reduction of notes will develop judgment on the reasonableness of answers—a critical factor in surveying and engineering practice.

An unfamiliar table should always be investigated by substituting values in it that will give known answers. For example, angles of 5, 10, and 15° can be used to check tabular results. Assuming a vertical angle of 15°00′ (zenith angle 75°), a rod intercept of 1.00 ft, and a stadia constant of 1 ft, the following results are secured. By Table E-1,

$$H = 93.30 \times 1.00 + 1 = 94.3, \quad \text{or } 94 \text{ ft}$$

Bu Eq. (15-8),

$$H = 100 \times 1.00 - 100 \times (0.259)^2 + 1 = 94.3, \quad \text{or } 94 \text{ ft}$$

EXAMPLE 15-1

Assume that in Figure 15-2, the elevation of M is 268.2 ft; h.i. = EM = 5.6 ft; rod intercept $AB = R = 5.28$ ft; vertical angle a to point D read at 5.6 ft on the rod is +4°16′; and $C = 1$ ft. Compute distance H, elevation difference V, and the elevation of point O.

SOLUTION

From Table E-1, for an angle of 4°16′ (zenith angle 85°44′) and a rod intercept of 1 ft, the horizontal and vertical distances are 99.45 and 7.42 ft, respectively. Then

$$H = (99.45 \times 5.28) + 1 = 525.1 + 1 = 526 \text{ ft}$$

$$V = (7.42 \times 5.28) + 0.08 = 39.18 + 0.08 = 39.3 \text{ ft}$$

The elevation of point O is

$$\text{elev } O = 268.2 + 5.6 + 39.3 - 5.6 = 307.5 \text{ ft}$$

The complete expression for determining the difference in elevation between M and O in Figure 15-2 is

$$\text{elev}_O - \text{elev}_M = \text{h.i.} + V - \text{rod reading} \qquad (15\text{-}10)$$

From Eq. (15-10), the advantage of sighting on the h.i. to read the vertical angle is evident. Since the rod reading and h.i. are opposite in sign, if equal in magnitude they cancel each other and can be omitted from the elevation computation. If the h.i. cannot be seen because of obstructions, any rod reading can be sighted and Eq. (15-10) used. Setting the middle cross wire on a full foot mark just above or below the h.i. simplifies the arithmetic.

Determination of elevation differences by stadia can be compared with differential leveling. The h.i. corresponds to a plus sight, and the rod reading to a minus sight. On these is superimposed a vertical distance which may be either plus or minus, its sign depending on the angle of inclination. On important sights to control points and hubs, instrumental errors will be reduced by good field procedures utilizing the principle of reversion—that is, reading vertical angles with the telescope in normal and plunged positions.

Direct rod readings with the line of sight horizontal (as in leveling), rather than vertical angles, are taken when possible to simplify the reduction of notes. Inspection of Table E-1 shows that for vertical angles smaller than about 4°, the difference between the slope and horizontal distance is negligible except on long sights (where the distance-reading error is also greater). Hence inclining the telescope by several degrees is permissible for the stadia distance reading after taking a level "foresight" to get the vertical angle.

15-4. STADIA RODS. Various types of markings are used on stadia rods but all have bold geometric figures designed for legibility at long distances. Most stadia rods have been graduated in feet and tenths (hundredths are interpolated), but metric ones are becoming more common. Different colors aid in distinguishing the numbers and graduations.

One-piece, folding, and sectional rods having lengths of 10 or 12 ft are typical. Longer ones increase the sight-distance limit but are heavy and awkward to handle. Often the lower foot or two of a 12-ft rod will be obscured by weeds or brush, leaving perhaps only 10 ft visible. The maximum length of sight would then be about 1000 ft. On longer shots, the half-interval (intercept between the middle cross hair and upper or lower stadia hair) can be read and doubled for use in the standard stadia reduction equations. With a quarter-hair between the middle cross line and upper stadia hair, theoretically a distance of almost 4000 ft can be estimated. On short sights, up to perhaps 200 ft, ordinary leveling rods such as the Philadelphia type are satisfactory.

15-5. BEAMAN ARC. The Beaman arc (Figure 15-3) is a device placed on some transits and alidades to facilitate stadia computations. It may be part of the vertical circle or a separate plate. The H and V scales of the arc are graduated in percent. The V scale shows the difference of elevation per 100 ft of slope distance, whereas the H scale gives the *correction* per 100 ft to be subtracted from the stadia distance. Since V is proportional to $\frac{1}{2}\sin 2a$ and the correction for H depends on $\sin^2 a$, spacing of the graduations decreases as the vertical angle increases. Therefore a vernier cannot be used, and an exact reading can be made only by setting the arc to read a whole number.

The V-scale indicator (index) is set to read 50 (perhaps 30 or 100 on some instruments) when the telescope is horizontal to eliminate minus values. Readings greater than 50 are obtained for sights above the horizon, smaller than 50 below it. The arithmetic required in using the Beaman arc is simplified by setting the V scale on a whole number and letting the middle cross wire fall somewhere near the h.i. The H scale generally then will not read a whole number and the values must be interpolated. This is unimportant, since the arithmetic remains simple.

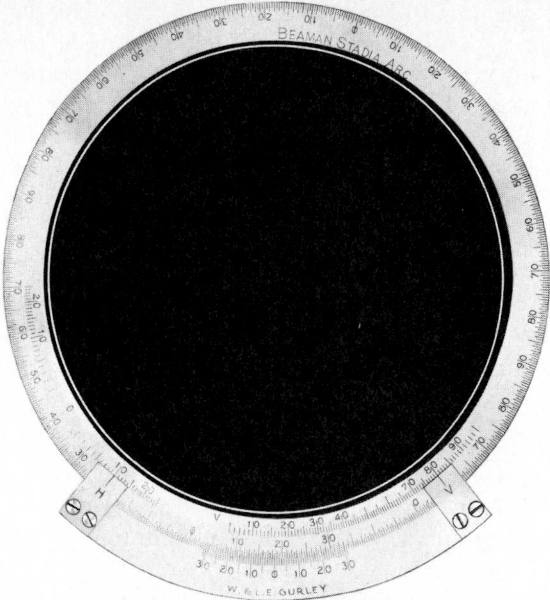

Figure 15-3. Beaman arc (and vertical circle). (Courtesy W. & L. E. Gurley.)

The elevation of a point B sighted with the transit set up over point A is found by the following formula:

$$\text{elev } B = \text{elev } A + \text{h.i.} + (\text{arc reading} - 50)(\text{rod intercept}) - \text{rod reading of middle cross wire} \tag{15-11}$$

Careful attention must be given to signs. Other instruments have a similar arc called a stadia circle with the same V scale, but the H scale gives a multiplier instead of a percentage correction.

EXAMPLE 15-2

Assume that in Figure 15-2 the V-scale reading of a Beaman arc is 56; H-scale reading, 0.4; rod intercept, 6.28 ft.; h.i. = 4.2 ft; rod reading of the middle wire, 7.3 ft; $C = 0$; and the elevation of point M is 101.5 ft. Compute the elevation of point B.

SOLUTION

$$\text{elev } O = 101.5 + 4.2 + (56 - 50)(6.28) - 7.3 = 136.1 \text{ ft}$$

$$H = (100)(6.28) - (0.4)(6.28) = 625 \text{ ft}$$

15-6. SELF-REDUCING TACHEOMETERS. Self-reducing tacheometers (Figure 15-4) and alidades have been developed in which curved stadia lines appear to move apart or closer together as the telescope is elevated or depressed. Actually, the lines are engraved on a glass plate which turns around a center (situated outside the telescope) as the telescope is transited.

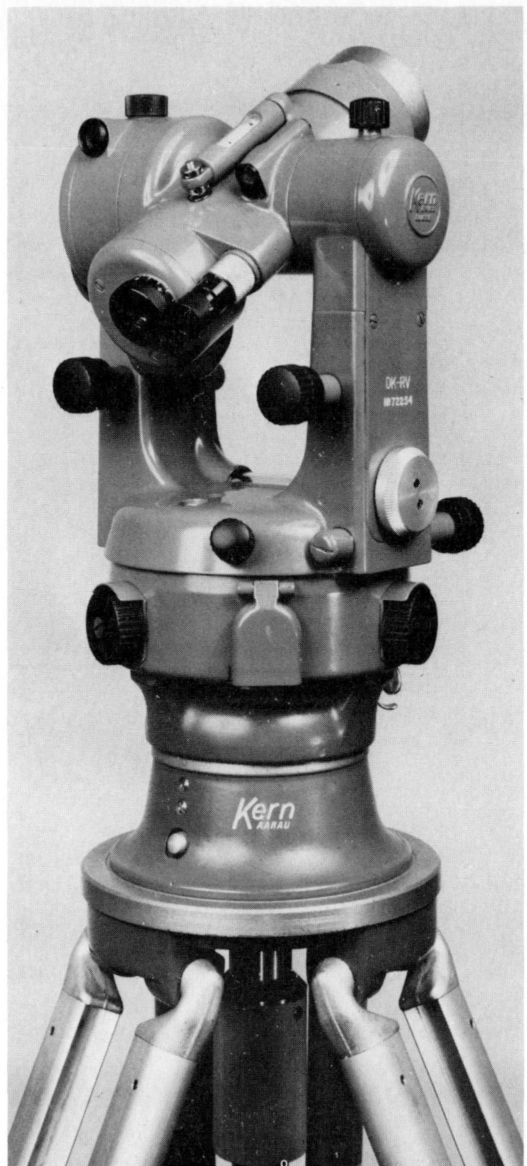

Figure 15-4. DK-RV reduction tacheometer. (Courtesy Kern Instruments, Inc.)

In Figure 15-5 (which does not apply to the DK-RV of Figure 15-4), the upper and lower lines (two outer lines) are curved to correspond to the variation in trigonometric function $\cos^2 a$ and are used for distance measurements. Two inner lines determine differences in elevation and are curved to represent the function $\sin a \cos a$. A vertical line, center cross, and short stadia lines are marked on a second, but fixed, glass plate, which is in focus simultaneously with the curve lines.

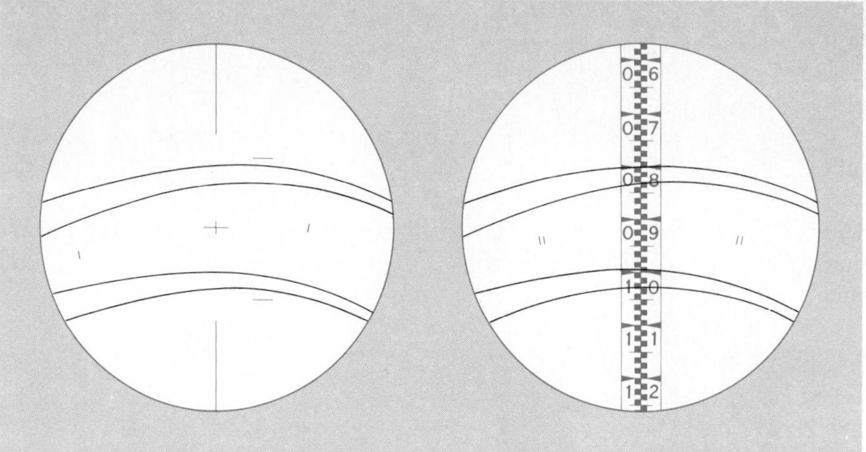

Figure 15-5. Curved stadia lines. **Left:** Field of telescope without rod. **Right:** Field of telescope with rod (horizontal distance = 100 × 0.227 = 22.7 m; difference in height = 20 × 0.167 = 3.30 m). (Courtesy Kern Instruments, Inc.)

A constant stadia-interval factor of 100 is used for horizontal distances. A factor of 20, 50, or 100 is applied to measurements of elevation differences. Its value depends on the angle of slope and is indicated by short lines placed between the elevation curves.

Other "diagram" tacheometers have basically worked on the same principle: Vertical angles are automatically compensated for by varying the stadia-wire separation. One self-reducing tacheometer utilizes a fixed horizontal line on one reticle and another horizontal line on a movable second reticle geared to a cam that operates with changes in vertical angle. Most planetable alidades employ some type of tacheometric reduction procedure.

A special movable-foot Topo Staff with zero set at the h.i. value is usually recommended to make tacheometric instruments fully self-reading.

15-7. FIELD NOTES. An example of stadia field notes is shown in Plate D-9. The transit or theodolite is usually oriented by a meridional azimuth, and clockwise angles are taken to desired points.

In stadia work, many shots may be taken from one hub. *Orientation should therefore be checked by sighting on the control line after every 10 or 20 topographic points and before leaving the hub.*

As previously stated, on nearly level ground horizontal sights should be used in place of inclined ones if possible. Rod readings then become minus sights, as in profile leveling. This procedure reduces chances for mistakes and simplifies calculations.

Columns in the notes are arranged in the order of reading—point sighted, rod intercept, azimuth, and vertical angle. A sketch and pertinent data are placed on the right-hand page as usual. Numbering topographic points begins with 1 at the first hub and continues successively clockwise through all others to eliminate possible duplication of numbers on any sketch. Note reduction is done in the office, unless information is needed immediately for control or plotting on a planetable sheet.

Newer methods use data collectors to expedite notekeeping and reduce tabulations. The notekeeper still must make sketches, however.

15-8. FIELD PROCEDURES. Proper field procedures save time and reduce the number of mistakes in all surveying operations. The order of taking readings best suited *for stadia work involving vertical angles* is as follows:

1. Bisect the rod with the vertical wire.
2. With the middle wire approximately on the h.i., set the lower wire at a full foot mark, or decimeter on metric rods.
3. Read the upper wire, and mentally subtract the lower wire reading from it to get the rod intercept. Record (or call out to the notekeeper) only the intercept.
4. Move the middle wire to the h.i. by using the vertical tangent screw.
5. Release the rodperson for movement to the next point by giving the proper signal.
6. Read and record the horizontal angle.
7. Read and record the vertical angle.

This procedure enables an instrument operator to keep two or three rodpersons busy in open terrain where points to be located are widely separated. The same order can be followed when the Beaman arc is used, but in step 4 the V scale is set to a whole number, and in step 7 the H- and V-scale readings are recorded.

When reading the stadia distance after the lower wire has been moved to a full foot mark, the middle wire is not precisely on the h.i. or the graduation read for the vertical angle. This does not normally cause a significant error in the reduction process except on long sights and steep vertical angles. Not having the rod plumb will, of course, cause appreciable errors, but use of a rod level eliminates this problem.

15-9. STADIA TRAVERSES. In a transit-stadia traverse, distances, horizontal angles, and vertical angles are measured at each point. Reduction of notes as the survey progresses provides elevations to be carried from hub to hub. Average values of stadia distances and differences in elevation are obtained from a foresight and backsight on each line. An elevation check should be secured by closing on the initial point, or a nearby bench mark for an open traverse. Although not as accurate as a taped traverse, a party of three members—an instrument operator, notekeeper, and rodperson—is typical. An added rodperson speeds work where many details are widespread.

The horizontal angles should also be checked for closure. Any angular misclosure should be adjusted, latitudes and departures computed, and traverse precision checked by the methods of Chapter 13.

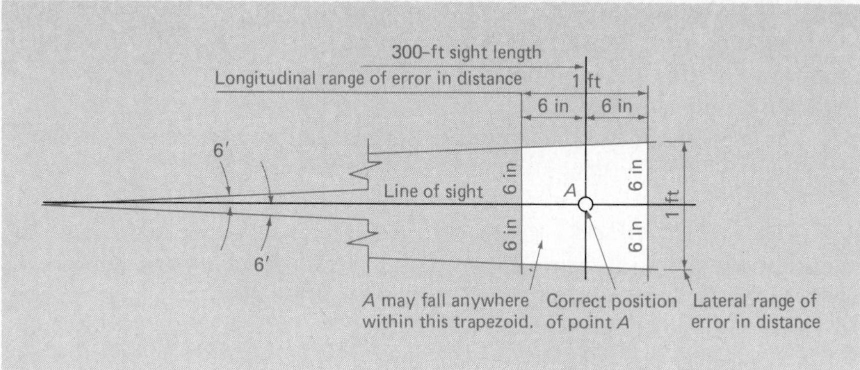

Figure 15-6. Comparable precision of angles and stadia distances.

15-10. TOPOGRAPHY. The stadia method is most useful in locating numerous topographic details, both horizontally and vertically, by transit or planetable. In urban areas, angle and distance readings can be taken faster than a notekeeper is able to record the measurements and prepare a sketch.

Use of stadia in topographic work is covered in more detail in the chapters on topography and the planetable.

15-11. STADIA LEVELING. The stadia method is adaptable to trigonometric leveling. The HI (height of instrument above datum) is determined by sighting on a station of known elevation, or by setting the instrument over such a point and measuring the height of the horizontal axis above it with a stadia rod. The elevation of any point can then be found by computation from the rod intercept and vertical angle. If desirable, a leveling circuit can be run to establish and check the elevations of two or more points.

15-12. PRECISION. A ratio of error of $\frac{1}{300}$ to $\frac{1}{500}$ can be obtained for a transit-stadia traverse run with ordinary care and reading both foresights and backsights. Shorter sights, a long traverse, and special procedures can give better ratios. Errors in stadia work are usually the result of poor rod readings rather than incorrect angles. An error of 1 min in reading a vertical angle does not appreciably affect the horizontal distance. The same 1-min error develops a difference in elevation smaller than 0.1 ft on a 300-ft sight for usual-sized vertical angles.

Figure 15-6 shows that if stadia distances are determined to the nearest foot (the usual case), horizontal angles *to topographic points* need be read only to within 5 or 6 min for comparable precision on 300-ft sights. A stadia distance given to the nearest foot is assumed to be correct to within about $\frac{1}{2}$ ft. Allowing the same $\frac{1}{2}$ ft error laterally, the direction can be off about 5 min (readily computed from sin 1 min = 0.00029). If an American transit is used, angles therefore can be read without using the vernier merely by estimating the position of the vernier index.

Accuracy of trigonometric leveling by stadia depends on sight lengths and size of vertical angles required.

15-13. SOURCES OF ERROR IN STADIA WORK. Errors that occur in transit and theodolite operations are inherent in stadia work, too. Additional sources include the following:

INSTRUMENTAL ERRORS

1. Improper spacing of the stadia wires.
2. Index error.
3. Incorrect length of rod graduations.
4. Axis of sight of a transit not parallel with the telescope-bubble axis.

PERSONAL ERRORS

1. Rod not held plumb. (Avoid by using a rod level.)
2. Incorrect rod readings resulting from long sights.
3. Careless leveling for vertical-arc readings.

Most errors in stadia work can be eliminated by (a) properly manipulating the instrument, (b) limiting lengths of sights, (c) using a good rod and rod level, and (d) averaging readings in the forward and backward directions. Line-of-sight errors cannot be corrected by field procedures—the instrument must be adjusted.

15-14. MISTAKES. Some typical mistakes in stadia work are:

1. Index error applied with wrong sign.
2. Confusion of plus and minus vertical angles.
3. Arithmetic mistakes in computing the rod intercept.
4. Use of an incorrect stadia interval factor.
5. Waving the rod. (*The rod should always be held plumb.*)

PROBLEMS

15-1. Upon what principle is the stadia method based?
15-2. When is it advantageous to use a transit as a level in stadia work?
15-3. What assumption is made in the stadia formula for inclined sights to simplify it?
15-4. Explain why the stadia "constant" is almost zero for an internal-focusing telescope.
15-5. Why should the line of sight through the lower cross line be kept several feet above any intervening ground if possible?
15-6. What stadia interval factor was used in some old "precise" levels?
15-7. An external-focusing telescope has a 10-in focal length. What spacing error of stadia lines changes the interval from 100.00 to 99.70?
15-8. Why is it more difficult to get good elevation closures than distance ones in transit-stadia surveying?
15-9. Is the subtense-bar method stadia in a horizontal plane? Explain.
15-10. Why is it very important in stadia work to have the instrument line of sight parallel with the telescope-bubble axis?
15-11. List the advantages of a Beaman stadia arc in reading distance and elevation differences?
15-12. Why should the scales of a Beaman arc be carefully checked before use?
15-13. Discuss the importance of paint colors on stadia rods.
15-14. Should both A and B transit verniers be read in stadia work? Explain.
15-15. Why should sights be taken both forward and backward on each course in running stadia traverses?

15-16. How can an open stadia traverse be checked?

15-17. How can a vertical angle reading be checked in the field when using a transit that has a stadia interval factor of 100?

15-18. Compute the error in distance if the top of a 12-ft stadia rod is inclined 6 in toward the observer and the rod intercept is 6.00 ft on a horizontal sight.

15-19. An instrument operator on a transit-stadia sight cannot set the center hair on the h.i. or see the upper hair because of overhanging branches. How can distance and elevation difference be found?

Compute the horizontal distances and elevation differences for the data in Problems 15-20 and 15-21.

15-20. Rod intercept = 5.54 ft, $a = -4°10'$, $K = 99.0$, $C = 0$.

15-21. Rod intercept = 6.17 ft, $a = +6°12'$, $K = 100$, $C = 1.2$ ft.

Calculate the rod intercept for the notes in Problems 15-22 and 15-23.

15-22. Horizontal distance = 200 ft, $a = 5°00'$ (on h.i.), $K = 98.0$, $C = 0$.

15-23. Horizontal distance = 135.0 ft, $a = +7°00'$ (on h.i.), $K = 101.4$, $C = 0$.

15-24. Specifications for a stadia traverse require an accuracy limit of $\frac{1}{500}$. Above what size vertical angle must inclined sights be reduced to the horizontal?

15-25. Similar to Problem 15-24, except for an accuracy of $\frac{1}{800}$.

At hub X, a sight is taken on BM Plug, elevation 175.0 ft. Compute the elevation of hub M in Problems 15-26 and 15-27.

15-26. $S = 400$ ft, $a = +1°30'$ to 7.00 on rod, h.i. = 4.8 ft, $C = 1.0$.

15-27. $S = 299$ ft, $a = +2°20'$ to 9.6 ft on rod, h.i. = 5.0 ft, $C = 0$.

Calculate the horizontal distance AB and the elevation of B for the Beaman arc readings taken at A, elevation = 345.6 ft, h.i. = 5.3 ft, for Problems 15-28 and 15-29. The V-scale index base is 50.

15-28. $R = 2.80$, V scale reads 78 to 6.7 ft on rod, and H scale = 9.

15-29. $R = 3.20$, V scale reads 72 to 5.8 ft on rod, and H scale = 5.

How close must angles be read for consistent precision of angles and stadia distances in Problems 15-30 through 15-32.

15-30. A sight of 650 ft, $C = 0$.

15-31. A sight of 200 m, $C = 0$.

15-32. A sight of 800 ft (estimated to nearest 5 ft), $C = 1.0$ ft.

At hub E, elevation 640.0, an FS is taken on hub F; the instrument is moved to F, and a BS taken on E. Compute the distance EF and the elevation of hub F in Problems 15-33 and 15-34.

15-33. At E, $R = 4.52$, $a = -3°15'$ to 7.2 on the rod, h.i. = 5.1 ft.
At F, $R = 4.50$, $a = +3°18'$ to 3.5 on the rod, h.i. = 5.2 ft.

15-34. At E, $R = 2.82$, $a + 1°52'$ to 7.6 on the rod, h.i. = 5.4 ft.
At F, $R = 2.82$, $a = -1°46'$ to 3.3 on the rod, h.i. = 4.9 ft.

BIBLIOGRAPHY

Colcord, J. E. 1971. "Tacheometry in Survey Engineering." *ASCE Journal of the Surveying and Mapping Division* 97(no. SU1): 39.

Harrington, E. L. 1955. "The Stepping Method of Stadia." *Surveying and Mapping* 15 (no. 4): 4.

Mussetter, W. 1953. "Stadia Characteristics of the Internal-Focusing Telescope." *Surveying and Mapping* 13(no. 1): 15.

Wolf, P. R., B. Wilder, and G. Mahun. 1978. "An Evaluation of Accuracies and Applications of Tacheometry." *Surveying and Mapping* 38(no. 3): 231.

16 TOPOGRAPHIC SURVEYS

16-1. INTRODUCTION. Topographic surveys are made to determine the configuration (*relief*) of the earth's surface and to locate *natural* and *cultural* features on it. By means of various lines and conventional symbols, topographic maps are produced from survey data. A topographic map is a large-scale representation of a portion of the earth's surface showing *culture, relief, hydrography*, and perhaps vegetation. Cultural (artificial) features are the products of people, such as roads, trails, buildings, bridges, canals, and boundary lines. Names and legends on maps identify the features.

Topographic maps are made and used by engineers to determine the most desirable and economical locations of highways, railroads, canals, pipelines, transmission lines, reservoirs, and other facilities; by geologists to investigate mineral, oil, water, and other resources; by foresters to locate fire-control roads and towers; by architects in housing and landscape design; by agriculturists in soil conservation work; and by archeologists, geographers, and scientists in numerous fields.

A *planimetric map* depicts natural and cultural features in plan only. A *hypsometric map* shows relief by conventions such as *contours, hachures, shading*, and *tinting*.

16-2. METHODS FOR TOPOGRAPHIC SURVEYING. Topographic surveys are conducted by either aerial (photogrammetric) or ground (field) methods, and often a combination of both. Refined equipment and procedures available today have made photogrammetry accurate and economical; hence, almost all topographic mapping projects covering large areas now employ this method. Ground surveys are still frequently used, however, especially for preparing large-scale maps of small areas. Even when photogrammetry is utilized, ground surveys are necessary to establish control and to field-check mapped features for accuracy. This chapter concentrates on ground methods. Several field procedures for locating topographic features, both horizontally and vertically, will be described. Photogrammetry is discussed in Chapter 28.

16-3. CONTROL FOR TOPOGRAPHIC SURVEYS. The first requirement of any topographic survey is good control, whether the survey is done by ground or aerial methods. Control, discussed in Chapter 20, is classified as either horizontal or vertical.

Horizontal control is provided by two or more points on the ground, precisely fixed in position horizontally by distance and direction. It is the basis for map scale and locating topographic features. Horizontal control is usually established by traversing, triangulation, trilateration, or inertial and satellite methods, and can be filled in photogrammetrically for large areas.

For small areas, horizontal control for topographic work is generally established by a traverse, although a single line may suffice in some cases. Triangulation and trilateration furnish the most economical basic control for surveys extending over a state or the entire United States. These techniques may, however, give way in the future to inertial systems and satellite doppler receivers. Monuments of the state plane coordinate systems are excellent for all types of work, but unfortunately more are needed in most areas.

Specified maximum allowable closure errors for both horizontal and vertical control should be established in advance of field work.

Vertical control is provided by bench marks in or near the tract to be surveyed. It becomes the foundation for correctly portraying relief on a map. A vertical control net is established by lines of levels starting from and closing on bench marks. Elevations are ascertained for all traverse hubs, with provision in some cases for marks set nearby and out of the way of construction. A lake surface is a continuous turning point or bench mark and may sometimes be used. Even a gently flowing stream may serve as supplementary control. Trigonometric and barometric leveling can be employed to extend vertical control in rugged terrain, but the latter is less accurate.

Topographic details are usually built upon a framework of traverse hubs whose positions and elevations have been established. Any errors in the hub positions or their elevations are reflected in the location of topography. It is advisable, therefore, to run, check, and adjust the traverse and level circuits before a topographic detail survey is begun, rather than carry on both processes simultaneously. This is particularly true in planetable work, where an error in elevation or position of an occupied station will displace the plotted locations of all cultural features and contours.

Field measurement errors are more difficult to eliminate than those in mapping procedures. Even though plotting measurements that have been made to the nearest 0.01 ft and 1 sec is not possible, the data may be used for other purposes. Thus special care is required in taking and recording field data.

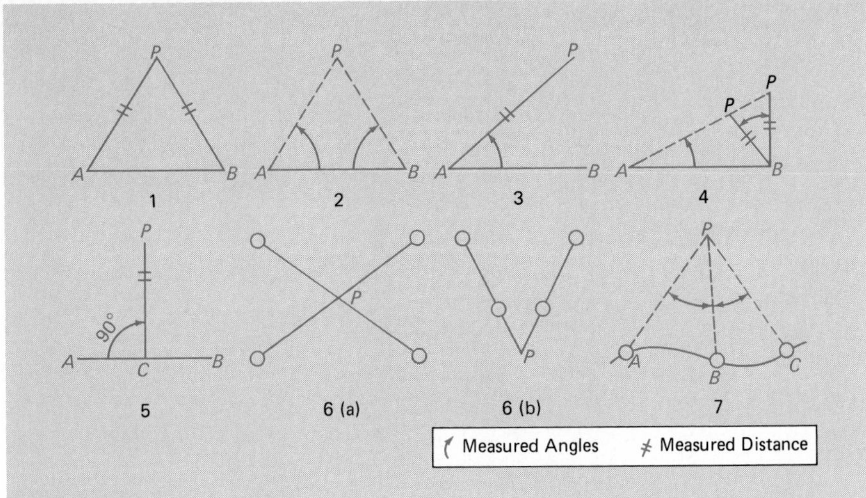

Figure 16-1. Locating a point P.

The kind of control (traverse, triangulation, or trilateration) and method selected to get topographic details govern the speed, cost, and efficiency of a topographic survey. A theodolite–EDMI combination, total-station instrument with tracking system, data storage accessory, and voice recorder plus a communication system for descriptions simplify the notekeeper's job and reduce field errors. Sketches still must be made in field books, however. Various equipment combinations can be selected for the location methods discussed in succeeding sections.

16-4. METHODS OF LOCATING TOPOGRAPHIC DETAILS IN THE FIELD. Objects to be located in a survey can range from single points to meandering streams and complicated geological formations. The process of tying topographic details to the control net is called *detailing*.

Seven methods used to locate a point P in the field are illustrated in Figure 16-1. All are based on horizontal control. One line, AB, must be fixed in each of the first four methods, and its length known in methods 1, 2, and 4. Positions of three points must be known or identifiable to apply the seventh method called *resection* or the *three-point problem.* Quantities to be measured in the respective diagrams are:

1. Two distances.
2. Two angles.
3. One angle and the adjacent distance.
4. One angle and the opposite distance (two possible points P).
5. One distance and a right-angle offset.
6. The intersection of string lines from straddle hubs.
7. Two angles at the point to be located.

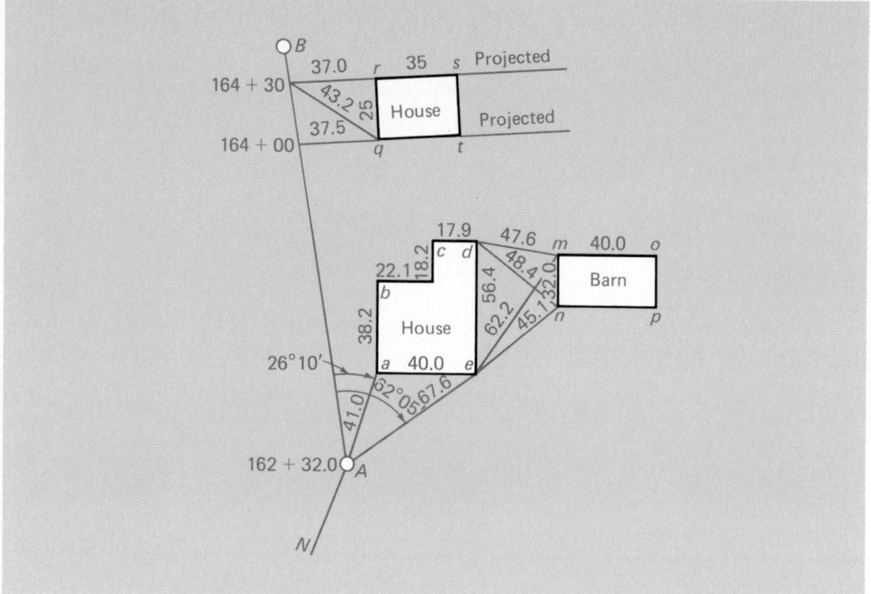

Figure 16-2. Location of details.

Method 3 is used most often, but an experienced party chief employs whichever method is appropriate in a given situation, considering both field and office (computation and map) requirements.

16-5. LOCATION OF LINES. Most objects are located by considering them as composed of straight lines, with each line being determined by two points. Irregular or curved lines can be assumed straight between points sufficiently close together; thus, detailing becomes a process of locating points. Two examples will illustrate the measurements made to locate straight and curved lines.

In Figure 16-2, house *abcdea* is to be tied to traverse line *AB*. The building shape requires defining only two main corners, such as *a* and *e*. Corner *a* could be located by any of the first five methods of Figure 16-1, but an angle and distance are used as shown. Good practice adds a third corner, if possible, to provide a check. All house sides are measured by taping and the lengths recorded on a sketch. Location of the barn by measurements from the house is satisfactory if only a tape is available, but determining two angles and distances from the traverse hubs provides a quicker and independent result. Stations projected on the center-line traverse from the second house, *q*, *r*, *s*, and *t*, are practical in route surveys, but again angles and distances are a strong alternative.

If a transit or theodolite is used to measure the angles, distances to *a* and *e* can be secured by taping or stadia. When a combination theodolite–EDMI is used for detailing, distances can be measured electronically—a convenience where long lines are involved. All measurements may be shown on a sketch, but generally angles and distances from traverse points are tabulated on the left-hand page of a field book to avoid crowding the drawing. Topographic

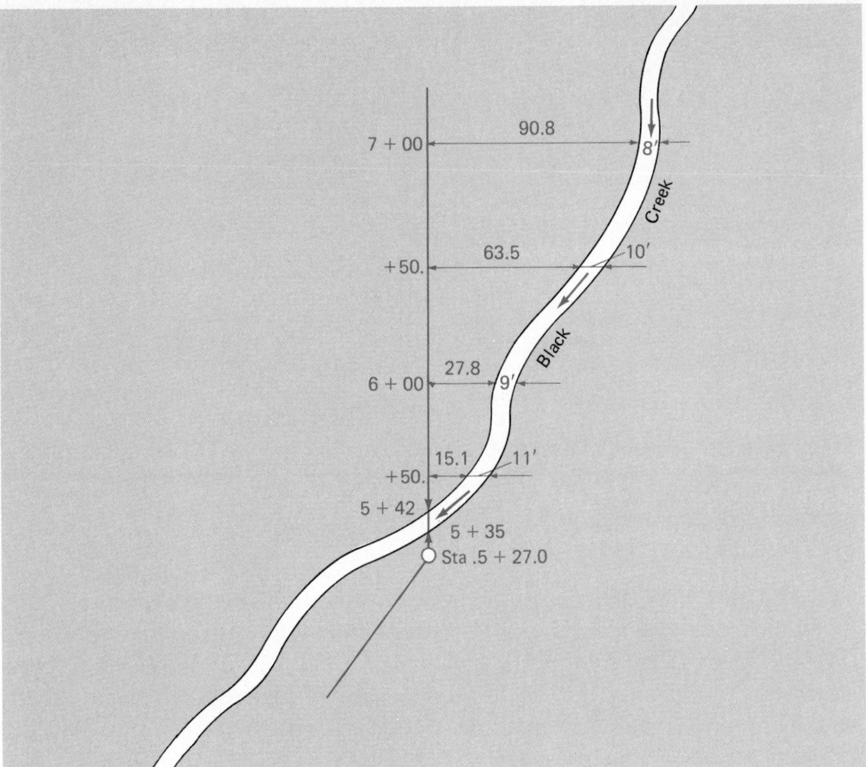

Figure 16-3. Location of creek by offset method.

points are then identified by consecutive numbers, rather than letters, as on Plate D-9.

Location of a crooked creek by using the offset method is shown in Figure 16-3. At intervals along the traverse line, offsets to the edge of the creek are measured. They can be taken at regular intervals or spaced at distances that permit the line to be considered straight between successive offsets.

16-6. LOCATION OF LINES FROM A SINGLE POINT. The single-point method can be used to locate lines of a closed figure, such as the boundaries of a field. Some point, O, is chosen from which all corners can be seen, as in Figure 16-4(a). The direction to each corner is found by measuring all central angles, or by azimuths from point O. Lengths of all radiating lines, such as OA and OB, are determined by stadia, taping, or EDM, and the perimeter lengths computed by trigonometry, since two sides and the included angle for each triangle are known. As a check, coordinates of each corner can be calculated from the lengths and directions of the radiating lines, and distances AB, BC, and so on, computed and verified by measurements.

A more rigid solution of the boundary lengths is obtained with the method presented in Figure 16-4(b). A line OO' is chosen as the base and its length carefully measured. Angles are taken from each end of the base to all corners, and all radiating lines measured as in the single-point method. Since this procedure

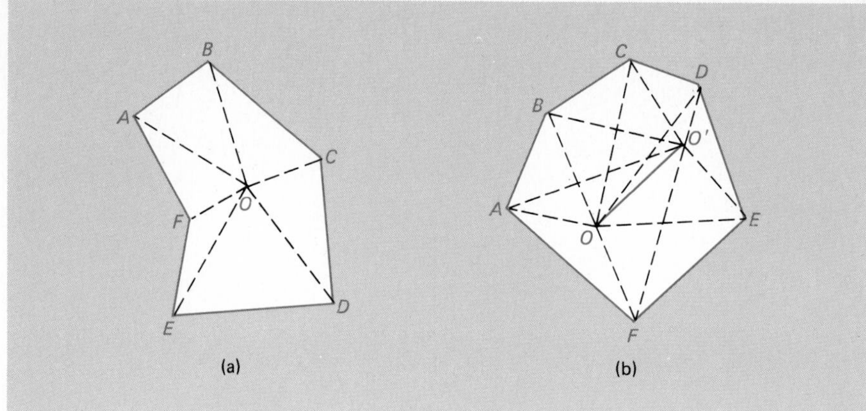

Figure 16-4. (a) Location of lines from single point. (b) Location of lines from base line.

provides redundant measurements, an adjustment can be made—for example, by least squares.

16-7. CONTOURS. The best method of quantitatively representing hills, mountains, depressions, and ground-surface undulations on a two-dimensional sheet of paper is by contours. *A contour is a line connecting points of equal elevation.* Contours may be visible, as in the case of a lake shoreline, but usually on the ground, elevations of only a few points are located and contours sketched between these controls.

Contours are shown on maps as the traces of level surfaces of different elevations (Figure 16-5). Thus level surfaces cutting a vertical cone form circular

Figure 16-5. Contour lines.

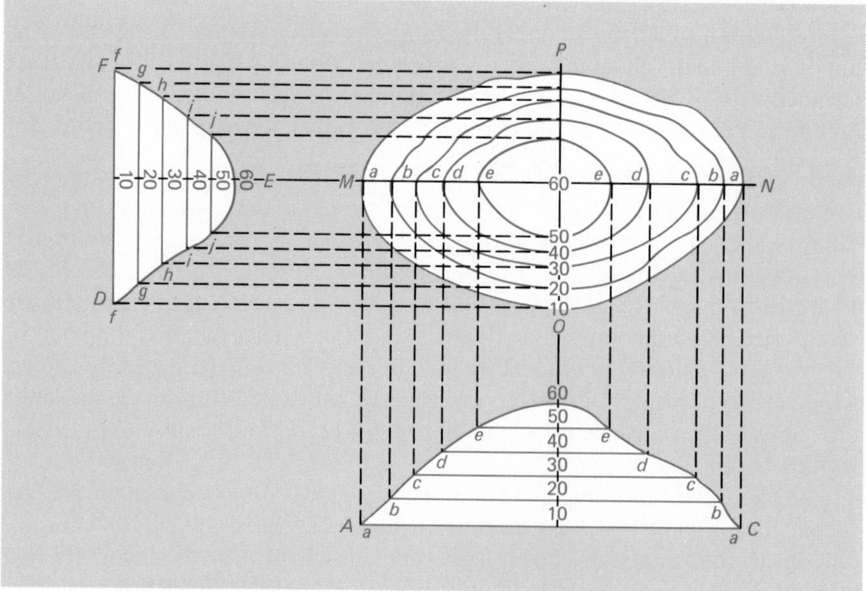

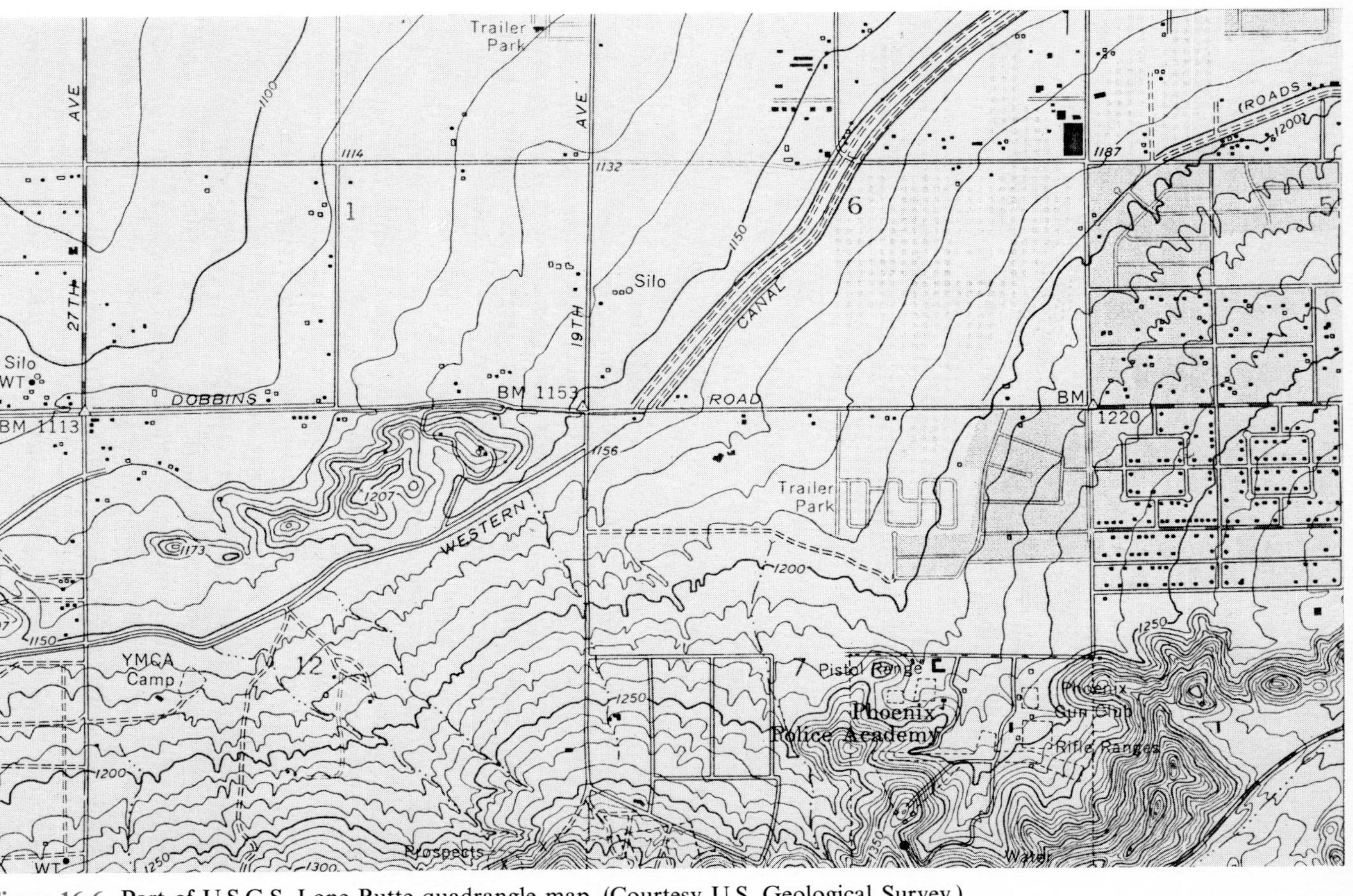

Figure 16-6. Part of U.S.G.S. Lone Butte quadrangle map. (Courtesy U.S. Geological Survey.)

contours and intersect a sloping cone to produce elliptical ones. On uniformly sloping surfaces, such as those in highway cuts, contours are straight lines.

Most contours are irregular lines like the closed loops for the hill in Figure 16-5. The vertical distance between level surfaces forming the contours is called the *contour interval.* For topographic quadrangles at 1:24,000 scale, the U.S. Geological Survey uses one of the following contour intervals: 5, 10, 20, 40, or 80 ft. Contour intervals in meters sometimes replace foot units.

The contour interval selected depends on a map's purpose, scale, and diversity of relief in the area. Reducing the interval requires more costly and precise field work. In regions where both flat coastal areas and mountainous terrain are included in a map, supplementary contours at one-half or one-fourth the basic interval are drawn (and shown in dashed lines).

Figure 16-6 is a contour map showing 10-ft contours. *Spot elevations* are given for critical point such as peaks, sags, streams, and highway crossings. Sketching ridge, valley, and drainage lines (dashed) prior to drawing contours is desirable.

16-8. PROPERTIES OF CONTOURS. Certain properties of contours are fundamental in their location and plotting.

1. Contour lines must close upon themselves, either on or off a map. They *cannot* deadend.
2. Contours are perpendicular to the direction of maximum slope.
3. The slope between contour lines is assumed to be uniform. If it is not, all breaks in grade should be located in topographic mapping.
4. The distance between contours indicates the steepness of a slope. Wide separation denotes gentle slopes; close spacing, steep slopes; even and parallel spacing, uniform slope.
5. Irregular contours signify rough, rugged country. Smooth lines imply gradual slopes and changes.
6. Concentric closed contours that increase in elevation represent hills. A contour forming a closed loop around lower ground is called a *depression contour.* Hachures inside the lowest contour and pointing to the bottom of a hole or sink with no outlet make map reading easier. Contour elevations are shown on the uphill side of lines or in breaks to avoid confusion, and for at least every fifth contour.
7. Cuts and fills for earth dams, levees, highways, railroads, canals and so on, produce straight or geometrically curved contour lines with uniform, or uniformly graduated, spacing.
8. Contours of different elevations never meet except on a vertical surface such as a wall, cliff, or natural bridge. They cross only in the rare case of a cave or overhanging shelf. Knife-edge conditions are seldom, if ever, found in natural formations.
9. A contour cannot branch or wye into two contours of the same elevation.
10. Controlling features in locating contours are usually drainage lines.
11. Contour lines crossing a stream point upstream and form V's; they point down the ridge and form U's when crossing a ridge crest.

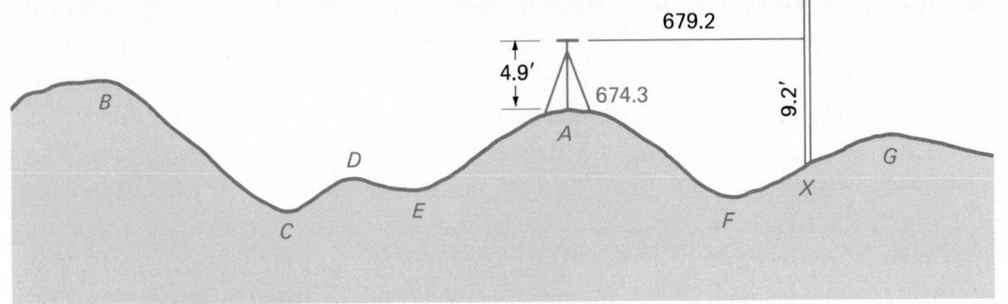

Figure 16-7. Direct method of locating contours

12. Contour lines go in pairs up valleys and sides of ridges. They tend to parallel streams.
13. Contours cross sloping, crowned streets in typical U-shaped curves.
14. The shoreline of a small lake forms a contour if inflow, outflow, and wind effects are negligible.

Keeping these principles in mind will make it easy to visualize contours when looking at an area, and prevent serious mistakes in sketching. Numerous points may be necessary to locate a contour in certain types of terrain. For example, in the unusual case of a level field that is at contour elevation, the exact location of a single unique contour would be time-consuming or perhaps impossible.

16-9. DIRECT AND INDIRECT METHODS OF LOCATING CONTOURS. Contours can be established by either the *direct method* (*trace-contour method*) or the *indirect method* (*controlling-point method*). In the direct method, the rod reading (foresight) that must be subtracted from the HI to give the contour elevation is determined. The rodperson then selects trial points believed to give this minus sight and is directed uphill or downhill by the instrument operator until the required reading is actually secured (within 0.1 to 0.5 ft, the allowable discrepancy depending on the terrain and specified accuracy).

In Figure 16-7 the instrument is set up at point *A*, elevation 674.3 ft, h.i. of 4.9 ft, and HI of 679.2 ft. If 5-ft contours are being located, a reading of 4.2 or 9.2 with the telescope level will place the rod on a contour point. For example, in Figure 16-7, the 9.2-ft rod reading means that point *X* lies on the 670 ft contour. After this point has been located by trial, distance and azimuth are read and the process repeated. Work is speeded by using a piece of red cloth perhaps 0.2 ft wide, which can be moved up and down on the stadia rod to mark the required reading and eliminate searching for a number.

The maximum distance between contour points is determined by the terrain and accuracy required. The tendency for beginners is to take more sights than necessary in ordinary terrain. Contours are sketched between the located points as part of the drafting-room work, but they may also be drawn in the field book to clarify unusual conditions.

Unless the rod can be read by using a transit or theodolite as a level, the direct method of locating contours is impractical. Too much time is wasted juggling the combination of a vertical angle and stadia distance to get the required difference in elevation. Self-reducing tacheometers, or EDMIs that can automatically resolve slope distances to their horizontal and vertical components, are effective in this work, however.

For the indirect method, a rod is set on critical points where changes in ground slope occur, such as *B*, *C*, *D*, *E*, *F*, and *G* in Figure 16-7. Elevations are procured with the telescope level whenever possible to save time in reducing notes and increase accuracy. Points are selected at random or along pertinent azimuth lines. The rodperson moves clockwise to facilitate notekeeping and plotting. Contours are interpolated between the high and low points whose elevations are found.

The direct method is advantageous in gently rolling country; the indirect method better in rough, rugged terrain.

Another system, a *radial grid*, can be laid out with a total-station instrument to locate contour lines accurately by the indirect method in rolling terrain. Radial lines at 20° (or other) intervals are projected from a traverse hub as far as desired, horizontal distances of 50 or 100 feet staked, and elevations determined. Supplementary readings can also be made on random selected radial-line points. The direct method of locating contours can be employed in a similar manner.

If a hill or ridge nearby overlooks a large part, or all, of the area to be surveyed, a stub traverse line point on it permits more and longer radial lines. Intersecting radial lines from adjacent traverse hubs insure thorough coverage and provide checks on contour locations.

16-10. FIELD METHODS OF OBTAINING TOPOGRAPHY.

Location of topographic details is usually accomplished by one of the following methods: (a) radiation, (b) stadia, (c) planetable, (d) coordinate squares (grid), or (e) offsets from a center line (cross-sectioning or cross-profiling). A brief explanation of the uses, advantages, and disadvatanges of each system will be given.

16-10.1. RADIATION METHOD.

In the radiation method, traverse stations are occupied with a transit or theodolite, and angles to desired contour points and features measured. Distances are found by taping or EDM. After corners of buildings, bridges, and other details have been located, their lengths, widths, and projections are taped and sketched in the field book. The radiation procedure is accurate and efficient if a combination theodolite–EDMI (total-station instrument) with self-reduction capability is used. If distances must be taped, however, it may be too slow and costly for ordinary work.

16-10.2. STADIA METHOD.

The stadia method is similar to the radiation process, except that distances are secured by stadia. This procedure is rapid and sufficiently accurate for most topographic surveys. Stadia distances, azimuths, and vertical angles are read for lines radiating from the transit, theodolite, or planetable to required points. Plate D-9 shows sample notes for a transit-stadia survey.

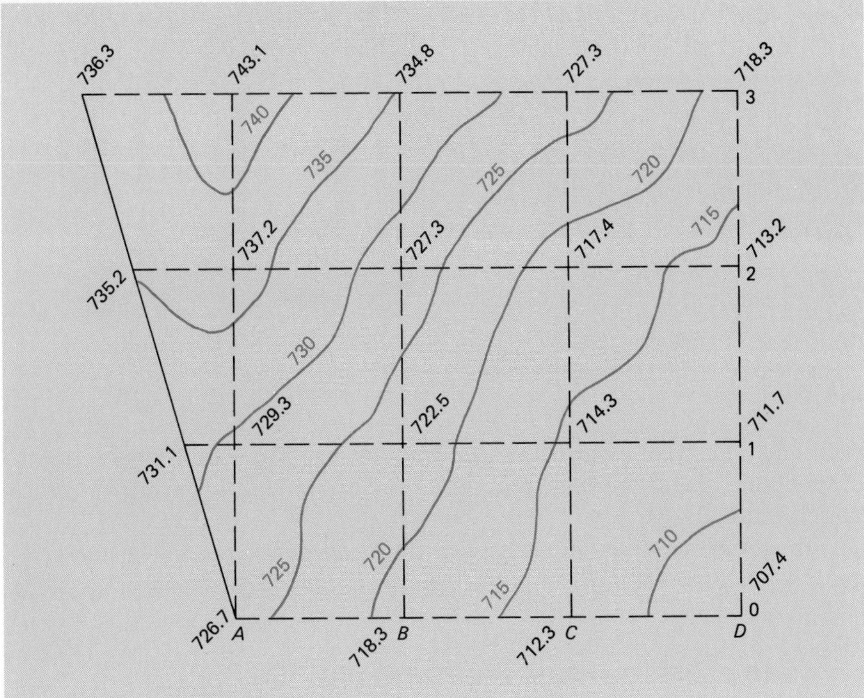

Figure 16-8. Coordinate squares.

16-10.3. PLANETABLE METHOD. In this method an alidade is sighted on a rod held at the point to be located, and a stadia distance and vertical angle (or Beaman arc) read. Direction of the line is drawn along the alidade ruler, thus eliminating the need to measure or record any horizontal angles. Vertical angles are also avoided, if possible, by using the alidade as a level.

The instrument operator sketches contours by either the direct or indirect method while looking at the area. Since the map is plotted in the field, coverage can be checked by observation. Planetable use is discussed in Chapter 18 in more detail.

16-10.4. COORDINATE SQUARES METHOD. The method of coordinate squares (grid method) is better adapted to locating contours than cultural features but can be used for both. The area to be surveyed is staked in squares 10, 20, 50, or 100 ft (5, 10, 20, or 40 m) on a side, the size depending on the terrain and accuracy required. A transit or theodolite can be used to lay out lines at right angles to each other, such as *AD* and *D*3 in Figure 16-8. Grid lengths are marked and the other corners staked by intersections of taped lines. Corners are identified by the number and letter of intersecting lines. If a transit is not available, all layout work can be done with a tape.

To obtain elevations of the corners, a level is set up in the middle of the area, or in a position from which level sights can be taken on each point. Contours are interpolated between the corner elevations (along the sides of the

Figure 16-9. Double pentagonal prism (on side). (Courtesy Kern Instruments, Inc.)

blocks) by estimation or by calculated proportional distances. Except for plotting contours, this is the same procedure as that used in the borrow-pit problem in Section 27-10.

In plotting contours by the grid method, *elevations obtained by interpolation along diagonals will generally not agree with those from interpolation along the four sides because the ground's surface is warped.*

16-10.5. OFFSETS FROM THE CENTER LINE. After the center line for a route survey has been established, a profile is run to get elevations at regular stations and critical points. Details such as fences and buildings are located by right-angle offsets. The right angle can be measured by a *pentagonal prism* of the type shown in Figure 16-9, or estimated by standing on the center line, pointing the arms in opposite directions along the line, then bringing the palms of the hands together with arms outstretched in front of the body (eyes should be closed to prevent unconsciously aiming at a prominent object near the perpendicular line).

Cross-sectioning consists in taking a vertical section on the surface of the ground at right angles to the center line of a route survey. In effect it is profiling normal to the center line. Rod readings are secured at all breaks in the ground surface and recorded with their respective distances out. Plate D-10 is a sample set of cross-section notes.

Cross-section data can be used for compiling contours, but more commonly it is the basis for making earthwork volume computations. In this application, the distance from centerline and elevations are plotted on special paper that is most commonly ruled in 1-in squares divided decimally with lighter lines 0.1 in apart horizontally and vertically. Cross-section notes—when plotted along with the design template (base width and side slopes) for a proposed highway or canal—outline the areas of cuts and fills, which are readily measured or computed. Excavation quantities can be calculated from cross-section areas and their known distances apart as described in Chapter 27.

In recent years, electronic computers have generally eliminated the need to manually plot cross sections and planimeter areas. Instead, cross-section field notes and design-template values are read into a computer, together with an appropriate program that calculates and lists cut and fill areas, also excavation and fill volumes. Cross sections with superimposed design templates can be plotted automatically by a fast electronic computer accessory.

On some surveys, contour points are directly located along with any important changes in ground slope. For example, if 5-ft contours are being delineated, a typical set of notes on the right-hand page of a field book would be in the following form:

L				CL	R			
$\frac{85}{63}$	$\frac{80}{45}$	$\frac{75}{28}$	$\frac{70}{9}$	$\frac{72.4}{0}$	$\frac{75}{12}$	$\frac{80}{27}$	$\frac{85}{38}$	$\frac{90}{56}$

Numbers above the line represent contour elevations; those below it are distances out from the center line.

Contour points can be plotted on cross-section paper to get the equivalent of a set of cross-section notes, or the contours may be drawn on the plan view of a proposed highway.

Readings are taken with a level, transit, theodolite, or, in some cases, a hand level.

16-10.6. CONTOURS BY HAND LEVEL. A hand level can be used for contouring when a high order of precision is not required. In this procedure, the observer's HI (eye height above datum) is determined first by taking a plus sight (or standing) on a point of known elevation, as in differential leveling. In going uphill, the point at which a level line of sight strikes the ground will have the same elevation as the observer's HI. After identifying this point on the ground, the observer moves there and occupies it to compute a new HI.

In leveling downhill, an observer finds by trial the location required to make a level backsight strike the ground at the previously occupied position. Elevation of the trial point is then found by subtracting the observer's h.i. (height of eye above ground), and the process repeated. In locating 5-ft contours, adoption of an exact 5-ft h.i. by using a rod or forked stick is desirable to speed work and simplify arithmetic. Horizontal positions of contour points thus located must be determined by one of the methods described in previous sections.

16-11. SELECTION OF FIELD METHOD. Selection of the field method to be used on any topographic survey depends on many considerations, including (1) purpose of the survey, (2) map use (accuracy required), (3) map scale, (4) contour interval, (5) size and type of area involved, (6) cost, (7) equipment and time available, and (8) experience of the survey personnel.

Items 1–5 are interdependent. The cost, of course, will be a minimum if the most suitable method is selected for a project. On large-scale work, personnel cost rather than equipment investment will govern (except perhaps in photogrammetric offices). The method chosen by a private surveyor making a topographic survey of 50 or 100 acres, however, may be governed by the equipment owned.

Special training is necessary before the average surveyor can do photogrammetric work. Likewise, relatively few persons have had enough planetable experience to be efficient in its operation.

16-12. SPECIFICATIONS FOR TOPOGRAPHIC SURVEYS. *National map standards of accuracy* specifications for the maximum errors permitted in horizontal positions and elevations shown on maps are as follows:

Horizontal Accuracy. For maps to scales larger than 1:20,000, not more than 10% of the points tested shall be in error by more than $\frac{1}{30}$ in (0.8 mm). On smaller-scale ones, the limit of error is $\frac{1}{50}$ in (0.5 mm), or approximately 40 ft on the ground at a map scale of 1:24,000. These limits of accuracy shall apply in all cases to positions of well-defined points only, such as monuments, bench marks, highway intersections, and building corners which can be plotted to 0.1 in at map scale.

With normal vision, distances can be plotted within $\frac{1}{50}$ or $\frac{1}{100}$ in (or 0.5 to 0.25 mm). At a scale of 1 in = 100 ft, map positions would then be within 1 or 2 ft. Greater field accuracy requirements for topography can be set accordingly.

Vertical Accuracy. Not more than 10% of the elevations tested shall be in error by more than one-half the contour interval. Published maps meeting these accuracy requirements usually note in their legends: "This map complies with the National Map Accuracy Standards."

The accuracy of any map may be tested by comparing the positions of points whose locations or elevations are shown on it with corresponding positions determined by surveys of a higher accuracy. Plotted horizontal positions of objects are checked by an independent traverse, triangulation, or trilateration run to points selected by the person or organization for whom the survey is made. A profile run from any point in any arbitrary direction is compared with one made from the plotted contours; thus, both field work and map drafting are checked.

16-13. SOURCES OF ERROR IN TOPOGRAPHIC SURVEYS. Some sources of error in topographic surveys are:

1. Instrumental errors, especially the line of sight being maladjusted in stadia work.
2. Errors in reading instruments.
3. Control not established, checked, and adjusted before topography is taken.
4. Control points too far apart and poorly selected for proper coverage of an area.
5. Sights taken on detail points that are too far away.
6. Poor selection of points for contour delineation.

16-14. MISTAKES. Some typical mistakes in topographic surveys are:

1. Unsatisfactory equipment or field method for the particular survey and terrain conditions.
2. Mistakes in instrument reading.

3. Failure to check azimuth orientation when many detail points are located from one instrument station.
4. Too few (or too many) contour points taken.
5. Omission of some topographic details.

PROBLEMS

16-1. List five topographic details classified as "cultural" features not mentioned in Section 16-1.
16-2. What advantages does a lake have in serving as a turning point or bench mark?
16-3. Describe the best methods to determine the water surface elevation of a lake disturbed by light winds and small waves?
16-4. Sketch a good application of each of the seven field methods for locating points.
16-5. Prepare a set of field notes to locate the topographic details in Figure 16-2. Scale distances and angles.
16-6. Give examples and sketch terrain conditions for which the best method of locating contours would be by the (a) grid method, (b) direct method, and (c) indirect method.
16-7. What is the purpose of preparing a topographic map for a new portion of a city? Of a city park? For the site of an industrial park?
16-8. Explain how the map in Figure 16-6 can be tested for accuracy by using an overlay sketch.
16-9. A contour map has an interval of 2 ft and a scale of 1:1200. If two adjacent contours are 80 ft apart, what is the average slope of the ground between the contours?
16-10. On a map whose scale is 1 in = 500 ft, how far apart would 1-ft contours be on a uniform slope (grade) of 4%?
16-11. On a map drawn to a scale of 1 in = 200 ft, contour lines are $\frac{3}{8}$ in apart at a certain place. Contour interval is 5 ft. What is the ground slope in percent between adjacent contours?
16-12. Same as Problem 16-11, except for a 2-m interval, 30-mm spacing, and map scale 1 cm = 50 m.
16-13. List and give examples of the pertinent factors considered in selecting the proper scale for a transit-stadia topographic mapping project.
16-14. Sketch a contour that crosses a 40-ft-wide street having a +5.00% grade, an 8-in parabolic crown, and 4-in-high curbs.
16-15. Same as Problem 16-14, except an 80-ft four-lane divided highway including a 20-ft median strip, on a +3.00% grade, 1.00% side slope, and 4-ft-wide shoulders with a minimum slope.
16-16. For a 25-ft contour interval, what is the greatest error in elevation expected of any point read from a map if it complies with National Map Accuracy Standards?
16-17. When should points located for contours be connected by straight lines? When by smooth curves?

The following are elevations at the corners of 50-ft coordinate squares and apply to Problems 16-18 through 16-21.

57	56	57	56	60	64
60	53	58	61	63	66
58	59	56	58	59	67

16-18. About which number in the table can a 5-ft closed contour be drawn?
16-19. Draw 5-ft contours for the area. Compare elevations at locations of several square centers by interpolating on diagonals.

16-20. Same as Problem 16-19, except add a fourth line of elevations: 55, 51, 56, 60, 68, and 71 (from left to right).

16-21. Same as Problem 16-20, except add a fifth line of elevations: 52, 54, 58, 61, 69, and 70.

16-22. Give an additional property of contours not mentioned in Section 16-8.

16-23. In your opinion, which is the most important of the 14 properties of contours listed in Section 16-8? Why?

16-24. How are "form lines" used on some maps or charts instead of contours?

16-25. On what kinds of surveys is the hand level suitable for contour location?

16-26. Why is the offsets-from-the-center-line method used more than any other ground system on route surveys to locate points?

16-27. Is the direct method of locating contours faster than the indirect method for ordinary terrain? More accurate? Explain.

16-28. A city area has many details to be located by transit stadia. Explain the best method to make notekeeping and plotting easier.

16-29. Is a person who makes a topographic survey to landscape an estate, using the property boundary lines as the control, required to be an LS in your state?

BIBLIOGRAPHY

Brown, R. L. 1980. "Proposed Manual on Selection of Map Uses, Scales, and Accuracies for Engineering and Associated Purposes: Map Availability—Chapter VI. "*ASCE Journal of the Surveying and Mapping Division* 106(no. SU1):149.

Crombie, B. W. 1977. "Contour Design and the Topographic Map User." *Canadian Surveyor* 31:34.

Eliel, L. T., et al. 1952. "Selection of Contour Intervals." *Surveying and Mapping* 12 (no. 4):344.

Feldscher, C. B. 1980. "A New Manual on Map Uses, Scales and Accuracies." *ASCE Journal of the Surveying and Mapping Division* 106(no. SU1):143.

Hotine, M. 1975. "Rapid Topographic Surveys of New Countries." *Surveying and Mapping* 25(no. 4):557.

Keates, J. S. 1961. "Techniques of Relief Representation." *Surveying and Mapping* 21 (no. 4):459.

Lee, M. P. 1981. "Automated Slope—a New Technique."*ACSM Bulletin* 74:43.

Lyddan, R. H. 1954. "How Much Topographic Detail?" *Surveying and Mapping* 14 (no. 1):29.

Thompson, M. M., and G. H. Rosenfeld. 1971. "On Map Accuracy Specifications." *Surveying and Mapping* 31(no. 1):57.

Wolf, P. R., B. Wilder, and G. Mahun. 1978. "An Evaluation of Accuracies and Applications of Tacheometry." *Surveying and Mapping* 38(no. 3):231.

17 MAPPING

17-1. INTRODUCTION. Throughout the ages, maps have had a profound impact on human activities, and today the demand for them is perhaps greater than ever. They are of utmost importance in engineering, resource management, urban and regional planning, management of the environment, construction, conservation, geology, agriculture, and many other fields. Maps show various features—for example, topography, property boundaries, transportation routes, soil types, vegetation, landownership for tax purposes, and mineral and resource locations. Maps are especially important in engineering for planning project locations, designing facilities, and estimating contract quantities.

The military services have always depended heavily on a steady flow of up-to-date maps and charts. During World War II, the Army Map Service, now the Topographic Center, Defense Mapping Agency (DMA), prepared more than 40,000 maps of all types covering approximately 400,000 mi^2 of the earth's surface; 500 million copies were printed. The Normandy invasion alone required 70 million copies of 3000 different maps. In the first four weeks of the Korean conflict, the Army Map Service and the Far East Command printed and distributed 10 million copies of maps—more than printed during all of World War I. Because of the size and dispersion of military forces in Vietnam, the Army Map Service issued an estimated 500 million copies of maps to support that conflict. Complete country coverage was made in 1:50,000-scale maps, and most was also shown at 1:25,000. Numerous special products for riverine operations and photo maps were prepared.

17-2 MAPPING AGENCIES. Maps are prepared by private surveyors, industries, cities, counties, states, and several agencies of the federal government. Unfortunately, many mapping activities have not been coordinated; hence, some duplication of effort has occurred, and the existence of much valuable information is unknown and thus unavailable to prospective users. A start has been made to improve this situation by setting up depositories in some cities and states, where every obtainable map of the area is filed for use by interested persons. The National Cartographic Information Center provides information helpful to surveyors, engineers, cartographers, and other technical map users, as well as the general public.[1] This office is a central source of information on surveys and maps and provides comprehensive data on topographic maps, aerial photography, and geodetic control surveys, plus basic facts needed for engineering and construction programs, such as irrigation projects, highway and railroad location, urban and rural development, transmission and pipelines, airports, and the location of radio and television facilities.

The Geological Survey began publishing topographic maps in 1886 as an aid to scientific studies. Standard sheets cover $7\frac{1}{2}$- or 15-min quadrangles and show cultural features in black, contours in brown, water features in blue, urban regions in red, and woodland areas in green. An index map giving the status of topographic mapping in the United States and its territories and possessions is available free of charge from the Geological Survey. Other index maps showing the status of aerial photography and aerial mosaics in the United States are published by the same agency.

The Geological Survey is proceeding toward metrication with due consideration for the needs of map users. Map elements shown in the metric system are the Universal Transverse Mercator (UTM) coordinate grid, contours, elevations, distances, bathymetric contours and soundings. Seven map scales are used. The basic contour intervals for them are 1, 2, 5, 10, 20, 50, and 100 m. Elevations are given in meters; distances in kilometers. For large-scale site plans, intervals of 0.1, 0.2, or 0.5 m have been recommended. As noted in Chapter 1, other agencies also produce maps for different purposes.

17-3. MAP SCALE. The choice of map scale depends on the purpose, size, and required precision of the finished map. Dimensions of a standard sheet, type and number of topographic symbols, and accuracy requirements for scaling distances from it are some considerations.

Map scales are given in three ways: (a) by *ratio* or *representative fraction*, such as 1:2000 or $\frac{1}{2000}$; (b) by an *equivalence*, for example, 1 in = 200 ft; and (c) *graphically*. Two graphic scales placed at right angles to each other and in diagonally opposite corners of a map sheet permit accurate measurements to be made even though the paper changes dimensions.

Map scales are generally classified as *large*, *medium*, and *small*. Their respective scale ranges are as follows:

[1] Requests for information should be addressed to National Cartographic Information Center, Reston, VA 22092. Telephone: (703) 860-7000.

Large scale, 1 in = 100 ft (1:1200) or larger.
Medium scale, 1 in = 100 to 1000 ft (1:1200 to 1:12,000).
Small scale, 1 in = 1000 ft (1:12,000) or smaller.

Surveyors and mappers could express all map scales by ratios as done now for metric scales. The common 1 in = 40 ft scale (1:480) is close to the metric ratio 1:500.

Discussion in this chapter will be primarily confined to large-scale maps. A map drawn to any scale can be enlarged or reduced by means of a pantograph, an opaque projector, or photographically. However, it is important to note that for enlarged maps, the errors are also magnified and the resulting product may no longer meet accuracy standards.

17-4. MAP DRAFTING. Map drafting generally consists of two steps: preparing the manuscript and drafting the final map. The manuscript is usually compiled in pencil. It should be carefully prepared to locate all features and contours as accurately as possible and be complete in every detail, including placement of symbols and letters. Lettering on the manuscript need not be done with extreme care, for its major purpose is to ensure good overall map design and proper placement. A well-prepared manuscript goes a long way toward achieving a good-quality final map.

The completed version is drafted in ink, or *scribed*. Either process involves tracing from the manuscript. If inked, the manuscript is placed on a "light table" and features are traced on a stable-base transparent overlay material. Lettering is usually done first; then planimetric features and contours are traced.

Scribing is executed on sheets of transparent stable-base material coated with an opaque emulsion. Manuscript lines are transferred to the coating in a laboratory process. Special scribing tools are used to vary line weights and make standard symbols. Lines representing features and contours are prepared by cutting and scraping to remove the coating. Scribing is generally easier and faster than inking, and its use is increasing.

Overlay drafting to update changed topographic conditions and produce an accurate composite map is now practical and economical. Pin bar registration using a metal strip with projecting pins and stable Mylar film properly punched on one edge at corresponding spacing provide precise "layering." Revisions and development stages are readily shown.

The process of preparing a pencil manuscript can be divided into four parts: (1) plotting the control; (2) plotting the details; (3) drawing the topography and special data; and (4) finishing the map, including labeling and lettering.

17-5. PLOTTING THE CONTROL. The method selected for plotting control depends on the surveying procedures employed to establish it and/or the form in which the control data are available. A traverse control survey can be plotted as a series of angles (using one of the methods described in Section 17-6) and distances laid out at the selected scale for the map (e.g., 1 in = 10, 20, 40, 50, or 100 ft if the English system is employed, or 1:100, 1:200, 1:500, 1:1000, and so on, in metric units). An engineer's scale is satisfactory but should be supplemented by steel scales and dividers for marking control points accurately to perhaps 0.02 or 0.01 in, or better.

For a traverse plotted by angles and distances, bearings and lengths of the courses are lettered parallel to the lines so they can be read easily when the user looks at the sheet *from the bottom or right-hand side*. Bearings are shown in the forward direction and continuous around the traverse. When a bearing is read from left to right, but actually runs from right to left, *an arrow is used to note the correct direction*, as in Figure 17-7.

Instead of angles and distances, the coordinate method can be adopted for plotting traverses after calculating X and Y values for the points as described in Section 13-8. If control was established by triangulation or trilateration, point locations will probably be computed in coordinates ready for the most accurate and convenient method of plotting. The map sheet is first laid out precisely in a grid pattern with unit squares of appropriate size, such as 100, 400, 500, or 1000 ft, and checked by measuring the diagonals. Each grid line is labeled with its coordinate value. The origin of coordinates may be the most westerly or southerly station of the traverse, or somewhere off the sheet to assure all plus values.

Control points are plotted by measuring their X and Y coordinates from the ruled grid lines. Small circles, $\frac{1}{8}$ in or smaller in diameter, are drawn to mark the hubs. Any errors in plotting are detected by comparing the scaled distance and bearing of each line with the length and direction measured in the field or computed. On most topographic maps, hub locations are omitted from the finished drawings. If shown, they may be drawn in light blue ink to make them less prominent on a print.

Special instruments called *coordinatographs* are also available for drawing the grid and plotting points by coordinates. Index marks set to coordinate values on perpendicular graduated X and Y rails permit rapid and accurate plotting to be performed. Computer-driven automated coordinatographs are now rapidly gaining prominence in map drafting. These systems are described in Section 17-17.

17-6. PLOTTING ANGLES.

Angles can be plotted by the tangent method, chord method, or by protractor. These procedures are explained in the following sections.

17-6.1. TANGENT METHOD.

To lay off an angle by the tangent method, a convenient distance is measured along a reference line to serve as a base. Thus in Figure 17-1, to plot a 12°14′ deflection angle at point A, length AB equal to 10 in is first marked on the prolongation of the back line. A perpendicular with length equal to the distance AB times the natural tangent of 12°14′ (2.17 in) is erected at B to locate point C. The line connecting A and C makes the desired angle with AB. Any length of base can be used, but a distance of 10 or 100 units requires only movement of the decimal place in natural tangents.

The tangent method is employed extensively for plotting deflection angles. It is not as practical for large direct angles.

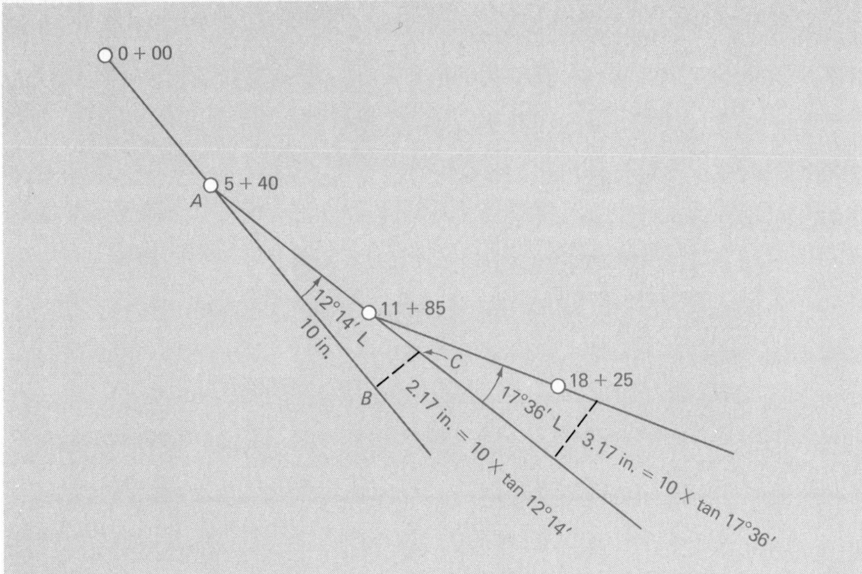

Figure 17-1. Tangent method.

17-6.2. CHORD METHOD. To lay off an angle by the chord method, as indicated in Figure 17-2, a convenient base length *BD* of 10 units is first marked on one side, *BA*, giving point *D*. With the vertex *B* as the center and a radius of 10 units, an arc is swung. Then with *D* as the center and a radius equal to the chord for the desired angle, another arc is drawn. The intersection of the two arcs locates *E*. A line connecting points *B* and *E* forms the angle's other side.

The chord for an angle is found by multiplying twice the distance *BD* by the sine of half the angle. Thus in Figure 17-2 the chord is $2 \times 10 \times \sin 16°14'$. By making the base length 10, 50, or 100 units to some scale, the chord is readily computed using natural sines taken from a calculator.

Figure 17-2. Chord method.

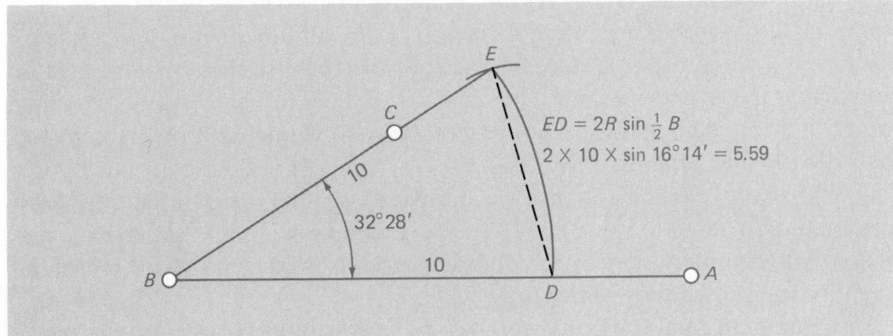

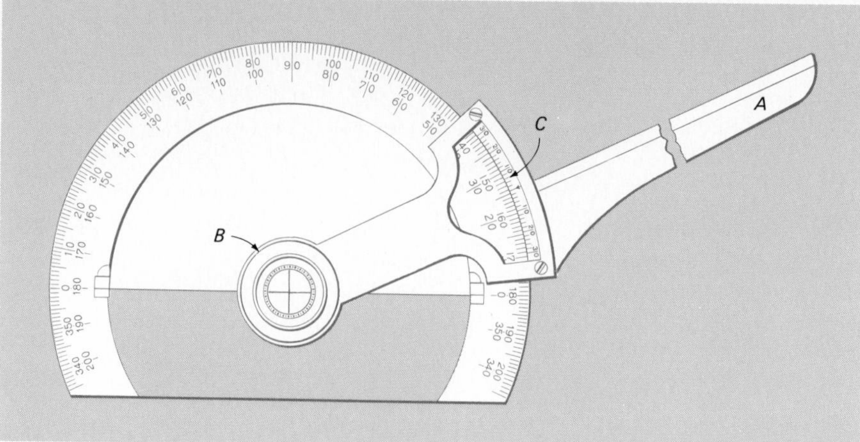

Figure 17-3. Protractor.

17-6.3. PROTRACTOR METHOD. A protractor is a device made of paper, plastic, or metal, cut in a full circle or semicircle, with angle graduations along the circumference. A fine point identifies the circle center. The protractor is centered at the angle vertex with the zero line along one side and the proper angle point marked at its edge.

Protractors are available in sizes having radii from 2 to 8 in or more. A metal circle with a movable arm extending beyond the edge, as illustrated in Figure 17-3, is often used. The arm *A* rotates about the center *B* and has a vernier *C* reading to minutes. A similar arrangement on drafting machines makes plotting fast and accurate.

Protractors are universally employed for plotting details but are not generally suitable for high-precision work on traverses or control.

17-7. ADVANTAGES AND DISADVANTAGES OF THE DIFFERENT METHODS.

With the coordinate method, any large errors in plotting are readily found by scaling. Correction usually involves replotting only one point. For example, if the scaled lengths of lines *CD* and *DE* of a plotted traverse are not equal to their field measurements, point *D* is off. The independence of each point in the plotting procedure is a definite advantage.

The tangent method is accurate and probably the best way to lay out a single angle precisely. In plotting a traverse, however, an error in any angle or distance is carried through the remainder of the traverse. If a traverse fails to close, each angle and line must be checked. If the closure is due to drafting rather than the field work, it may be necessary to rotate each line, except the first, slightly and progressively.

The chord method also has the disadvantage that errors in any traverse line are passed along to succeeding courses. Erection of a perpendicular is eliminated, but determination of the chord lengths is somewhat more laborious than finding perpendicular offsets.

Laying out angles by protractor is the fastest but least accurate method.

17-8. PLOTTING DETAILS. As in plotting the control traverse, methods used to position details on a map depend on the survey procedures used to locate them, and the form in which the data are available. If the field information is given by angles and distances, boundary corners and important points upon which construction work may depend are plotted by coordinates or the tangent method, but a protractor is used for most details. Orientation of the protractor zero line is by meridian for details secured by azimuths or bearings, and by the backsight line if direct angles were measured. Angles are marked along the protractor edge, and distances scaled from the vertex to plot details. To avoid obliterating the vertex with numerous lines that must later be erased, an engineer's scale can be laid beside the vertex in the direction to a topographic detail and its location plotted directly at the required distance. Or, a short line may be drawn from the angle point at only the estimated distance and the exact length scaled on it.

Some protractors have a graduated scale on the arm to make detail plotting easier. For greatest accuracy, the distance to a detail should be smaller than the protractor radius unless an extended-arm type is used.

If a total-station instrument has been used to locate details in the field, their coordinates will likely have been computed and recorded and are ready for plotting. Coordinatographs, especially computer-driven ones, are exceptionally fast and accurate.

17-9. CONTOUR INTERVAL. As noted in Section 16-7, the choice of contour interval to be used on a topographic map depends on its intended use, required accuracy, type of terrain, and scale. If, according to National Standards of Map Accuracy, elevations can be interpolated from a map to within one-half the contour interval, then for an accuracy of 1 ft, a 2-ft maximum interval is necessary. However, if only 10-ft accuracy is required, a 20-ft contour interval will suffice.

Terrain type and map scale combine to regulate the contour interval needed to produce a suitable density (spacing) of contours. Rugged terrain requires a larger contour interval than gently rolling country, and flat ground mandates a relatively small one to portray the surface adequately. Also, if map scale is reduced, the contour interval must be increased; otherwise, lines are crowded, confuse the user, and possibly obscure other important details.

For average terrain, the following large and medium map scales and contour interval relationships generally provide suitable spacing:

ENGLISH SYSTEM		METRIC SYSTEM	
SCALE (ft/in)	CONTOUR INTERVAL (ft)	SCALE	CONTOUR INTERVAL (m)
50	1	1:500	0.5
100	2	1:1000	1
200	5	1:2000	2
500	10	1:5000	5
1000	20	1:10,000	10

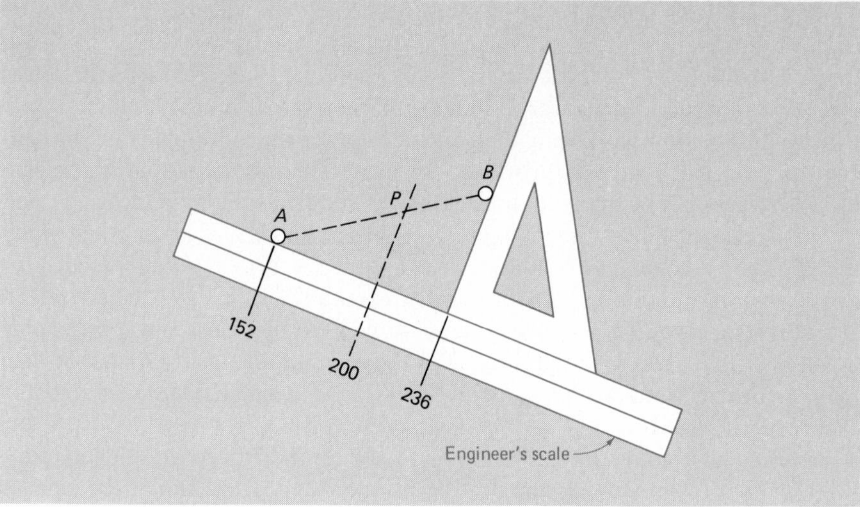

Figure 17-4. Interpolating using engineer's scale and triangle.

17-10. PLOTTING CONTOURS. Points to be used in plotting contours are located in the same manner as those for details. Contours found by the *direct method* are sketched through the points. Interpolation between plotted points is necessary for the *indirect method.*

Interpolation to find contour locations between points of known elevation can be done in several ways:

1. Estimating.
2. Scaling the distance between points of known elevation and locating the contour points by proportion.
3. Using a rubber band graduated to some scale and stretching it to make convenient marks fall on the known elevations. Special devices called *variable scales* are available which contain a graduated spring. The spring may be stretched to make suitable marks fall on the known elevations.
4. Using a triangle and scale, as indicated in Figure 17-4. To interpolate for the 420-ft contour between point *A* at elevation 415.2 and point *B* at elevation 423.6, first set the 152 mark on any of the engineer's scales opposite *A*. Then, with one side of the triangle against the scale and the 90° corner at 236, the scale and triangle are pivoted together around *A* until the perpendicular edge of the triangle passes through point *B*. The triangle is then slid to the 200 mark and a dash drawn to intersect the line from *A* to *B*. This is the interpolated contour point *P*.
5. Using a thin plastic, converging-line device, such as that in Figure 17-5, which can be pivoted and adjusted to fit the difference in elevation between any two points. The procedure is illustrated by the following example: Assume two plotted points *A* and *B* have elevations of 17.6 and 25.9. Draw a straight line between these points. Determine the

Figure 17-5. Contour finder for interpolating contours between plotted elevations, assuming uniform slope. (Courtesy Keuffel & Esser Company.)

difference between the given elevations to the nearest whole number. Since 25.9 − 17.6 = 8.3, use 8. Place the contour finder over the map so that its horizontal lines are parallel to line *AB* and eight intervals nearly fill the space between the plotted points. Adjust the finder until the 17.6 (*A*) point lies 0.6 into the first interval and the 25.9 (*B*) point lies 0.9 beyond the eighth interval. With a needle point, punch through the contour finder at the contour points desired. An obvious disadvantage is that the manuscript (and expendable finder) are perforated.

Contours are drawn only for elevations evenly divisible by the contour interval. Thus, for a 20-ft interval, elevations of 800, 820, and 840 are shown, but 810, 830, and 850 are not. To improve legibility, every fifth line, evenly divisible by 5 times the contour interval, is made heavier. So for a 20-ft interval, the 800, 900, and 1000 lines would be heavier. On U.S. Geological Survey maps, the standard color for inked contour lines is sepia (dark reddish brown), but on large-scale engineering maps, they are usually black.

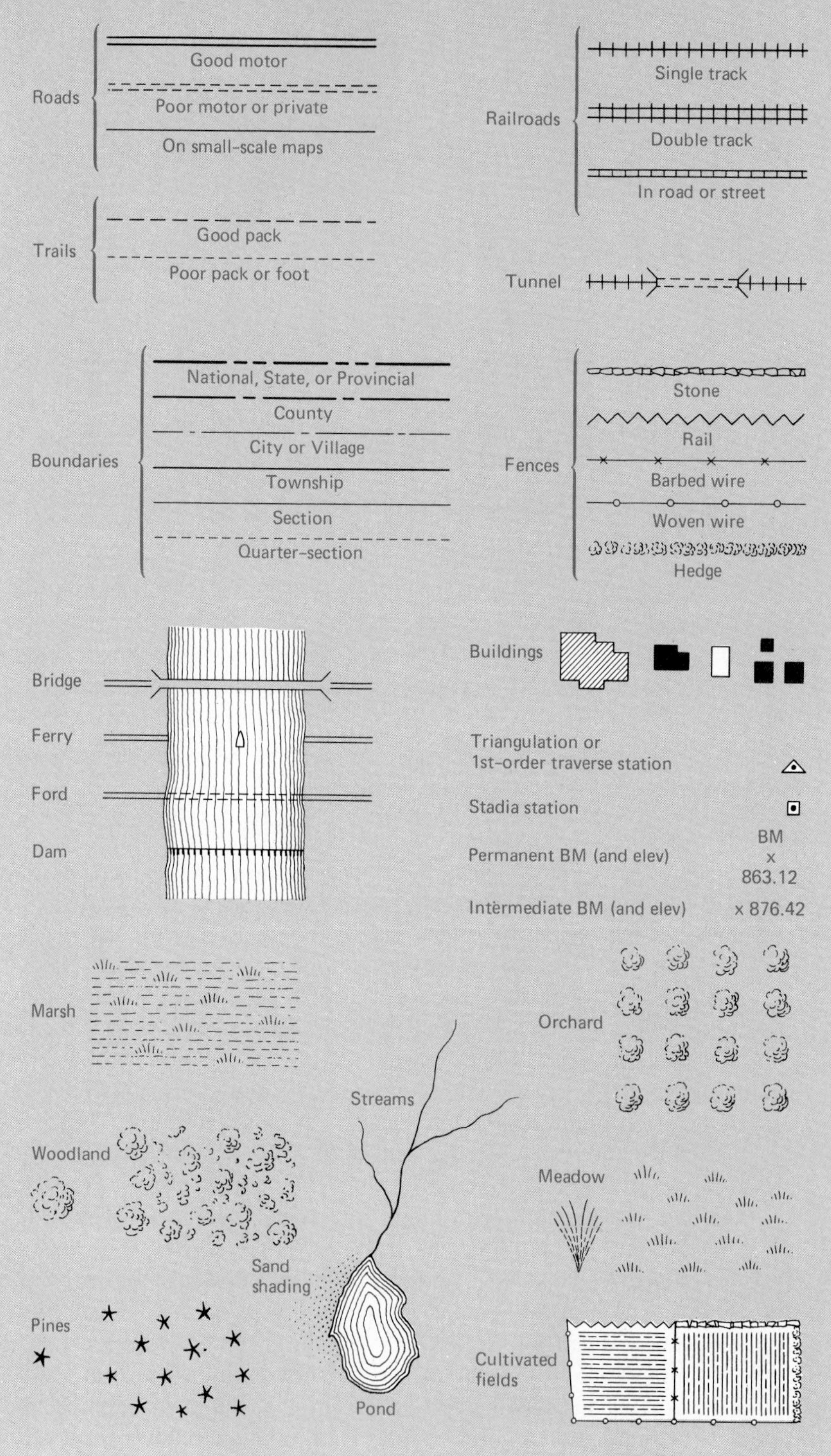

Figure 17-6. Topographic symbols.

Hills, mountains, depressions, and ground-surface undulations can also be represented by hachures and other forms of hill shading. *Hachures* are short lines drawn in the direction of slope. They are heavy and closely spaced for steep slopes, fine and widely separated for gentle slopes. They show only the general ground shape rather than actual elevations and thus are not suitable for most engineering work.

The *Quick Topo Plotter*, a series of Mylar charts having different scales and comparable to a contour finder, accelerates plotting topographic information without sacrificing accuracy. The charts eliminate the need for a scale.

17-11. TOPOGRAPHIC SYMBOLS. Standard symbols are used to represent special topographic features, thereby making it possible to show many details on a single sheet. Figure 17-6 gives a few of the hundreds of symbols employed on topographic maps. Considerable practice is required to draw these symbols well at a suitable scale. Before they are placed on a map, such things as buildings, roads, and boundary lines are plotted and inked. The symbols are then drawn, or cut from standard "stickup" sheets with an adhesive on the back, and pasted on the map. A fully detailed map with coloring and shading is a work of art.

17-12. PLACING THE MAP ON A SHEET. The appearance of a finished map has considerable bearing on its acceptability and value. A map that is poorly arranged, carelessly lettered, and unfinished looking does not inspire confidence in its accuracy. A border line somewhat heavier than all other lines improves a sheet's appearance.

The first step in map arrangement is to determine the best position for control and topography to properly "balance" the sheet. Figure 17-7 shows a traverse (with no topography outside it) suitably placed.

Before any plotting is done, the proper scale for a sheet of given size must be selected. Assume an 18 × 24-in sheet with a 1-in border on the left (for possible binding) and $\frac{1}{2}$-in borders on the other three sides. If the most westerly station *A* has been chosen as the origin of coordinates, then divide the total departure to the most easterly point *C* by the number of inches available for plotting in the east-west direction. The maximum scale possible in Figure 17-7 is 774.25 divided by 22.5, or 1 in = 34 ft. The nearest standard scale that will fit is 1 in = 40 ft.

This scale must be checked in the *Y* direction by dividing the total difference in *Y* coordinates, 225.60 + 405.57 = 631.17, by 40 ft, giving 15.8 in required in the north-south direction. Since $16\frac{1}{2}$ in are usable, a scale of 1 in = 40 ft is satisfactory, although 1 in = 50 ft provides a better border margin. If a scale of 1 in = 40 ft is not suitable for the map's purpose, a sheet of different size should be selected, or perhaps more than one sheet must be employed to map the required area.

In Figure 17-7 the traverse is centered on the sheet in the *Y* direction by making each distance *m* equal to $\frac{1}{2}[17 - (631.17/40)]$, or 0.61 in, and the same 0.61 in used for the left margin. "Weights" of the title, notes, and north arrow compensate for the traverse being to the left of the sheet center.

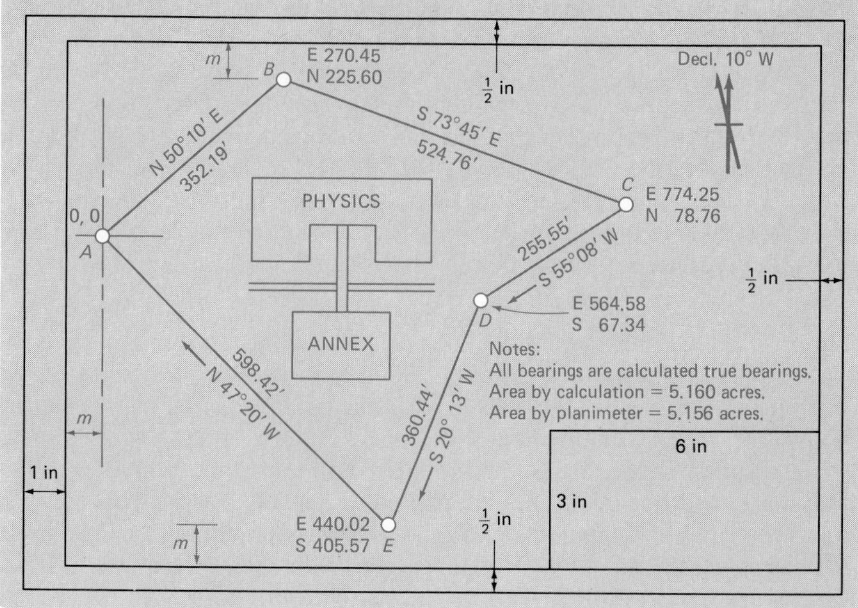

Figure 17-7. Map layout.

If topography is to be plotted outside the traverse, the maximum north-south and east-west distances to topographic features must be added to the traverse coordinates before computations are made for the scale and centering distances.

On some types of maps it is permissible to skew the meridian with respect to the borders of the sheet in order to better accommodate a traverse relatively long in the NE–SW or NW–SE direction, or to make street lines parallel with the borders. When a traverse with topography is to be plotted by any means other than coordinates, it is advisable to first make a sketch showing controlling features. If this is done on tracing paper, orientation for the best fit and appearance is readily determined by rotating and shifting the tracing.

17-13. MERIDIAN ARROW. Every map must display a meridian arrow for orientation purposes. It should preferably be near the top of the sheet, although it may be moved elsewhere for balance. An arrow should not be so large, elaborate, or heavily blacked in that it becomes the focal point of a sheet, as was true on maps of 50 years ago.

True, grid, or magnetic north or all three may be shown. A true-meridian arrow is identified by a full head and full feather; a grid and/or magnetic arrow by a half-head and half-feather. The half-head and half-feather are put on the side away from the true north arrow to avoid touching it. Figure 17-8 shows dimensions of an arrow suitable for routine maps. In practice, an arrow is traced from a sheet of standards, or cut out as a "stickup" for pasting on the map.

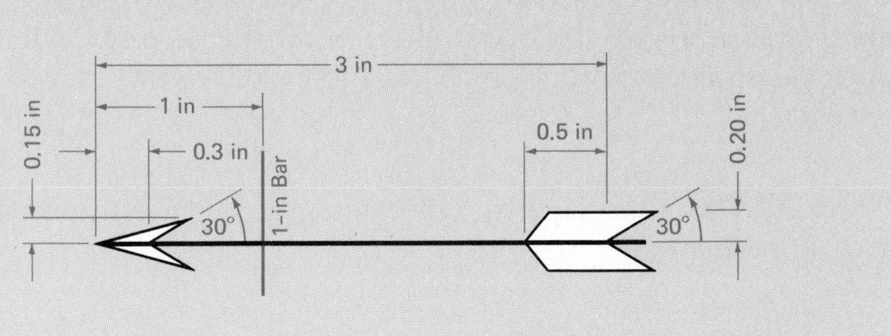

Figure 17-8. Dimensions for simple arrow.

17-14. TITLE. The title may be placed wherever it will best balance the sheet, but is always kept outside the property lines on a boundary survey. Usually the title occupies the lower right-hand corner, with any pertinent notes just above or to the left of it. Search for a particular map in a bound set or loose pile of drawings is facilitated if all titles are in the same location. Since sheets are filed flat, bound on the left border, or hung from the top, the lower right-hand corner is the most convenient position.

The title should state the type of map, name of property or project and its owner or user, location or area, date completed, scale, contour interval, horizontal and vertical datums used, and name of the surveyor with his license number on property surveys. Additional data may be required on special-purpose maps. Lettering should be simple in style rather than ornate, and conform in size with the individual map sheet. Emphasis is placed on the most important parts of the title by increasing their height and using uppercase (capital) letters.

Perfect symmetry of outline about a vertical center line is necessary, since the eye tends to exaggerate any defection. Also, an appearance of stability is obtained by having a full-width bottom line.

An example of a title and its arrangement for the 18 × 24-in sheet of Figure 17-7 is given in Figure 17-9. Heights of letters in lines 1 and 2 could be $\frac{1}{4}$ in; in line 3, $\frac{5}{16}$ in; and in the last two lines, $\frac{1}{8}$ in.

Figure 17-9. Title arrangement.

UNIVERSITY OF HAWAII
CIVIL ENGINEERING DEPARTMENT
TRANSIT-STADIA SURVEY OF ENGINEERING CAMPUS
SCALE 1 INCH = 20 FEET — DATE 8 NOV. 1983
SURVEY BY L. HILL, T. HALL, A. AKER — MAP BY J. JONES

No part of a map better portrays the artistic ability of a draftsperson than a neat, well-arranged title. Today, however, most companies and government agencies use sheets with preprinted title forms to be filled in with individual job data.

17-15. NOTES. Notes cover special features pertaining to the individual map, such as the following:

All bearings are true bearings (or magnetic, or grid, or calculated).
Area by calculation is X acres or hectares.
Area by planimeter is Y acres or hectares.
Legend (explanation of usual symbols; for example, $*$ represents cooling towers).

Notes must be in a prominent place where they are certain to be seen upon even a cursory examination of the map. The best location is just above or to the left of the title block in the lower right-hand corner of the map. A user then finds the desired map by title and checks any special notes before examining the drawing.

17-16. DRAFTING MATERIALS. Polyester film and tracing and drawing papers are the materials commonly used for preparing maps in surveying and engineering offices. Polyesters such as Mylar are by far most frequently employed because they are dimensionally stable, strong, durable, and waterproof; take pencil, ink, and stickup items; and withstand erasing. Tracing papers are available in a variety of grades, and good ones also are stable, take pencil, ink, and stickups, and endure some erasing. Both Mylar and tracing paper are transparent so blueprints can be made from them.

Drawing papers of different types and grades are utilized for map drawings; however, they have the disadvantages of not tolerating much erasing, tearing more easily, cracking with age, and not being transparent. Cloth and aluminum-backed papers are superior to other grades.

Considering the time and expense required to gather field data and draft a map, the small extra cost for excellent-quality drafting material is certainly justified.

Surveys of city lots, normally drawn on small sheets, and other maps may be reproduced in a surveyor's office by an ozalid-type process which does not entail wetting and drying the printing paper.

17-17. AUTOMATED MAPPING. More data are needed today, faster and better, for individual and combined projects. Examples include property location and ownership, transportation data, and soil and mineral investigations. Numerous varied electronic machines have been developed to automatically draw maps and extract and analyze data from them. Human errors are minimized or eliminated. Special symbols can be automatically located by touching a cursor.

The required input to a computer for an automatic mapping system includes control data, topographic detail information, map scale, and contour interval. Appropriate programs direct the computer to solve for the positions of points using the survey data and to plot contours and other features.

Advantages of the computer-driven systems are their speed, accuracy, and a consistently uniform final product. Moreover, all data can be stored in a databank or on magnetic tape with different numerical codes for the various kinds of features, and recalled later for plotting in total or in parts for special-purpose maps. As an example, only roads and utilities might be needed by a city engineering department, whereas an assessor may want only the property boundaries mapped. Scale and contour interval can be varied readily, and either English or metric units specified.

With contour information stored in the computer, profiles along selected lines can be plotted automatically. By including grade lines and design templates, survey stakeout information and earthwork quantities are obtained for projects such as highways and canals.

Interactive drafting systems which include a cathode ray tube (CRT) and automatic drafting machine interfaced with a computer are extremely versatile. An operator can examine a visual display of a map on the CRT as it is being compiled by the computer and make any changes needed. As examples, lines can be added, deleted, or their styles altered; placement of symbols and lettering modified; lettering sizes and styles varied; and essentially the map designed and checked for completeness and accuracy. When the operator is satisfied that the map meets requirements and is the optimum design, the automatic plotter is actuated to draft the final product.

Modern surveying instruments such as the total-station type and photogrammetric stereoplotters produce large quantities of terrain data in coordinate form rapidly and economically. The data, recorded digitally in the form of X, Y, and Z coordinates of a grid of points, produce a *digital terrain model* (DTM). This information can be fed to a computer, and with an appropriate program reduced automatically to contours and drafted.

Computer-assisted cartography is revolutionizing map making. The United States has begun digitizing the 40,000 quadrangle maps covering almost the entire country. One device, the *raster scanner*, scans printed material, line by line, and converts the information to numerical form.

Another short-cut to map production being tested is a laser platemaker which uses digital data and an ultraviolet laser to expose press plates. The high cost of silver halide film is then eliminated.

Descriptions of competing products for the new market are available from surveying literature and their manufacturers.

Figure 17-10 is a computerized planimetric subdivision plat.

17-18. SOURCES OF ERROR IN MAPPING. Sources of error in mapping are:

1. Not checking (scaling) distances when plotting by coordinates.
2. Plotting by protractor.
3. Using a soft pencil, or one with a blunt point, for plotting.

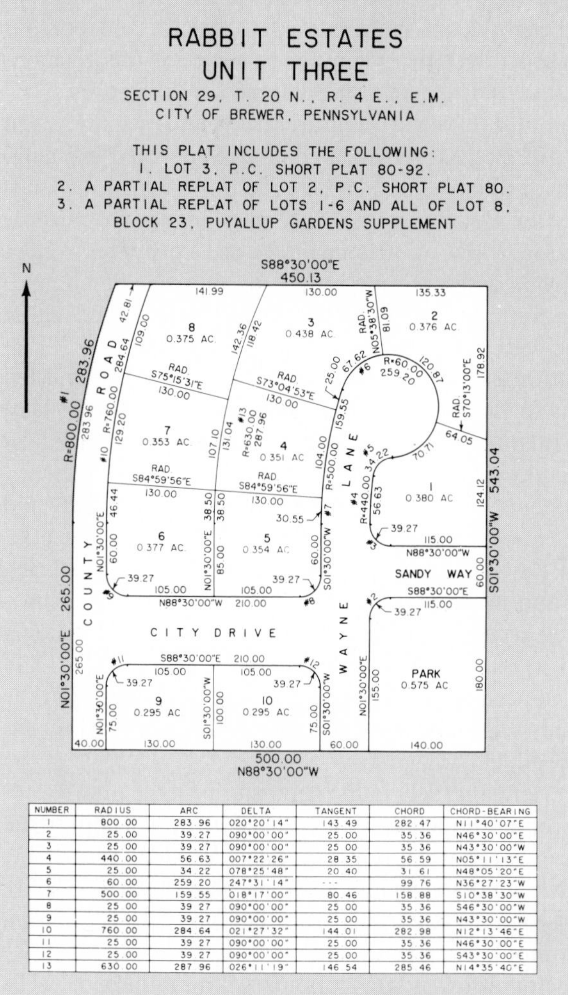

NUMBER	RADIUS	ARC	DELTA	TANGENT	CHORD	CHORD-BEARING
1	800.00	283.96	020°20'14"	143.49	282.47	N11°40'07"E
2	25.00	39.27	090°00'00"	25.00	35.36	N46°30'00"E
3	25.00	39.27	090°00'00"	25.00	35.36	N43°30'00"W
4	440.00	56.63	007°22'26"	28.35	56.59	N05°11'13"E
5	25.00	34.22	078°25'48"	20.40	31.61	N48°05'20"E
6	60.00	259.20	247°31'14"	---	99.76	N36°27'23"W
7	500.00	159.55	018°17'00"	80.46	158.88	S10°38'30"W
8	25.00	39.27	090°00'00"	25.00	35.36	S46°30'00"W
9	25.00	39.27	090°00'00"	25.00	35.36	N43°30'00"W
10	760.00	284.64	021°27'32"	144.01	282.98	N12°13'46"E
11	25.00	39.27	090°00'00"	25.00	35.36	N46°30'00"E
12	25.00	39.27	090°00'00"	25.00	35.36	S43°30'00"E
13	630.00	287.96	026°11'19"	146.54	285.46	N14°35'40"E

Figure 17-10. Subdivision map compiled automatically by a computer-driven plotter. (Courtesy Technical Advisors, Inc.)

4. Variations in dimensions of map sheet due to temperature and moisture.
5. Selecting an inappropriate scale or contour interval for the map.

17-19. MISTAKES. Some common mistakes in mapping are:

1. Improperly orienting topographic notes in field and office.
2. Using wrong edge of engineer's scale.
3. Making the north arrow too large, complex, or black.
4. Omitting the scale or necessary notes.
5. Failing to balance the sheet by making a preliminary sketch.
6. Drafting the map on a poor-quality medium.

17-1. What is the status of metrication in surveying?

17-2. Specifications require plotting errors to be kept within 5 ft using an engineer's scale. What is the smallest feasible map scale? Minimum scaling accuracy is $\frac{1}{50}$ in.

17-3. On a planetable sheet having a scale of 1:3600, what is the smallest distance that can be plotted with an engineer's scale? Minimum scaling accuracy is $\frac{1}{100}$ in.

17-4. Classify the maps in Figures 16-2, 16-6, and 17-10 on the basis of scale.

17-5. Describe three conditions for which overlay drafting is useful.

17-6. Why should an arrow be added to a bearing read from left to right but actually running from right to left?

17-7. List side by side the various methods of plotting control and traverses. Give the relative advantages and disadvantages of each.

17-8. Similar to Problem 17-7, except for plotting topographic details.

17-9. What ratio scales are suitable to replace the following equivalent scales: 1 in = 8 ft, 1 in = 20 ft, 1 in = 40 ft, 1 in = 80 ft, 1 in = 200 ft, and 1 in = 800 ft.

17-10. Why are traverse lines shown in light blue on a survey map? Should the hubs be light blue or black? Why?

17-11. An area that varies in elevation from 626 to 715 ft is being mapped. Which contour intervals will be numbered if a 10-ft interval is used? Which lines drawn heavier?

17-12. For mapping an average terrain, what contour interval would you recommend for map scales of 1 in = 200 ft, 1:1000, and 1:24,000?

17-13. List factors that influence the selection of a contour interval.

17-14. When are form lines used on a map? How do they differ from contours?

17-15. What is meant by the statement: A contour map lacks expression?

17-16. List four convenient classifications of items that can be represented on general-purpose maps.

17-17. Draw the conventional symbols for national, state, and county lines.

17-18. Draw 1-ft contours for the data in Plate D-6.

17-19. Sketch 5-ft contours for the notes in Plate D-10.

17-20. Plot the topography covered in Plate D-9.

17-21. If elevations on a map must be interpolated to the nearest ± 2 ft, what contour interval is necessary according to the National Map Accuracy Standards?

17-22. Describe the first two steps in drawing a survey map and how they are best accomplished.

17-23. How do the features shown on a large-area topographic map differ from those on a small-area map?

17-24. List items that should be in the notes or title of a profile map. Do the same for a property survey map.

17-25. Why are titles usually placed in the lower right-hand corner of most map sheets, and the meridian arrow on the top or right side?

17-26. Give another name or names for the term "Legend" on a map.

17-27. Why are stickups placed on maps?

17-28. List three essentials of a good meridian arrow on a survey map.

17-29. What additional items might be helpful in the map title of Figure 17-9?

17-30. List the advantages of computerized mapping systems, and one disadvantage.

17-31. What do Spk., Stk., Tel., and W. L. mean on a surveying map?

17-32. Is there any difference between a map and a chart? Between a map and an aerial photograph?

17-33. Describe two mapping conditions where it may be desirable to skew the meridian arrow with respect to the border, and one case where it must be parallel.

17-34. List the important items that are usually in uppercase letters and those in lowercase on a (a) topographic map and (b) subdivision map of a property bordering a river or lake.

17-35. For a student survey and map similar to Figure 17-7, how close would you expect the computed and planimetered area within the traverse to check?

17-36. Are copyright laws for maps the same as those for textbooks?

BIBLIOGRAPHY

Briefs. 1981. "Lasers Add Zap to Map Making." *ACSM Bulletin* 74:74.

Dix, W. 1981. "Mapmakers Shift to 'Digitization.'" *ACSM Bulletin* 72:43.

Greulich, G. 1982. "Metric Standards in Subdivision and Zoning." *Surveying and Mapping* 42(no. 3):257.

Hebrank, A. J. 1981. "Overlay Drafting for Surveyors." *Surveying and Mapping* 41(no. 2): 151.

Iler, W. H. 1981. "The Use of Computer Mapping in Planning and Public Works Management: The Case for Chicago." *The American Cartographer* 8:115.

Jacober, R. P., Jr. 1980. "Map Content and Symbols: Chapter V of Proposed Manual on Map Uses, Scales and Accuracies for Engineering and Associated Purposes." *ASCE Journal of the Surveying and Mapping Division* 106(no. SU1):41.

———1981. "Standard for Symbology of Engineering Scale Maps." *ASCE Journal of the Surveying and Mapping Division* 107(no. SU1):21.

McKelvey, V. E. 1977. "USGS Statement on the Preparation of Base Maps for the National Mapping Program." *ACSM Bulletin* 58:23.

Monteith, W. J. 1970. "Chart Compilation—Field and Office." *ASCE Journal of the Surveying and Mapping Division* 96(no. SU2):157.

Quick Topo Plotter. 1982. "Manufacturer's Announcement." *Point of Beginning*, Dec.–Jan., p. 62.

Robinson, A. H., R. D. Sale, and J. L. Morrison. 1978. *Elements of Cartography*, 4th ed. New York: Wiley.

Sloane, R. C., and J. M. Montz. 1943. *Elements of Topographic Drawing*. New York: MacGraw-Hill.

Thompson, M. M. 1972. "Water Features on Topographic Maps." *ASCE Journal of the Surveying and Mapping Division* 98(no. SU1):1.

Urban, L. J. 1978. "An Overview of Maps, Their Use and Importance." *Proceedings of the Fall Meeting of the American Congress of Surveying and Mapping*, p. 400.

U.S. Geological Survey. 1977. "Statement on the Preparation of Metric Base Maps for the National Mapping Program." *American Congress of Surveying and Mapping Bulletin* 58:23.

18 THE PLANETABLE

18-1. INTRODUCTION. The planetable is one of the oldest surveying instruments, now used more extensively overseas than in the United States. In its present form, a planetable outfit consists of a tripod, a drawing board which fastens on the tripod, an alidade equipped with stadia wires, a stadia rod, and tape. A sheet of drawing paper, Mylar, or other material is fastened to the board and a map made in the field by plotting directions and distances obtained by sighting with the alidade.

Mapping topographic features while they are in full view is advantageous in many types of surveys in civil and mining engineering, forestry, geology, agriculture, archeology, and military operations. Geologists use the Brunton compass and planetable almost exclusively in their surveys. A small board, light tripod, and peep-sight alidade, collectively known as a *traverse table*, have been standard equipment in military mapping. The table is leveled and oriented by movement of the tripod legs.

Photogrammetry is now the principal method employed for large mapping projects, but the planetable is still used for various classification and supplemental surveys. It is impossible to extract *everything* mapworthy from photographs. Extra surveys on the ground will therefore always be required, although they may be called "ground-truth" surveys or something else. The U.S. Geological Survey has indicated that there is no real substitute for the alidade and planetable in surveying small tracts at large scales.

18-2. DESCRIPTION OF THE PLANETABLE. The drawing board of a planetable (usually 24 × 31 in) is carefully made to resist warping and other damage from

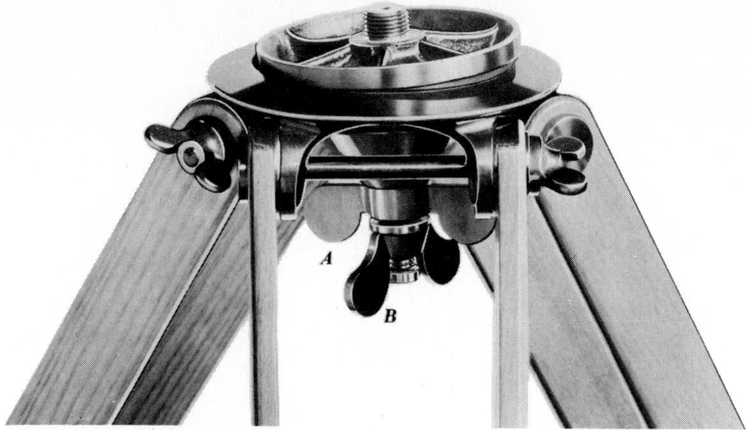

Figure 18-1. Johnson-type head.

weathering. The upper side is smooth but has some means such as brass screws for attaching a map sheet to the board. A socket at the center of the underside of the board has threads that fit a head fastened to the tripod top.

Two radically different tripod heads, the *Coast and Geodetic Survey type* and the *Johnson head*, are available for leveling and orienting the board. The Coast and Geodetic Survey type has four leveling screws, a clamp, and a tangent screw like those of a transit. Leveling and orientation are accomplished easily. The Johnson head (Figure 18-1) has a ball-and-socket arrangement to hold the board in position after leveling and prevent its turning in a horizontal direction. Clamp *A* regulates the motion of the table as it moves on the larger ball joint. After leveling the board by pressing or lifting on one edge, it is tightened. Lower clamp *B* controls movement of the board about its vertical axis and is tightened after orientation.

Keeping the table level is the most difficult part of planetable operation for beginners. Even light pressure on an edge of the board applies a strong turning moment on the relatively small supporting area of the leveling head.

Older-type alidades (Figure 18-2) consist of a telescope supported by a pedestal rigidly attached to a base or blade that is up to 18 in long on some instruments. The telescope, similar to that of a transit, is equipped with one vertical and three horizontal cross hairs. A sensitive level vial, bull's-eye level, vertical arc and/or Beaman arc, compass needle, and lifting knobs are provided. The telescope may be centered over the blade or offset to place the line of sight along the edge.

Accessories used with the planetable include a scale, triangles, compass, magnifying glass, French curve, erasers, protractor, declinator, drafting tape, and hand calculator. A *declinator* is a brass plate on which a compass box and one or two level vials are mounted. Two edges of the plate are parallel with the north-south line of the compass circle. The declinator is used, in the absence of a compass on the blade, to determine bearings and orient the alidade by placing an edge against the base.

An alidade of European design, shown in Figure 18-3, is a self-reducing stadia instrument and has a parallel-ruler plotting device. Some have the same

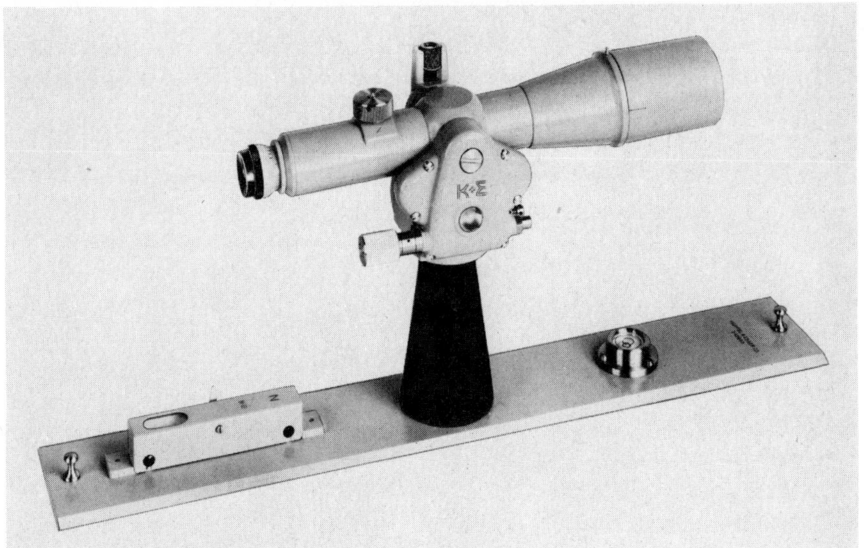

Figure 18-2. Self-indexing alidade. (Courtesy Keuffel & Esser Company.)

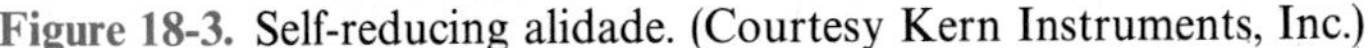

Figure 18-3. Self-reducing alidade. (Courtesy Kern Instruments, Inc.)

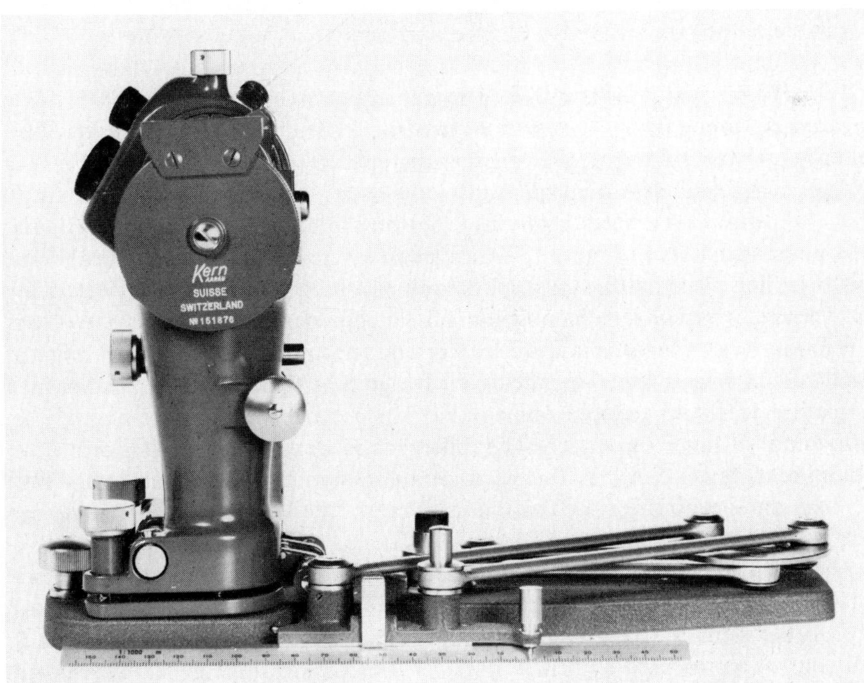

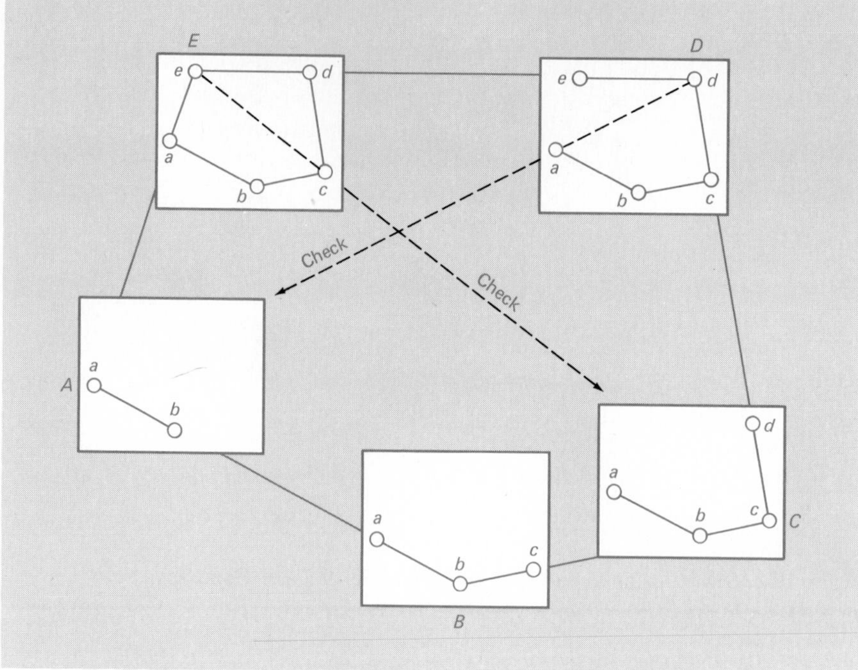

Figure 18-4. Traversing with planetable.

features as certain theodolites—microscopes for reading the enclosed glass circle, self-indexing, and other improvements. Electronic alidades are also available.

18-3. USE OF THE PLANETABLE. The planetable is best suited for traversing and taking topography. It is seldom used on boundary or route surveys but often employed to add details on maps prepared from theodolite, EDM, or photogrammetric measurements.

With a suitable sheet firmly fastened on the board, a table is set up over any point such as *A* in Figure 18-4 and leveled. The beveled edge of the alidade blade is then placed at the corresponding point *a* on the sheet. This point may be selected in a convenient position on the board if not previously plotted. Direction to a point *B* is laid off by sighting through the telescope to align it and then drawing a line along the beveled edge. The stadia distance and vertical angle are read and reduced immediately, so *b* can be located by scaling the horizontal distance on line *ab*. The difference in elevation computed from the stadia reading is applied to the elevation of *A* (assuming a reading on the h.i.) to find the elevation of *B*. This value is noted beside the plotted point for ready reference.

Specified maximum closure errors for both horizontal and vertical control should be established in advance of field work. On precise work, horizontal control in the form of triangulation, trilateration, or traverse stations, and vertical control represented by bench marks scattered over the area to be mapped, must be plotted carefully on the sheet in advance of any field mapping. Stadia readings are reduced and plotted as soon as they are taken, since notes are not recorded. *Sketching and interpolation for contours should be done along with the plotting while the ground location of all mapped points is still fresh in the observer's mind.*

A planetable sheet must be resistant to moisture and temperature changes because it may be used in the field for extended periods. A hard pencil, 6-H or 8-H, is necessary to avoid smudges. The board can be covered with a waterproof material to protect the drawing. A small "window" is left open over the working area. Mylar sheets, which can be marked with a stylus, are often employed.

18-4. SETTING UP AND ORIENTING THE PLANETABLE. After the board is screwed on the tripod head and the sheet fastened, the tripod is set up with a plotted point over the corresponding hub on the ground. Generally this relationship is estimated by eye for small-scale maps. If greater accuracy is required, a plumb bob is used or a pebble dropped from the underside of the board to check the position. The table must be carefully leveled by means of the leveling screws (or by the ball-and-socket movement of the Johnson head).

For convenience in drafting and to avoid pressure on the table which will disturb its level, the board should be placed at a height about 1 in lower than the observer's elbows when standing in a comfortable position.

The table may be oriented by (1) compass, (2) backsighting, or (3) resection.

1. To orient by compass, the alidade is laid on the sheet in the direction deemed most desirable for the north-south line, and the board turned until the compass on the blade, or the declinator against the ruler, reads north. The board is then clamped and a line drawn along the blade edge for future reference.

If a line *AB* of known bearing—say, S28°30′E—has already been plotted, the alidade is placed along *AB*, the table turned until the compass reads S28°30′E, and then clamped. Orientation by compass is not recommended if the backsighting or resection method can be employed readily.

2. The backsighting method is most commonly used in traversing, although the table may initially be oriented by compass to get north at the top or left side of the sheet. After a line *AB* has been drawn by sighting from a setup at *A* to point *B*, the table is moved to *B*, positioned, and leveled. With the alidade blade along *BA*, the table is rotated to line-in point *A* and the board clamped. Directions of other sights taken with the table in this position will automatically be referred to the same reference line or meridian as *AB*.

3. Resection is discussed in Section 18-8.

After the table has been oriented by any of the three methods, prominent distant points should be sighted and short lines drawn near the edge of the sheet in their direction. At intervals the alidade can be laid on the occupied station point and then along these lines to check the orientation. Observing on such *permanent backsights* reduces a rodperson's walking time. The term *effective eccentricity* is used to describe the combined effects of both the rod and planetable being off their corresponding ground points.

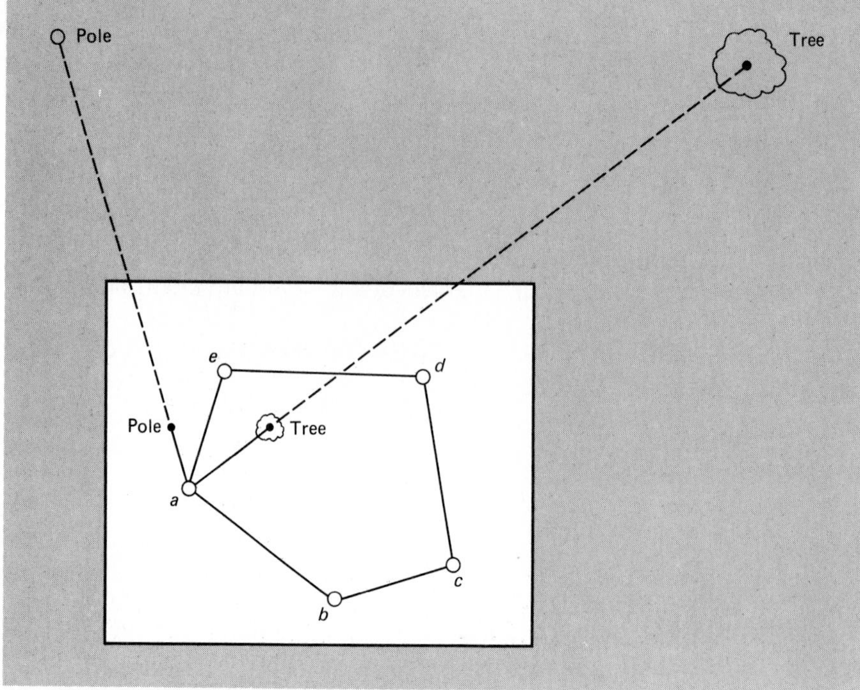

Figure 18-5. Radiation method with planetable to locate details.

18-5. TRAVERSING. To run a traverse, the table is set up over initial survey point *A*, leveled, and clamped. Point *a* on the planetable sheet (see Figure 18-4) is marked to represent this hub. With the ruler's edge on *a* and the alidade sighted at *B*, line *ab* is drawn. Distance *AB* can be determined by tape or stadia and its length marked off at map scale to locate *b*. It is essential that this first course be plotted accurately, since it serves as the base line for all other measurements.

The table can now be moved to hub *B*, set up, and leveled. It is oriented by placing the blade edge on line *ba* and turning the board on its vertical axis until the alidade sights point *A*. Distance *BA* is measured and the average of *AB* and *BA* used in laying out *ab*. The next hub *C* is observed with the blade touching *b*, distance *BC* read, and length *bc* plotted. In similar fashion, succeeding points can be occupied and traverse lines plotted. Whenever possible, check sights should be taken on previously occupied hubs. Small discrepancies are adjusted, but if a plotted point is missed by an appreciable distance, some or all measurements must be repeated.

Details can be located while the traverse is being run, or later. It is desirable to close and adjust the traverse before taking topography, however, since all plotting work done at an incorrectly located station will be wasted. Two methods are used to obtain details: radiation and intersection.

18-6. RADIATION METHOD. With the table oriented at any traverse station, radiating lines can be drawn to points whose locations are desired, as indicated in Figure 18-5. Generally, distances are measured by stadia or an EDMI but a

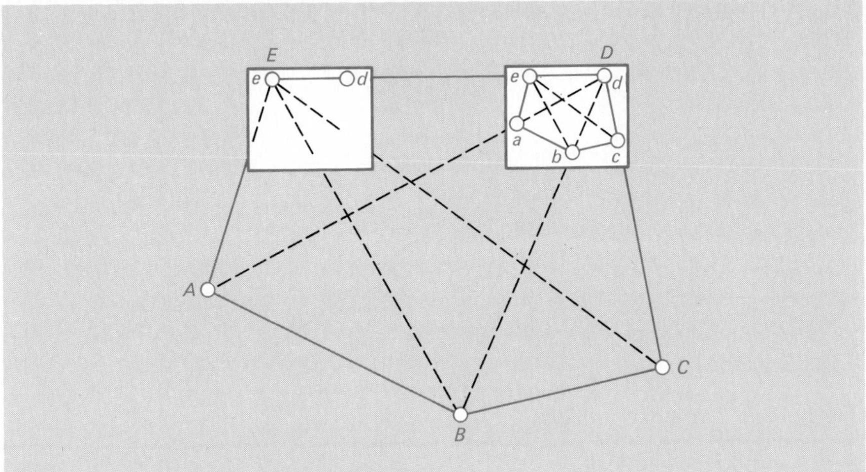

Figure 18-6. Intersection with planetable.

cloth tape is suitable for short sights. Radiating lines, or *rays*, are drawn along the blade edge and the distance plotted to locate points.

18-7. INTERSECTION, OR GRAPHICAL TRIANGULATION METHOD. In the intersection method, shown in Figure 18-6, rays of indefinite length are drawn toward the same point from at least two setups of the planetable. The intersection of rays locates the point desired. One measured line serves to establish the scale; no other distances are required.

The term *graphical triangulation* is sometimes applied to this procedure, which corresponds to triangulation by transit intersections. *The method is particularly important in locating inaccessible and distant points.* In actual field practice, both radiation and intersection may be used at the same setup. Most shots are taken by radiation, however.

The elevation of an inaccessible point located by intersection can be found by reading the vertical angle, scaling the plotted map distance to the point, and computing by trigonometric formula.

18-8. RESECTION. Resection is a method of orientation used when the table occupies a position not yet located on the map. Solutions for two field conditions will be described. In the first, called the *two-point problem*, the length of one line is known. In the second, the *three-point problem*, locations of three fixed points must be known.

18-8.1. THE TWO-POINT PROBLEM. In Figure 18-7, *ab* represents the known length of line *AB* on the ground. With the planetable set up and oriented at *B* by sighting on *A*, a line of indefinite length *bx* is drawn in the direction of *C*. Distance *bc* is not measured, since the observer may wish to reserve judgment on the most advantageous position for point *c*, or may want the rodperson to remain at point *A*. The table is moved to hub *C* (any point on line *bx*), oriented by backsighting along *xb*, and clamped. The alidade blade is set on *a* and pivoted

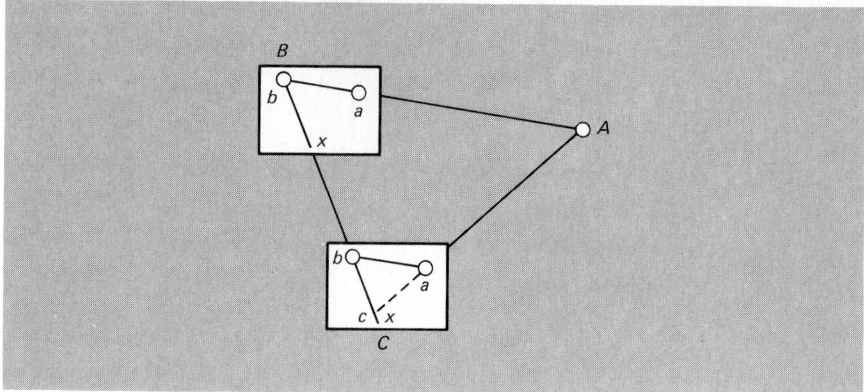

Figure 18-7. Two-point problem.

around this point until the corresponding station A is sighted. The line drawn along the ruler from point a to intersect line bx gives the location c of the occupied point. Line ac is called a resection line.

18-8.2. THE THREE-POINT PROBLEM. The three-point method of location has many uses. It permits the topographer to set up the planetable at any favorable position for taking details and then determine its location on the map by sighting three plotted points, such as church steeples, water towers, flagpoles, radio masts, lone trees, jutting cliff formations, or other prominent objects. The three-point method has long been employed in navigation to ascertain a ship's position by observing with a *sextant* on three shore features recognizable on a coastal chart.

The three-point problem is discussed frequently in technical literature. Trigonometric, mechanical, and graphic solutions have been devised. Only the tracing-paper and three-arm-protractor solutions will be described here.

If the planetable is on a *great circle* (the circumscribed circle) through the three known points, its location is indeterminate. A strong solution results when (a) the table is well inside the great triangle formed by the three points (A, B, and C in Figure 18-8), or (b) the table is *not* near the great circle passing through the three points.

Tracing-Paper Method. A sheet of tracing paper is fastened to the table as shown in Figure 18-8. From any point p' on the sheet, three radiating lines $p'a'$, $p'b'$, and $p'c'$ are drawn toward signals A, B, and C observed through the alidade. The tracing paper is then moved until the three radiating lines pass through the corresponding points a, b, and c previously plotted on the map. The vertex of these lines, p'', is the correct location of the table. This point is marked and the board turned to make the lines radiate to the signals A, B, and C on the ground. The board is then clamped.

Three-Arm-Protractor Method. A similar solution can be obtained mechanically by means of a device called a three-arm protractor. This instrument has a center (somewhat like that of a drafting machine) around which two arms

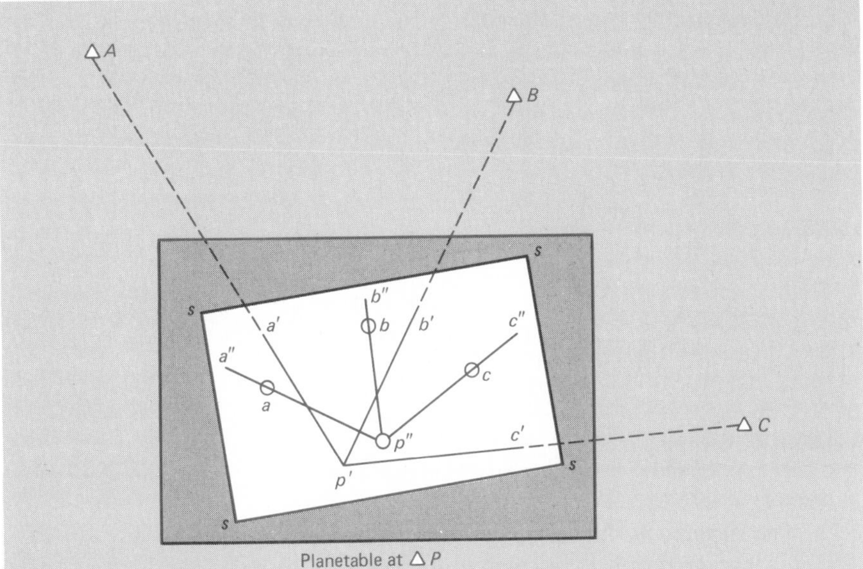

Figure 18-8. Tracing-paper method of three-point resection.

can be rotated. The desired angle between each rotating arm and the fixed arm is set by a vernier on a 360° graduated circle.

Angles *Ap′B* and *Bp′C* in Figure 18-8 are read with a sextant, transit, or theodolite. The protractor arms are clamped to these angles and then made to pass through plotted points *a*, *b*, and *c* by trial. The three-arm protractor is seldom used on planetable work but is a valuable tool in hydrographic surveying and coastal navigation.

Lehmann's Method. A different trial solution of the three-point problem by *Lehmann's method* was widely used by the U.S. Geological Survey when a planetable was the basic equipment for preparing extensive topographic maps. The fifth and preceding editions of this textbook describe that method in detail.

18-9. LEVELING. Elevations are obtained by using the alidade as a level, or by reading vertical angles and slope distances, and performing stadia reductions. After carefully leveling the alidade, a sight is taken on a rod held on a bench mark to get the HI. Or the HI can be secured by adding the h.i. (found by standing a rod beside the table and measuring the telescope's height above the point occupied) to the elevation of that point. With the HI known, rod readings are taken as in ordinary stadia or leveling to determine elevations of other points.

Trigonometric leveling is commonly used where there are considerable differences of elevation. This method requires measuring vertical angles and stadia distances from which differences of elevation are calculated. A Beaman arc on the alidade facilitates computations, but a stadia-reduction chart, stadia table, or hand-held calculator can be used.

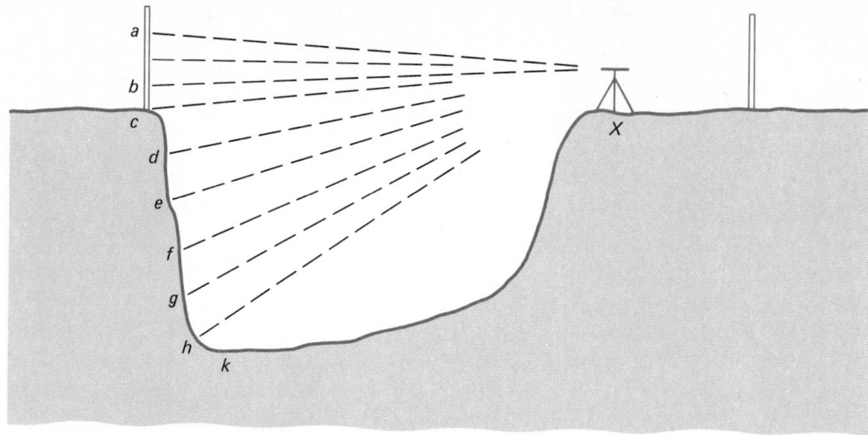

Figure 18-9. Stepping method.

The *stepping method* to acquire elevations is adequate for some purposes, and in rough terrain it saves time and effort for a rodperson. In Figure 18-9, assume the elevation of point k at the bottom of a gorge is needed on a survey that uses two rodpersons, one on each side of the canyon. After the rod intercept ab (say, 5.2 ft) between stadia hairs is found, the upper hair is set on the rod base at c. The position d where the lower cross hair strikes the ground is noted, and the upper cross hair depressed to hit this point. The process is continued until point h is found and the remaining drop to the lowest point k estimated at 2 ft. The difference in elevation between c and k is then $5(5.2) + 2 = 28$ ft.

Another version of the stepping method relies on the fact that the axis of sight through the upper (or lower) cross hair is on a $\frac{1}{2}\%$ grade from the axis through the middle hair.

Differences of elevation are usually computed in the field and plotted on the planetable map. The topographer may compute (if a self-reducing tacheometer is not being used) as well as plot, or one party member may make the reductions. Contours are drawn through points established by the direct method, or interpolated if breaks in slope have been taken. Spot elevations are shown for grade crossings, peaks and depressions that do not fall at contour levels, and all other critical points.

The number of points taken need be only 50 to 60% of the total on a comparable transit-stadia survey to locate contours with the same degree of accuracy. Furthermore, the features shown on a map can be compared with the terrain as work progresses, and any discrepancies quickly discovered.

One of the greatest difficulties arising in the use of a planetable is keeping it level. Pressing on one corner of the board is a practical way to center the bubble and complete a few sights with the Johnson head. This procedure eliminates loosening the ball-and-socket joint, which usually disturbs the orientation.

18-10. PLANETABLE USE IN CONSTRUCTION WORK. A new approach employs a planetable and contoured models for construction layout of earthwork and paving. In this context, a contoured model simply means the contoured plan of a construction site drawn on a conventional planetable sheet or other dimensionally stable medium to a suitable scale and contour interval, as opposed to a three-dimensional architectural model. The contoured model must be accurately drafted to fit the exact geometry of contract drawings. The maximum allowable error in all horizontal and vertical positions mandated by the project specifications determines the model's horizontal scale and contour interval.

The technique described is based on the concept that if a portion of the earth's surface can be mapped by planetable methods, then the process can be reversed. Therefore, the same equipment may be used for layout provided that a suitable site model is available. As usual in any layout situation, horizontal and vertical control are necessary.

In practice, the planetable is set up and oriented in the customary way at a convenient location for the layout requirements. As for any survey procedure, accurate correlation betweeen the ground features to be constructed and the survey control system is essential.

The planetabler selects the points to be set, such as at toe of a slope, top of cut, intersection of two slopes at a valley or ridge line, curved surfaces below bridge abutments, ditches, and so on. Distances from a setup station to selected points are accurately scaled from the model and recorded in a noteform. The alidade straightedge is placed beside the plotted setup station mark and aligned in the direction to a selected point. A rodperson is directed along the line; the stadia distance to a trial point on the line is read; then the rodperson is guided by hand signals forward or backward to the selected point (a tracking mode instrument would be helpful). When the proper location has been accurately established and a hub or lath set, the elevation is computed and recorded. The cut or fill is lettered on the lath for the contractor's use.

Some advantages of this system are:

1. The geometry shown in the contract drawing is checked and any abnormal conditions detected.
2. The stake-setting process is fast and requires only two persons.
3. A planetable operator can see the objectives without abstract visualizations.
4. The procedure can be used regardless of traffic conditions and heavy equipment operating around and between members of the layout party.
5. Moderate elevation differences of about 50 ft, up or down, are no deterrent.
6. Since only selected points are set, it is not necessary to stake or replace the project survey centerline or base line.
7. In complex configurations such as interchange ramps or channel relocations, the slopes can be readily checked during progress of construction, thus allowing the contractor an option to make mid-course corrections.
8. If a site is prone to vandalism or any other condition where stakes may have a rather short life, they can be economically reset.

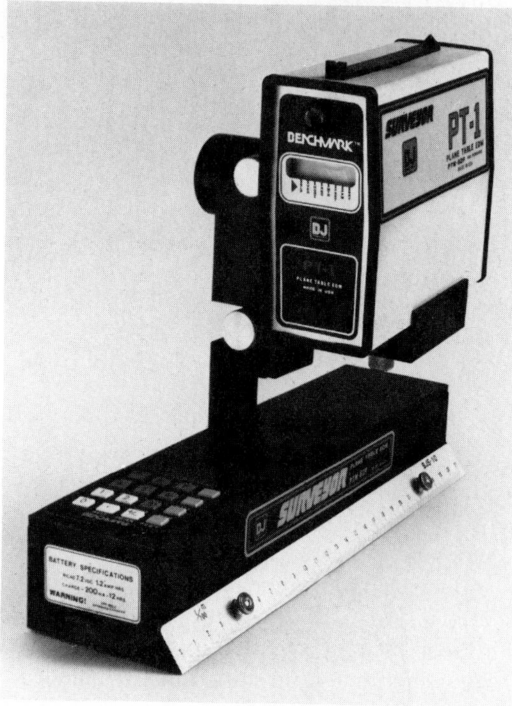

Figure 18-10. Survey PT-1 EDMI mounted on an alidade. (Courtesy Benchmark Company.)

An important stimulant to increased planetable usage is the introduction of the first and smallest EDMI to be mounted on an alidade. The Benchmark Surveyor PT-1 shown in Figure 18-10 measures 2 × 4 × 4 in; weighs only 1.6 lb (thus does not make the alidade topheavy); has a vertical angle sensor for automatic slope reduction; range of 1 km; stakeout, tracking and averaging modes; accuracy of $\pm$(5 mm + 10 ppm); angular resolution of 16 sec; distance resolution of 1 mm; and a readout to 7 digits.

18-11. ADVANTAGES AND DISADVANTAGES OF THE PLANETABLE. Some advantages of a planetable over the transit-stadia method follow.

1. The map is made while looking at the area.
2. Irregular lines, such as stream banks and contours, can be sketched.
3. Notes are not taken.
4. Fewer point locations are required for the same precision in drawing contours and filling in data properly spaced in the field.
5. A map is produced in shorter overall time (field plus office).

Some disadvantages of a planetable compared with the transit-stadia method are:

1. More field time is necessary.
2. Bad weather may halt field work.

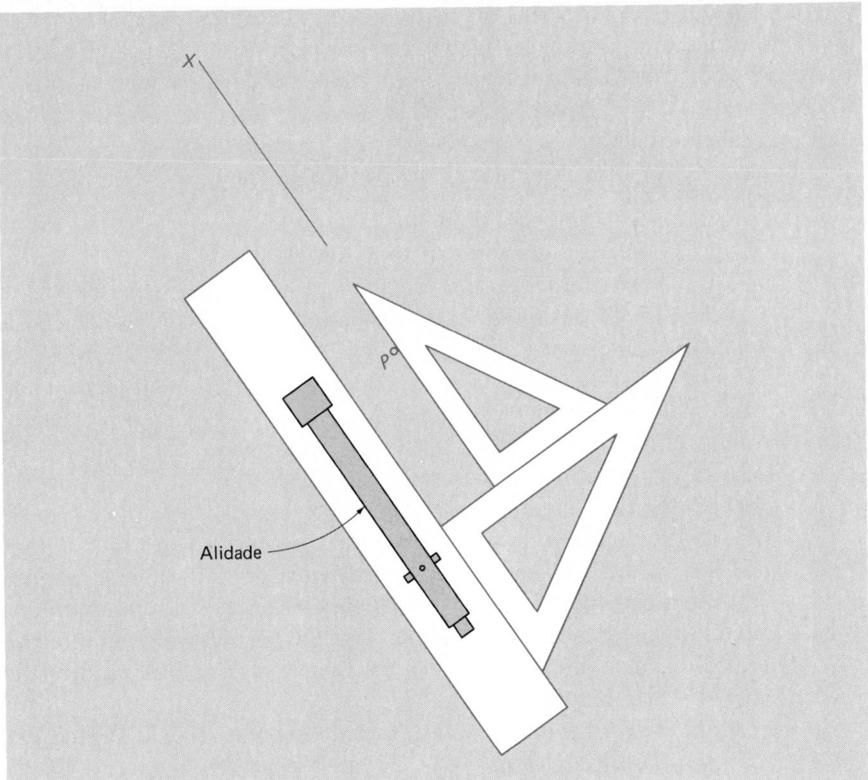

Figure 18-11. Transfer of pivot point.

3. Control must be plotted in advance for precise work.
4. Much more time is required to become proficient with a planetable than with a transit or theodolite.
5. The table is normally set lower than a transit, thus providing reduced sight clearance.
6. The planetable is unsuited for heavily wooded country.
7. Distances must be scaled if lengths are to be taken from the map and areas computed.
8. Many awkward items must be carried.

18-12. PLANETABLE POINTERS. One troublesome problem in operating a planetable is keeping the alidade blade on the plotted position of an occupied point, such as *P* in Figure 18-11. As the alidade is moved to sight a detail, the edge moves off point *P*. A solution sometimes tried is to stick a pin at *P* and pivot around it, but a progressively larger hole is gouged in the sheet with each sight.

Some alidades, like that in Figure 18-3, have a parallel ruler attached to the blade; this provides the best solution. The next best is to place the alidade within an inch or two of point *P*, pivot as necessary for the sight, and then transfer the line through *P* by using two triangles, as illustrated in Figure 18-11. The small error resulting from the eccentric sight is no greater than that from not being exactly over ground point *P*, or even the small one caused by the telescope axis not being over the blade edge on most alidades.

Other helpful pointers on use of the planetable are:

1. Use buff or green detail paper to lessen glare.
2. Plot and ink the traverse in advance of detailing, giving lengths of traverse lines, coordinates of triangulation stations, and useful signals available.
3. Have at least one vertical control point for every three traverse hubs, and show all known elevations.
4. Carry all needed accessories.
5. Cover the portion of a map not being used.
6. Set up the table slightly below elbow height.
7. Verify orientation on two or more lines, if possible.
8. Check the distance and elevation differences in both directions when setting a new hub. Read vertical angles with the telescope in normal and plunged positions, if possible. Use the average value in plotting.
9. Read distance first, then the vertical angle; with a Beaman arc read the H scale, then the V scale.
10. Lift the forward end of the alidade blade to pivot, instead of sliding it, to keep the sheet cleaner.
11. Clean the blade frequently to remove graphite.
12. Reduce stadia readings in the following preferred order: self-reducing tacheometer, Beaman arc, hand-held calculator, charts, and tables.
13. Check hub locations by resection and *cutting-in* (sighting and plotting) prominent points.
14. Draw short lines at the estimated distance on the map to plot points. *Do not start lines at the hub occupied.*
15. Keep in mind comparable limits of accuracy for direction (angle) and distance.
16. Identify points by consecutive numbers, or names, as they are plotted. Number the contours.
17. Have the rodperson make independent sketches on long shots for later transfer to the planetable map.
18. Carry radios to help rodpersons describe topographic features when the observer cannot identify them because of distance and obstacles.
19. Use the same points to locate details and contours whenever possible.
20. Locate contours by the direct method in gently sloping and rolling terrain. Employ the indirect procedure in rugged country and areas that have uniform slopes.
21. Tie a piece of colored cloth on the stadia rod at the required reading to speed work in locating contours by the direct method.
22. Sketch contours after three points have been plotted. Points on the map lose their value if they cannot be identified on the ground.

23. Show spot elevations for summits, sags, bridges, road crossings, and all other critical points.
24. Utilize the stepping method in areas where access is difficult.
25. Draw lines at 1-in intervals on all four margins of the planetable sheet at the time the first plotting is done. The lines serve as a means to determine any expansion and contraction of the sheet.
26. Use a 6-H or harder pencil to avoid smudging.

18-13. SOURCES OF ERROR IN PLANETABLE WORK. Some sources of error in planetable operation include the following:

1. Table not level.
2. Orientation disturbed during detailing.
3. Sights too long for accurate sketching.
4. Insufficient or poor-quality control.
5. Traversing and detailing simultaneously.
6. Too few points taken for good sketching.

18-14. MISTAKES. Some typical mistakes made in planetable use are:

1. Mistakes in reading rod intercepts and angles.
2. Mistakes in stadia reduction.
3. Mistakes in plotting contours.
4. Mistakes in orientation.

PROBLEMS

18-1. Compare the respective advantages and disadvantages of the Johnson-type and Coast and Geodetic Survey tripod heads.

18-2. List three methods of orienting a planetable, their advantages and disadvantages, and an example of the practical use of each method.

18-3. Upon what factors does accurate planetable orientation depend?

18-4. Why is it dangerous to run a planetable traverse and locate details at the same time?

18-5. Does a mark on the wall of a brick building having the sun shine on it provide a good permanent backsight? Explain.

18-6. Describe a situation in which effective eccentricity may enter.

18-7. It may be difficult or impossible to measure a vertical angle of 40° with some planetable alidades? Why?

18-8. For what field conditions is the planetable preferred to transit-stadia in making a topographic survey?

18-9. Compare the radiation and intersection methods for planetable operations. Give an example of the most desirable applications of each.

18-10. Locate the position of a point *C* on your desk by the two-point method. Mark point *A* on a blackboard and point *B* on another desk. Measure distance *AB* and plot it to scale as *ab* on a sheet fastened to the desk. Locate point *c* on the sheet over *C*. Sight over one edge of an engineer's scale to get the resection line.

18-11. Similar to Problem 18-10, except using the three-point tracing-paper method. Mark three points: *A* in one corner, *B* and *C* on the blackboards. Measure distances *AB* and *AC*. Plot the points to scale. Sight over an engineer's scale to determine resection lines.

18-12. Examine planetable literature or *Elementary Surveying*, 5th edition. Discuss advantages of the Lehmann three-point location method.

18-13. Select two field situations that make planetable use preferable to photogrammetric mapping.

18-14. How can the planetable be used in connection with topographic maps prepared photogrammetrically?

18-15. List the pertinent factors to be considered in selecting the proper scale for any specific planetable project.

18-16. On what maximum sight length can the effects of curvature and refraction be neglected in getting elevations with a planetable?

18-17. Are spur traverses desirable with a planetable? Explain.

18-18. What is the biggest problem in leveling, trig leveling, and stadia work with a planetable?

18-19. A planetable at hub D is oriented by compass to start a survey. As a consequence, the first line is off by 30 min from the assumed direction. If extended five miles and hub E set, what is the plotted position error of E if the scale is 1:24,000?

18-20. How is a planetable map checked for accuracy?

18-21. What factors cause errors in the "stepping method"?

18-22. In leveling a planetable, which condition demands the most care: short sights, large vertical angles, or a small scale?

18-23. What main troubles arise in using a planetable? How are they corrected?

18-24. The alidade on a planetable is placed 4 in off the proper point when sighting on an object 90 ft away. Map scale is 1 in = 50 ft. Is the error significant? Explain.

18-25. The telescope of an alidade is centered over a blade 3 in wide. What error is introduced in plotting a 50-ft shot if the map scale is 1:2000? If the shot is 600 ft and map scale 1 in = 500 ft?

18-26. Total length of a closed planetable traverse is 7800 ft. When plotted it fails to close by 0.30 in. If map scale is 1 in = 200 ft, what misclosure is obtained?

18-27. Similar to Problem 18-26, except the linear closure is 0.12 in and the scale 1 in = 100 ft.

18-28. The stepping method is employed to determine the elevation of point k in Figure 18-9. A planetable is set up over X, elevation 820.0 and h.i. = 5.4 ft. With the telescope level, an intercept ab = 3.56 ft is obtained on a rod at c and 5.2 ft read on the middle cross line. If five steps are taken from the rod bottom at c, and an estimated vertical distance of 0.5 ft remains to point k, compute its elevation.

18-29. Similar to Problem 18-28, except the rod intercept is 4.65 ft, six steps are taken, and the estimated vertical distance to k is 2 ft.

18-30. A rectangular lot running N-S and E-W is 75 × 50 ft. To locate contours it is divided into 25-ft square blocks and the following readings taken successively along the E-W lines at the corners: 6.4, 5.8, 3.8; 6.0, 6.7, 5.2; 7.9, 6.8, 6.2; and 9.5, 8.2, 6.7. The Hl is 180.3 ft. Sketch 1-ft contours.

BIBLIOGRAPHY

Harrington, E. L. 1955. "The Stepping Method of Stadia." *Surveying and Mapping* 15(no. 4):460.

Loew, J. W. 1952. *Planetable Mapping*. New York : Harper & Row.

19 ASTRONOMICAL OBSERVATIONS

19-1. INTRODUCTION. Astronomical observations in surveying consist of measuring positions of the sun or certain stars; the principal purpose of this in plane surveying is to determine the direction of the true meridian (astronomic north). True bearings and true azimuths can then be calculated using this meridian. These are needed to establish directions of new property lines so parcels can be properly described; to retrace old property boundaries whose descriptions include bearings; to specify directions of tangents on route surveys; and for many other purposes. Other important but less frequently performed astronomical observations find latitudes and longitudes of points.

To expand on the definition of the true meridian given in Section 8-4, at any point it is a line tangent to, and in the plane of, the great circle which passes through the point and the earth's north and south geographic poles. This is illustrated in Figure 19-1, where P and P' are the poles located on the earth's axis of rotation. Line PAP' is the great circle through A, and line AN the true meridian (tangent to the great circle at A in plane $POP'A$). With a true meridian established, the true azimuth α of any line, such as AB of Figure 19-1, can readily be obtained by determining horizontal angle NAB.

Astronomical observations are not necessarily required on every project where true bearings or true azimuths are needed. If a pair of intervisible control monuments from a previous survey exist in the area, and the true bearing or azimuth is known for that line, new directions can be referenced to it. Also, north-seeking gyros are now available which can automatically and quickly

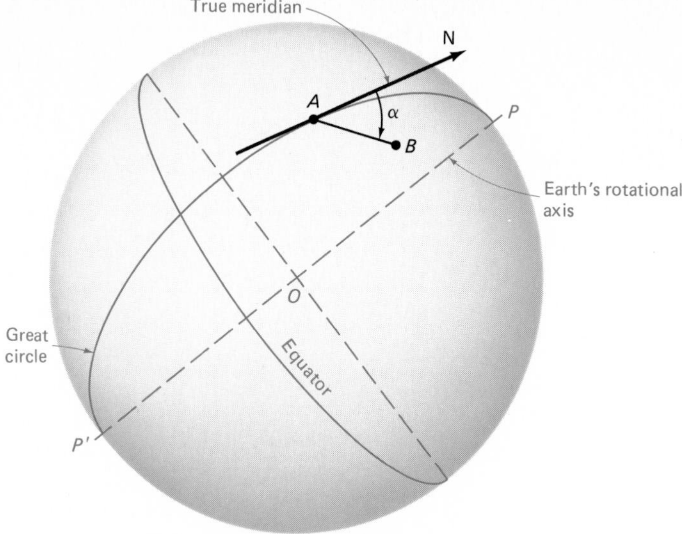

Figure 19-1. True meridian and true azimuth.

ascertain the direction of true north. This is achieved when the axis of the spinning gyro has aligned itself with the earth's axis of rotation. The line of sight of the theodolite, to which the gyro is attached (Figure 19-2), can then be turned to north as determined by the gyro. The process takes only a few minutes and can yield accuracies of ± 20 sec or better. Gyros are especially useful for projects where astronomical observations cannot be made—for example, in mine surveys.

Existing control monuments are often not available for reference, and north-seeking gyros are relatively expensive and limited exclusively in their use

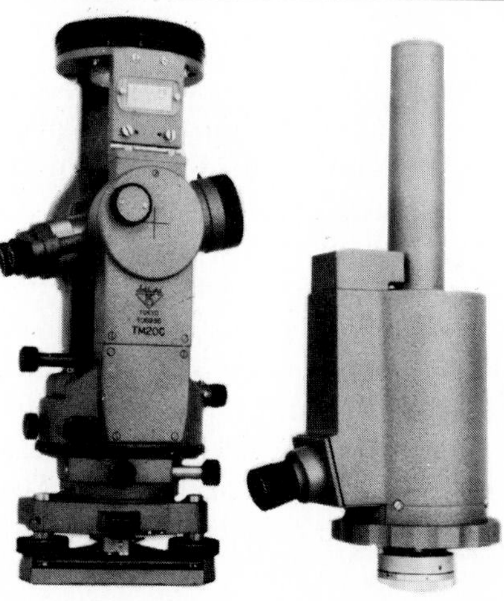

Figure 19-2. TM-20C theodolite and GP-1 gyro attachment. (Courtesy Lietz Co.)

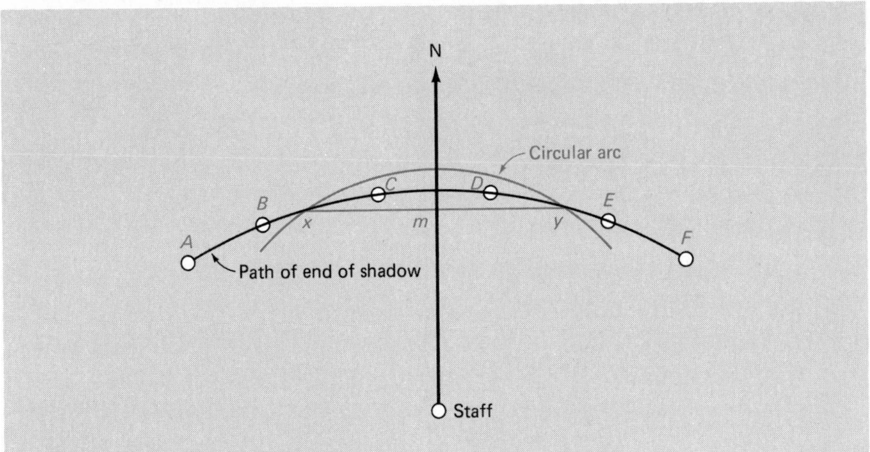

Figure 19-3. Determination of azimuth by shadow method.

to azimuth determination. Thus, astronomical observations that can be made with standard surveying equipment are extremely important. Field and office procedures involved in this process are the subjects of this chapter.

19-2. SIMPLE METHODS OF DETERMINING THE MERIDIAN.

To introduce the subject of meridian determination, two simple methods requiring no computation are described. Both are based on the knowledge that the sun appears to move in essentially a circular path and achieves its highest altitude as it crosses the meridian. Thus if the direction to the sun at its maximum altitude can be defined, the true meridian is located.

19-2.1. SHADOW METHOD.

The true meridian can be established by the shadow method using only a straight rod or pole and a piece of string. In Figure 19-3, points *A*, *B*, *C*, *D*, *E*, and *F* mark the end of the shadow of a plumbed staff or a telephone pole at intervals of perhaps 30 min throughout the period from 9 A.M. to 3 P.M. A smooth curve is sketched through the marks. With the staff or pole as a center and any appropriate radius, a circular arc is swung to obtain two intersections, *x* and *y*, with the shadow curve. A line from the staff through *m*, the midpoint of *xy*, approximates the meridian. If the ground is level, the pole plumb, and the shadow points carefully marked, the angle between the line established and the true meridian can be obtained to an accuracy of within about ± 30 min.

19-2.2. EQUAL ALTITUDES OF THE SUN.

Determination of the meridian by equal altitudes of the sun requires a transit or theodolite, but the method is similar in principle to the shadow method. Assume that a meridian is to be passed through point *P* in Figure 19-4, over which the instrument is set up. At some time between 8 and 10 A.M., say about 9 A.M., with a dark glass over the eyepiece or

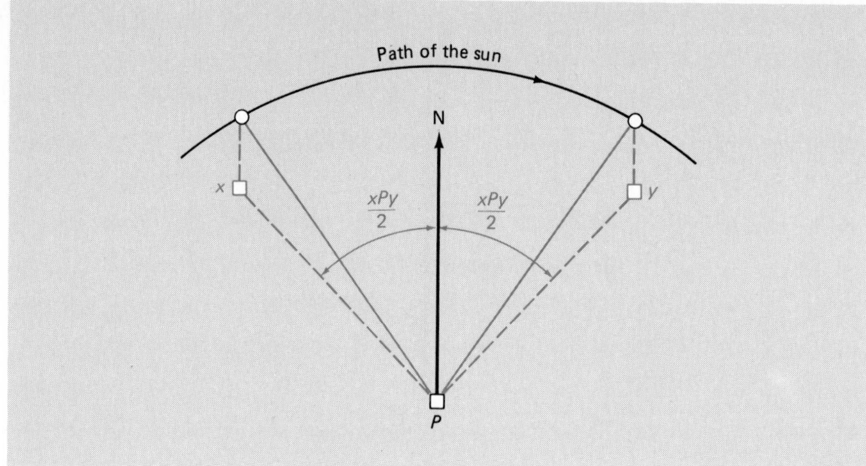

Figure 19-4. Azimuth by equal altitudes of the sun.

objective lens, the sun's disk is bisected by both the horizontal and vertical cross hairs. The vertical angle is read, the telescope depressed, and point x set at least 500 ft from the instrument. Shortly before 3 P.M., with the vertical angle previously read placed on the arc, the sun is followed until the vertical and horizontal cross hairs again simultaneously bisect the sun. The telescope is depressed and a point y set at approximately the same distance from P as x. The bisector of angle xPy is the true meridian.

Perfect results cannot be attained with either of these simplified methods, because the sun's changing declination (see Section 19-5 for the definition of declination) gives it a path slightly skewed to the equator instead of parallel with it. Other disadvantages are the time and delay required, and the possibility of clouds obscuring the sights. Because of their drawbacks, these procedures are seldom used by surveyors, but they do illustrate uncomplicated methods of meridian determination.

19-3. OVERVIEW OF USUAL PROCEDURES FOR ASTRONOMICAL AZIMUTH DETERMINATION. The usual field procedures employed by surveyors to define the direction of true north consist of the following steps: (1) a transit or theodolite is set up and leveled at one end of the line whose azimuth is to be determined, like point A of Figure 19-5; (2) the horizontal and vertical circles are read when pointing on a celestial body S; (3) the precise time of pointing is recorded; and (4) a horizontal angle is measured from the celestial body to a point on the other end of the line, like angle θ of Figure 19-5 from S to B. Office work involves (a) obtaining the precise location of the body in the heavens at the instant sighted from an *ephemeris* (almanac of celestial body positions); (b) computing the celestial body's azimuth (angle Z of Figure 19-5) based on observed and ephemeris data; and (c) calculating the line's azimuth by applying the measured horizontal angle to the computed azimuth of the body ($\alpha = Z + \theta$ in Figure 19-5).

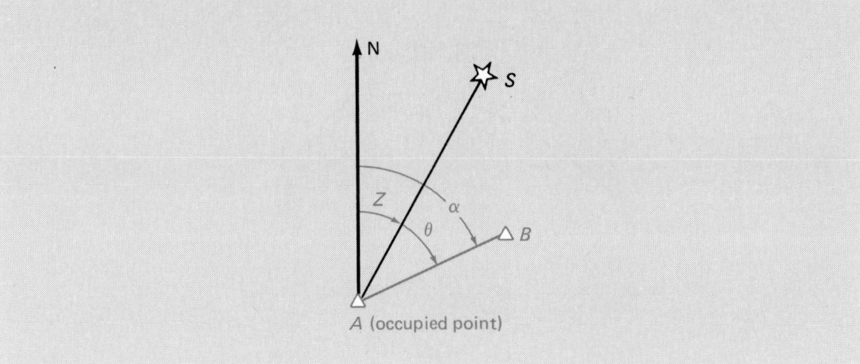

Figure 19-5. Azimuth determination from observation on a celestial body.

Any celestial body for which ephemeris data are available can be employed in the procedures outlined; however, the sun and, in the United States, Polaris (north star) are almost always selected. The sun permits observations to be made in lighted conditions during normal daytime working hours; Polaris is preferred because of the observing convenience and accuracy afforded due to its favorable location in the heavens.

Accuracies attainable in astronomical azimuths depend on many variables, including (a) precision of transit or theodolite used, (b) ability and experience of the observer, (c) weather conditions, (d) quality of the clock or chronometer used to measure the time of sighting, (e) celestial body sighted and its position when observed, and (f) accuracy of ephemeris and other data available. Polaris observations provide the best results and, with several repetitions of measurements utilizing first-order instruments, accuracies to within 1 sec are possible. Sun observations yield a lower order of accuracy, but values accurate to within ± 10 to 15 sec can be obtained if careful repeated measurements are made.

19-4. EPHEMERIDES. A variety of ephemerides are available to surveyors for obtaining sun and star positions. Some of the most useful ones are: *The Nautical Almanac*, published by the U.S. Naval Observatory; *The Ephemeris of the Sun, Polaris and Other Selected Stars*, also published by the U.S. Naval Observatory but specifically for surveyors (both are available from the U.S. Government Printing Office, Washington, D.C. 20402); *Apparent Places of Fundamental Stars*, published by Astronomisches Rechen-Institute, Heidelberg, Germany; and the pocket-sized *Solar Ephemeris*, published by and available from the Keuffel & Esser Company, Morristown, N.J. 07960.

Values tabulated in ephemerides are given for Greenwich civil time, so before extracting values from them, standard times normally recorded for observations must be converted to Greenwich time. This topic is discussed further in Section 19-6.

Possession of an ephemeris is assumed in the following discussion. The 1983 Keuffel & Esser *Solar Ephemeris*, with values carried out to the nearest

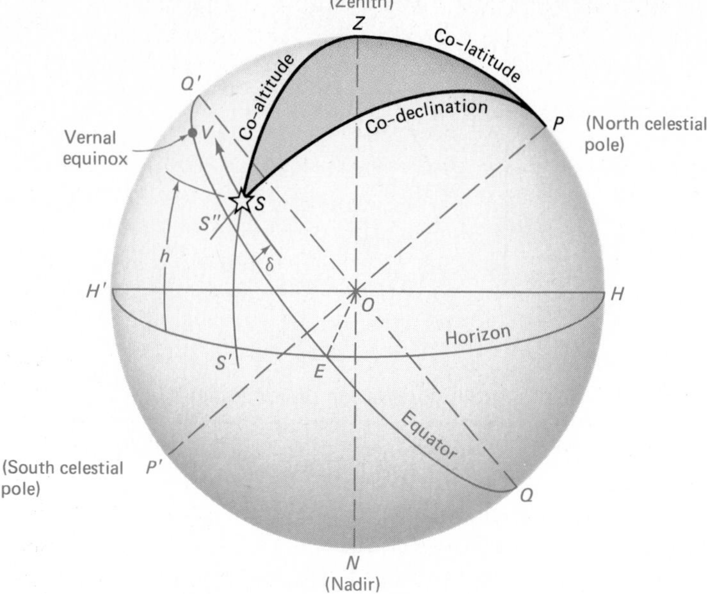

Figure 19-6. Celestial sphere.

0.1 min (far enough for ordinary accuracy), is used for reference in examples given in this chapter. Table and page numbers are noted for values taken from it. (These page numbers usually change slightly in successive years.)

19-5. DEFINITIONS. In making and computing astronomical observations, the sun and stars are assumed to lie on the surface of a *celestial sphere* of infinite radius having the earth as its center. Due to the earth's rotation on its axis, all stars appear to move around centers that are on the north-south axis of the celestial sphere. Figure 19-6 is a sketch of the celestial sphere and illustrates some terms used in field astronomy. Here *O* represents the earth, and *S* a heavenly body, as the sun or a star whose apparent direction of movement is indicated by an arrow. Students find it helpful to sketch the various features on a true sphere or globe.

The *zenith* is located where a plumb line projected above the horizon meets the celestial sphere. On a diagram it is usually designated by *Z*. Stated differently, it is the point on the celestial sphere vertically above the observer.

The *nadir* is that point on the celestial sphere vertically beneath the observer and exactly opposite the zenith.

The *north celestial pole* is point *P* where the earth's rotational axis, extended from the north geographic pole, intersects the celestial sphere.

The *south celestial pole* is point *P*′ where the earth's rotational axis, extended from the south geographic pole, intersects the celestial sphere.

A *vertical circle* is any great circle of the celestial sphere passing through the zenith and nadir, and represents the line of intersection of a vertical plane with the celestial sphere.

The *celestial equator* is the great circle on the celestial sphere whose plane is perpendicular to the axis of rotation of the earth. It corresponds to the earth's equator enlarged in diameter. Half of the equator is represented by $Q'S''EQ$ in Figure 19-6.

An *hour circle* is any great circle on the celestial sphere which passes through the north and south celestial poles. Therefore, hour circles are perpendicular to the plane of the celestial equator. They correspond to meridians (longitudinal lines) and are used to measure hour angles.

The *horizon* is a great circle on the celestial sphere whose plane is perpendicular to the direction of the plumb line. In surveying, the plane of the horizon is determined by a spirit level. Half of the horizon is represented by $H'S'EH$ in Figure 19-6.

A *celestial meridian* is the hour circle containing the zenith. It is also defined as the vertical circle that passes through the celestial poles. The intersection of the celestial meridian plane with the horizon plane is line $H'OH$ of Figure 19-6, which defines the direction of true north. Thus, it is the astronomic meridian line used in plane surveying. Since east is 90° clockwise from true north, line OE in the horizon plane is a true east line. The celestial meridian is composed of two branches; the *upper branch* contains the zenith and is the semicircle $PZQ'H'P'$ of Figure 19-6, and the *lower branch* includes the nadir and is arc $PHQNP'$.

A *diurnal circle* is the complete path of travel of the sun or a star in its apparent daily orbit about the earth. Four terms describe specific positions of heavenly bodies in their diurnal circles: (1) *lower culmination*—the body's position when it is exactly on the lower branch of the celestial meridian; (2) *eastern elongation*—when the body is farthest east of the celestial meridian with its hour circle and vertical circle perpendicular; (3) *upper culmination*—when it is on the upper branch of the celestial meridian; and (4) *western elongation*—when the body is farthest west of the celestial meridian with its hour circle and vertical circle perpendicular.

An *hour angle* is the angle between the celestial meridian plane and the hour circle plane passing through a celestial body. It is measured by the angle at the pole between the meridian and hour circle, or by the arc of the equator intercepted by those circles. Hour angles are measured westward (in the direction of apparent travel of the sun or star) from the upper branch of the meridian of reference.

The *Greenwich hour angle* of a heavenly body at any instant of time is the angle, measured westward, from the upper branch of the meridian of Greenwich to the meridian over which the body is located at that moment.[1] In the ephemeris it is designated by GHA. *Local hour angle* (LHA) is similar to GHA except it is measured from the observer's meridian.

The *declination* of a heavenly body is the angular distance measured along the hour circle between the body and the equator; it is plus when the body is north of the equator, and minus when south of it. Declination is usually denoted by δ in formulas, and represented by $S''S$ in Figure 19-6.

[1] The meridian of Greenwich, England is internationally accepted as the reference meridian for specifying longitudes of points on earth and for giving positions of celestial bodies.

The *position* of a heavenly body with respect to the earth at any moment may be given by its declination and Greenwich hour angle.

The *polar distance* of a body is its angular distance from the pole measured along an hour circle. It is equal to 90° minus the declination. In Figure 19-6, polar distance is arc *PS*, labeled co-declination.

The *altitude* of a heavenly body is the angular distance measured along a vertical circle above the horizon, *S′S* in Figure 19-6. It is measured with the vertical arc of a transit, theodolite, or sextant and usually denoted in formulas by *h*.

The *co-altitude* or *zenith distance* is arc *ZS* in Figure 19-6 and equals 90° minus the altitude.

The *astronomical* or *PZS triangle* (darkened in the figure) is the spherical triangle whose vertices are the pole (*P*), zenith (*Z*), and astronomical body (*S*). Because of the body's apparent movement through its diurnal circle, the three angles in this triangle are constantly changing.

The *azimuth* of a heavenly body is the arc of the horizon clockwise from either the north or south point to the vertical circle through the body. An azimuth from north is represented by *HS′* of Figure 19-6, and it equals the *Z* angle of the *PZS* triangle.

The *latitude* of an observer is the angular distance, measured along the meridian, from the equator to the observer's position. In Figure 19-6 it is arc *Q′Z*. It is also the angular distance between the polar axis and horizon, or arc *HP*. Latitude is measured north or south of the equator. Formulas in this book denote it as ϕ.

The *vernal equinox* is the point of intersection of the celestial equator and the hour circle through the sun at the instant it reaches zero declination (about March 21 each year). For any calendar year it is a fixed point on the celestial sphere (the astronomer's zero-zero point of coordinates in the sky) and moves with the celestial sphere just as the stars do. On Figure 19-6 it is designated by *V*.

The *right ascension* of a heavenly body is the angular distance *VS″* measured eastward from the hour circle through the vernal equinox to the hour circle of a celestial body. Right ascension frequently replaces Greenwich hour angle as a means of specifying the position of a star with respect to the earth. In this system, however, the Greenwich hour angle of the vernal equinox must also be given.

Refraction, as illustrated in Figure 19-13, is the angular increase in the apparent (observed) altitude of a heavenly body due to bending of light rays that pass obliquely through the earth's atmosphere. It varies from zero for an altitude of 90° to a maximum of about 35 min at the horizon, and for intermediate positions its value, in minutes, is roughly equal to the natural cotangent of the observed altitude. Small adjustments must also be made for temperature and pressure variations. Refraction makes observations on heavenly bodies near the horizon less reliable than those taken at high altitudes. Most ephemerides list refraction corrections for varying values of altitude, temperature, and pressure. The correction is always subtracted from observed altitudes.

Parallax, also illustrated in Figure 19-13, results from observations being made from the surface of the earth instead of at its center. It causes a small an-

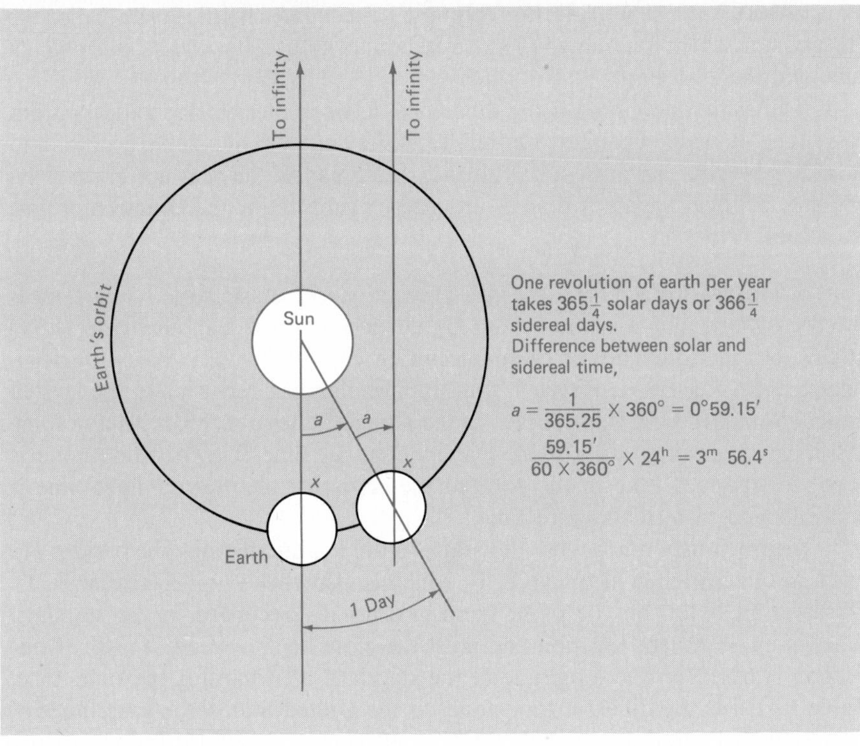

Figure 19-7. Earth's orbit.

gular decrease in the apparent altitude; hence, the correction is always added. Parallax is insignificant when a star is observed but must be added on sun altitudes. The ephemeris also contains tables of parallax corrections.

19-6. TIME. Four kinds of time are used in making and computing an astronomical observation.

Sidereal Time. A sidereal day is the interval of time between two successive upper culminations of the vernal equinox over the same meridian. Sidereal time is star time. At any location for any instant, it is equal to the local hour angle of the vernal equinox.

Apparent Solar Time. An apparent solar day is the interval of time between two successive lower culminations of the sun. Apparent solar time is sun time, and the length of a day varies somewhat because the rate of travel of the sun is not constant. Since the earth revolves about the sun once a year, there is one fewer day of solar time in a year than sidereal time. Thus the length of a sidereal day is shorter than a solar day by approximately 3 min 56 sec. The relationship between sidereal and solar time is illustrated in Figure 19-7.

Mean Solar, or Civil, Time. This time is related to a fictitious sun, called the "mean" sun, which is assumed to move at a uniform rate. It is the basis for watch time and the 24-h day.

The *equation of time* is the difference between mean solar and apparent solar time. Its value changes continually as the apparent sun gets ahead of and then falls behind the mean sun. Values for each day of the year are given in an ephemeris. Local apparent time is obtained by subtracting the equation of time from local civil time.

Standard Time. This is the mean time at meridians 15° or 1 h apart, measured eastward and westward from Greenwich. Eastern standard time (EST) at the 75th meridian differs from Greenwich civil time (GCT) by 5 h (earlier, since the sun has not yet traveled from the meridian of Greenwich to the United States). Standard time was adopted in the United States in 1883, replacing some 100 local times previously used. Daylight saving time (DST) in any zone is equal to standard time in the zone to the *east*; thus, central daylight time is equivalent to eastern standard time.

As previously noted, sun and star positions tabulated in ephemerides are given in Greenwich civil time (GCT), which is also called universal time (UT). Observation times, on the other hand, are usually recorded in the standard times of an observer's location and must therefore be converted to GCT. Conversion is based on the longitude of the standard meridian for the time zone. Table 19-1 lists the different time zones in the United States, the longitudes of their standard meridians, and the number of hours to be added for converting standard and daylight time to GCT.

In making civil time conversions based on longitude differences, the following relationships are helpful:

$$360^\circ \text{ of longitude} = 24 \text{ h}$$
$$15^\circ \text{ of longitude} = 1 \text{ h}$$
$$1^\circ \text{ of longitude} = 4 \text{ min (of time)}$$

To make precise timings of astronomical observations, a check of the timepiece used should be made against accurate signals broadcast on shortwave

TABLE 19-1. LONGITUDES OF STANDARD MERIDIANS IN THE UNITED STATES AND TIME DIFFERENCES FROM GREENWICH

STANDARD TIME ZONE (AND ABBREVIATION)	LONGITUDE OF STANDARD MERIDIAN	CORRECTION IN HOURS, TO ADD TO OBTAIN GCT	
		STANDARD TIME	DAYLIGHT TIME
Eastern (EST)	75°	5	4
Central (CST)	90°	6	5
Mountain (MST)	105°	7	6
Pacific (PST)	120°	8	7
Yukon (YST)	135°	9	8
Alaska/Hawaii (AHST)	150°	10	9
Bering Sea (BST)	165°	11	10

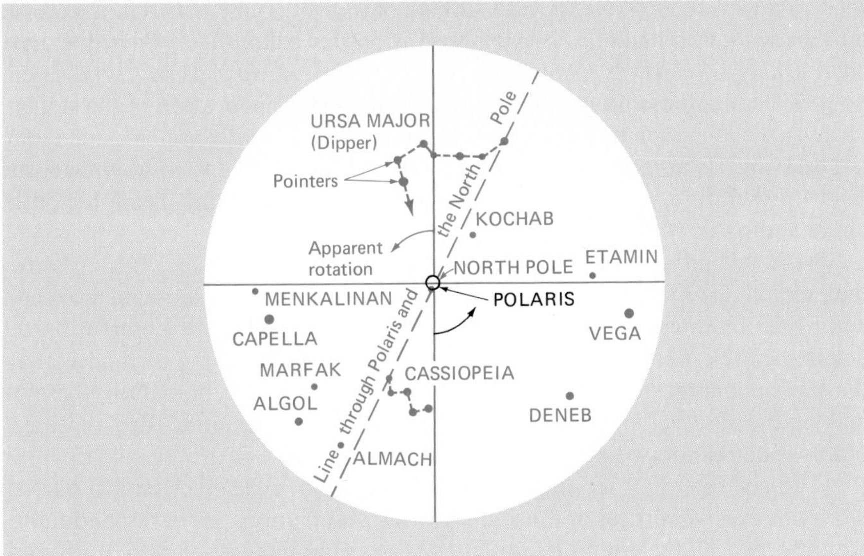

Figure 19-8. Position of Polaris.

radio. In the United States, the National Bureau of Standards broadcasts time signals from station WWV in Ft. Collins, Colorado, on frequencies of 2.5, 5, 10, and 15 MHz. To broaden coverage, signals are also transmitted from station WWVH in Hawaii on the same frequencies. These signals are broadcast as audible ticks with a voice announcement of Greenwich civil time at each minute.

19-7. STAR POSITIONS. If the pole could be seen as a definite point in the sky—that is, if a star were there—a meridian observation would require only a simple sighting. Since the pole is not so marked, observations must be made on stars—preferably those close to the pole—whose positions at given times are listed in an ephemeris.

If the polar distance of a star is less than the latitude of an observer's position, that star will never set; that is, if *PS* in Figure 19-6 is less than *PH*, the star will never move below the horizon plane and be invisible. Thus it can always be seen during nighttime hours. Such stars are called *circumpolar stars.*

The visible star nearest the north pole is *Polaris,* part of the constellation *Ursa Minor,* also called the *Little Dipper.* Its polar distance changes slightly from year to year and in 1983 was approximately 0°48.5′. Polaris is a circumpolar star for all latitudes in the United States. In most of the southern hemisphere, stars of the constellation *Southern Cross* are circumpolar (their diurnal circles are close to the south celestial pole) and thus they are most frequently used for azimuth observations there.

Polaris is located in the sky by first finding the Big Dipper in Ursa Major. The two stars of the dipper farthest from the handle are the Pointers, as shown in Figure 19-8. Polaris is the nearest bright star along the line through the Pointers, and it is also on the line through the easternmost star of Cassiopeia and the end star of the Big Dipper handle.

19-8. AZIMUTH FROM POLARIS OBSERVATIONS. Three different methods have been used in making Polaris observations for azimuth: (1) *Polaris at any hour angle*, (2) *Polaris at culmination*, and (3) *Polaris at elongation*. As indicated by their names, these procedures differ principally in the location of the star in its diurnal circle when observed. The method most frequently used by surveyors is observing Polaris at any hour angle, since it can be done at a convenient time. Furthermore, repeated observations can be made with the theodolite both direct and reversed, which increases precision.

The advantage of observing Polaris at culmination, just when crossing the meridian, is that its direction is true north then, and no subsequent calculations are required. However, computations must precede the observation to determine the exact time of culmination. A disadvantage of the method is that the star's apparent horizontal movement is maximum at culmination, thereby making the observations slightly more difficult and less accurate. Also, repeated observations cannot be made.

Advantages of observing Polaris at elongation are (1) the star's apparent motion is exactly vertical at those times, thus observations are easier and more accurate, and (2) the computations needed to determine its azimuth are somewhat simplified. Again, the time of elongation must be calculated prior to observing. Because the star's azimuth remains nearly constant for 15 to 20 min before and after elongation, repeated sightings can be made.

One serious disadvantage of both culmination and elongation methods is that observations are restricted to the fixed times of these events, which often occur during the daytime or at inconvenient hours late at night or early in the morning. Thus these methods are seldom used, so discussion in this book concentrates on Polaris at any hour angle.

19-9. COMPUTATIONS FOR AZIMUTH FROM OBSERVATIONS OF POLARIS AT ANY HOUR ANGLE. In this method, which can be done at any convenient time beginning about dusk, only the horizontal circle reading and precise time have to be recorded when the star is sighted. Readings of the vertical circle are recommended, however, to ensure that the correct star has been sighted and provide a check on the observations (see Section 19-11). After observing the star, a horizontal angle is measured to the point on a line whose azimuth is desired.

Computations after field work require the solution for angle Z in the astronomical (PZS) triangle (see Figure 19-6). The formula for Z, derived from laws of spherical trigonometry, is

$$Z = \tan^{-1}\left(\frac{\sin t}{\cos\phi\tan\delta - \sin\phi\cos t}\right) \tag{19-1}$$

The geometry upon which Eq. (19-1) is based is shown more clearly in Figure 19-9, where the PZS triangle is again darkened. The latitude ϕ of the observer's position is arc HP; thus, arc PZ is $(90^\circ - \phi)$ or *co-latitude*. Declination δ of the star is arc $S''S$, so SP is $(90^\circ - \delta)$ or polar distance. Angle ZPS of Figure 19-9 is t, the so-called *meridian angle*.

The latitude of an observer's position can be scaled from a U.S. Geological Survey (USGS) quadrangle map and, with reasonable care, obtained to within

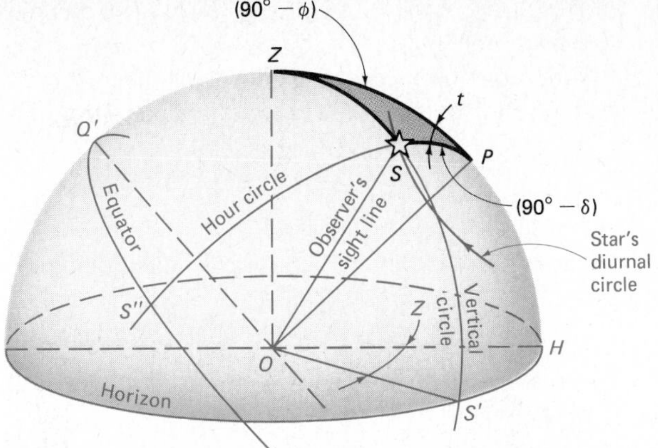

Figure 19-9. The *PZS* triangle for Polaris at any hour angle.

±2 sec. The "Topo Aid," a transparent template placed over a USGS quad sheet, can also be used to measure latitude and longitude. Alternatively, latitude can be obtained from astronomical observations as explained in Section 19-17. Declination is extracted from an ephemeris for the instant of sighting.

Meridian angle t equals the local hour angle (LHA) of the star if it is west of north, or (360° − LHA) if east of north. The local hour angle is between 0 and 180° when the star is west of north, and between 180 and 360° if the star is east of north. Diagrams such as those of Figure 19-10(a) and (b) are helpful in determining t angles. These diagrams show the north celestial pole at the center of the star's diurnal circle as viewed from the observer's position within the sphere. The apparent motion of stars is counterclockwise. Angle λ between the meridian of Greenwich G and the local meridian L through the observer's

Figure 19-10. Computation of meridian angle (t).

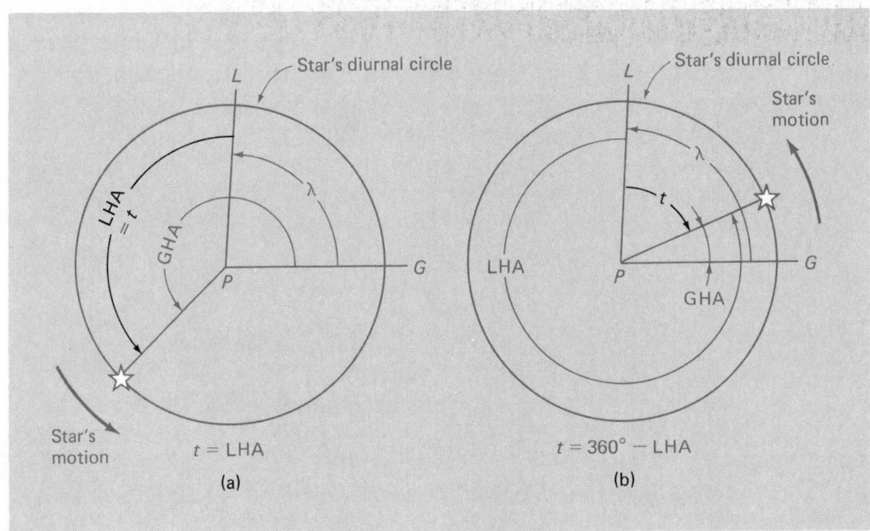

position is the longitude of the station occupied. The Greenwich hour angle (GHA) of the star is taken from an ephemeris for the observation time, and longitude scaled from a USGS map or observed as described in Section 19-17. Plotting these two values to scale on a diagram immediately makes clear the star's position.

EXAMPLE 19-1

For the following set of field notes obtained on June 3, 1983, compute the azimuth of line *AB*.[2]

FIELD DATA

POLARIS OBSERVATION FOR AZIMUTH
STATION OCCUPIED = *A*
LATITUDE *A* = N43°05.4′;
LONGITUDE *A* = W89°26.0′

OBSN TIME (P.M. CDT)	INST D/R	STA STD	HORIZ CIRCLE	VERT CIRCLE
8:30:49	D	*B*	0°00.0′	
		Polaris	49°36.2′	47°36.8′
8:39:00	R	*B*	180°00.0′	
		Polaris	229°38.3′	312°22.6′
8:44:33	D	*B*	0°00.0′	
		Polaris	49°39.7′	47°38.0′
8:49:46	R	*B*	180°00.0′	
		Polaris	229°41.1′	312°21.7′

When the observations occur over a short period of time, average values for observation time and horizontal angle can be used and one reduction made. This assumes the star moves linearly throughout the time between the first and last observation. For more accurate results, or if the elapsed time exceeds more than about 20 min, each observation should be reduced independently and the resulting azimuths averaged. In this example, the reductions are performed independently.

Solution for the first observation:

1. GCT of observation:

8:30:49	P.M. CDT 6/3/83
+12	corr. for P.M.
+ 5	corr. to Greenwich
25:30:49	
1:30:49 =	GCT of observation, 4 June 1983

[2] A computer program written in BASIC is given in Appendix C. To illustrate its use, this problem has been solved.

2. GHA of Polaris (from K&E ephemeris):

$$
\begin{array}{rl}
218^\circ 31.8' = & \text{GHA at } 0^{\text{h}} \text{ GCT, 4 June 1983 (Table 1, p. 58)} \\
+\ 22^\circ 33.7' & \text{incr. for } 1^{\text{h}}\ 30^{\text{m}} \\
+\ \quad 12.3' & \text{incr. for } 49^{\text{s}} \\
\hline
241^\circ 17.8' = & \text{GHA of Polaris at observation}
\end{array}
\left.\vphantom{\begin{array}{l} a \\ b \end{array}}\right\} \text{Table 5, p. 68}
$$

3. Declination of Polaris (from K&E ephemeris):

$$
\left.\begin{array}{l}
0^\circ 48.94' = \text{polar distance at } 0^{\text{h}} \text{ GCT, 31 May 1983} \\
0^\circ 48.98' = \text{polar distance at } 0^{\text{h}} \text{ GCT, 10 June 1983}
\end{array}\right\} \text{Table 3, p. 67}
$$

By interpolation for approx $3\frac{1}{4}$ days beyond 31 May:

$$(0.04)3\tfrac{1}{4}/10 = 0.01'$$

$$\text{polar distance at observation} = 0^\circ 48.95'$$

$$\text{declination} = 90^\circ - 0^\circ 48.95' = 89^\circ 11.05' \text{ north}$$

4. Local hour angle of Polaris:

$$
\begin{array}{rl}
\text{GHA} & 241^\circ 17.8' \\
-\lambda & \ \ 89^\circ 26.0' \\
\hline
\text{LHA} = & 151^\circ 51.8' = t \text{ (Polaris is west of north)}
\end{array}
$$

[Note: A sketch similar to Figure 19-10(a) applies.]

5. Azimuth of Polaris, by Eq. (19-1):

$$Z = \tan^{-1}\left[\frac{\sin 151^\circ 51.8'}{\cos 43^\circ 05.4' \tan 89^\circ 11.05' - \sin 43^\circ 05.4' \cos 151^\circ 51.8'}\right]$$

$$= \tan^{-1}\left[\frac{0.47158}{(0.73028)(70.22501) - (0.68315)(-0.88183)}\right]$$

$$= \tan^{-1}(0.0090887)$$

$$= 0^\circ 31.2' \text{ (west of north)}$$

$$\text{azimuth of star} = 360^\circ - 0^\circ 31.2' = 359^\circ 28.8'$$

6. Azimuth of line AB:

$$
\begin{array}{rl}
359^\circ 28.8' & \text{azimuth of Polaris} \\
-\ \ 49^\circ 36.2' & \text{horizontal angle } B \text{ to Polaris} \\
\hline
\alpha_{AB} = 309^\circ 52.6' & = \text{azimuth of line } AB
\end{array}
$$

(Note: A sketch similar to Figure 19-5 will aid this calculation.)

The other three observations were calculated in the same way, and all provide an identical azimuth for line *AB*.

19-10. REDUCTION OF POLARIS OBSERVATIONS BY DOUBLE INTERPOLATION. The parameters in Eq. (19-1) that have the greatest variation are the meridian angle t and observer's latitude ϕ. Declination (90° − polar distance) is also a variable, but it changes only very slightly throughout the year. For Polaris in 1983, it fluctuated from a minimum value of 0°48.26′ to a maximum of 0°49.02′. Equation (19-1) has been solved repeatedly using increments of meridian angles from 0 to 180°, and increments of latitudes from 10 to 70°. Polar distance was held constant at 0°48.50′. Azimuths resulting from these repeated solutions are listed in Table 10 of the K&E ephemeris.

To obtain the azimuth of Polaris at any hour angle, a simple double interpolation can be made from Table 10 after determining the observer's latitude and computing the star's local hour angle. A small correction must be applied, however, for the difference between the actual polar distance of Polaris and the value of 0°48.50′ used to obtain the data listed in Table 10. These corrections are given in Table 11.

EXAMPLE 19-2

Determine the azimuth of Polaris for the first observation of Example 19-1 by double interpolation.

SOLUTION

In Example 19-1, the observer's latitude was 43°05.4′N, and the local hour angle of the star was 151°51.8′. The following values (underlined) were extracted from Table 10, p. 78, of the K&E ephemeris:

LHA	LATITUDE		
	42°	43°05.4′	44°
150°00′	0°32.3′	0°32.8′	0.33.3′
151°51.8′		0°30.9′	
155°00′	0°27.3′	0°27.7′	0°28.1′

Interpolating first between latitudes 42 and 44° to obtain azimuth Z at latitude 43°05.4′ (which is 1.09° greater than 42°) gives at LHA = 150°:

$$Z = 0°32.3' + \frac{1.09}{2}(33.3' - 32.3') = 0°32.8'$$

At LHA = 155°

$$Z = 0°27.3' + \frac{1.09}{2}(28.1' - 27.3') = 0°27.7'$$

Now a second interpolation is made between local hour angles of 150 and 155° to obtain Z at 151°51.8′ (which is 1.86° greater than 150°):

$$Z = 0°32.8' - \frac{1.86}{5}(32.8' - 27.7') = 0°30.9'$$

A correction must finally be made for the declination of Polaris, since Table 10 is based on a polar distance of 0°48.50′, and in Example 19-1, the polar distance of Polaris at the time of observation was 0°48.95′. In Table 11 (p. 78 of the K&E ephemeris), this correction is also obtained by double interpolation between bearing angles of 0°20′ and 0°40′ and polar distances of 0°49.0′ and 0°48.9′. The correction is +0.3′, which, when added to Z obtained previously, yields:

$$Z = 0°30.9' + 0.3' = 0°31.2'$$

Note that this checks with the value computed in Example 19-1.

19-11. VERIFICATION OF FIELD OBSERVATIONS. Before performing an azimuth reduction, the correctness of the field observations should be verified. Obvious mistakes, or values containing large errors, can then be eliminated from the solution or, if too many are rejected, the observations repeated. If some observations are rejected, equal numbers of direct and reversed measurements should always be retained in the computations to eliminate the effects of instrumental errors.

A good method of verifying observations is to plot both horizontal and vertical angles versus time. The results should lie along a straight line for observations taken over a relatively short period of time. To illustrate, the values of Example 19-1 have been plotted in this manner on Figure 19-11. Note that the horizontal angles do fit a straight line as they should, and the computations prove this was an excellent set of observations. The vertical angles should also plot as a straight line, but instead direct and reversed readings yield two parallel straight lines. This indicates the values were correctly read, but the vertical spacing between the lines shows there was an approximate instrumental error of 0.1 min.

19-12. LOCATING POLARIS IN THE TELESCOPE. If Polaris is observed at any hour angle, field operations can begin at dusk and be completed when some daylight still exists. This greatly assists in handling the instrument and recording notes. Polaris cannot be seen with the naked eye before dark, but is visible through the telescope of a transit or theodolite. To bring the star into the field of view, it is very helpful to compute its altitude for the anticipated time of observation. This can be done using the following formula derived for the PZS triangle from the cosine law of spherical trigonometry:

$$h = \sin^{-1}(\sin \phi \sin \delta + \cos \phi \cos \delta \cos t) \tag{19-2}$$

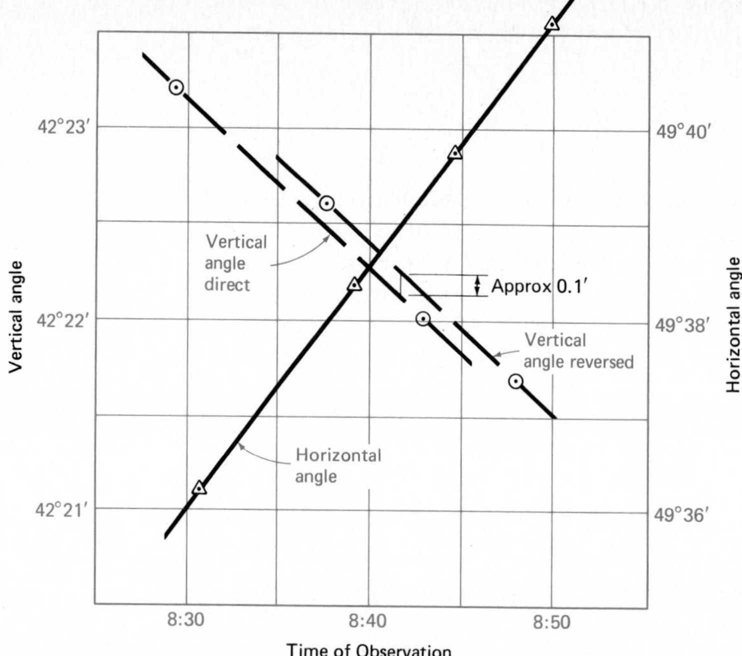

Figure 19-11. Plot of horizontal and vertical angles versus time for verifying correctness of observations.

Equation (19-2) will yield the true altitude of the star, but a refraction correction must be added to actually find it. This can be obtained from tables for varying altitudes.

EXAMPLE 19-3

An observation on Polaris is planned to begin at 5:00 P.M. CST on November 17, 1983, at a station whose latitude and longitude are 42°35′N and 89°28′W, respectively. Determine the altitude angle at which the vertical circle should be set to find the star.

1. GCT of planned observation:

5:00	CST
+12:00	corr. to 24-h system
+ 6:00	corr. to GCT
23:00	GCT, 17 November 1983

2. GHA Polaris (from K&E ephemeris):

21°24.4′ at 0^h GCT (Table 1, p. 63)
345°56.7′ corr. for 23^h (Table 5, p. 72)
367°21.1 − 360° = 7°21.1′ = GHA Polaris at 5:00 P.M. CST

3. Meridian angle of Polaris:

$$\begin{array}{r} 89°28.0' = \lambda \\ -\ \ 7°21.1' = \text{GHA} \\ \hline 82°06.9' = t \end{array}$$

[Note: A sketch similar to Figure 19-10(b) applies.]

4. Declination of Polaris (from K&E ephemeris):

$p = 48.47'$ at 0^{h} GCT, 17 November 1983 (Table 3, p. 67)
(change in 23^{h} negligible)

$$\delta = 90° - 0°48.47' = 89°11.53'$$

5. True altitude of Polaris, by Eq. (19-2):

$$\begin{aligned} h &= \sin^{-1}(\sin 42°35' \sin 89°11.53' + \\ &\quad \cos 42°35' \cos 89°11.53' \cos 82°06.9') \\ &= 42°41.3' \end{aligned}$$

6. Refraction correction, C (from K&E ephemeris):

$C = +0°01.1'$ (Table 2, p. 65, for altitude of 42°41.3')

7. Vertical angle required to sight the star:

$$h = 42°41.3' + 0°01.1' = 42°42.4'$$

Calculation of h should be done before going into the field. Then when an instrument has been set up in preparation for observing, the telescope is focused on some faraway object (the moon is excellent if visible; otherwise any distant point will suffice). The computed vertical angle is set on the instrument, the telescope turned in the approximate direction of north (which can be determined by reading a compass and applying magnetic declination), and the sky slowly scanned until the tiny speck of light, Polaris, is found. Scanning need span only a few degrees in each direction from north as measured by compass, since Polaris is never more than 2° from the meridian for latitudes up to nearly 70°. Once the star has been found, the telescope is plunged and a mark set on-line a few hundred feet from the instrument to provide a reference direction for resighting the star.

19-13. PRACTICAL SUGGESTIONS FOR POLARIS OBSERVATIONS. Although different individuals may prefer other techniques, with repeating theodolites and transits a recommended observing procedure consists of setting the horizontal circle to 0°00′; then with the telescope direct, sight a mark with the lower motion. Open the upper motion, sight the star, and record the precise time of pointing and the horizontal and vertical circle readings. Plunge the scope and with the upper motion resight the mark. (The reading should be exactly 180°; any difference represents instrumental and/or personal errors.) The upper motion is again used to sight the star, whereupon similar data are recorded. Several repetitions of direct and reversed readings are made, the number being a function of the required accuracy. The notes of Example 19-1 should clarify the procedure.

The following additional suggestions make observations on Polaris easier to perform:

1. Precompute the star's altitude for the anticipated time of beginning observations.
2. Prepare the noteforms in advance of starting field work.
3. Set up the instrument in daylight if possible. With the altitude of Polaris known, observations can begin before dark.
4. Have a flashlight, reflector, and pencils. The accuracy of transit vernier readings is lowered at night, because lighting from the side produces parallax if the plates are not at the same level. Hold the flashlight over the compass box behind the ground-glass upright piece and let the light diffuse through it.
5. The mark must be visible at night, recognizable and definite in the daytime. It should be at least 1 mi away if possible to avoid instrumental errors in refocusing between mark and star, star and mark sights.
6. A good watch, either digital or one with a sweep second hand, should be available for timing and checked against WWV radio time signals before and after field operations.
7. To enable pointing after dark, it is necessary to illuminate the cross hairs. This can be done by directing a flashlight obliquely into the objective end of the telescope. The light must not be too bright or the star becomes invisible, and it cannot be too dim or the cross hairs fade out. Many theodolites have a special internal battery-operated illumination system to light the cross wires, horizontal circle, micrometer scale, and vertical arc.

19-14. AZIMUTH FROM SUN OBSERVATIONS. When the sun is observed to determine azimuth, the following equation, derived from the cosine law of spherical trigonometry, is used to compute the Z angle of the PZS triangle:

$$Z = \cos^{-1}\left(\frac{\sin\delta - \sin h \sin\phi}{\cos h \cos\phi}\right) \tag{19-3}$$

An alternate form of Eq. (19-3) is:

$$Z = \cos^{-1}\left(\frac{\sin\delta}{\cos\phi\cos h} - \tan\phi\tan h\right) \tag{19-4}$$

where h is the true altitude of the sun and all other terms are as previously defined.

Again, latitude can be taken from a map or observed, and declination extracted from an ephemeris for the time of observation. Because declination changes slowly (never more than about 0.1 min in 6 min of time), precise timing is not needed, which is an advantage of the method. True altitude is the measured value corrected for various systematic errors, which are explained in the following section. In solving Eq. (19-3) or (19-4), either a positive or negative value can be obtained for cos Z. For morning observations the azimuth of the sun is equal to Z; in the afternoon, $360° - Z$.

19-15. METHODS OF MAKING AND CORRECTING SUN OBSERVATIONS.

Observations of the sun can be made directly by placing an optically plane and parallel dark glass over the eyepiece, or indirectly by focusing the sun's image on an unlined white card held behind the eyepiece. In the latter approach, the telescope is rotated in a horizontal plane until the shadow of transit standards and telescope axle (or the theodolite outline) is symmetrical; then the telescope is turned up and down until the sun's image is visible and sharp at infinite distance focus. The eyepiece must also be adjusted to obtain a clear image of the cross hairs on the card. *Looking at the sun directly through the telescope without a dark glass will result in permanent eye injury.*

At least four observations on the sun, two normal and two plunged, should be made in a minimum of time on the sun's disk. The Roelof solar prism, when properly aligned, produces four intersecting suns, permitting sighting its center to within perhaps 5 sec. The sun's diameter as viewed from the earth is approximately 32 min. If a Roelof prism is not available, bisecting such a large object moving both horizontally and vertically is difficult, but the average observer can do it with an accuracy of perhaps 1 min. A better method is to sight on the sun's edges (limbs). In Figure 19-12(a) the disk is brought tangent to both cross lines, first in one quadrant and then in the diagonally opposite one. Averaging the two readings eliminates consideration of the semidiameter. To avoid coordinating the movement of both tangent screws with the sun, it

Figure 19-12. Observation on the sun.

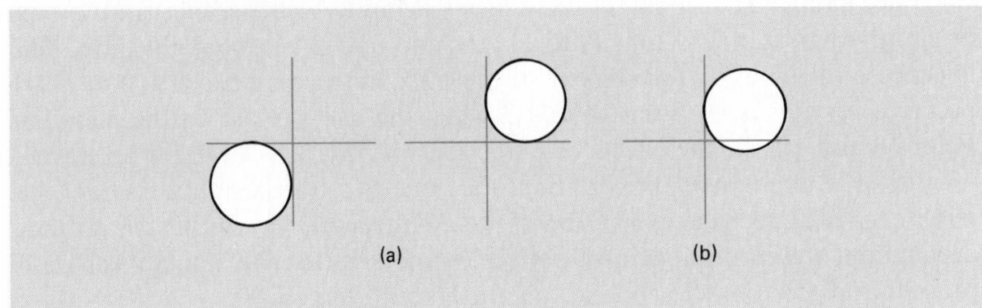

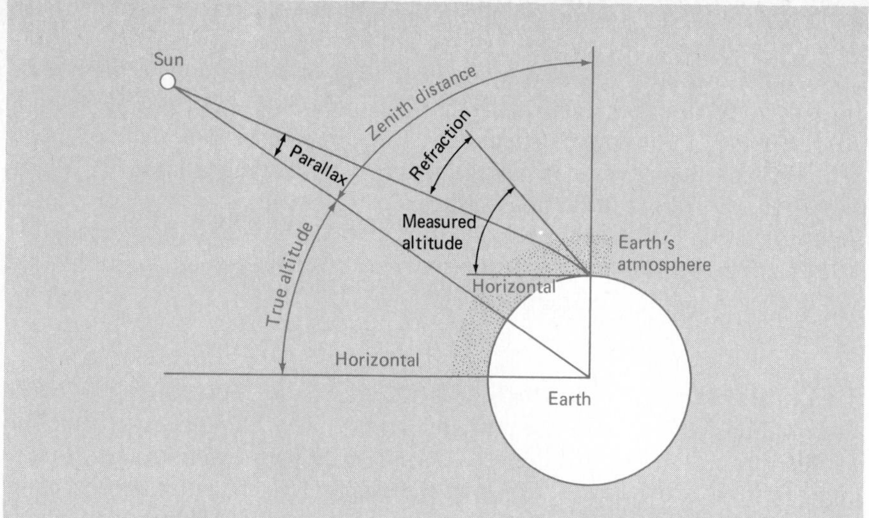

Figure 19-13. Altitude, refraction, and parallax. True altitude = measured altitude − refraction + parallax. (Courtesy Keuffel & Esser Company.)

is simpler to follow the disk by keeping the vertical cross hair tangent to it and letting the sun come tangent to the horizontal hair, as in Figure 19-12(b).

The apparent altitude of the sun is its angular distance above the horizon as measured by a transit or theodolite. To obtain true altitude, the observed value must be corrected for index error, refraction and parallax as indicated in Figure 19-13, and for semidiameter if the border is sighted in only one quadrant. Refraction and parallax corrections can be obtained from tables. Errors of adjustment of the standards and cross hairs are eliminated by sighting with the telescope both normal and plunged. Careful leveling is required.

In observing the sun, the transit or theodolite is oriented with zero or a known value on the horizontal circle by sighting along a fixed line from the observer's position to a mark. After recording the times and vertical and horizontal angles for four or more observations, the mark is resighted to be certain the circle still reads the starting figure. The field notes of Example 19-4 illustrate this procedure. The average horizontal angle to the sun's center is combined with its mean computed azimuth to obtain the line's azimuth.

The best time to observe the sun for azimuth is between 8 and 10 A.M. or 2 and 4 P.M. standard time (9 to 11 A.M. and 3 to 5 P.M. daylight time). This is because the solution for angle Z of the PZS triangle by Eq. (19-3) or (19-4) becomes very weak between 10 and 2 when the sun is close to the meridian. Before 8 and after 4 the effects of refraction on observed altitudes are severe.

A solar attachment, priced about the same as a transit or lower-order theodolite, gives a mechanical solution of the PZS triangle. By setting the latitude, declination, and hour angle on three scales and sighting the sun, a mechanical procedure places the telescope in the north-south direction.

19-16. COMPUTATIONS FOR AZIMUTH FROM SUN OBSERVATIONS. The following example is presented to illustrate procedures in computing azimuth from sun observations.

EXAMPLE 19-4

For the following set of field notes, calculate the azimuth of line Rover-Ridge.[3]

⩚ at △ Rover, Latitude 42°45′N, Longitude 73°56′W
30 July 1983, Temp 80°F, Pressure 28.7 in Hg

POINT* SIGHTED	TELESCOPE	WATCH TIME (EST)	VERTICAL ANGLE	HORIZONTAL ANGLE (CLOCKWISE)
△ Ridge				0°00′
♀ Sun	Direct	$3^h33^m10^s$	49°44′	214°08′
♀ Sun	Direct	$3^h34^m20^s$	49°32′	214°28′
♄ Sun	Plunged	$3^h35^m37^s$	49°52′	215°19′
♄ Sun	Plunged	$3^h36^m49^s$	49°40′	215°37′
△ Ridge				0°00′
Mean		$3^h34^m59^s$	49°42.0′	214°53.0′

* Erecting telescope used, and position of sun's image with respect to cross hairs as viewed on card.

SOLUTION

Mean vertical angle	= 49°42.0′
Index correction	= 00.0′
Refraction and parallax correction (temp = 80°F, press = 28.7 in) −0.80′(0.97)(0.94) + 0.09′ (pp. 65 and 66, K&E ephemeris)	= −00.6′
1. True altitude	49°41.4′
EDT of observation, P.M.	= $3^h34^m59^s$
Correction 0^h GCT to noon	$+12^h$
Correction for EDT	$+4^h$
2. Greenwich civil time of observation	= $19^h34^m59^s = 19.58^h$
Declination 0^h GCT, 31 July 1983 (p. 59, K&E ephemeris)	= N18°27.8′
Correction for 4.42^h earlier = 0.61×4.42^h	+ 02.7′
3. Declination at time of observation	= N18°30.5′

[3] A computer program written in BASIC is given in Appendix C. To illustrate its use, this problem has been solved.

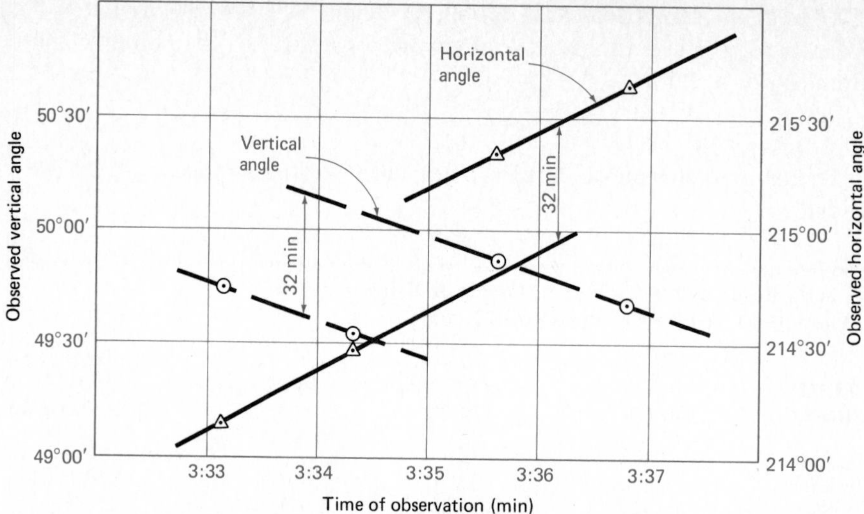

Figure 19-14. Plot of vertical and horizontal angles versus time to verify observations.

By Eq. (19-4):

$$\cos Z = \frac{\sin 18°30.5'}{\cos 42°45'/\cos 49°41.4'} - \tan 42°45' \tan 49°41.4'$$

$$= -0.421387$$

4. $Z = 114°55.4'$

Since the observation was made in the afternoon, angle Z is counterclockwise from north, and the minus sign indicates an angle greater than 90°. Thus, the sun's azimuth at observation time is:

$$360°00.0' - 114°55.4' = 245°04.6'$$

Horizontal angle, Ridge to sun = 214°53.0′

5. Azimuth of line Rover-Ridge = 30°11.6′

Sun observations can be verified by plotting horizontal and vertical angles versus time in the same manner described for Polaris observations. The values should define a straight line. Figure 19-14 illustrates this plot for the values of Example 19-4. Note they produce two sets of parallel straight lines separated by 32 min in both horizontal and vertical angles. This is correct, since observations were taken on opposite limbs of the sun, which has a 32-min diameter.

19-17. DETERMINATION OF LATITUDE AND LONGITUDE. If maps are unavailable for scaling latitude and longitude, they can be determined astronomically. Several procedures are available, and textbooks that specialize in

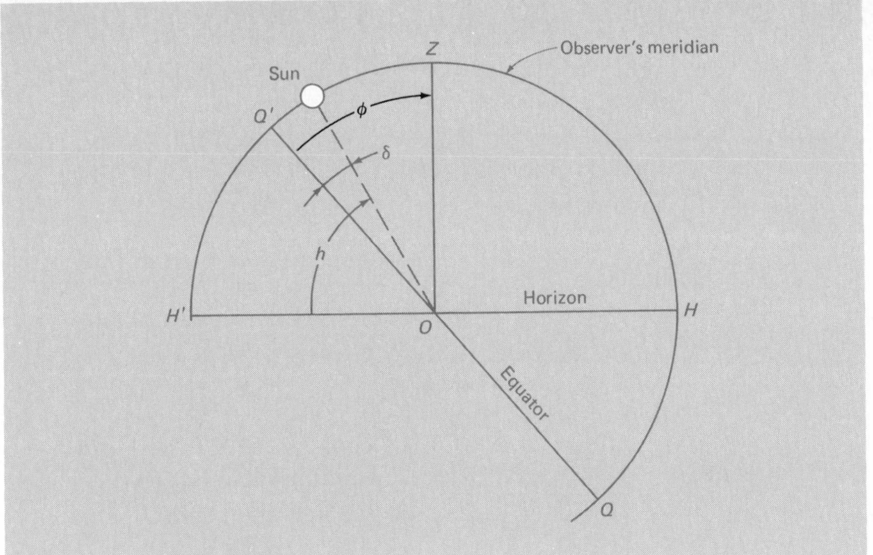

Figure 19-15. Determination of latitude by sun observation at culmination.

field astronomy describe them in detail. In this book only one simple method of determining each is presented to introduce students to these topics.

19-17.1. DETERMINATION OF LATITUDE. The easiest method of observing for latitude is to measure the sun's altitude at culmination. An instrument can be set up on a station shortly before noon and carefully leveled. The sun is then followed as its altitude increases to a maximum, at which position it is on the meridian. Vertical circle reading and time are recorded. Time noted to the nearest few minutes is sufficient, since only declination, which changes slowly, is computed from it. Figure 19-15 illustrates the geometry, and from it the following equation for latitude is evident:

$$\phi = 90^\circ - h + \delta \tag{19-5}$$

where h is the sun's true altitude corrected for index error, refraction, parallax, and semidiameter if a limb is observed. Declination δ is taken from an ephemeris for the observation time. If it is southerly, a negative sign must be affixed in computations.

EXAMPLE 19-5

On December 1, 1983, a theodolite was set up before noon at a point of unknown latitude and carefully leveled. Using a Roelof's prism, the zenith angle to the center of the sun at its maximum altitude was measured as 61°02.7′ at 12:03 P.M. EST. Compute the latitude of the place.

1. True altitude:

Observed altitude = 90° − 61°02.7′ =	28°57.3′
Refraction (Table 2, p. 65)	− 1.7′
	28°55.6′
Parallax (Table 2, p. 65)	+ 0.1′
True altitude	28°55.7′

2. Sun's declination:

Observation time = 12^h03^m EST
Correction to GCT = + 5^h GCT
17^h03^m

$\delta = -21°40.7'$ at 0^h GCT, 1 December 1983 (Table 1, p. 64)
Change = $-6.6'$ (17.05 h × 0.39 min/h)(Table 1, p. 64)
$-21°47.3'$ = declination at observation time

3. Latitude calculation by Eq. (19-5):

$$\phi = 90° - 28°55.7' + (-21°47.3') = 39°17.0'$$

Accuracies of latitudes determined by this method depend largely on instrument precision, but values to within ±1 min are possible.

19-17.2. DETERMINATION OF LONGITUDE. A simple method of determining longitude is to measure the exact standard time at which the sun's center crosses the true meridian. This is converted to Greenwich civil time; the equation of time applied to get Greenwich apparent time (GAT); and GAT then converted to longitude.

EXAMPLE 19-6

The sun's center crossed the meridian of a place at 11:57:19 CST on December 17, 1983. Determine the station's longitude.

1. Compute GAT:

Observation time, CST	11:57:19
Correction to GCT	6:00:00
GCT of observation	17:57:19
Eq. of time, 0^h GCT, 17 December (Table 1, p. 64)	+ 4:20.2
Change in eq. = 17.96(1.22) (Table 1, p. 64)	− 21.9
GAT of observation	18:01:17

2. Convert GAT to longitude:

GAT of observation	18:01:17
Correction, midnight to noon	−12:00:00
	6:01:17 = 6.021389^h

$15°/h \times 6.021389h = 90°19'15''$ west longitude

This method can produce longitudes accurate to within ±1 min, but timing must be extremely precise, since an error of 4 sec causes approximately a 1-min error in longitude.

19-18. SOURCES OF ERROR IN ASTRONOMICAL OBSERVATIONS. Sources of error in astronomical observations are:

1. Instrument not perfectly leveled.
2. Horizontal axis of the instrument not truly horizontal.
3. Index error not corrected.
4. Sun not bisected by both cross hairs.
5. Time not correct, or not read exactly at the moment of observation.
6. Parallax in readings taken at night.

19-19. MISTAKES. Some more common mistakes that occur in astronomical observations are:

1. Sighting on the wrong star.
2. Using a poor signal on the reference line.
3. Computational mistakes.

PROBLEMS

19-1. For what purposes are astronomical observations made in surveying?
19-2. Is it necessary to make astronomical observations on every project where true azimuths are needed? Explain.
19-3. Describe two simple methods of determining the true meridian that require no computations. Discuss the advantages and disadvantages of these methods.
19-4. The coordinates of a point on the earth's surface are latitude and longitude. What are the coordinates of the sun in a similar system?
19-5. Why is sundial time not accurate?
19-6. What is the approximate date of the autumnal equinox? Winter solstice?
19-7. When is the sun's declination 0°00′? When is it positive?
19-8. (a) What is the declination of a star that passes through the zenith of an observer at latitude 36°25′N? (b) Of a star that rises due east of an observer at latitude 29°15′N?
19-9. Why is the moon not commonly used for meridian observations?
19-10. Do standard time zones follow 15° intervals exactly? Why?
19-11. Explain the gain or loss of one day in crossing the international date line.
19-12. What is the reason for having both apparent and mean solar time?

19-13. At 12 noon Greenwich civil time, what are the standard times at the following longitudes:
(a) 91°15′W (b) 74°17′W
(c) 167°10′W (d) 120°50′W

19-14. Express the following differences in longitude in terms of mean solar time:
(a) 63°15′ (b) 97°38′
(c) 113°20′30″ (d) 81°17′23″

19-15. Explain what is meant by the equation of time. How is it used?

19-16. At what central standard time on 10 May of this year will the center of the apparent sun be on the meridian of 90°W longitude?

19-17. At what eastern standard time on 5 March of this year will the center of the apparent sun be on the meridian of 71°30′W longitude?

19-18. For an observer in the northern hemisphere, what are the azimuth and altitude of the north celestial pole?

19-19. What are the standard times for the following locations and circumstances?
(a) EST in New York (longitude 73°57′30″W) when it is 10:03 A.M. PST in San Francisco (longitude 122°45′54″W)
(b) PST in San Francisco (longitude 122°45′54″W) when it is 8:57 P.M. CST in Chicago (longitude 87°51′12″W)

19-20. The longitude of Cincinnati is 84°25′21″W. What is the central standard time there when local time is 6:18:15 P.M.?

19-21. Describe and show on a sketch a method for determining, without use of a table or any equipment, whether Polaris is on the meridian of the observer.

19-22. List the advantages and disadvantages of making an azimuth observation on Polaris at (a) culmination and (b) elongation.

19-23. What is the local hour angle of Polaris at a place of 92°39′W longitude when its Greenwich hour angle is 157°28′?

19-24. Compute the local hour angle of Polaris at 7:30 P.M. CST on 15 December of this year at latitude 34°45′N and longitude 88°36′W. What is the star's meridian angle t at that instant of time?

19-25. Compute the altitude angle that must be put on a transit's vertical circle in order to sight Polaris at 8:00:00 P.M. PST on November 12 of this year at a place of latitude 42°13′N and longitude 119°47′W.

19-26. Assume eight observations (4D and 4R) have been taken on Polaris for an azimuth determination. List the data that should be recorded and explain what checks are available to verify correctness of the observations.

19-27. Compute the true bearing of line AB for the following situations:
(a) Polaris was observed at upper culmination from station A with 0°00′ on the horizontal circle. After turning a clockwise angle to station B, the horizontal circle reading was 163°27′.
(b) Polaris was observed at lower culmination from station A with 45°15′ on the horizontal circle. After turning a clockwise angle to station B, the horizontal circle reading was 318°52′.
(c) Polaris was observed at any hour angle from station A with 0°00′ on the horizontal circle. After turning a clockwise angle to station B, the horizontal circle reading was 213°48′. Computations yielded a bearing of Polaris of N 0°33′W for the instant sighted.

19-28. Determine the sun's declination for the following times this year:
(a) 9:30 A.M. CST, Feb. 10 (b) 10:15 A.M. EDT, June 13
(c) 2:08 A.M. PDT, Aug. 25 (d) 3:40 P.M. MST, Dec. 2

19-29. Direct and reversed observations on a fixed point yielded vertical angle readings of +28°10′ and +28°06′. With the same instrument, an azimuth observation made in the direct mode on the sun's center produced an altitude angle of 37°42′. If the temperature and pressure at the time of observation were 80°F and 28.7 in Hg, respectively, what was the true altitude of the sun at the instant sighted?

19-30. Cite the advantages of observing Polaris for azimuth rather than the sun.

19-31. What data must be recorded in making observations on the sun for azimuth?

19-32. List the sides of the *PZS* triangle used in computing solar observations for azimuth, and give the angles opposite them.

19-33. Why are the hours of 8 to 10 A.M. and 2 to 4 P.M. standard time best to make solar observations for azimuth?

19-34. The vertical angle to the sun's center, measured in the northern hemisphere as it crosses the observer's meridian, is +42°18′ after all corrections have been made. If the sun's declination was −2°12′, what is the observer's latitude?

19-35. Using a Roelof prism, the zenith angle to the sun at its maximum altitude was measured in a direct mode as 59°10.5′ at 11:57 A.M. MST on November 15 of the current year. The temperature and pressure were recorded as 38°F and 29.0 in Hg, respectively. A check on the instrument revealed an index error of +0°00.4′. Calculate the observer's latitude.

19-36. At what central standard time on March 10 of this year will the apparent sun cross an observer's meridian at longitude 91°05′W?

19-37. What is the latitude of an observer if the measured altitude of Polaris at upper culmination is 41°07.5′ when the polar distance to the star is 0°49.3′? (At the time of observation the pressure and temperature were 29.0 in Hg and 70°F, respectively.)

19-38. Assume the center of the sun crossed the meridian of a place at 12:10:25 EST on January 6 of this year. What is the longitude of the place?

19-39. Similar to problem 19-38, except the crossing occurred at 12:46:41 CDT on May 26.

BIBLIOGRAPHY

Buckner, R. B. 1975. "Reasons and Methods for Accurate Direction in Land Surveying." *Surveying and Mapping* 35(no. 4):305.

Flowe, R. T. 1962. "Gyroscopic Azimuth Instruments for Surveying." *Surveying and Mapping* 22(no. 2):271.

Mackie, J. B. 1978. *Elements of Astronomy for Surveyors*, 8th ed. London: Charles Griffin.

McDonnell, P. W., Jr. 1982. "Radio Time Signals for Astronomic Observations." *Point of Beginning* 7(no. 2):44.

Mueller, I. 1969. *Spherical and Practical Astronomy as Applied to Geodesy*. New York: Ungar.

Rish, R. F. 1980. "Rate of Change Methods in Astronomical Surveying." *ASCE Journal of the Surveying and Mapping Division* 106(no. SU1):137.

Schwartz, W. M. 1973. "Astronomic Azimuth Calculations on a Desk-Top Computer." *Canadian Surveyor* 27 (no. 1):32.

20 CONTROL SURVEYS

20-1. INTRODUCTION. Control surveys establish precise horizontal and vertical positions of reference monuments to serve as the basis for originating or checking subordinate surveys for projects such as topographic and hydrographic mapping; property boundary delineation; and route and construction planning, design, and layout.

There are two general types of control surveys: *horizontal* and *vertical.* Horizontal surveys over large areas generally establish geodetic latitudes and longitudes of stations. From these values, plane rectangular coordinates, usually in a state plane or UTM coordinate system (see Chapter 21), can be computed. On control surveys of smaller areas, plane rectangular coordinates may be determined directly without obtaining geodetic latitudes and longitudes.

To explain geodetic latitude and longitude, it is necessary to define the *spheroid,* which is a mathematical surface obtained by revolving an ellipse about the earth's polar axis. The ellipse dimensions are selected to give a good fit of the spheroid to the *geoid* over a large area. The geoid is the earth's mean sea level surface, and it is everywhere perpendicular to the direction of gravity. Because of variations in the earth's mass distribution, the geoid is irregular.

A two-dimensional view of the geoid and spheroid is shown in Figure 20-1. As illustrated, the geoid contains nonuniform undulations (which are exaggerated in the figure for clarity) and is therefore not readily defined mathematically. The Clarke Spheroid of 1866 fits the geoid in North America very well, and currently is used as a reference surface for specifying geodetic positions of points in the United States, Canada, and Mexico. Its equatorial semiaxis a

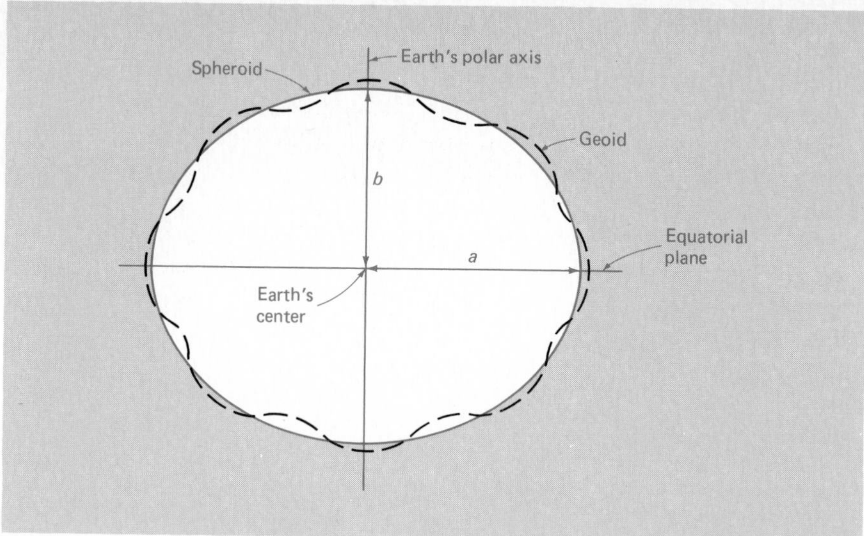

Figure 20-1. Spheroid and geoid.

and polar semiaxis b are 6,378,206.4 m and 6,356,583.8 m, respectively. Thus, the polar semiaxis is only about 21.6 km (13.4 mi) shorter than the equatorial semiaxis, so for some calculations involving moderate lengths (usually up to about 50 km) a true sphere can be assumed.

Figure 20-2 shows a three-dimensional view of the spheroid and illustrates geodetic latitude ϕ_A and geodetic longitude λ_A for any point A on the spheroid. Great circles on the circumference of the spheroid that pass through the north

Figure 20-2. Geodetic latitude, ϕ_A, and geodetic longitude, λ_A.

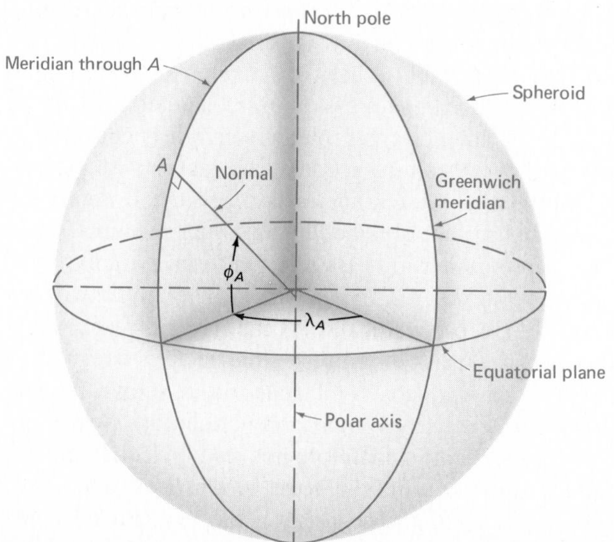

and south poles are called *meridians*; any plane containing a meridian and the polar axis is a *meridian plane*. Geodetic latitude is the angle, in the meridian plane containing A, between the equatorial plane and the *normal* to the spheroid at A. Geodetic longitude is the angle, in the equatorial plane, between the planes of the Greenwich meridian and the meridian through A. Precise latitudes and longitudes that accurately specify the relative positions of widely spaced points are determined by using geodetic field and computational techniques.

Field procedures used in horizontal-control surveys include *triangulation, precise traversing, trilateration,* and combinations of these basic methods. Rigorous photogrammetric techniques have also been used to densify control in limited areas. Newer techniques employ inertial and satellite doppler systems. Astronomical observations are also made to determine azimuths, latitudes, and longitudes. Terrain in the area, project requirements, available equipment, and relative economy normally dictate the system selected.

Vertical-control surveys establish elevations for a network of monuments called *bench marks*. Depending on accuracy requirements, vertical-control surveys may be run by *barometric, trigonometric,* or *differential leveling*. Inertial and satellite doppler systems are now also used to establish vertical control. The most accurate and widely applied method is precise differential leveling.

20-2. REFERENCE DATUMS. Horizontal and vertical datums are necessary to provide a frame of reference for specifying the positions of control monuments and elevations of bench marks. The current horizontal datum used in the United States consists of (1) an initial point ("Meades Ranch" in Kansas) having a known geodetic latitude and longitude, (2) a fixed azimuth from "Meades Ranch" to an intervisible point "Waldo," and (3) a spheroid of known dimensions. As noted in Section 20-1, the Clarke Spheroid of 1866 is used. With this datum as a frame of reference, a general adjustment was performed in 1927 (the most recent one) in which all the then-existing control monuments were incorporated. This datum is called the *North American Datum of* 1927 (NAD27). Control monuments established subsequent to that date have been adjusted to stations included in the 1927 adjustment.

A new general adjustment, to be called the *North American Datum of 1983* (NAD83), is now in progress. It will utilize a new spheroid called the Geodetic Reference System 1980 (GRS80), whose equatorial and polar semiaxes are approximately 6,378,137 m and 6,356,752 m, respectively. All horizontal observations used to establish more than 260,000 stations in the United States, Canada, Mexico, and Central America will be simultaneously adjusted by least squares — a huge undertaking! The results will be published in the form of revised values for geodetic latitudes, longitudes, and state plane and UTM coordinates in metric units. This adjustment is deemed necessary to eliminate inaccuracies and distortions in the current NAD27.

Vertical datums for referencing bench-mark elevations are based on mean sea level. The current vertical datum in use in the United States is the *National Geodetic Vertical Datum of 1929* (NGVD29), since that was the year of the most recent general adjustment of bench marks. The NGVD29 was obtained from a best fit of mean sea level observations taken at 26 gauging stations in the United States and Canada. Since 1929, more than 625,000 km of additional

TABLE 20-1. **HORIZONTAL-CONTROL SURVEY ACCURACY STANDARDS**

ORDER AND CLASS	RELATIVE ACCURACY REQUIRED BETWEEN DIRECTLY CONNECTED ADJACENT POINTS
First order	1 part in 100,000
Second order	
Class I	1 part in 50,000
Class II	1 part in 20,000
Third order	
Class I	1 part in 10,000
Class II	1 part in 5000

control leveling has been run. Crustal movements and subsidence have changed the elevations of many bench marks. To incorporate the additional leveling and correct elevations of erroneous bench marks, a new general vertical adjustment is now also under way. It should be completed by 1988 and will be referred to as the *North American Vertical Datum of 1988* (NAVD88).

20-3. ACCURACY STANDARDS AND SPECIFICATIONS FOR CONTROL SURVEYS. The required accuracy for a control survey depends primarily on its purpose. Some major factors that affect accuracy are type and condition of equipment used, field procedures adopted, and experience and capabilities of available personnel. To guide surveyors, the Federal Geodetic Control Committee (FGCC) has prepared and published a two-volume set of detailed classifications, standards of accuracy, and specifications.[1] The rationale for them is twofold: (1) to provide a uniform set of standards specifying minimum acceptable accuracies of control surveys for various purposes, and (2) to establish specifications for instruments, field procedures, and misclosure checks to ensure that the intended level of accuracy is achieved.

Standards and specifications developed by the FGCC establish the following three distinct so-called *orders of accuracy*, given in descending order: *first order*, *second order*, and *third order*. For horizontal-control surveys, second order and third order each have two separate accuracy categories, *class I* and *class II*. For vertical surveys, first order and second order also each have class I and class II accuracy divisions. Thus a total of five classes are defined in the specifications for both horizontal- and vertical-control surveys.

Triangulation, traverse, and trilateration surveys are included in the FGCC horizontal-control standards and specifications, but only differential leveling is covered in the vertical-control section.

Tables 20-1 and 20-2 give *relative accuracies* between directly connected adjacent points required by the FGCC for the various orders and classes.

The ultimate success of any engineering or mapping project depends on appropriate survey control. The higher the order of accuracy demanded, the

[1] "Classification, Standards of Accuracy, and General Specifications of Geodetic Control Surveys" and a companion document, "Specifications to Support Classification, Standards of Accuracy, and General Specifications of Geodetic Control Surveys," are available from the Superintendent of Documents, U.S. Government Printing Office, Washington, D.C. 20402.

TABLE 20-2. **VERTICAL-CONTROL SURVEY ACCURACY STANDARDS**

ORDER AND CLASS	RELATIVE ACCURACY (STANDARD ERROR) REQUIRED BETWEEN DIRECTLY CONNECTED BENCH MARKS
First order	
Class I	$0.5 \text{ mm} \times \sqrt{K}$
Class II	$0.7 \text{ mm} \times \sqrt{K}$
Second order	
Class I	$1.0 \text{ mm} \times \sqrt{K}$
Class II	$1.3 \text{ mm} \times \sqrt{K}$
Third order	$2.0 \text{ mm} \times \sqrt{K}$

where K is the distance between bench marks in kilometers.

more time and expense required. It is therefore important to select the proper order of accuracy for a given project and carefully follow the specifications. Note that no matter how accurately a control survey is conducted, errors will exist in the computed positions of its stations, but a higher order of accuracy presumes smaller errors.

20-4. THE NATIONAL CONTROL NETWORK. To meet the various local needs of surveyors, engineers, and scientists, the federal govenment has established a National Control Network consisting of more than 250,000 horizontal-control monuments and approximately 500,000 bench marks throughout the United States. The National Geodetic Survey (NGS) began control surveying operations as the Survey of the Coast in 1807, changed to Coast Survey in 1836, to Coast and Geodetic Survey in 1878, and to a division of the National Ocean Survey (NOS) in 1970. It continues to send field parties to all states to establish new control stations and upgrade and maintain existing ones.

The National Control Network is split into *horizontal* and *vertical* divisions. All control within each part is classified in a ranking scheme based on purpose and order of accuracy.

20-5. HIERARCHY OF THE NATIONAL HORIZONTAL CONTROL NETWORK. The hierarchy of control within the National Horizontal Control Network, from highest to lowest order, follows.

Primary Control. Consists principally of east-west arcs of triangulation spaced at about 100 km, crossed by north-south arcs having similar spacing. In addition to triangulation, traverse and trilateration have been used, and more recently, *doppler* methods (see Section 20-15). Primary control is established by first-order methods.

Secondary Control. Densifies the network within areas surrounded by primary control, especially in high-value land areas. Secondary-control surveys are executed to second-order, class-I standards; employ triangulation, traverse,

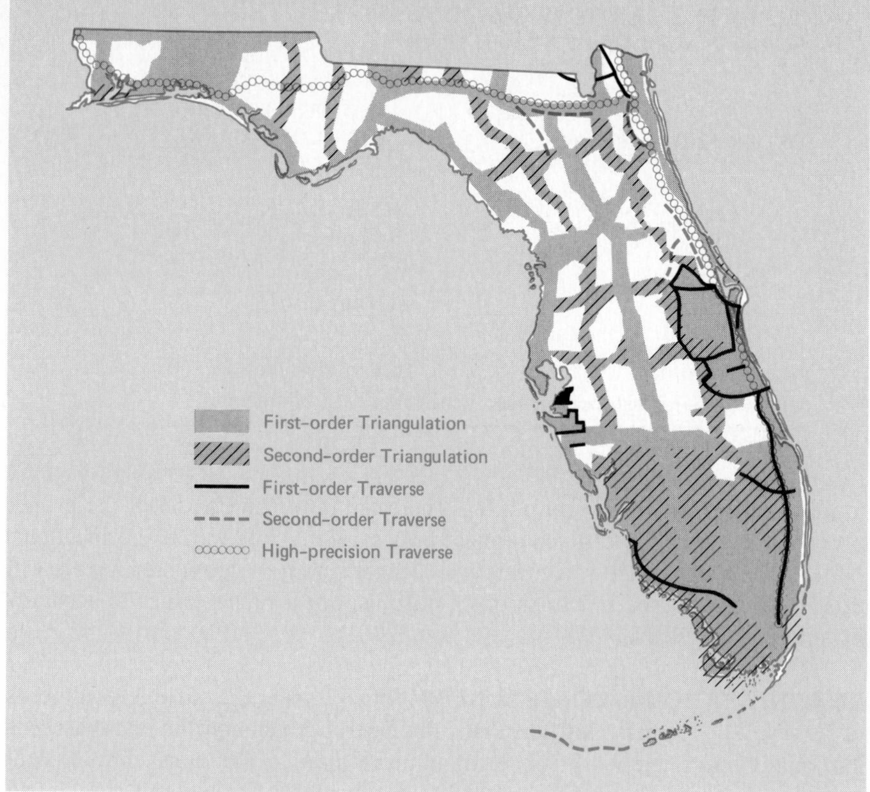

Figure 20-3. Primary and secondary triangulation arcs and traverse network in Florida. (Adapted from National Geodetic Survey map.)

and trilateration; and are adjusted simultaneously with, and thus strengthen, the primary network.

Supplemental Control. Serves in general to densify control between the primary network in lightly developed areas. It is also placed along coastlines and on extensive mapping and construction projects. Supplemental control surveys originate at stations of the primary and, occasionally, secondary network, and are executed to second-order, class-II standards.

Local Control. Provides reference points for local construction projects and small-scale topographic mapping. These surveys are referenced to higher-order control monuments and, depending on accuracy requirements, may be third order, class I, or third order, class II.

Figure 20-3 shows the locations of primary triangulation arcs and densifying secondary, supplement surveys in Florida.

20-6. HIERARCHY OF THE NATIONAL VERTICAL CONTROL NETWORK. The scheme of bench marks within the National Vertical Control Network may be classified as follows:

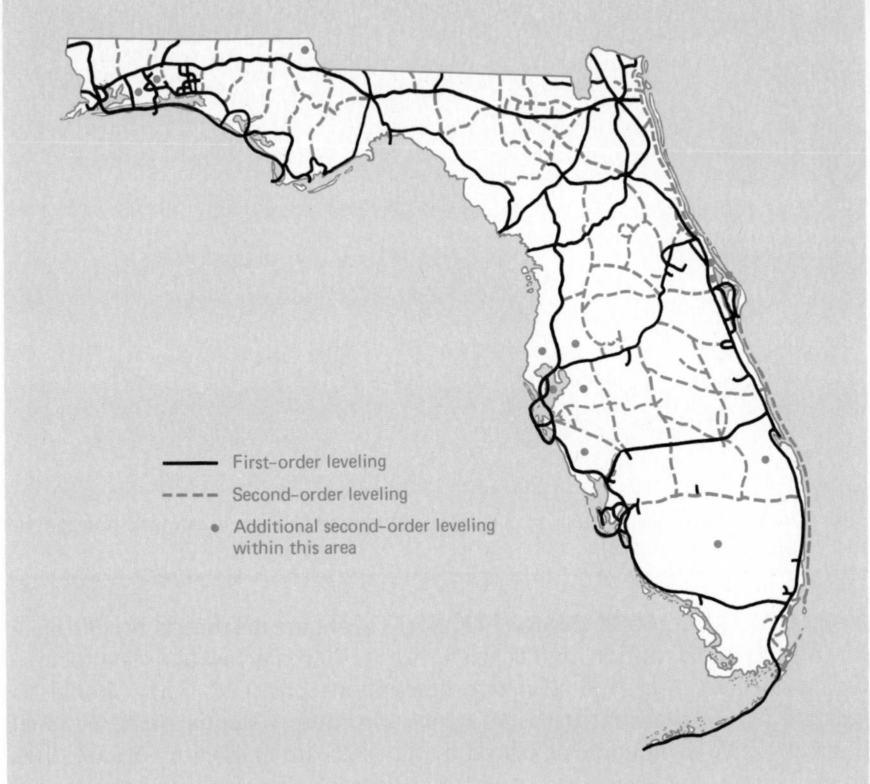

Figure 20-4. Basic framework and secondary nets of the National Vertical Control Network in Florida. (Adapted from National Geodetic Survey map.)

Basic Framework. A uniformly distributed nationwide network of bench marks whose elevations are determined to the highest order of accuracy. It consists of nets *A* and *B*. In net *A*, adjacent bench marks are spaced an average of about 100 to 300 km apart using first-order, class-I standards; in net *B* the average separation is about 50 to 100 km and first-order, class-II standards are used.

Secondary Network. Densifies the basic framework, especially in metropolitan areas and for large engineering projects. It is established to second order, class-I standards.

General Area Control. Vertical control for local engineering, surveying, and mapping projects. It is established to second-order, class-II standards.

Local Control. Serves as vertical reference for minor engineering projects and small-scale topographic mapping. Bench marks in this category satisfy third-order standards.

Figure 20-4 shows the basic framework of the National Vertical Control Network in Florida.

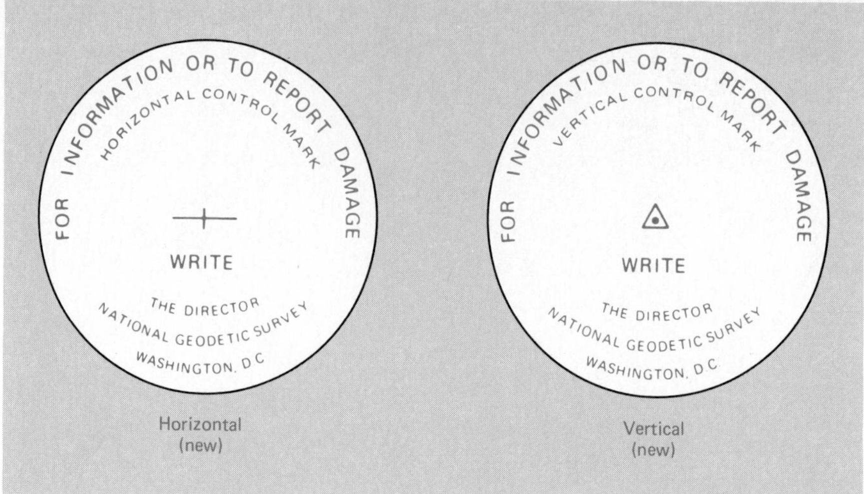

Figure 20-5. Bronze disks used by the National Geodetic Survey to mark horizontal- and vertical-control stations.

20-7. CONTROL-POINT DESCRIPTIONS. To obtain maximum benefit from control surveys, all stations and bench marks are placed in locations favorable to their subsequent use, and adequate descriptions provided. They should be permanently monumented to ensure recovery by future potential users. National Control Network monuments placed by the NGS are marked by bronze disks about $3\frac{1}{2}$ in in diameter set in concrete or bedrock. Figure 20-5 shows two of these disks.

The NGS publishes and makes available to local surveyors location diagrams and complete descriptions of all their control. These give general placement in relation to nearby towns, and specific positions by means of distances and directions to several nearby reference monuments.[2] For horizontal control, the descriptions generally include the station's geodetic latitude and longitude, state plane coordinates, approximate elevation, and geodetic and plane azimuths to a nearby station or stations. Geodetic and plane azimuths differ as a result of convergence of meridians, and therefore the appropriate azimuth must be selected for the particular surveying methods used. Published benchmark data include station locations and adjusted elevations in both meters and feet. All descriptions are likely to contain notes on successes or failures of earlier station recovery attempts. Upon completion of the new general horizontal and vertical adjustments, revised data will be published and distributed.

Besides control within the national network set by the NGS, additional marks have also been placed in various parts of the United States by other federal agencies such as the U.S. Geological Survey, Corps of Engineers, and Tennessee Valley Authority. Also state, county, and municipal organizations have added control. This work is frequently coordinated through the NGS, and descriptions of stations are distributed by that agency.

[2] Requests for control data in a given area should be made to the National Geodetic Information Center, National Ocean Survey, 6001 Executive Blvd., NOAA, Rockville, MD 20852.

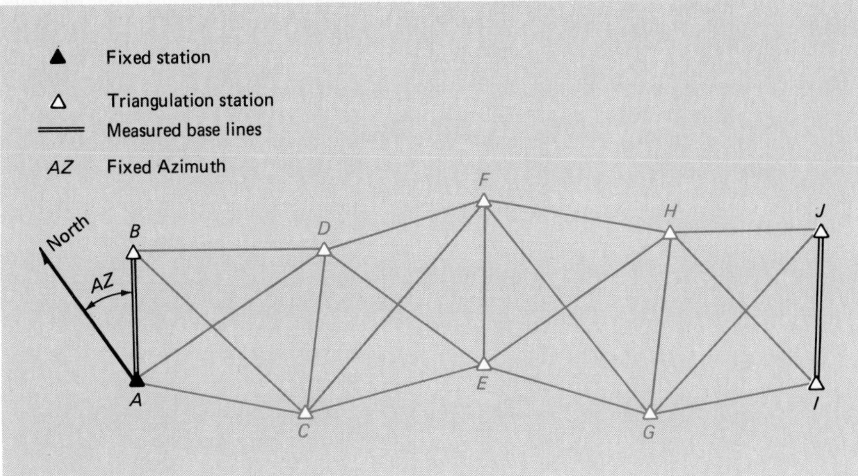

Figure 20-6. Chain of quadrilaterals.

20-8. TRIANGULATION. Prior to the emergence of electronic distance-measuring equipment, triangulation was the preferred and principal method for horizontal-control surveys, especially if extensive areas were to be covered. Angles could be more easily measured compared with distances, particularly where long lines over rugged and forested terrain were involved, by erecting the very versatile Bilby towers. The method possesses a large number of inherent checks and closure conditions which help detect blunders and errors in field data and increase the possibility of meeting a high standard of accuracy.

As implied by its name, triangulation utilizes geometric figures composed of triangles. Horizontal angles and a limited number of sides called *base lines* are measured. Using the angles and base-line lengths, triangles are solved trigonometrically and positions of stations (vertices) calculated.

Different geometric figures have been employed for control extension by triangulation, but chains of quadrilaterals called *arcs* (Figure 20-6) are most common. They are the simplest geometric figures permitting rigorous closure checks and adjustments of field observational errors, and they enable point positions to be calculated by two independent routes for computational checks. More complicated figures like that illustrated in Figure 20-7 are frequently used to establish horizontal control by triangulation in a metropolitan area.

Arcs of triangulation originate from one or more stations of known or fixed position and require the azimuth of at least one line. If two or more stations are fixed, azimuth orientation of the network is automatically determined. Today, fixed starting stations and initial azimuths are normally available from other previous higher-order control surveys. The NGS established beginning positions and azimuths for the national network from astronomical observations, which also are made at various intervals throughout extensive arcs to check and supplement angle and base-line measurements and help maintain true azimuth orientation. In Figure 20-6, the arc of triangulation originates from fixed station *A* and employs the known azimuth of line *AB*. All horizontal

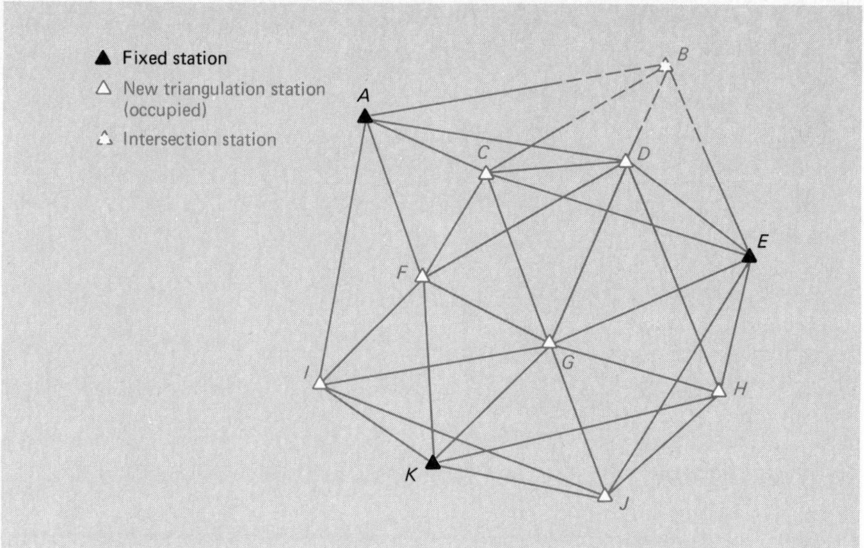

Figure 20-7. Triangulation network for a metropolitan control survey.

angles within the arc, and base lines *AB* and *IJ* have been measured. From this information, positions of stations *B* through *J* were calculated.

In executing triangulation surveys, it is general practice to locate a number of *intersection stations* as part of the project. They can be tall prominent objects in the area, such as church spires, smokestacks, or water towers visible from several triangulation stations. Angles to them are measured from as many occupied points as possible, but the intersection stations themselves are not occupied. Their positions are calculated and thus become available as local reference points. An example is station B in Figure 20-7.

20-9. TRIANGULATION RECONNAISSANCE. One of the most important aspects of any triangulation survey is the reconnaissance and selection of station locations. Factors to be considered are (1) strength of figure, (2) station intervisibility, (3) station accessibility for the original triangulation observing party and surveyors who will subsequently use the stations, and (4) overall project efficiency. Careful attention must be given to each factor in planning and designing the optimum triangulation network for a given project.

Strength of figure deals with the relative accuracies of computed station positions that result from use of angles of various sizes in calculations. Triangulation computations are based on the trigonometric *law of sines*. The sine function changes significantly for angles near 0° and 180°, so a small observational error in an angle close to these values produces a comparatively large difference in position calculations. Conversely, sines of angles near 90° change very slowly; thus, a small observational error in that region causes little change in the computed position. Since similar observational errors are expected for each angle, design of triangulation figures having favorable angle sizes increases overall triangulation accuracy.

Figure 20-8. Kern DKM-3 directional theodolite suitable for first-order control surveying. (Courtesy Kern Instruments, Inc.)

Rigorous procedures beyond the scope of this book have been developed for evaluating relative strengths of geometric figures used in triangulation. In general, angles approximately 90° are optimum, and if no angles smaller than 30° or larger than 150° are included in calculations, the figure should have sufficient strength. Locations of triangulation stations fix the angle sizes so they must be planned carefully for maximum strength of figure. If local terrain or other conditions preclude use of figures having strong angles, more frequent base-line measurements are necessary.

Station intervisibility is vital in triangulation because lines of sight to all stations within each figure must be clear for measuring angles. Preliminary decisions on station placement can be resolved from available topographic maps. Intervening ridges that might obstruct sight lines are checked by plotting profiles of the lines between stations. Trees on line and, for long lengths, the combined effects of earth curvature and refraction are additional factors affecting station intervisibility. After making a preliminary decision on station locations, a visual test should be made by visiting each proposed site. Stations are normally placed on the highest points in an area and, if necessary, towers erected to elevate the theodolite, observer, and targets above the ground stations. Because of the uncertainty of refraction near the ground, lines of sight should be kept at least 10 ft above it and not graze intervening ridges.

20-10. FIELD MEASUREMENTS FOR TRIANGULATION. As previously stated, the basic field measurements for triangulation are horizontal angles and base-line lengths. Angles can be measured using repeating instruments, or more likely directional theodolites such as the Kern DKM-3 (Figure 20-8) having a plate bubble sensitivity of 10-sec/2-mm division, or the Wild T-3 with a 7-sec bubble sensitivity. Both theodolites are suitable for first-order work and permit angles to be read by estimation to the nearest 0.1 sec. Other lower-order theodolites were discussed in Chapter 10.

To reduce effects of atmospheric refraction on high-order triangulation, observations are made at night with lights for targets. At each station, several "positions" are read; a position consists of angles or directions distributed around the horizontal circle of the instrument in both the direct and plunged modes. With directional theodolites, to compensate for possible circle graduation errors, the circle is advanced by approximately $180°/n$ for each successive position (as described in Section 11-9), where n is the number of positions at the station. Angles should be computed in the field from the directions, checked for acceptable misclosure, and any rejected ones repeated before leaving the station. The average of all satisfactory values for each angle is used in the triangulation calculations.

Base lines now preferably are measured by electronic methods, which produce excellent accuracies. Precise Invar tapes may also be used. Several measurements should be made in both directions. Slope distances must be reduced to horizontal, and mean sea level lengths calculated. If computations involve state plane coordinates, sea level distances are converted to grid lengths by applying scale factors (see Section 21-5).

20-11. TRIANGULATION ADJUSTMENT. Errors that occur in angle and distance measurements require an adjustment. The most rigorous method utilizes least squares. In that procedure, all angle measurements plus distance or azimuth observations can be simultaneously included in the adjustment, and any configuration of quadrilaterals or more complicated figures handled to get station positions having maximum probability. The theory is beyond the scope of this text.

Other approximate methods for triangulation adjustment, easily applied to standard figures such as quadrilaterals, also give satisfactory results and are described in advanced surveying books.

20-12. PRECISE TRAVERSE. Employment of precise traversing is common among local surveyors for horizontal-control extension especially for projects of limited size. Field work consists of two basic parts: reading horizontal angles at the traverse hubs and measuring distances between stations. Angles can be secured with either a repeating or directional instrument, and distances measured with EDM equipment (*electronic traversing*) or by taping. Precise traverses are always the closed type.

The FGCC has defined standards and specifications for five orders of accuracy for traverses. First-order and second-order classes supplement the National Horizontal Control Network, particularly where a greater density of control is needed than that afforded by triangulation. Second- and third-order traverses are run extensively to solidify control in metropolitan areas for engineering and construction projects, property surveys, aerial photogrammetric surveys, and numerous other projects.

Unlike triangulation, in which stations are normally widely separated and placed on the highest ridges and peaks in an area, traverse routes generally follow the cleared rights-of-way of highways and railroads with stations located relatively close together. Besides easing field work, this provides a secondary benefit in accessibility of the stations. Traverses lack the automatic checks inherent in triangulation, and extreme observational caution must therefore be applied to avoid blunders. Also, since traverses generally run along single lines, they are not as good as triangulation for establishing control over large areas.

Procedures for precise traverse computation vary, depending on whether a geodetic or plane reference system is used. In either case, it is necessary first to adjust angles and distances for observational errors. Closure conditions are enforced for (1) azimuths or angles, (2) latitudes, and (3) departures. The most rigorous process, the least squares method, should be used because it simultaneously satisfies all three conditions and gives residuals having the greatest likelihood according to the theory of probability. Other less precise systems such as the compass rule (Bowditch Method) (see Section 13-7) are sometimes employed.

In calculating state plane coordinates, it is necessary to reduce observed horizontal distances to their grid lengths before making the calculations, as described in Section 21-5.

In recent years the NGS has conducted special "ultra-high-accuracy transcontinental traverses" with results approaching one part per million. These surveys were originally begun to provide base lines for a worldwide satellite triangulation program performed in the 1960s and 1970s, and have subsequently been used to upgrade the horizontal-control network. Theodolites and EDM equipment of highest precision are required for transcontinental traverses, and rigorous field procedures adopted to meet the extremely stringent specifications. In addition to showing triangulation, Figure 20-3 illustrates locations of high-precision transcontinental traverses in part of Florida.

20-13. TRILATERATION. Trilateration, a method for horizontal-control surveys based exclusively on measured horizontal distances, has gained acceptance because of EDM instrumentation. Both triangulation and traversing require horizontal angle measurement; hence, trilateration surveys often can be executed faster and produce equally acceptable accuracies.

The geometric figures used in trilateration, although not as standardized, are similar to those employed in triangulation. Stations should be intervisible and therefore placed on the highest peaks, perhaps with towers to elevate instruments and observers.

Strength of figure in trilateration is less quantified than for triangulation; however, slender figures are weakest in the direction transverse to their long dimensions. Hence networks covering essentially square areas are better, since they give stronger overall uniform accuracy.

Because of intervisibility requirements and the desirability of having essentially square networks, trilateration is ideally suited to densify control in metropolitan areas and on large engineering projects. In special situations where topography or other conditions require elongated narrow figures, the network can be strengthened by reading some horizontal angles. Also, for long trilateration arcs, astronomic azimuth observations prevent the network from deforming in direction. A complete and accurate estimate of strength of

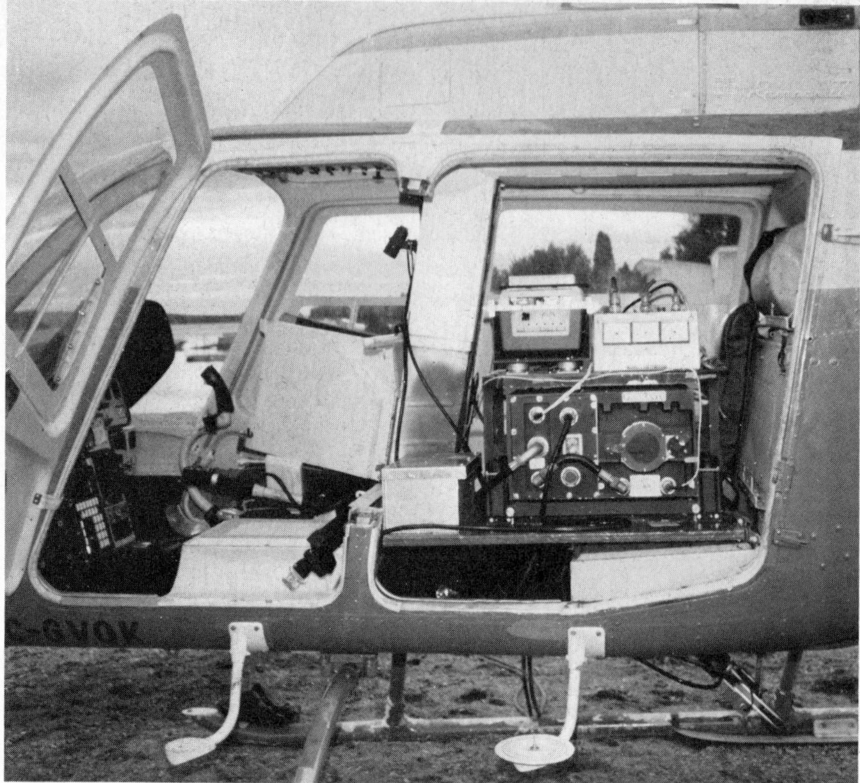

Figure 20-9. Inertial surveying system mounted in a helicopter. (Courtesy Shell Canada Resources, Ltd.)

figure for trilateration networks is secured by presurvey least squares analysis of trial configurations, but the procedure is beyond the scope of this book.

As in triangulation, surveys by trilateration can be extended from one or more monuments of known position. If only a single station is fixed, at least one azimuth must be known or observed.

Trilateration computations consist of reducing measured slope distances to horizontal lengths; then to mean sea level equivalents; and finally to grid lengths if the calculations are being done in state plane coordinate systems. Observational errors in trilateration networks must be adjusted, preferably by the least squares method.

20-14. INERTIAL SURVEYING SYSTEMS. New inertial surveying systems have recently been introduced which have the potential to completely revolutionize current control surveying practices. These systems, carried in helicopters or land vehicles, incorporate precise gyroscopes, accelerometers, and a computer. They are spinoffs from the military guidance systems used on aircraft and missiles, and can display latitude and longitude (or X and Y coordinates) and elevation at any position. An inertial surveying system mounted in a helicopter is shown in Figure 20-9.

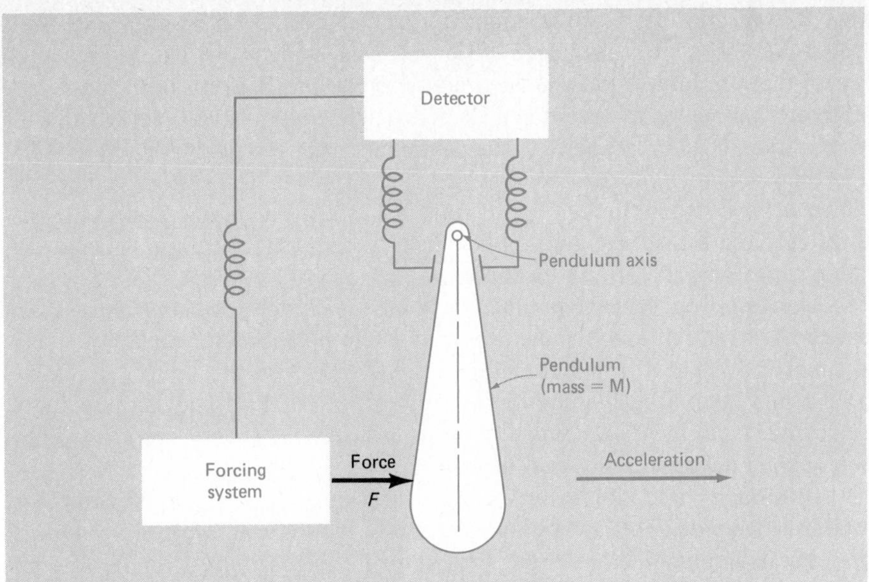

Figure 20-10. Simplified schematic diagram of operation of an accelerometer.

Inertial surveying systems are oriented by a computer-controlled process called *gyrocompassing*. Gyros sense the earth's rotation and orient themselves orthogonally, facing north-south and east-west. The gyros maintain their orientation, while accelerometers (three are required) measure components of movement in the cardinal directions, and in elevation, as they are moved from point to point. The system is initialized at a point of known position and elevation, and by applying the components of change, positions and elevations of new points occupied are determined.

The method of conducting surveys with inertial systems follows the general procedure used in traversing, that is, closed loops or circuits are established. After occupying and determining preliminary positions and elevations of all desired intermediate new points in an area, the circuit is closed by reoccupying either the initial point or another station of known position and elevation. Any misclosure in running the circuit can be observed at this final station, and then preliminary positions and elevations adjusted for the known misclosure.

The operating principle of the accelerometers of inertial surveying systems can be illustrated with a simplified diagram like that of Figure 20-10. In this figure, the accelerometer consists of a pendulum of mass M, equipped with a feedback system. If the accelerometer is at rest, or moving at a constant rate, the pendulum will hang in its rest position. The pendulum's inertia resists changes in velocity. Thus when acceleration or deceleration occurs as a result of movement, the pendulum will have a tendency to swing away from its rest position. A detector senses this tendency to move and sends feedback signals to a forcing system that applies torque forces that are just sufficient to prevent the pendulum from swinging.

The magnitudes of forces applied, together with the lengths of time over which they occur, are measured. In accordance with Newton's fundamental law of motion, force equals mass times acceleration. Because both mass and force are known, acceleration can be computed, that is, $a = F/M$, where a is acceleration, F is force, and M is the pendulum mass. Finally, with acceleration and time available, the distance traveled is determined. The system's computer converts the combinations of forces and time to movement components in the three required directions, and applies them to the initial position to yield the equipment's position at any location.

As inertial surveying equipment is refined, accuracies are improving. Tests conducted over calibrated routes show absolute position and elevation errors as small as a few centimeters for short circuits, and precisions up to $\frac{1}{10,000}$ have been achieved on long courses. Inertial surveying systems do not require direct angle or distance measurements; thus, clear lines of sight are not necessary. They can operate day and night, rain or shine.

Although the initial cost of inertial surveying systems is high, control is established rapidly, so they are comparatively economical, especially on large projects. In an integrated mapping project for DuPage County, Illinois, 68 control points were established in seven days at a cost estimated to be one-fifth of that needed to perform the same work by conventional methods. The Bureau of Land Management is now using inertial systems extensively for establishing section corners in Alaska. In this operation, coordinates of required section corners are entered into the system's computer, whereupon the device leads the operator to that location.

20-15. SATELLITE DOPPLER SYSTEMS. In satellite doppler surveying, receivers located at ground stations measure the frequencies of radio signals transmitted from satellites operating in a polar orbit about the earth. The altitude of the satellites is approximately 1100 km, and they encircle the earth every 105 minutes. Development of the system began in 1958 under the Navy Navigation Satellite System (NNSS) to provide guidance for Polaris submarines. It became operational in 1964 and in 1967 was made available for commercial use. Because of its many advantages (including global coverage, all-weather operation, and no intervisibility required between points), its use in surveying has increased dramatically in recent years.

The principle of operation of satellite doppler systems is illustrated in the simplified diagram of Figure 20-11. A precisely controlled radio frequency is transmitted from the satellite as it passes in its orbit above an observing station [See Figure 20-11(a)]. When the transmitter approaches a receiver, the received signal will have a frequency higher than that transmitted, but it will be decreasing. Then as the satellite moves away from the station, the frequency decreases below the transmitted level. This phenomenon, illustrated in Figure 20-11(b), can be appreciated by anyone who has listened to the change in pitch of a train whistle as it approaches and then moves past. The magnitude of frequency change (doppler shift), which is a function of the range (distance to the satellite), is measured by the receiver. With the transmitting frequency, satellite orbit, and precise timing of observations known, the position of a receiving station can be computed from the measured doppler shifts.

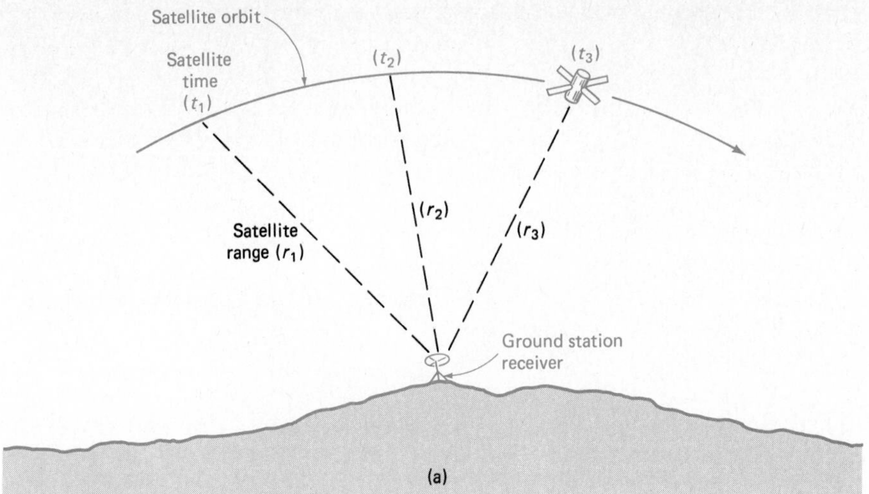

Figure 20-11(a). Satellite and doppler receiver geometry.

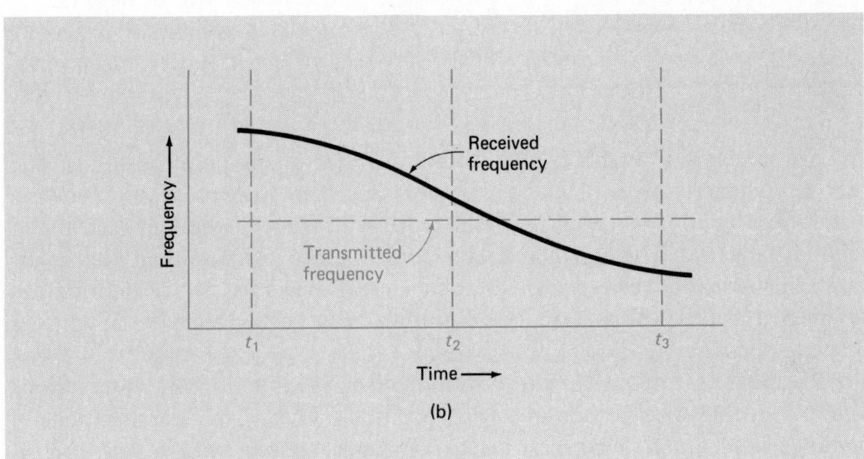

Figure 20-11(b). Frequency change (doppler shift) measured by receiver.

Most receivers, such as the one shown in Figure 20-12, are compact and portable, use a standard 12-V battery, and can be operated in any weather conditions. Doppler stations must be established in open areas providing a relatively unobstructed line of sight, or the area must be cleared. They have to be in locations free from radio interference.

As noted earlier, knowledge of the satellite's orbit is an essential ingredient for computing positions of doppler receiving stations. This orbital information can be obtained either from the so-called *broadcast ephemeris*, or a *precise ephemeris.* The broadcast ephemeris consists of a prediction of what the satellite's orbit will be at a given time, made by extrapolating ahead from tracking data recently measured at four stations located in Hawaii, California, Minnesota and Maine. The data are transmitted from the satellite in real time,

Figure 20-12. Receiver of a satellite doppler surveying system. (Courtesy Magnavox.)

and are therefore available for immediate use in reducing point positions. The precise ephemeris gives orbital parameters based on computations involving actual observations made on the satellite from 20 tracking stations distributed around the world. These orbital data are much more accurate than the broadcast ephemeris, but there is a time lag of several weeks before the information becomes available. Also a security clearance may be necessary to obtain the precise ephemeris.

Different techniques are applied in making satellite doppler observations and in reducing them to obtain station positions. Three basic classifications of methods used are (1) *point-positioning*, (2) *translocation*, and (3) *short-arc*. In point-positioning, a receiver at a single station of unknown position collects data from multiple satellite passes. Based upon the measured data, the position of the receiver is first reduced to its location in a coordinate system relative to the satellite's position. Then the point is transformed into one of the conventional coordinate systems used by surveyors. For a single pass, and using the broadcast ephemeris, position to within about ± 30 m can be obtained. By observing succeeding passes, and using the precise ephemeris, accuracy increases to an ultimate of less than $\frac{1}{2}$ m in horizontal position and elevation for 30 or 40 passes. The time required to observe this many passes would be on the order of from 2 to 8 days. This time variation occurs because only 5 satellites are currently operating, and depending upon the latitude of the station, several hours can elapse between observable passes.

In the translocation and short-arc methods, receivers located at two or more stations simultaneously track a satellite. The position of at least one of the stations must be known. In translocation, coordinates of the known point are computed using the point-positioning technique just described. These are compared to the station's known coordinates, which enables detection of errors in the system. Based upon these errors, corrections are determined and applied to the positions of the unknown stations, whose locations have also been determined using point-positioning methods of computation.

The short-arc method is fundamentally the same as translocation, except that it is more rigorous because corrections are made for the satellite's orbital parameters. This is not done in the translocation procedure. Positions accurate to within a few tenths of a meter have been obtained by observing 25 to 30 satellite passes, and then reducing by short-arc methods. A similar number of passes, reduced by translocation, would produce errors slightly larger.

Satellite doppler systems are being used in many surveying applications, but their principal application is in control surveying. As an example, they are being used to augment and strengthen the national geodetic network. In the conterminous United States, positions of approximately 150 stations, spaced from 50 to 300 km apart, have been determined. An additional 100 stations are being established in Alaska, Hawaii, and Puerto Rico. The stations, located about 300 km apart, are considered by the NGS to have relative accuracies on the order of 1 part in 200,000! Relative accuracies, of course, decrease as stations become closer together.

A satellite surveying program planned for the future called the *Global Positioning System* (GPS) is expected to yield relative positional accuracies on the order of a few centimeters. Anticipated to be operational by 1987, the system will use a minimum of 18 satellites in 20,000-km-altitude orbits. With this many satellites, observing time will be drastically reduced. This system has tremendous potential as a means of upgrading and densifying the national control network.

20-16. VERTICAL-CONTROL SURVEYS. Vertical-control surveys are run in a variety of ways, depending on required accuracy. Barometric leveling, described in Section 6-4, is used to get approximate elevations suitable for reconnaissance surveys. Trigonometric leveling (see also Section 6-4) can provide a higher order of accuracy, suitable for example, to control photogrammetric mapping. Differential leveling, described in Section 7-4, produces different accuracies depending on the precautions taken. In this section only *precise differential leveling* is considered.

As noted in Section 20-3 and Table 20-2, the FGCC has established accuracy standards and specifications for various orders of differential leveling, but the same basic principles are employed. To achieve higher orders, however, special care must be exercised to minimize errors.

For the most accurate work, precise tilting levels such as the Wild N-3 of Figure 20-13 are used. This instrument has a split image or coincidence bubble with a sensitivity of 10 sec/2-mm division that is centered by means of a tilting screw. It has a parallel-plate glass micrometer which elevates or depresses the line of sight parallel to itself to set precisely on the nearest rod graduation. A micrometer shows the distance raised or lowered and permits rod readings to about $\frac{1}{100}$ of the smallest graduation, instead of $\frac{1}{10}$ division by estimation. The instrument reticle is shown in Figure 20-14. It has a single

Figure 20-13. Wild N-3 precise tilting level. (Courtesy Wild Heerbrugg Instruments, Inc.)

horizontal line on the right for making rod readings in the normal way, and two lines that form a wedge on the left to straddle a reading for utmost accuracy.

Special level rods are needed for precise work. They have scales graduated on Invar strips, which are only slightly affected by temperature variations. Nevertheless thermometers attached to the rod are read and corrections applied for any shrinkage or expansion of the scale. Precise level rods are equipped with rod bubbles to facilitate plumbing, and special braces aid in holding the rod steady. They usually have two separate graduated scales. One type is divided in centimeters on an Invar strip on the rod's front side, and a scale in feet painted on the back for checking readings and minimizing blunders. A

Figure 20-14. Reticle of the Wild N-3 precise level shown with dual metric-scale precise leveling rod. (Courtesy Wild Heerbrugg Instruments, Inc.)

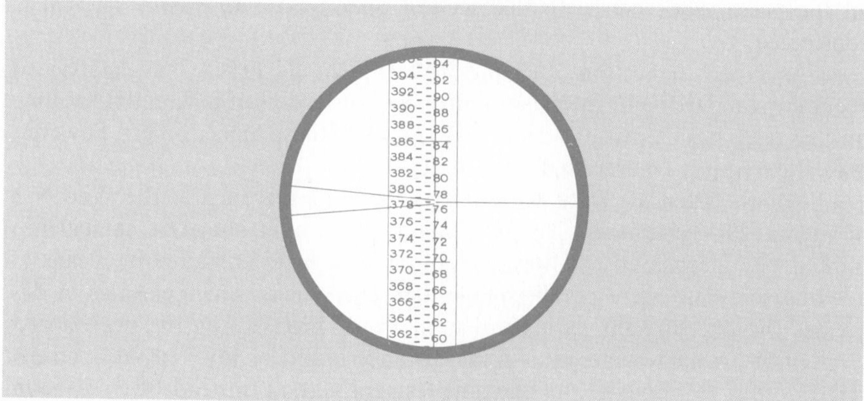

second kind, shown in Figure 20-14, has two sets of centimeter graduations on the Invar strip, with the right one precisely offset from the left by a constant, thereby giving checks on readings.

Cloudy weather is preferable for precise leveling, but an umbrella can be used on sunny days to shade the instrument and prevent uneven heating which causes the bubble to run (one design encases the vial in a styrofoam shield). Automatic levels are not as susceptible to errors caused by differential heating. Precise work should not be attempted on windy days. For best results, short equal backsight and foresight distances up to a maximum of 250 ft (75 m) are recommended and their lengths balanced to within 6 ft (2 m) at each setup. Rodpersons pace or count rail lengths or highway slab joints to set sight distances, which are checked for accuracy by three-wire stadia. Precise leveling demands good-quality turning points. Lines of sight should not pass closer than about 2 ft from the ground to avoid refraction. Readings at any setup must be completed in rapid succession; otherwise changes in atmospheric conditions might significantly alter refraction characteristics between them.

Three-wire leveling has been employed for much precise surveying in the United States. The method has the advantages of (a) providing checks against rod reading blunders, (b) producing greater accuracy because averages of three readings are available, and (c) furnishing stadia measurements of sight lengths. In the three-wire procedure, rod readings at the upper, middle, and lower cross hairs are taken and recorded for each backsight and foresight. The difference between the upper and middle readings is compared with that for the middle and lower values, and must agree within one or two of the smallest units being recorded (usually 0.1 of the least count of the rod graduations); otherwise the readings are repeated. An average of the three readings is actually used, but as a computational check it must be very close to the middle wire figure. The difference between the upper and lower wire readings multiplied by the instrument stadia constant gives the sight distances.

A sample set of field notes for the three-wire method is illustrated in Figure 20-15 Backsight readings on BM *A* of 0.718, 0.633, and 0.550 taken on the upper, middle, and lower wires, respectively, give upper and lower differences (multiplied by 100) of 8.5 and 8.3 m, which agree within acceptable tolerance. Stadia measurement of the backsight length (the sum of the upper and lower differences) is 16.8 m. The average of the three backsight readings on BM *A*, 0.6337 m, agrees within 0.0007 m of the middle reading. The stadia foresight length of 15.9 m at this setup is within 0.9 m of the backsight length, and satisfactory. The HI for the first setup is found by adding the average backsight reading to the elevation of BM *A*. Subtracting the average foresight reading on TPI gives its elevation. This process is repeated for each setup.

A second technique in precise leveling employs the parallel-plate micrometer attached to a precise leveling instrument, and a pair of precise rods like those described earlier. While this method has been used in Europe for more than 50 years, it was not adopted in the United States until the late 1960s.

It is generally advisable to design large level networks so that several smaller circuits are interconnected to supply checks which isolate blunders

THREE-WIRE LEVELING
TAYLOR LAKE ROAD

Sta.	+ Sight	Stadia	− Sight	Stadia	Elev.
BM A					103.8432
	0.718		1.131		
	0.633	8.5	1.051	8.0	+0.6337
	0.550	8.3	0.972	7.9	104.4769
	3)1.901	16.8	3)3.154	15.9	−1.0513
	+0.6337		−1.0513		
TP 1					103.4256
	1.151		1.041		
	1.082	6.9	0.969	7.2	+1.0820
	1.013	6.9	0.897	7.2	104.5076
	3)3.246	13.8	3)2.907	14.4	−0.9690
	+1.0820		−0.9690		
TP 2					103.5386
	1.908		1.264		
	1.841	6.7	1.194	7.0	+1.8410
	1.774	6.7	1.123	7.1	105.3796
	3)5.523	13.4	3)3.581	14.1	−1.1937
	+1.8410		−1.1937		
BM B					104.1859
	Σ +3.5567		Σ −3.2140		Check
Page Check:					
	103.8432	+3.5567	−3.2140	= 104.1859	

Figure 20-15. Sample field notes for three-wire leveling.

or large errors. In Figure 20-16 for example, it is required to determine the elevations of points X, Y, and Z by commencing from BM A and closing on BM B. As a minimum, this could be done by running level lines 1 through 4, but if an unacceptable misclosure was obtained at BM B, it would be impossible to discover in which lines the blunder occurred. If additional lines 5, 6, and 7 are run, calculating differences in elevation by other routes through the network might isolate the blunder. Furthermore, by including supplemental measurements, precision of the resulting elevations at X, Y, and Z is increased.

Regardless of precautions taken in field observations, errors accumulate in leveling and must be adjusted to provide perfect mathematical closure of all loops. For simple level circuits, adjustment procedures presented in Sections 7-14 and 7-15 can be followed; for interconnected level networks such as that of Figure 20-16, the method of least squares is preferable.

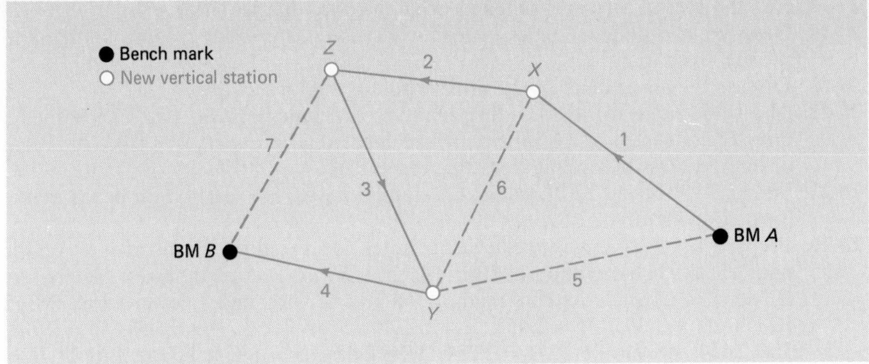

Figure 20-16. Interconnecting level circuit.

Precise leveling is very tedious and requires extreme care. Field personnel must heed minute details to minimize systematic errors which are always present. Probably no aspect of control surveying has been studied for so many years (over a century).

PROBLEMS

20-1. Describe what is meant by the terms *geodetic latitude* and *geodetic longitude.*

20-2. What different field methods are used in horizontal-control surveying? Which factors influence the choice of field method to use?

20-3. Discuss the reference datums used for horizontal- and vertical-control surveys.

20-4. List the orders and classes of accuracy of horizontal-control surveys and give their relative accuracy requirements.

20-5. Name the orders and classes of accuracy of vertical-control surveys and give their relative accuracy requirements.

20-6. Describe the hierachy of the National Horizontal Control Network.

20-7. Discuss the hierachy of the National Vertical Control Network.

20-8. Explain why it is important to permanently monument and adequately describe control stations. State the contents of a good control station description.

20-9. Why are quadrilaterals the most commonly used geometric figure in triangulation?

20-10. Analyze the factors that must be considered in triangulation reconnaissance.

20-11. Obtain a U.S. Geological Survey quadrangle map of your area. On the map, lay out a quadrilateral having sides from 3 to 5 mi long. Check all lines for intervisibility by plotting their profiles, and revise station positions if necessary to provide unobstructed lines of sight.

20-12. Explain the reason for establishing triangulation figures that contain angles between 30 and 150°.

20-13. Discuss the special precautions taken in observing angles on high-order triangulation.

20-14. Outline the advantages and disadvantages of traverse control surveys as compared to triangulation for horizontal networks.

20-15. Explain the differences between triangulation and trilateration. List their similarities.

20-16. Describe the advantages of inertial surveying systems. Explain why these systems could potentially revolutionize current control surveying practice.

20-17. Briefly describe satellite doppler surveying systems. What is meant by doppler shift?

20-18. List the special precautions taken on high-order differential leveling.

20-19. Describe two different types of level rods used to prevent reading blunders in precise leveling.

20-20. Discuss the advantages of the three-wire leveling procedure.

20-21. The standard error of the difference in elevation between two bench marks directly connected in a level circuit and located 27 mi apart is ± 0.008 m. What order and class of leveling does this represent?

20-22. Similar to Problem 20-21, except the standard error is ± 0.014 ft for bench marks located 40 km apart.

20-23. Prepare a set of three-wire leveling notes for the data given, and make the page check. The elevation of BM X is 149.387 m. Rod readings (in meters) are (H denotes upper cross wire reading, M middle wire, and L lower wire): BS on BM X: $H = 1.683$, $M = 1.453$, $L = 1.224$; FS on TP 1: $H = 2.959$, $M = 2.707$, $L = 2.454$; BS on TP 1: $H = 2.254$, $M = 2.054$, $L = 1.854$; FS on BM Y: $H = 1.013$, $M = 0.817$, $L = 0.620$.

20-24. Similar to Problem 20-23, except the elevation of BM X is 859.307 ft, and rod readings (in meters) are: BS on BM X: $H = 2.573$, $M = 2.321$, $L = 2.070$; FS on TP 1: $H = 1.949$, $M = 1.653$, $L = 1.356$; BS on TP 1: $H = 1.470$, $M = 1.195$, $L = 0.921$; FS on BM Y: $H = 1.674$, $M = 1.453$, $L = 1.231$.

BIBLIOGRAPHY

Berry, R. M. 1976. "History of Geodetic Leveling in the United States." *Surveying and Mapping* 36(no. 2):137.

———. 1977. "Observational Techniques for Use with Compensator Leveling Instruments for First Order Levels." *Surveying and Mapping* 37(no. 1):17.

Bomford, G. 1971. *Geodesy*, 3rd ed. London: Oxford University Press.

Bossler, J. D. 1981. "A Note on Global Positioning System Activities." *Bulletin, American Congress on Surveying and Mapping*, no. 74, p. 39.

———. 1982. "New Adjustment of North American Datum." *ASCE Journal of the Surveying and Mapping Division* 108(no. SU2):47.

Colcord, J. E. 1981. "The Surveying Engineer and NAD-83." *ASCE Journal of the Surveying and Mapping Division* 107(no. SU1):25.

Hanson, R. H. 1976. "The New Adjustment of the North American Datum: The Network Adjustment." *Bulletin, American Congress on Surveying and Mapping*, no. 55, p. 21.

Haug, M. D., et al. 1980. "A Simplified Explanation of Doppler Positioning." *Surveying and Mapping* 40(no. 1):47.

Hoar, G. J. 1982. Satellite Surveying. Torrance, Calif.: Magnavox Advanced Products and Systems Co.

Hothem, L. P., et al. 1978. "Doppler Satellite Surveying System." *ASCE Journal of the Surveying and Mapping Division* 104(no. SU1):79.

Kulp, E. F. 1970. "High Precision Levels with Automatic Instruments." *ASCE Journal of the Surveying and Mapping Division* 96(no. SU2):121.

Lippold, H. R., Jr. 1980. "Readjustment of the National Geodetic Vertical Datum." *Surveying and Mapping* 40(no. 2):155.

Mezera, D. F. 1979. "Geodetic Surveying: The Next Decade." *ASCE Journal of the Surveying and Mapping Division* 105(no. SU1):93.

National Geodetic Survey. 1982. "Bulletin to Users of Compensator-Type Leveling Instruments." *Bulletin, American Congress on Surveying and Mapping*, no. 79, p. 35.

Treftz, W. H. 1981. "An Introduction to Inertial Positioning as Applied to Control and Land Surveying." *Surveying and Mapping* 41(no. 1):59.

Vanicek, P., and E. Krakiwsky. 1982. *Geodesy, the Concepts.* Amsterdam: North-Holland.

Whalen, C. T. 1982. "The New Adjustment of the North American Vertical Datum." *Bulletin, American Congress on Surveying and Mapping*, no. 78, p. 39.

Wolf, P. R., and S. D. Johnson. 1974. "Trilateration with Short Range EDM Equipment and Comparison with Triangulation." *Surveying and Mapping* 34(no. 4):337.

21 STATE PLANE COORDINATES

21-1. INTRODUCTION. Most surveys of small areas are based on the assumption that the earth's surface is a plane. As explained in Chapter 20, however, for large-area surveys, it is necessary to consider earth curvature. This is done by computing the horizontal positions of widely spaced stations in terms of geodetic latitudes and longitudes. Unfortunately, the calculations necessary to determine geodetic positions from survey measurements and get distances and azimuths from them are lengthy. Practicing surveyors often are not familiar with this procedure. Clearly a system for specifying positions of geodetic stations using plane rectangular coordinates is desirable, since it allows computations to be made using relatively simple coordinate geometry formulas, such as those presented in Appendix B. The National Geodetic Survey met this need by developing a state plane coordinate system for each state. The first such system evolved in 1933 for the state of North Carolina.

A state plane coordinate system provides a common datum of reference for horizontal control of all surveys in a large area, just as mean sea level furnishes a single datum for vertical control. It eliminates having individual surveys based on different assumed coordinates, unrelated to those used in other adjacent work. At present, state plane coordinates are widely used in all types of surveys, including those for photogrammetric mapping, highway construction projects, and property boundary delineation. In many states new subdivisions must include state plane coordinates.

As discussed in Chapter 20, the earth's curved or mean sea level surface closely approximates a *spheroid* (derived by mathematically revolving an ellipse

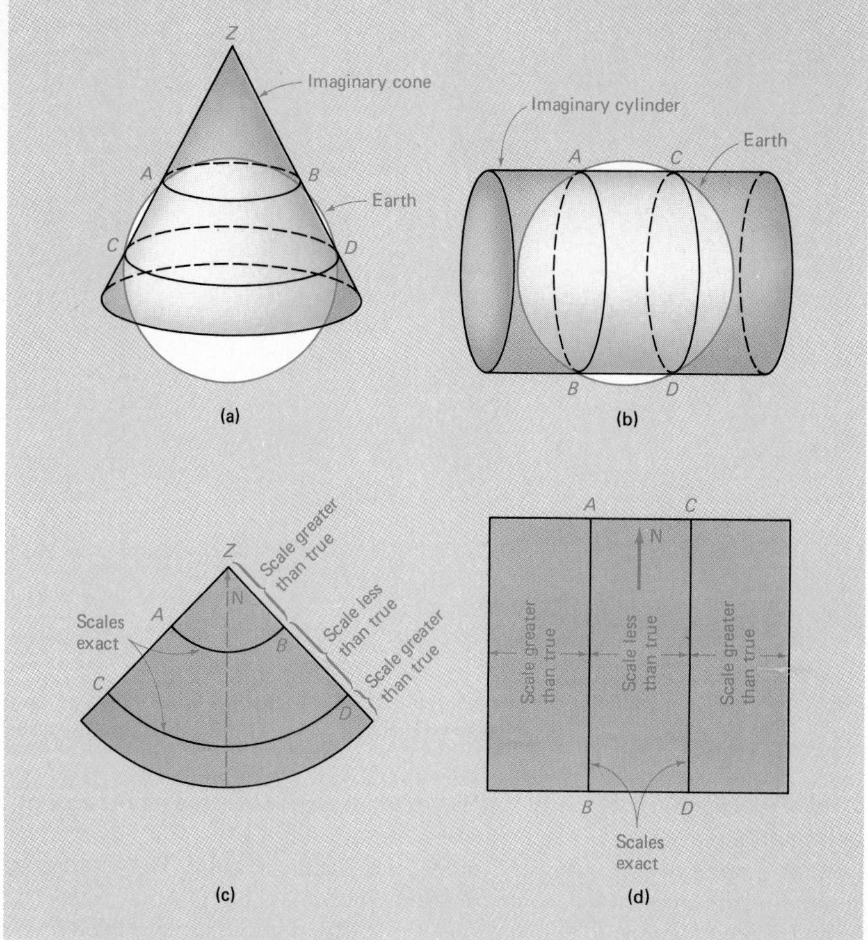

Figure 21-1. Surfaces used in state plane coordinate systems.

about the earth's polar axis). To convert geodetic positions of a portion of the earth's surface to plane rectangular coordinates, points are projected mathematically from the spheroid to some imaginary *developable surface*—a surface that can be developed or "unrolled and laid out flat" without distortion of shape or size. A rectangular grid can be superimposed on the *developed* plane surface and the positions of points in the plane specified with respect to X and Y grid axes. A plane grid so developed is called a *map projection.*

Two basic projections are used in state plane coordinate systems: the *Lambert conformal conic projection* and the *transverse Mercator projection.* The former utilizes an imaginary cone, the latter a fictitious cylinder, as their developable surfaces. These are shown in Figure 21-1(a) and (b), respectively. The cone and cylinder are *secant* to the spheroid in the state plane coordinate systems; that is, they intersect the spheroid along two small arcs AB and CD as shown. Figure 21-1(c) and (d) illustrate plane surfaces developed from the cone and cylinder, respectively.

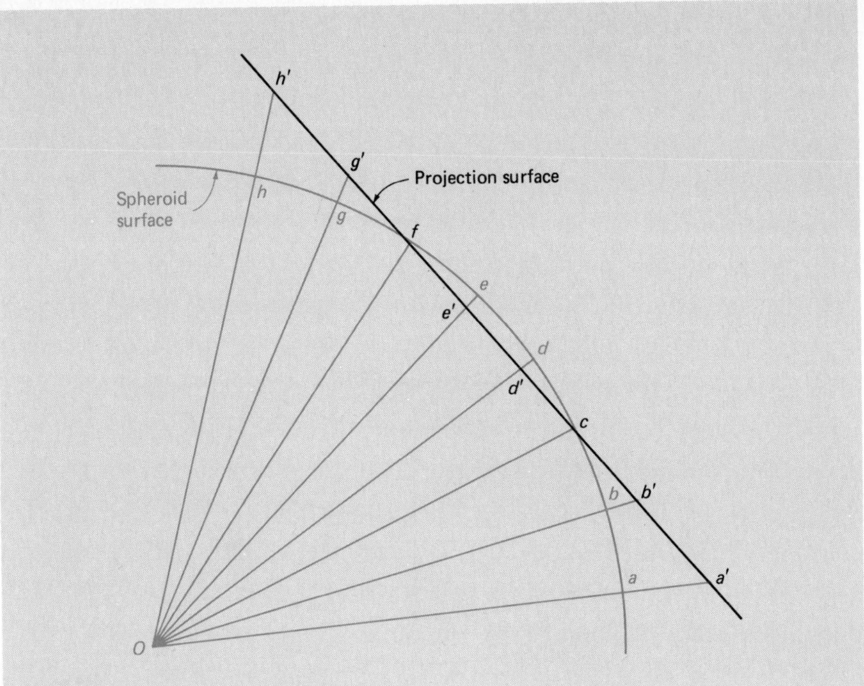

Figure 21-2. Method of projection.

In computing state plane coordinate systems, points are projected mathematically from the spheroid to the surface of the imaginary cone or cylinder. For discussion purposes this may be considered to be a radial projection from the earth's center. Figure 21-2 illustrates this process diagrammatically and displays the relationship between the length of a line on the spheroid and its extent when projected onto the surface of either a cone or cylinder. Note that distance *a′b′* on the projection surface is greater than *ab* on the spheroid, and similarly *g′h′* is longer than *gh*. From this observation it is clear that map projection scale is larger than true spheroid scale where the cone or cylinder is outside the spheroid. Conversely, distance *d′e′* on the projection is shorter than *de* on the spheroid, and thus map scale is smaller than true spheroid scale when the projection surface is inside the spheroid. Points *c* and *f* occur at the intersection of projection and spheroid surfaces, and therefore map scale equals true spheroid scale along the lines of intersection. These relationships of map scale to true spheroid scale for various positions on the two projections are indicated in Figure 21-1(c) and (d).

From the foregoing discussion, it should be clear that points cannot be projected from the spheroid to developable surfaces without introducing distortions in the lengths of lines or shapes of areas. These distortions are held to a minimum, however, by selected placement of the cone or cylinder secant, by choosing a *conformal projection* (one that preserves true angular relationships around points), and also by limiting the zone size or extent of coverage on the earth's surface for any one projection. If the width of zones is held to a maxi-

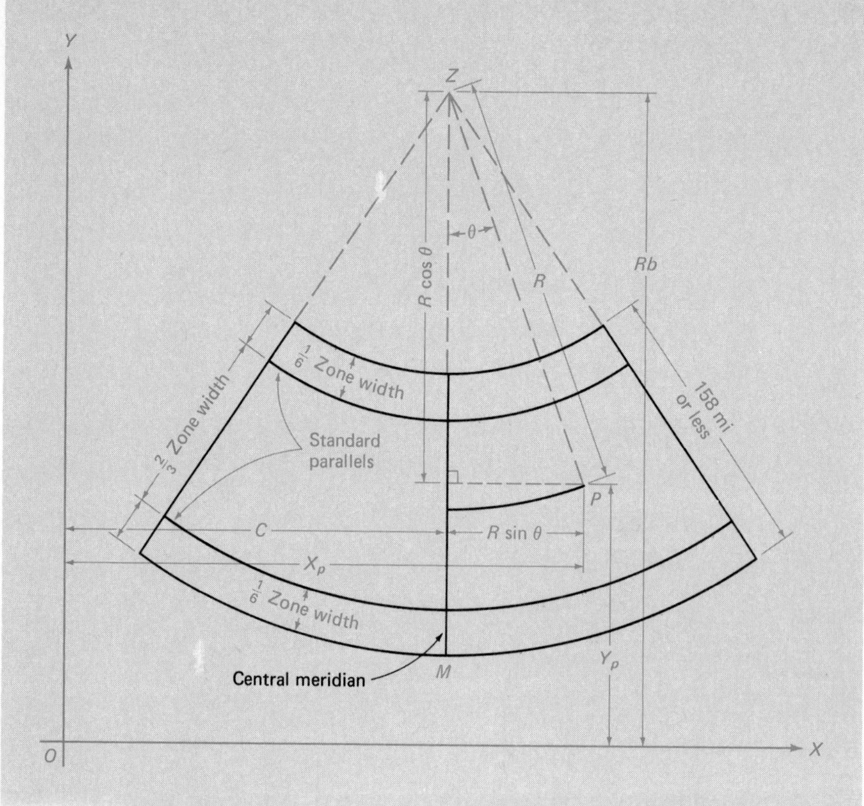

Figure 21-3. The Lambert conformal conic projection.

mum of 158 mi, and if two-thirds of this zone width is between the secant lines, distortions (differences in line lengths on the two surfaces) are kept to 1 part in 10,000 or less. This is the accuracy intended by the NGS in its development of the state plane coordinate systems. One zone is sufficient to cover an entire small state—for example, Connecticut or Delaware. Larger states, such as California and Texas, require more zones; California has seven, Texas five.

21-2. LAMBERT CONFORMAL CONIC PROJECTION. The Lambert conformal conic projection, as its name implies, is a projection onto the surface of an imaginary cone. The term *conformal*, as earlier noted, means true angular relationships are retained around all points. This projection is used in 31 of the 50 states. Scale on a Lambert projection varies from north to south but not from east to west, as shown on Figure 21-1(c). This projection is therefore ideal for mapping areas extending great distances in an east-west direction—for example, Kentucky, Pennsylvania, and Tennessee.

In the Lambert projection, shown in Figure 21-3, the cone intersects the spheroid along two parallels of latitude, called *standard parallels*, at one-sixth of the zone width from the north and south zone limits. On the projection, all meridians are straight lines converging at Z, the apex of the cone, and all parallels of latitude are arcs of concentric circles having centers at the apex of the cone. The projection is located in a zone in an east-west direction by selecting a *central meridian* whose longitude is near the middle of the area to be covered. The direction of the central meridian on the projection establishes *grid north.* All lines parallel with the central meridian point in the direction of grid north. Except at the central meridian, therefore, directions of true and grid north do not coincide.

Given the latitude and longitude of any point *P*, its *X* and *Y* state plane coordinates in the Lambert projection are readily calculated. Consider the developed plane of the Lambert projection illustrated in Figure 21-3. Point *Z* is the apex of a cone, and point *O* the origin of rectangular coordinates. Line *ZM* is the central meridian of the projection. A constant *C*, usually 2,000,000 ft, is adopted to offset the central meridian from the *Y* grid axis and make *X* coordinates of all points positive. Line *ZP* represents a portion of the meridian through point *P* with its length designated as *R*. Angle θ between the central meridian and the meridian *ZP* is termed the *mapping angle.*

The NGS has computed and published projection tables for every state.[1] For any point *P*, the value of *R* is listed in the tables versus the latitude of *P*, and θ recorded versus longitude *P*. A constant R_b (the *Y* coordinate of the cone's apex *Z*) is also given for any particular zone. From Figure 21-3 and appropriate projection tables, the following equations can be solved for the *X* and *Y* coordinates of *P*:

$$\begin{aligned} X_p &= R \sin \theta + C \\ Y_p &= R_b - R \cos \theta \end{aligned} \tag{21-1}$$

Note that if θ is to the left of the central meridian, its sign is negative; if to the right, it is positive. Except where a line of reference azimuth exceeds 5 mi in length, grid azimuth may be calculated with sufficient accuracy from geodetic azimuth using the following equation:

$$\text{grid azimuth} = \text{geodetic azimuth} - \theta \tag{21-2}$$

21-3. TRANSVERSE MERCATOR PROJECTION. The transverse Mercator projection is also a conformal projection based on an imaginary secant cylinder as its developable surface. Because scale varies in an east-west direction, but not from north to south, it is used to map areas of 22 states that are long in a north-south direction, such as Illinois and Indiana.[2]

The axis of the imaginary cylinder of a transverse Mercator projection lies in the plane of the earth's equator. The cylinder cuts the spheroid along two small circles equidistant from the central meridian. On the developed plane surface (Figure 21-4) all parallels of latitude, and all meridians except the central meridian, are the curves shown in light broken lines. A central meridian establishes the direction of grid north. *X* and *Y* coordinates of points are measured perpendicular to and parallel with the central meridian, respectively.

[1] Projection tables for every state are available at a nominal fee from the Superintendent of Documents, U.S. Government Printing Office, Washington, D.C. 20402.

[2] Both the Lambert conformal conic and the transverse Mercator projections are used in Alaska, Florida, and New York.

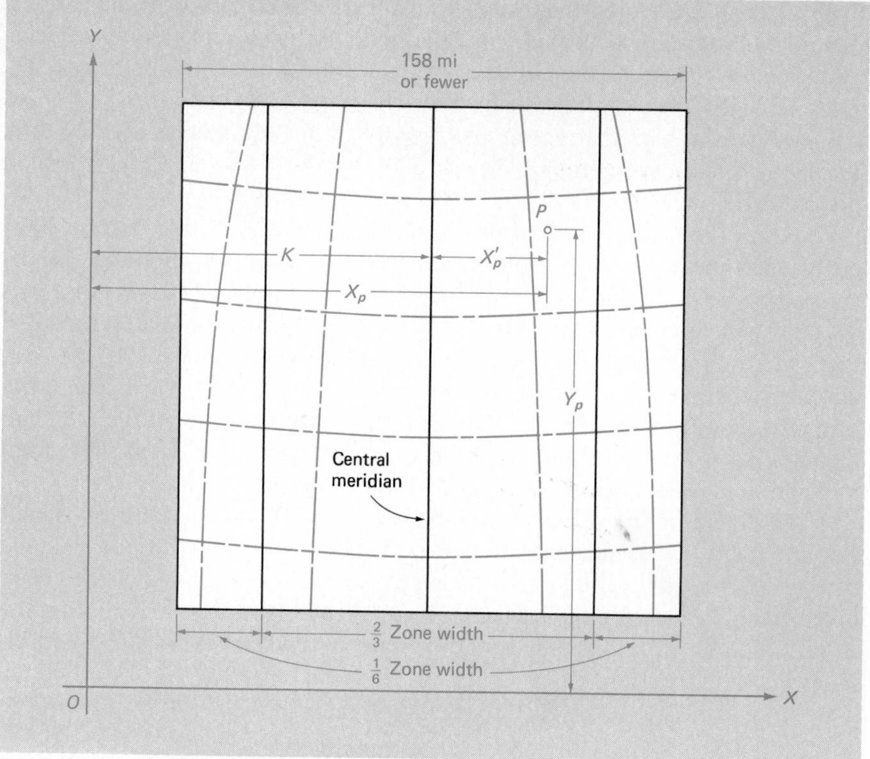

Figure 21-4. The transverse Mercator projection.

Referring to Figure 21-4 and appropriate transverse Mercator projection tables, the following equations may be solved for the X and Y coordinates of any point P:

$$X'_p = H \times \Delta\lambda'' \pm ab$$

$$X_p = X'_p + K \tag{21-3}$$

$$Y_p = Y_o + V\left[\frac{\Delta\lambda''}{100}\right]^2 \pm c$$

In the equations, X'_p is the distance to point P either east or west of the central meridian. The difference in seconds between longitudes of the central meridian and point P is $\Delta\lambda''$, its algebraic sign being negative if P is west and positive if P is east of the central meridian. Constant K offsets the Y axis from the central meridian, so all X coordinates are positive. Its value is 500,000 ft for most states. Values of H, a, Y_o, and V are tabulated versus the latitude of point P in projection tables, and b and c listed versus $\Delta\lambda''$. A negative sign for product ab decreases $H \times \Delta\lambda''$; a positive one increases it.

Except where a reference azimuth line exceeds 5 mi in length, grid azimuth can be calculated with sufficient accuracy from the geodetic azimuth using the following equation:

$$\text{grid azimuth} = \text{geodetic azimuth} - \Delta\alpha'' \tag{21-4}$$

In the equation, $\Delta\alpha'' = \Delta\lambda'' \sin \phi_P + g$ (where g is listed versus $\Delta\lambda''$ in projection tables), and ϕ_P is the latitude of point P.

State plane coordinates and the grid azimuth to a nearby reference mark are published by the NGS for all stations of the U.S. horizontal network and therefore need not be computed (or, if necessary, are determined readily using projection tables and a calculator). In addition to the data described, the tables provide detailed example problems. Projection tables for your state and adjacent ones should be obtained when working with state plane coordinates, since a survey may cover several zones (double sets of coordinates are tabulated near zone edges).

A new general control adjustment to be completed in 1983 by the NGS will replace the last general adjustment of 1927. It will result in revised geodetic latitudes and longitudes of all control stations, and thus produce new state plane coordinates for them. In some localities computed positions may shift as much as 50 ft or more. Upon completion of the general adjustment, new station descriptions will be published with coordinates given in the metric system. New tables giving zone parameters for use in Eqs. (21-1) through (21-4) will also be available from the NGS.

21-4. STATE PLANE COORDINATES BY AUTOMATIC DATA PROCESSING. To eliminate looking up values from tables and manually solve Eqs. (21-1) through (21-4), the NGS has published the complete formulas and constants used in computing tables for each state.[3] These formulas, when programmed for computer solution, provide state plane coordinates that agree exactly with those obtained from Eqs. (21-1) through (21-4), but with significant time savings. Furthermore, once the program has been prepared and tested, if a point's geodetic latitude and longitude are correctly entered into the computer, results free from calculation mistakes are assured.

21-5. COMPUTING STATE COORDINATES OF TRAVERSE STATIONS. Placing a survey on the state plane coordinate grid normally requires traversing (or triangulating or trilaterating) to start and end on existing stations that have known state plane coordinates and from which known grid azimuth lines have been established. Generally these data are available for immediate use, but if not they can be calculated as indicated when latitude and longitude are known.

It is important to note that if a survey begins with a given grid azimuth and ties into another, all intermediate ones will automatically be grid azimuths. Thus, corrections for convergence of meridians are not necessary when the state plane coordinate system is used throughout the survey.

A simple two-sided traverse computation in the state plane coordinate system is presented in this section to illustrate its simplicity. As shown in Figure 21-5, it starts at Irwin Triangulation Station in Ohio, which uses the Lambert conformal conic projection, and ends at BM 1705. Station A in between is the only new point in the survey and its coordinates will be determined.

[3] Coast and Geodetic Survey Publication 62-4, "State Plane Coordinates by Automatic Data Processing," is available from the Superintendent of Documents, U.S. Government Printing Office, Washington, D.C. 20402.

The first step is to reduce the traverse distances to mean sea level (spheroid surface),[4] and then to lengths according to their positions on the state plane coordinate grid. The following equation reduces measured lengths to mean sea level (MSL) distances:

$$L_s = L_m \frac{R_e}{R_e + h} \tag{21-5}$$

where L_s is the sea level line length, L_m the measured line length, R_e the mean radius of the earth (approximately 20,906,000 ft or 6,372,200 m), and h the average elevation of the measured line above MSL. The ratio $R_e/(R_e + h)$ is commonly called the *sea-level factor*. In this example problem, average elevation is 687 ft, and the factor is 0.999966.

The sea-level line length is next multiplied by a *scale factor* obtained from projection tables corresponding to a particular area of the zone in which the line falls. The range of this reduction or increase in sea-level length varies from zero along the two lines of exact scale, to maximum and minimum values determined by the zone size. In Connecticut, for example, the correction is never more than 1 part in 40,000, but it may reach 1 part in 10,000, or slightly more in a few cases for other states. For the illustrative problem, the average latitude of the traverse location obtained from published data on the two control stations is used as the argument to enter the table and find a scale factor of 0.999941.

The product of sea-level factor and scale factor is commonly called the *grid factor*–in this problem, 0.999966 × 0.999941 = 0.999907.

If a traverse extent is so short that the scale factor does not change appreciably, and uniform elevations throughout the survey area permit applying a single sea-level factor, a common grid factor simplifies calculations. It may be ignored if near 1.000000.

Figure 21-5 shows the given or fixed stations and azimuth lines for the example traverse. Dashed lines represent measured angles and courses. Published data for the two control stations are:

Station Irwin: $X = 1{,}367{,}887.24$ ft; $Y = 442{,}126.54$ ft
Azimuth to Irwin azimuth mark = 111°11′07″ from grid N
Approximate elevation = 883 ft
BM 1705: $X = 1{,}364{,}481.50$ ft; $Y = 437{,}001.53$ ft
Azimuth to BM 1704 = 298°23′09″ from grid N
Approximate elevation = 492 ft

The traverse computation has been performed in the following six steps:

1. Distribution of the angular misclosure to get corrected grid azimuths for all traverse sides. Clockwise angles were measured. Misclosure error (difference between given azimuth at BM 1705 and that computed by using the three measured angles and given azimuth at Irwin) is −0°00′15″. This error was distributed equally among the three angles as shown in Table 21-1. Final grid azimuths are listed in Table 21-2.

[4] Michigan is the only exception, using lengths converted to the elevation of 800 ft.

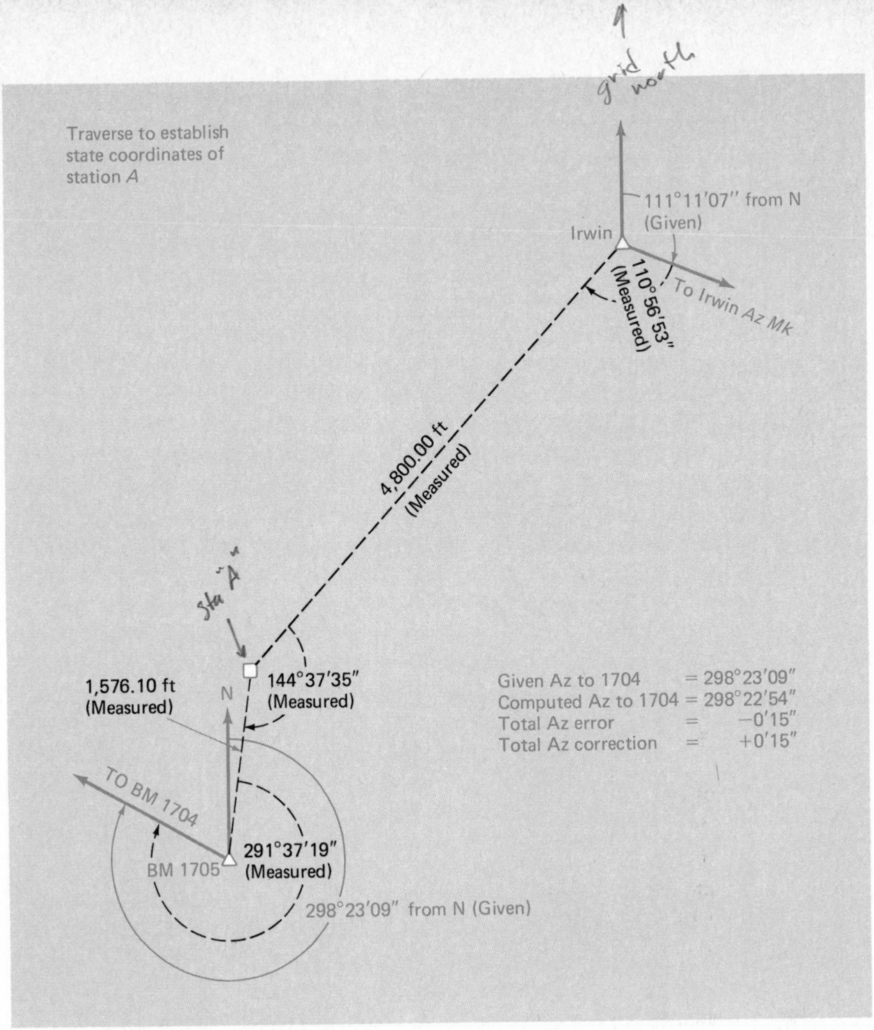

Figure 21-5. Example traverse.

TABLE 21-1. ANGULAR CLOSURE AND ADJUSTMENT

STATION	FROM	TO	PRELIM. AZIMUTH ANGLE TO RIGHT PRELIM. AZIMUTH	CORR. CUM. CORR.	FINAL AZIMUTH ANGLE TO RIGHT FINAL AZIMUTH
	Irwin	Irwin Az.	111°11′07″		111°11′07″ (given)
Irwin	Irwin Az.	*A*	110°56′53″	+5	110°56′58″
	Irwin	*A*	222°08′00″	+5	222°08′05″
	A	Irwin	42°08′00″		42°08′05″
A	Irwin	1705	144°37′35″	+5	144°37′40″
	A	1705	186°45′35″	+10	186°45′45″
	1705	*A*	6°45′35″		6°45′45″
BM 1705	*A*	1704	291°37′19″	+5	291°37′24″
	1705	1704	298°22′54″	+15	298°23′09″ (given)

TABLE 21-2. TRANSVERSE CLOSURE AND ADJUSTMENT (COMPASS OR BOWDITCH RULE)

STATION	GRID AZIMUTH	MEASURED DISTANCE	GRID FACTOR	GRID DISTANCE	SINE COSINE	DEPARTURE	LATITUDE	GRID COORDINATES PRELIM. X CORR. FINAL X	PRELIM. Y CORR. FINAL Y
Irwin							(Given) ⟶	1,367,887.24	442,126.54
	222°08′05″	4800.00	0.999907	4799.55	−0.6708761 −0.7415694	−3219.90	−3559.20	1,364,667.34 −0.20	438,567.34 −0.62
A								1,364,667.14	438,566.72
	186°45′45″	1576.10	0.999907	1575.95	−0.1177540 −0.9930428	−185.57	−1564.98	1,364,481.77 −0.27	437,002.36 −0.83
BM 1705							(Given) ⟶	1,364,481.50	437,001.53
		6376.10					Misclosures =	+0.27	+0.83

Linear misclosure = $\sqrt{(0.27)^2 + (0.83)^2} = 0.87$ ft
Precision = 0.87/6376 = 1/7330

Compass (Bowditch) rule:

X correction = −0.27/6.376 = −0.042 ft/1000 ft of cumulative distance
Y correction = −0.83/6.376 = −0.130 ft/1000 ft of cumulative distance

2. Reduction of measured distances to grid distances is done in Table 21-2 by multiplying values in the third column by those in the fourth column, or by subtracting a correction of 0.0093 ft/100 ft (which is less than 1 part in 10,000).
3. Computation of latitudes and departures.
4. Calculation of preliminary grid coordinates listed in the last column of Table 21-2. Misclosure in the X and Y directions is found by subtracting the given, or fixed, coordinates of BM 1705 from those obtained by traversing. The precision is 1:7330.
5. Traverse adjustment by the compass (Bowditch) rule. Corrections for preliminary coordinates are computed in proportion to the accumulated traverse distances up to any given station and applied in the last columns to get final state plane coordinates for station A. Full corrections (-0.27 and -0.83) applied at BM 1705 give the fixed values written in previously.
6. Based on final state plane coordinates, adjusted lengths and directions of the two traverse lines are obtained by inverse computations from their coordinates as follows:

Irwin–A

$$\Delta X = 1{,}364{,}667.14 - 1{,}367{,}887.24 = -3220.10 \text{ ft}$$

$$\Delta Y = 438{,}566.72 - 442{,}126.54 = -3559.82 \text{ ft}$$

$$\text{length} = \sqrt{(-3220.10)^2 + (-3559.82)^2} = 4800.14 \text{ ft}$$

$$\text{azimuth} = \tan^{-1}\left(\frac{-3220.10}{-3559.82}\right) = 222°07'53.4''$$

A–BM 1705

$$\Delta X = 1{,}364{,}481.50 - 1{,}364{,}667.14 = -185.64 \text{ ft}$$

$$\Delta Y = 437{,}001.53 - 438{,}566.72 = -1565.19 \text{ ft}$$

$$\text{length} = \sqrt{(-185.64)^2 + (-1565.19)^2} = 1576.16 \text{ ft}$$

$$\text{azimuth} = \tan^{-1}\left(\frac{-185.64}{-1565.19}\right) = 186°45'50.4''$$

21-6. SURVEYS EXTENDING FROM ONE ZONE TO ANOTHER. Surveys in border areas often cross into different zones or even abutting states. This presents no unusual problem, however, because adjacent zones overlap by approximately 50 mi.

To convert state plane coordinates of points from one zone to another, it is necessary to calculate latitudes and longitudes of two points using state plane coordinates from the zone where the survey originated. Depending on whether the coordinates are based on the Lambert conformal conic projection or the transverse Mercator projection, Eqs. (21-1) or (21-3) would be used but solved in reverse. From the latitude and longitude, state plane coordinates of the point are calculated using either Eq. (21-1) or (21-3) as appropriate for the zone entered. The grid azimuth of a line in the zone entered can then be obtained from the new state plane coordinates of the two points.

Suppose, for example, that a survey originates in southern Wisconsin, which uses the Lambert conformal conic system, and extends into northern Illinois, which uses the transverse Mercator grid. With the X and Y coordinates of two points inside the Illinois border known in Wisconsin's south zone, Eqs. (21-1) are rewritten for the first point as follows:

$$R \sin \theta = X_p - C \tag{21-6}$$

$$R \cos \theta = R_b - Y_p \tag{21-7}$$

Dividing Eq. (21-6) by Eq. (21-7),

$$\tan \theta = \frac{X_p - C}{R_b - Y_p} \tag{21-8}$$

All terms on the right side of Eq. (21-8) are known, with C and R_b coming from the projection tables for Wisconsin's south zone, and X_p and Y_p available from the survey. Thus, mapping angle θ can be obtained. R can then be found by substituting θ back into either Eq. (21-6) or (21-7). Finally, the geodetic latitude of the point is secured by interpolation from the Wisconsin tables for the value of R, and the geodetic longitude is obtained by interpolation using θ. The second point is handled in the same way. Using the geodetic latitudes and longitudes of the two points, Eqs. (21-3) are solved using constants for the appropriate Illinois zone to obtain their X and Y coordinates in that zone. An inverse computation using the coordinates yields the grid azimuth. Thus, the survey can continue into Illinois.

21-7. THE UNIVERSAL TRANSVERSE MERCATOR PROJECTION. The universal transverse Mercator (UTM) system is another important map projection. Originally developed by the military primarily for artillery use, it provides worldwide coverage from 80°S latitude, through the equator, to 80°N latitude (the polar caps are covered by polar stereographic systems; see Section 21-8). The UTM system is a modified transverse Mercator projection, and in military applications zone widths of 6° longitude are used. Adjacent zones overlap by 30 min.

The UTM projection has recently taken on added importance to surveyors, since this system has been adopted for computing and publishing the plane coordinates of all points adjusted in the new NAD83. UTM coordinates in metric units will be included along with state plane and geodetic coordinates for all station descriptions published thereafter, and UTM grids will be included on all maps of the national mapping program. Because the zone widths of 6° longitude used by the military yield accuracies as low as 1 part in 2500, zone widths for the new adjustment will be reduced to 2°, thereby attaining accuracies of 1 part in 10,000, to be consistent with state plane systems.

Equations for calculating X and Y UTM coordinates are the same as those for the transverse Mercator projection. As with state plane systems, the NGS will publish tables giving formulas and constants for each zone in the system. Advantages of the new UTM grid are that all points within the NAD83

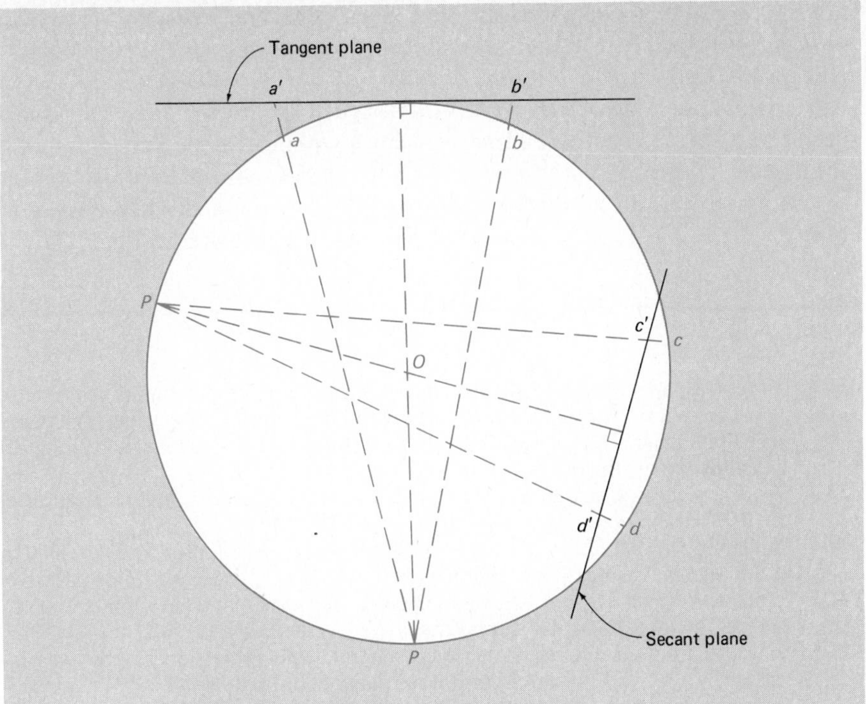

Figure 21-6. Tangent plane and secant plane horizon stereographic map projections.

will be on the same system and calculations between widely spaced points can readily be made. This is entirely consistent with current capabilities for conducting surveys of global extent with new devices such as satellite doppler systems (see Section 20-15).

21-8. OTHER MAP PROJECTIONS.

The Lambert conformal conic and transverse Mercator map projections are designed to cover areas extensive in east-west or north-south directions, respectively. These systems do not, however, conveniently cover circular areas or long strips of the earth that are skewed to the meridians. Two other systems, the *horizon stereographic* and *oblique Mercator* projections, satisfy these problems.[5]

The horizon stereographic projection can be divided into two classes: *tangent plane* and *secant plane*. In either class, as illustrated in Figure 21-6, the projection point P is on the spheroid where a line perpendicular to the map plane and passing through center point O intersects the spheroid. In the tangent plane system, spheroid points a and b are projected outward to a' and b', respectively, on the map plane. For the secant plane system spheroid points c and d are projected inward (if they were outside of the secant points, projection would be outward) to c' and d' on the map plane. Horizon stereographic pro-

[5] The oblique Mercator projection is also called the Hotine skew orthomorphic projection, named after the English geodesist Martin Hotine.

jections are not employed in the United States, but are used in Canada and other parts of the world. If point P is the north or south pole, the projection is called *polar stereographic*; if on the equator, *equatorial stereographic*.

The oblique Mercator projection is designed for areas that run obliquely such as northwest to southeast. It has gained acceptance in the United States in recent years, and is currently used as the state plane coordinate projection for the south-east portion of the state of Alaska, and by the U.S. Lake Survey to cover the area of Lakes Ontario and Erie, and the St. Lawrence River.

PROBLEMS

21-1. Discuss the advantages of placing surveys on state plane coordinate systems.

21-2. Name two basic projections used in state plane coordinate systems. What are their fundamental differences? Which one is preferred for states whose long dimensions are north-south? East-west?

21-3. Develop a table of sea-level factors for ground elevations ranging from sea level to 10,000 ft above sea level. Use increments of 500 ft.

21-4. Similar to Problem 21-3, except for ground elevations from sea level to 3000 m above sea level using 200-m increments.

21-5. Explain how surveys can be extended from one state plane coordinate zone to another, or from one state to another.

21-6. What accuracy in differences between spheroid and projection lengths was intended by the NGS in developing state plane coordinate systems? What maximum zone width is allowable to achieve desired accuracy?

21-7. The state plane coordinates of points A and B are as follows:

POINT	X	Y
A	2,179,431.75	895,784.92
B	2,184,960.42	892,576.67

Calculate the grid length and grid bearing of line AB.

21-8. Similar to Problem 21-7, except points A and B have the following state plane coordinates:

POINT	X	Y
A	1,574,206.86	773,438.20
B	1,569,831.22	770,207.84

21-9. Station A in Wisconsin's north zone (Lambert conformal conic) is at north latitude 45°21′32.507″ and west longitude 89°44′28.652″. Constants for this zone are $C = 2{,}000{,}000.00$ ft and $R_b = 20{,}489{,}179.67$ ft. R values tabulated for latitudes 45°21′ and 45°22′ are 20,422,328.19 ft and 20,416,250.84 ft, respectively, and θ values tabulated for longitudes 89°44′ and 89°45′ are +0°11′32.5160″ and +0°10′49.2337″, respectively. Compute the state plane coordinates for station A.

21-10. Point B, near station A of Problem 21-9, has state plane coordinates $X =$ 2,081,867.24 ft and $Y = 77{,}946.99$ ft. Calculate the grid length, grid azimuth, and geodetic azimuth of line AB.

21-11. Station Hill in Pennsylvania's south zone (Lambert conformal conic) has north latitude 40°10′36.205″ and west longitude 79°43′48.079″. Constants for this zone are $C = 2{,}000{,}000.00$ ft and $R_b = 24{,}984{,}826.43$ ft. R values tabulated for latitudes 40°10′ and 40°11′ are 24,681,260.43 ft and 24,675,189.10 ft, respectively, and θ values tabulated for longitudes 79°43′ and 79°44′ are −1°16′33.4556″ and −1°17′12.3832″, respectively. Determine the state plane coordinates for this point.

21-12. Station Vale, near station Hill of Problem 21-11, has state plane coordinates $X = 1{,}455{,}253.84$ and $Y = 315{,}987.55$. Calculate the grid length, grid azimuth, and geodetic azimuth of line Hill–Vale.

21-13. A station in Indiana's west zone (transverse Mercator) has north latitude 39°25′19.068″ and west longitude 87°40′58.795″. Constant K for this zone is 500,000.00 ft and the longitude of the central meridian is 87°05′00.000″. Values for Y_o, H, V, and a tabulated for latitudes 39°25′ and 39°26′ are as follows:

LATITUDE	Y_o	H	V	a
39°25′	697,992.23	78.482017	1.208187	−0.730
39°26′	704,062.75	78.463323	1.208326	−0.727

Values for b and c listed versus $\Delta\lambda''$ are:

$\Delta\lambda''$	b	c
2100	+3.218	−0.085
2200	+3.293	−0.091

Compute the state plane coordinates of this point.

21-14. What corrections must be made to measured slope distances prior to computing state plane coordinates?

21-15. In computing state plane coordinates for a project area whose mean elevation is 3480 ft above mean sea level, an average scale factor of 0.9999375 was used. The given distances between points in this project area were computed from state plane coordinates. What horizontal length would have to be measured to lay off these lines on the ground?
(a) 950.00 ft
(b) 2485.50 ft

21-16. Similar to Problem 21-15, except that the mean project area elevation was 1235 m, the scale factor 1.0000863, and the computed lengths of lines were:
(a) 675.821 m
(b) 1246.950 m

21-17. Which state utilizes an oblique Mercator projection for its state plane coordinates system? Why?

21-18. The horizontal ground lengths of a three-sided closed polygon traverse were measured as follows: $AB = 1754.93$, $BC = 2400.23$, and CA = 3147.54 ft. If the average elevation of the area is 2025 ft above sea level, calculate the sea-level lengths of the lines.

21-19. Assuming a scale factor for the traverse of Problem 21-18 to be 0.9999364, calculate grid lengths for the traverse lines.

21-20. For the traverse of Problem 21-18, the grid azimuth of a line from A to a nearby azimuth mark was 125°17′48″ and the clockwise angle measured at A from the azimuth mark to B 118°22′51″. The measured interior angles (to the right) were $A = 49°11'07''$, $B = 97°14'15''$, and $C = 33°35'02''$. Balance the angles and compute grid azimuths for the traverse lines.

21-21. Using grid lengths of Problem 21-19 and grid azimuths from Problem 21-20, calculate latitudes and departures, linear misclosure, and precision for the traverse.

21-22. If station A has state plane coordinates $X = 2{,}063{,}294.81$ and $Y = 563{,}042.46$, balance the latitudes and departures computed in problem 21-21 using the compass (Bowditch) rule, and determine the state plane coordinates of stations B and C.

21-23. The horizontal ground lengths of a four-sided closed polygon traverse were measured as follows: $AB = 384.707$ m, $BC = 666.343$ m, $CD = 523.797$ m, and $DA = 680.840$ m. If the average elevation of the area is 1545 m above sea level and the scale factor for the traverse 1.0000996, calculate grid lengths of the lines.

21-24. For the traverse of Problem 21-23, the grid bearing of the line BC is N57°29′18″W. Interior angles (to the right) were measured as follows: $A =$ 119°25′26″, $B =$ 72°47′54″, $C =$ 102°28′01″, and $D =$ 65°18′27″. Balance the angles and compute grid bearings for the traverse lines.

21-25. Using grid lengths from Problem 21-23 and grid bearings from Problem 21-24, calculate latitudes and departures, linear misclosure, and precision for the traverse. Balance the latitudes and departures by the compass (Bowditch) rule. If the state plane coordinates of point B are $X = 712{,}666.481$ m and $Y =$ 179,341.609 m, calculate state plane coordinates for points C, D, and A.

BIBLIOGRAPHY

Doyle, F. J. 1973. "Federal Mapping and the National Grid." *Surveying and Mapping* 33(no. 3):305.

Dracup, J. F. 1977. "The New Adjustment of the North American Datum: Plane Coordinate Systems." *Bulletin, American Congress on Surveying and Mapping,* no. 59, p. 27.

Howe, R. F. 1970. "Coordinates Versus Confusion." *ASCE Journal of the Surveying and Mapping Division* 96(no. SU2):191.

Hurlbert, D. D., and J. D. McDonald. 1976. "Horizontal Control in Kansas City by State Plane Coordinates." *ASCE Journal of the Surveying and Mapping Division* 102(no. SU1):9.

Lindsey, J. A. 1969. "Relocating Government Corners on the State Coordinate System." *Surveying and Mapping* 29(no. 3):401.

McDonnell, P. W., Jr. 1980. *Introduction to Map Projections.* New York: Dekker.

Meade, B. K. 1973. "Coordinate Systems for Surveying and Mapping." *Surveying and Mapping* 33(no. 3):337.

Pryor, W. J. 1973. "Plane Coordinates for Engineering and Cadastral Surveys." *Surveying and Mapping* 33(no. 3):317.

Read, J. R. 1981. "A Coordinate System for North America Based upon the 6° UTM Zone." *Surveying and Mapping* 41(no. 1):83.

22 BOUNDARY SURVEYS

22-1. INTRODUCTION. The earliest surveys were made to locate or relocate boundary lines of property. From Biblical times,[1] when the death penalty was assessed for destroying corners, to the colonial days of George Washington,[2] who was licensed as a land surveyor by William and Mary College of Virginia, and through the years to the present, trees and other natural objects, or stakes driven into the ground, have been used to identify corners.

As property increased in value and owners disputed rights to land, the importance of more accurate surveys, permanent monuments, and written records was obvious. "When Texas became a state in 1845, the public domain amounted to about 172,700,000 acres, which the United States government could have acquired by payment of approximately $13,000,000 in debts accumulated by the Republic of Texas. However, Congress allowed the Texans to retain their land and pay their own debts"[3]—a good bargain, even then, at roughly 7.6 cents/acre. Several years ago, land at Waikiki in Honolulu sold for well over $100/ft^2 or more than $5,000,000/acre. In Tokyo property can be even more expensive.

[1] "Cursed be he that removeth his neighbor's landmark. And all the people shall say Amen." Deut. 27:17.

[2] "Mark well the land, it is our most valuable asset." George Washington.

[3] Rupert F. Carroll, "Property Surveys Must Fit Their Titles," *ASCE Proceedings*, July 1949.

Land titles now are transferred by written documents called *deeds* (*grant*, *quitclaim*, *agreement*, or *warranty*) which contain a description of the property boundaries. The various methods of description include (a) metes and bounds, (b) block and lot number, (c) coordinate values for each corner, and (d) township, section, and smaller subdivision. The first three methods are discussed briefly in this chapter, the fourth one in Chapter 23.

Most property surveys today are wholly or partly resurveys rather than originals. In retracing old lines, a surveyor must exercise acute judgment based on education, practical experience, and a knowledge of land laws, and be accurate in making measurements. The necessary mathematics and proper use of a transit, theodolite, level, tape, and EDMI can be learned in a relatively short time. This background must be bolstered by tenacity in searching the records of all adjacent property as well as studying descriptions of the land in question. In field work, a surveyor must be untiring in efforts to find points called for by the deed. Often it is necessary to obtain testimony from people who have knowledge of accepted land lines and the location of corners, reference points, fences, and other evidence of the correct lines.

Modern-day land surveyors are confronted with a multitude of problems created over the past two centuries, under different technology and legal systems, that now require professional solutions. These include defective compass and chain surveys; incompatible descriptions and plats of common lines for adjacent tracts; lost or obliterated corners and reference marks; discordant stories by local residents; questions of riparian rights; and a tremendous number of legal decisions on cases involving property boundaries.

The responsibility of a professional surveyor is to sift all evidence and try to obtain a meeting of minds among persons involved in any property-line dispute, although without legal authority to force a compromise or settlement. Fixing title boundaries must be done by agreement of adjacent owners or court action. To serve as an expert witness in proceedings to establish boundaries, a surveyor should be registered.

Because of the complicated technical and judgment decisions that must be made, the increasing cost of land surveyors' professional liability insurance for "errors and omissions" has become a major part of operating expenses. Some clients demand that a surveyor have it for protection of all parties.

Many municipalities have rigid laws covering subdivisions. Regulations may specify the minimum lot size; allowable closures for surveys; types of corner marks to be used; minimum width of streets and the procedure for dedicating them; rules for registry of plats; and other matters. The mismatched street and highway layouts of today could have been eliminated by suitable subdivision regulations in past years.

Various problems associated with real estate titles and transfers have provided the impetus for research and development toward implementation of Modern Land Data Systems (MOLDS). Sometimes referred to as "modern cadasters," they will assemble computerized databanks at central locations. Information available for rapid retrieval may include parcel descriptions and identifiers, their geographic positions, records of ownership, easements, land use, soil types, and specialized data. Such systems can be invaluable to surveyors, lawyers, developers, planners, environmentalists, government officials, and others.

22-2. BASIS OF LAND TITLES. In the eastern part of the United States individuals acquired the first land titles by gifts or purchase from the English Crown. Surveys and maps were completely lacking or inadequate, and descriptions could be given in only general terms. Remaining land in the 13 colonies was transferred to the states at the close of the Revolutionary War. Later this land was parceled out to individuals, generally in irregular tracts. Boundary lines were described by metes and bounds (directions by magnetic bearings, and lengths in Gunter's chains, poles, or rods).

Many original transfers and subsequent ownerships and subdivisions were not recorded. Those that were legally registered usually had scanty or defective descriptions, since land was cheap and abundant. Trees, rocks and natural landmarks defining the corners, as in the first example metes-and-bounds description (see Section 22-3) were soon disturbed. The intersection of two property lines might be described only as "the place where John killed a bear" or "the bend in a footpath from Jones's cabin to the river."

Numerous problems in land surveying stem from the confusion engendered by early property titles, descriptions, and compass surveys. The locations of thousands of corners have been established by compromise after resurveys, or by court interpretation of all available evidence pertinent to their original or intended positions. Other corners have been fixed by *squatters' rights, adverse possession,* and *riparian changes.* Many boundaries still are in doubt, particularly in areas having marginal land where the cost of a good survey ("retracement of history") exceeds the property's value.

The fact that four corners of a field can be found, and that the distances between them agree with the calls in a description, does not necessarily mean they are in the proper place. Title or ownership is complete only when the land covered by a deed is positively identified and located on the ground.

Land law from the time of the Constitution has been held as a state's right, subject to interpretation by state court systems. Many millions of land parcels have been created in the United States over the past four centuries under different technology and legal systems. Some of the numberless problems passed on to today's professional surveyors, equipped with immensely improved equipment, are discussed in this chapter and in Chapter 23.

Land surveying measurements and analysis follow basic plane surveying principles. But years of experience in a given state are needed by a land surveyor to become familiar with local conditions, basic reference points, and legal interpretations of complicated boundary problems. Methods used in one state for prorating differences between recorded and measured distances may not be acceptable in another. Rules on when and how fences determine property lines are not the same in all states or even in adjacent ones.

The term *practical location* is used by the legal profession to describe an agreement, either explicit or implied, in which two adjoining property owners settle a boundary dispute or mark out an ambiguous boundary. Fixed principles enter the process and the boundary established becomes permanent.

Different interpretations are given locally to the (a) superiority or definiteness of one distance over another associated with it; (b) position of boundaries shown by occupancy; (c) value of corners in place in a tract and its subdivisions; and (d) many other factors. Registration of land surveyors is therefore required in all states to protect the public interest.

22-3. PROPERTY DESCRIPTION BY METES AND BOUNDS. Descriptions by metes (to measure, or assign by measure) and bounds (boundary lines or property limits) have a *point of beginning* (POB), such as a stake, fence post, road intersection, or some natural feature. Lengths and bearings of successive lines from the point of beginning are given. Values in chains, poles, and rods are being replaced by distances in feet and decimals, and slowly by metric units. A 1975 American Congress on Surveying and Mapping (ACSM) metric workshop recommended, among other things, that (a) surveyors immediately show equivalent values for areas in square meters or hectares, depending on parcel size, on all plans for recording; and (b) legal descriptions of existing deeds, record plans, or plats be converted to the metric system only if and when conveyancing or subdivision takes place.

Bearings may be assumed, magnetic, or true, the last being preferable. Care must be exercised to clearly indicate which meridian is the basis of bearings so no confusion arises. A West Virginia survey regulation calls for exterior lines of new subdivisions to be based on the true meridian, thus requiring an astronomical observation by a surveyor for each subdivision unless it adjoins another where an astronomical observation was made, or is near some other line of known true azimuth.

In relocating an old survey, precedence (weight of importance) is commonly assigned as follows: (1) marks or monuments in place; (2) calls for boundaries of adjoining tracts; and (3) courses and distances shown in the original notes or plat. If numbers are spelled out and also given as figures, words control unless other proof is available. There is a greater likelihood of transposing than misspelling—and lawyers prefer words!

Property descriptions are written by surveyors and lawyers. A single mistake in transcribing a numerical value, or one incorrect or misplaced word or punctuation mark, may result in litigation for more than a generation, since the intentions of *grantor* (person selling property) and *grantee* (person buying property) are not clearly fulfilled.

The importance of permanent monuments is evident; in fact, some states require pipes, iron pins, and/or concrete markers long enough to reach below the frost line at all property corners before surveys will be accepted for recording. Actually, almost anything can be called for as a monument. A map attached to the description clarifies it, and scaling provides a rough check on the angles and distances.

To increase precision of property surveys, large cities and some states have established a network of control monuments to supplement triangulation stations of the NGS. Property corners can be tied to these control points and boundary lines relocated with assurance.

Description of land in a deed should always contain the following information in addition to the recital:

1. *Point of beginning* (POB). This point must be identifiable, permanent, well referenced, and one of the property corners. Coordinates, preferably state plane, should be given if known or computable. Note that a POB is no more important than others and a called-for monument in place at the next corner establishes its position, even though bearing and distance calls to it do not agree.
2. *Definite corners.* Such corners are clearly defined points with coordinates if possible.
3. *Lengths and directions of the property sides.* All lengths in feet and decimals (or metric units), and directions by angles, true bearings, or azimuths must be stated to permit computation of any misclosure error. Omitting the length or bearing of a closing line to the POB and substituting a phrase "and thence to the point of beginning" is no longer acceptable. The date of the survey is required and particularly important if bearings are referred to magnetic north.
4. *Names of adjoining property owners.* These are helpful to show the intent of a deed in case an error in the description leaves a gap or creates an overlap. However, called-for monuments in place will control title over calls for adjoiners.
5. *Areas.* The included area is normally given as an aid in valuation and identification of a piece of property. Areas of rural land are given in acres or hectares; those of city lots in square feet or square meters. Because of differences in measurements, and depending on the adjustment method used for a traverse (compass, transit, least squares, etc.), one surveyor's calculated area, angles, and distances may differ from another's.

The expression "more or less" which may follow a computed area allows for only minor errors. It avoids "nuisance suits" for insignificant variations.

A partial metes-and-bounds description for the tract shown in Figure 22-1 is given as an example.

> That part of the SW$\frac{1}{4}$ of the NW$\frac{1}{4}$ of Section 28, T 22 N, R 11 E, Town of Little Wolf, Brock County, Wisconsin, described as follows: Commencing at a stone monument at the W$\frac{1}{4}$ corner of said Section 28; thence N45°00′E, 400.00 feet along the Southeasterly R/W line of Lake Street to a 1″ iron pipe at the point of beginning of this description, said point also being the point of curvature of a curve to the right having a central angle of 90°00′ and radius of 300.00 feet; thence Easterly, 471.24 feet along the arc of the curve, the long chord of which bears East, 424.26 feet, to a 1″ iron pipe at the point of tangency thereof, said arc also being the aforesaid Southerly R/W line of Lake Street; thence continuing along the Southwesterly R/W line of Lake Street, S45°00′E, 150.00 feet to a 1″ iron pipe; thence S45°00′W 200.00 feet to a 1″ iron pipe located N45°00″E, 20 feet, more or less, from the water's edge of Green Lake, and is the beginning of the meander line along the lake, thence West 141.42 feet along the said meander line to a 1″ iron pipe at the end of the meander line; said pipe being located N45°00′W, 20 feet, more or less from the said water's edge; thence N45°00′W, 350.00 feet to a 1″ iron pipe at the point of beginning... including all lands lying between the meander line herein described and the Northerly shore of Green Lake, which lie between true extensions of the Southeasterly and Southwesterly boundary lines of the parcel herein described, said parcel containing 2.54 acres. Bearings are based on astronomic north.

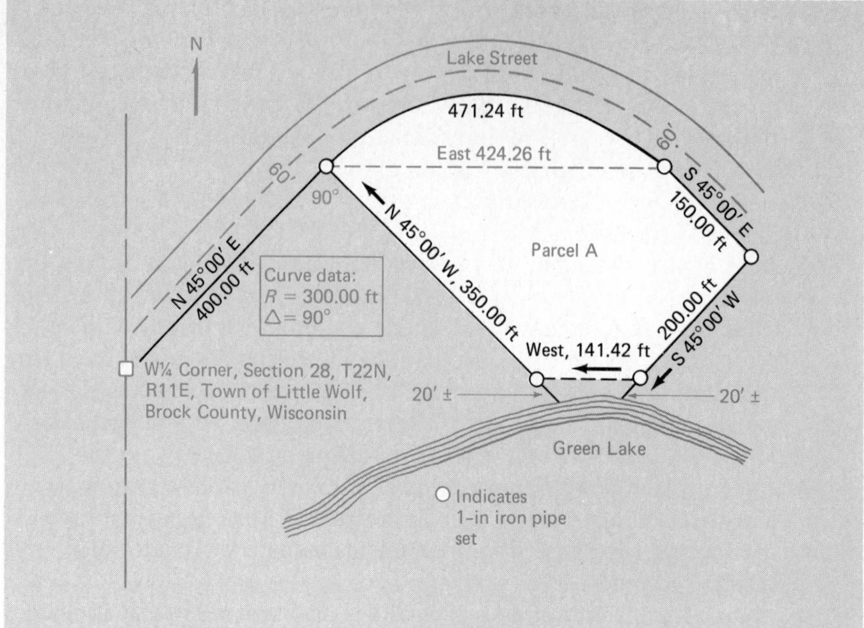

Figure 22-1. Metes-and-bounds tract.

Two examples of *old* metes-and-bounds descriptions from the eastern United States will be given. The first, part of an early deed registered in Maine, is:

> Beginning at an apple tree at about 5 minutes walk from Trefethens Landing thence easterly to an apple tree, thence southerly to a rock, thence westerly to an apple tree, thence northerly to the point of beginning.

With numerous apple trees and an abundance of rocks in the area, the dilemma of a surveyor trying to retrace the boundaries many years later is obvious.

The second, a more typical old description of a city lot showing lack of comparable precision in angles and distances, follows:

> Beginning at a point on the west side of Beech Street marked by a brass plug set in a concrete monument located one hundred twelve and five tenths (112.5) feet southerly from a city monument No. 27 at the intersection of Beech Street and West Avenue; thence along the west line of Beech Street S15°14′30″E fifty (50) feet to a brass plug in a concrete monument; thence at right angles to Beech Street S74°45′30″W one hundred fifty (150) feet to an iron pin; thence at right angles N15°14′30″W parallel to Beech Street fifty (50) feet to an iron pin; thence at right angles N74°45′30″E one hundred fifty (150) feet to place of beginning; bounded on the north by Norton, on the east by Beech Street, on the south by Stearns, and on the west by Weston.

22-4. PROPERTY DESCRIPTION BY BLOCK-AND-LOT SYSTEM. In subdivisions and in large cities it is more convenient to identify individual lots by *block and lot number*, by *tract and lot number*, or by *subdivision name and lot number*.

Examples are:

> Lot 34 of Tract 12314 as per map recorded in book 232, pages 23 and 24 of maps, in the office of the county recorder of Los Angeles County.
>
> Lot 9 except the North 12 feet thereof, and the East 26 feet of Lot 10, Broderick's Addition to Minneapolis. [Parts of two lots are included in the parcel described.]
>
> That portion of Lot 306 of Tract 4178 in the City of Los Angeles, as per Map recorded in Book 75, pages 30 to 32 inclusive of maps in the office of the County Recorder of said County, lying Southeasterly of a line extending Southwesterly at right angles from the Northeasterly line of said Lot, from a point in said Northeasterly line Southeasterly 23.75 feet from the most Northerly corner of said Lot.

Map books in the city or county recorder's office give the location and dimensions of all the blocks and lots. It is now standard practice to require subdividers to file a map with the proper office showing the type and location of monuments, size of lots, and other pertinent information such as the dedication of streets. It is evident that if the boundary lines of a tract are in doubt, the individual lot lines must be questioned also.

The block-and-lot system is a short and unique means of describing property for tax purposes as well as for transfer. Identification by street and house number is satisfactory only for tax-assessment records.

Figure 22-2 is an example of a small block-and-lot subdivision.

22-5. PROPERTY DESCRIPTION BY COORDINATES. The advantages of state plane coordinate systems in improving the accuracy of local surveys, and in facilitating relocation of lost and obliterated corners, have led to their legal acceptance in property descriptions. A coordinate description of corners may be used alone in various units, including metric, but usually is prepared in conjunction with an alternative method. Wider use of the coordinate system will be made as more reference points become available to the local surveyor.

A description by coordinates of a parcel in California follows.

> A parcel of tide and submerged land, in the State-owned bed of Seven Mile Slough, Sacramento County, California, in projected Section 10, T 3 N, R 3 E, Mt. Diablo Meridian, more particularly described as follows:
>
> BEGINNING at a point on the southerly bank of said Seven Mile Slough which bears S62°37′E, 860 feet from a California State Lands Commission brass cap set in concrete stamped "JACK 1969," said point having coordinates of $X = 2{,}106{,}973.68$ and $Y = 164{,}301.93$ as shown on Record of Survey of Owl Island, filed October 6, 1969, in Book 27 of Surveys, Page 9, Sacramento County Records, thence to a point having coordinates of $X = 2{,}107{,}196.04$ and $Y = 164{,}285.08$; thence to a point having coordinates of $X = 2{,}107{,}205.56$, $Y = 164{,}410.72$: thence to a point having coordinates of $X = 2{,}106{,}983.20$, $Y = 164{,}427.57$; thence to the point of beginning.
>
> Coordinates, bearings, and distances in the above description are based on the California Coordinate System, Zone II.

Earthquakes in Alaska, California, and Hawaii, and subsidence due to withdrawal of oil and groundwater in many states, have caused ground shifts that move corner monuments and thereby change their coordinates. The monuments, rather than the coordinates, then have greater weight in ownership rights.

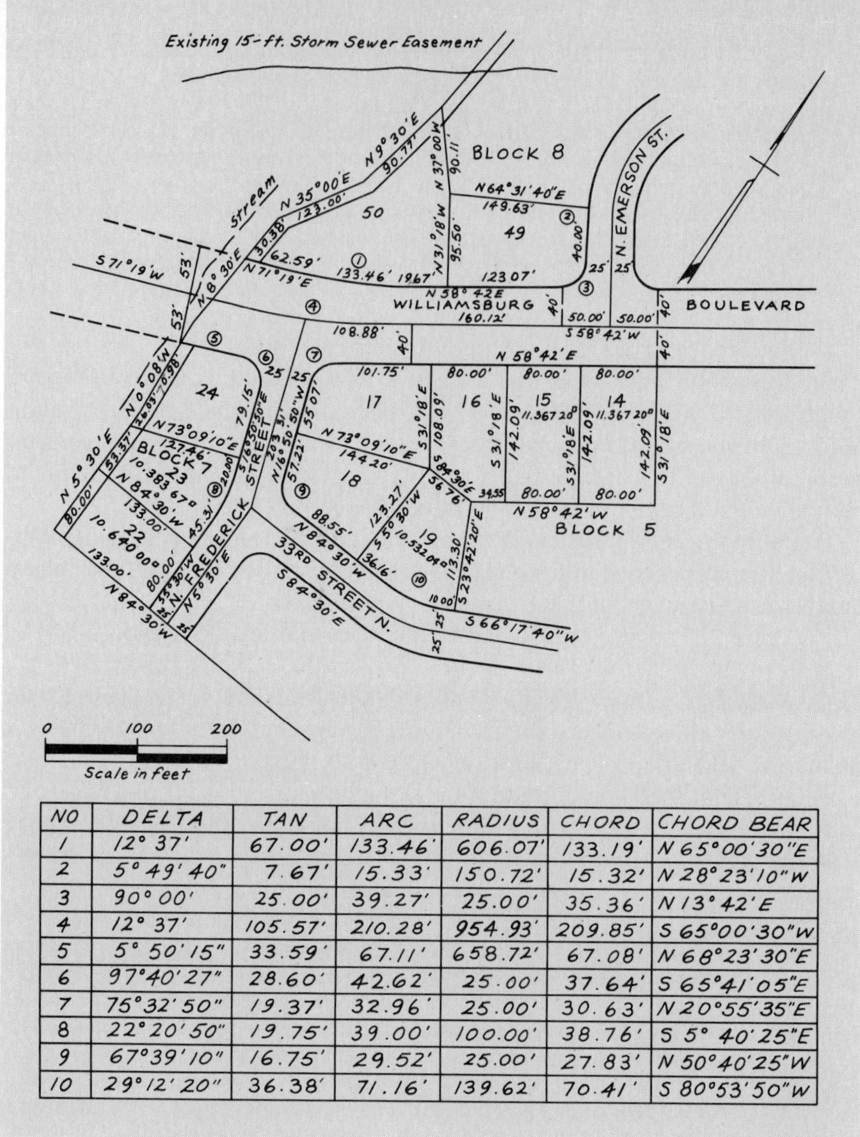

NO	DELTA	TAN	ARC	RADIUS	CHORD	CHORD BEAR
1	12° 37′	67.00′	133.46′	606.07′	133.19′	N 65°00′30″E
2	5° 49′ 40″	7.67′	15.33′	150.72′	15.32′	N 28°23′10″W
3	90° 00′	25.00′	39.27′	25.00′	35.36′	N 13° 42′ E
4	12° 37′	105.57′	210.28′	954.93′	209.85′	S 65°00′30″W
5	5° 50′ 15″	33.59′	67.11′	658.72′	67.08′	N 68°23′30″E
6	97°40′27″	28.60′	42.62′	25.00′	37.64′	S 65°41′05″E
7	75°32′50″	19.37′	32.96′	25.00′	30.63′	N 20°55′35″E
8	22°20′50″	19.75′	39.00′	100.00′	38.76′	S 5° 40′25″E
9	67°39′10″	16.75′	29.52′	25.00′	27.83′	N 50°40′25″W
10	29°12′20″	36.38′	71.16′	139.62′	70.41′	S 80°53′50″W

Figure 22-2. A small subdivision plat.

22-6. SUBDIVISIONS. A real estate *subdivision* is an unimproved tract of land, surveyed and divided into lots for sale purposes. It may be synonymous in some localities with a *development*, which implies improvements are made before sales. Any of the four description methods listed in Section 22-1 can be applied to the parcels in some states.

A small block-and-lot subdivision is shown in Figure 22-2. (Some lot areas have been deleted so their calculations can become end-of-chapter problems.) Large electronic computers with appropriate available programs greatly reduce

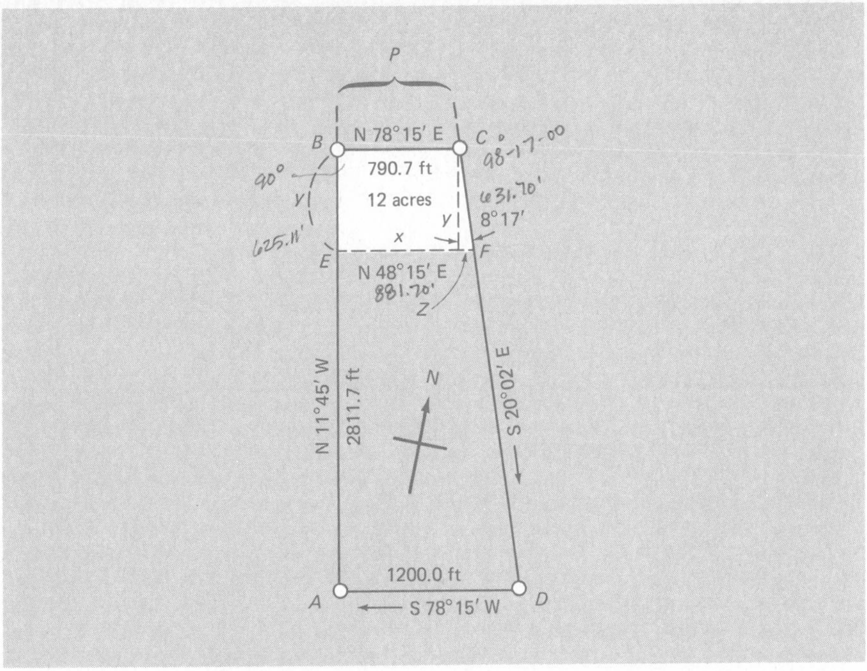

Figure 22-3. Land partitioning.

the labor of computing lot sizes and areas in subdivisions with curved streets. Automatic plotters using the printout results make platting accurate, simple, and fast.

Critical subdivision design and layout considerations include creation of good building sites, an efficient street layout, and assured drainage.

22-7. PARTITIONING LAND. A common problem in property surveys is partitioning land into two or more pieces for sale or distribution to family members, heirs, and so on. A boundary survey is run; latitudes and departures computed; the traverse balanced; and total enclosed area calculated. Some parcel shapes and requirements permit formula solutions, often by using analytical geometry. Others require trial-and-error methods.

EXAMPLE 22-1

Figure 22-3 illustrates cutting off 12 acres by means of a line *EF* parallel to base *AD*. The distances in this old farm deed were given and balanced only to the nearest 0.1 ft and thus, without exceeding the number of significant figures in the data, a "more or less" area of 12 acres = 522,720 ft² is required.

SOLUTION

$$\text{Required area} = 522{,}720 = xy + \frac{y^2 \tan 8^\circ 17'}{2} = 790.7y + \frac{y^2(0.1455872)}{2}$$

then

$$0.0727936y^2 + 790.7y - 522{,}720 = 0$$

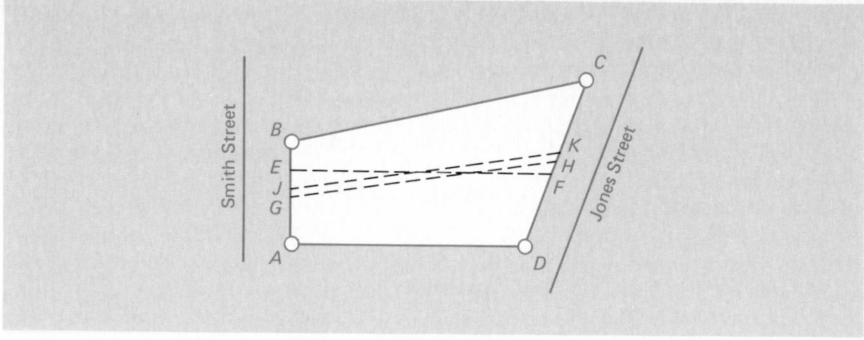

Figure 22-4. The "southerly half."

and

$$y = 625.1 \text{ ft}$$

$$z = y \tan 8°17' = 91.0 \text{ ft}$$

$$\text{base } EF = 790.7 + 91.0 = 881.7 \text{ ft}$$

$$CF = \frac{625.1}{\cos 8°17'} = 631.7 \text{ ft}$$

$$\text{area } EBCF = 625.1\left(\frac{790.7 + 881.7}{2}\right) = 522{,}700 \text{ ft}^2 \quad \text{(check)}$$

(With lengths of sides measured to only tenths of a foot, a computed area to more than four significant figures is not justified.)

Another approach is to extend AB and DC to meet at a point P, then use another triangle EPF in which simple proportions are readily visualized.

Cutoff lines to separate a certain area from the parcel may have (1) a specific starting point (distance from one corner of the tract polygon and run to the midpoint, or any other location on the opposite side); or (2) a required direction (parallel with, perpendicular to, or on a designated bearing angle from a selected line). These cases can often be handled by trial-and-error solutions involving an initial assumption such as the cutoff line direction or starting point. Certain problems are amenable to solution using coordinate formulas for the intersection of two lines (see Appendix B).

Figure 22-4 shows that a statement "the southerly half" of tract $ABCD$ can have a number of meanings—the *most* southerly half, half the frontage, or half the actual acreage. An important consideration is the final shape of each lot. Connecting midpoints G and H leaves the southerly "half" smaller than the northerly "half," but provides equal frontage for both parts on the two streets. Course EF, parallel with AD, produces one trapezoidal lot but a poorly shaped northerly parcel with meager frontage on Smith Street. The intent of the deed should therefore be clearly stated.

Before field work, a surveyor must spend considerable time searching courthouse records for descriptions, adjoiners, easements, and other pertinent facts.

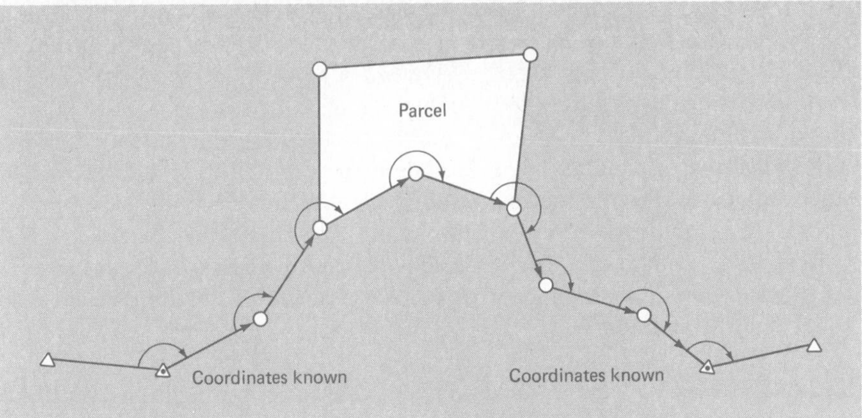

Figure 22-5. Transfer of coordinates to a parcel by traverse.

22-8. FIELD WORK. The first task is to locate all property corners. Here a most valuable piece of equipment to a land surveyor—a shovel—frequently comes into use along with a magnetic locater (particularly if magnetized metal survey caps have been set). A flux-gate-type metal detector, rather than the loop model all-metal detector, is now available. It is an important cost-saving tool for land surveyors.

In many cases, one or more lines must be run from control points some distance away to check or establish the location of a corner. If control monuments are available with a known direction and coordinates on a state or local system (Figure 22-5), occupying one of them, orienting the transit or theodolite, and running a traverse to a property corner will provide directions and coordinates for a boundary survey. A check is secured by closing the traverse back on the starting station or on a second nearby monument.

Generally a closed traverse is run around the property, with all corners being occupied if possible. Fences, trees, shrubbery, hedges, common (party) walls, and other obstacles may necessitate a traverse inside or outside the property. From stub (side shot) measurements to the corners, their coordinates, side lengths, and bearings are calculated (see Sections 13-9 and 13-10). All measurements should be made with a precision suited to the specifications and land values. If coordinates on a grid system are used in the description, the order of accuracy of the system itself should be maintained if practical.

Measurements made with an EDMI or standardized tape corrected for temperature and tension may not agree with distances on record or those between the marks. This situation provides a real test for surveyors. Perhaps the tape has a different length from that employed on the original survey; the marks may have been disturbed or are the wrong ones; monuments may not check with others in the vicinity believed or known to be correct; and perhaps several previous surveys disagree with each other.

Bearings and angles between adjacent sides may not fit those called for. The discrepancy could be due to a faulty original compass survey, incorrect corner marks, or other causes.

A surveyor's responsibility is to relocate or establish boundary markers in the exact position where they were originally set. If part of a surveyed area might be subject to claims, such as possession by other persons, the findings must be specifically noted on the surveyor's map.

If differences exist between distances of record and measured values for found monuments, the data should be used to get a "calibrated tape" length which relates to the original surveyed distances. In relocating monuments, calibration is employed to lay off the actual distances noted by the original surveyor. The method can vary depending on conveyances subsequent in time (the older has dominant boundary rights) or conveyances simultaneous in time (equal rights at boundary).

22-9. REGISTRATION OF TITLE. To remedy difficulties arising from inaccurate descriptions and disputed boundary claims, some states provide for registration of property titles under rigid rules. The usual requirements include marking each corner with standardized monuments referenced to established points, and recording a plat drawn to scale and containing specified items. Titles are then guaranteed by the court under certain conditions.

A number of states have followed Massachusetts's example and maintain separate *land courts* dealing exclusively with land titles. As the practice spreads, the accuracy of property surveys will be increased and transfer of property simplified.

A comparable service is offered by *title insurance companies*, which search, assemble, and interpret offical records, laws, and court decisions affecting ownership of land. Title companies then will insure purchasers against loss by guaranteeing that its findings regarding defects, liens, encumbrances, restrictions, assessments, and easements are correct. Defense against lawsuits is provided by the company against threats to a clear title from claims shown in public records and not exempted in the policy. Location of corners and lines is not guaranteed; hence, it is necessary to establish on the ground the exact boundaries called for by the deed and title policy. Close cooperation between surveyors and title insurance companies is necessary to prevent later problems for their clients.

Many technical and legal problems are considered before title insurance is granted. In some states, title companies refuse to issue a policy covering a lot if fences in place are not on the property line and exclude from the contract "all items that would be disclosed by a proper survey." Occupation and use of land belonging to a neighbor but outside his apparent boundary line as defined by a fence may lead to a claim of *adverse possession.*

Adverse rights are obtained against all except the public by occupying a parcel of land for a period of years specified by law and performing certain acts. Possession must be (a) actual, (b) exclusive, (c) open and notorious, (d) hostile, and (e) continuous. It may also be necessary for the property to be held under *color of title* (a claim to a parcel of real property based on some written instrument, though a defective one). In some states all taxes must be paid. The time required to establish a claim of adverse possession varies from a minimum of 7 years in Florida to a maximum of 60 years for urban property in New York. The customary period is 20 years. In Wisconsin it is 10 years under color of title, 20 years without.

Continuous use of a street, driveway, or footpath by an individual or the general public for a specified number of years results in establishment of a right-of-way privilege which cannot be withheld by the original owner.

An *easement* is a right, by grant or agreement, which allows a person or persons to use the land of another for a specific purpose. It always implies an interest in the land upon which it is imposed. Black's *Law Dictionary* lists and defines 18 types of easements; hence, the exact purpose of an easement should be clearly stated.

Discussion of property surveys has necessarily been condensed in this text but provides helpful information to readers while deterring inexperienced people from attempting to run boundary lines. For more extensive coverage, references are listed in the Bibliography.

22-10. SOURCES OF ERROR. Some sources of error in boundary surveys follow:

1. Corner not defined by a unique point.
2. Unequal precision of angles and distances.
3. Length measurements not made with a standardized tape or corrected for temperature, tension, and slope.

22-11. MISTAKES. Some typical mistakes in connection with boundary surveys are:

1. Use of wrong corner marks.
2. Failure to check deeds of adjacent property as well as the description of the parcel in question.
3. Ambiguous deed descriptions.
4. Omission of the length or bearing of the closing line.
5. Failure to close on known control.
6. Magnetic bearings not properly corrected to the date of the new survey.

PROBLEMS

22-1. From publications listed in the Bibliography, find the meaning of prescription, color of title, eminent domain, and escrow.
22-2. Define covenant, appurtenance, balk, and proration.
22-3. Approximately what year was a steel "ribbon" tape first used as a replacement for the Gunter's chain?
22-4. What must surveyors do in interpreting obscure descriptions?
22-5. Determine from the county courthouse or other local records the types of property descriptions used in your area.
22-6. Write a metes-and-bounds description for the house and lot where you live. Assume any necessary data. Sketch a map of the property.
22-7. Outline the size of strip and form of easement used by your local telephone and utilities companies. Submit an example.
22-8. Prepare a copy of a typical recital from a local property deed description.

22-9. A surveyor is employed to map a farm located in an unfamiliar area. What is the first step to be taken?

22-10. In your city's residential section, what setbacks are required from property lines in the front, rear, and on the sides?

22-11. In a description by metes and bounds, what purpose may be served by the phrase "more or less" following the acreage?

22-12. Plot the metes-and-bounds description of the city lot described in Section 22-3. Compute the lot's misclosure. Is the accuracy satisfactory for an average town of 2000 people?

22-13. Which method of property description do you consider best? Why?

22-14. Determine how the assignment of block, tract, subdivision, and lot numbers is made, and by whom, in your area.

22-15. In your town, can lots in a subdivision be designated by letters A, B, C, D, and so on, instead of numbers?

22-16. List in their order of importance the following items in deed descriptions: written and measured lengths; written and measured bearings; monuments in place or lost; witness corners and ties; areas; and testimony of witnesses. Justify your answer.

22-17. Sketch the property described as part of lots 9 and 10 in Section 22-4.

22-18. Plat the portion of lot 306 described in Section 22-4.

22-19. List all types of pertinent information or data that should appear on a completed property survey plat.

22-20. Secure a sample or copy the important items included in a title insurance company policy for a city lot.

22-21. Telephone poles are set on two corners of a rectangular lot. Explain how to survey the lot, locate boundaries, and find the area.

22-22. Two disputing neighbors employ a surveyor to check their boundary line. Discuss the surveyor's authority if (a) the line established is agreeable to both clients, and (b) the line is not accepted by one or both of them.

22-23. What time period is required to claim adverse rights in your area?

22-24. State the reason for a "hostile" requirement in acquiring adverse rights to real property.

22-25. A mistake was made on a subdivision plat and duly recorded. How can this defect be remedied and by whom?

22-26. Five items are required in a deed description. Which do you consider the most important? The least important? Why?

22-27. What is the main advantage of describing property by coordinate values for the corners?

22-28. Find two obvious mistakes in the plat in Figure 22-2.

22-29. Compute the area of lot 16 in Figure 22-2.

22-30. Calculate the area of lot 17 in Figure 22-2.

22-31. Compute the area of lot 18 in Figure 22-2.

22-32. Check the area of lot 19 in Figure 22-2.

22-33. Calculate the area of lot 24 in Figure 22-2.

22-34. Using a line perpendicular to AB through x, cut off one-third of the area to include corner B.

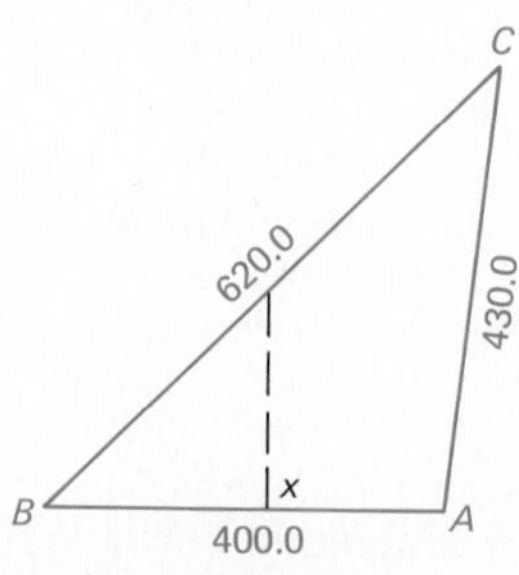

Problem 22-34

22-35. Side EF of lot $EFGH$ runs along a street. Compute the length of line GH parallel to EF that will cut off 18,000 ft^2, and lengths EH and FG.

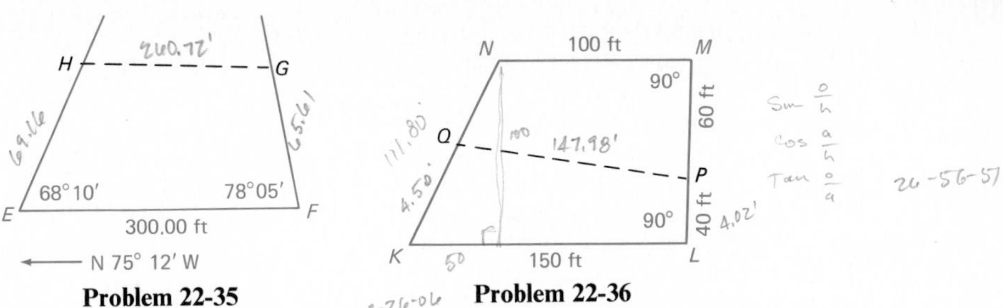

Problem 22-35

Problem 22-36

22-36. Determine the length and direction of line PQ that will cut off 600 ft^2 for a new lot $KLPQK$, and the length of KQ.

22-37. Is a surveyor who delivered an inaccurate survey to a builder liable to a third party (lot purchaser) for misrepresentation?

22-38. Is a recorded deed of a tract of land proof of ownership by the grantee named therein? Explain.

22-39. Can a railroad right-of-way survey crew be held liable for trespassing and crop damage to an abutting owner?

22-40. An intersection development leaves an odd-sized parcel described as follows: AB N28°04′20″W, 135.86 ft along Oakland Street; BC N76°55′31″E, 211.80 ft; CD S53°57′31″W, 88.02 ft; and DA S31°22′21″W, 136.11 ft. Compute the area.

22-41. How long after performing and recording a property survey in your state is a surveyor liable for damages due to mistakes in the work?

22-42. What does a surveyor do when the surveyed lines do not agree with the "use" or "occupied" lines?

22-43. A city block survey, starting and ending on proven corner monuments, determines the actual length to be 301.00 ft although the recorded length is 300.00 ft. There are six lots in the block. What should be done about the 1.00 ft excess if (a) the lot boundaries were never staked, and (b) four lots at one end have been staked but never occupied?

BIBLIOGRAPHY

Bauer, K. W. 1979. "Drainage and Subdivision of Land. "*ACSM Bulletin* 67:9.

Betts, D. N. 1979. "Certificates of Survey—What Are They?" *Surveying and Mapping* 39(no. 3):239.

Black, Henry C. 1979. Black's Law Dictionary, 5th ed. West Co.

Brown, C. M. 1969. *Boundary Control and Legal Principles*, 2nd ed. New York: Wiley.

Brown, C. M., W. G. Robillard, and D. A. Wilson, 1981. *Evidence and Procedures for Boundary Location*, 2nd ed. New York: Wiley.

Buckner, R. B. 1975. "Reasons and Methods for Accurate Direction in Land Surveys." *Surveying and Mapping* 35(no. 4):305.

Fant, J. E., A. R. Freeman, and C. Madson. 1981. "Metes and Bounds Descriptions." *Surveying and Mapping* 41(no. 2):222.

Greulich, G. 1982. "Metric Standards in Subdivision and Zoning." *Surveying and Mapping* 42(no. 3):257.

Horak, T. E. 1981. "Professional Dialogue." *Surveying and Mapping* 41(no. 4):438.

Kratz, K. E. 1981. "The Land Surveyor's Quasi-judicial Rights and Duties Regarding Boundary Location." *Surveying and Mapping* 41(no. 2):213.

Lampert, L. L. 1980. "The Retracement." *Surveying and Mapping* 40(no. 3):315.

Lowman, R. E. 1974. "Writing Deed Descriptions." *Surveying and Mapping* 34(no. 1):41.

Madson, C. M. 1980. "Over, Under and Across." *Surveying and Mapping* 40(no. 1):47.

Madson, T. S. 1978. "Practical Location and the Land Surveyor." *Proceedings of the Fall Meeting of the American Congress of Surveying and Mapping*, p. 247.

McCall, C. E. 1980. "Subdivision of Land Today." *Surveying and Mapping* 40(no. 4):415.

McEntyre, J. G. 1978. *Land Survey Systems.* New York: Wiley.

Robillard, W. G. 1974. "The Land Surveyor and the Law on Modernization." *American Congress of Surveying and Mapping Bulletin* 46:11.

U.S. Department of the Interior, Bureau of Land Management. 1973. *Manual of Surveying Instructions 1973.* Washington, D.C.: U.S. Government Printing Office.

Wattles, G. H. 1976. *Writing Legal Descriptions.* New York: Parker.

Zeman, R. L. 1971. "Improvement of Land Titles." *ASCE Journal of Surveying and Mapping Division* 97(no. SU1):113.

23 SURVEYS OF THE PUBLIC LANDS

23-1. INTRODUCTION. The term *public lands* is applied broadly to the areas which have been subject to administration, survey, and transfer of title under the public-land laws of the United States since 1785. These lands include those turned over to the federal government by the colonial states and the larger areas acquired by purchase from (or treaty with) the native Indians or foreign powers that had previously exercised sovereignty.

Thirty states, including Alaska, constitute the *public domain*, which has been, or will be, subdivided into rectangular tracts (see Figure 23-1). The area represents approximately 72% of the United States.

Title to the vacant lands, and therefore direction over the surveys within their own boundaries, was retained by the colonial states, the other New England and Atlantic coast states (except Florida), and later by the states of West Virginia, Kentucky, Tennessee, and Texas. In these areas the U.S. public-land laws have not been applicable.

The beds of navigable bodies of water are not public domain and not subject to survey and disposal by the United States. Sovereignty is in the individual states.

Survey and disposition of the public lands were governed originally by two factors:

1. A recognition of the value of grid-system subdivision based on experience in the colonies and another large-scale systematic boundary survey—the 1656 Down Survey in Ireland.

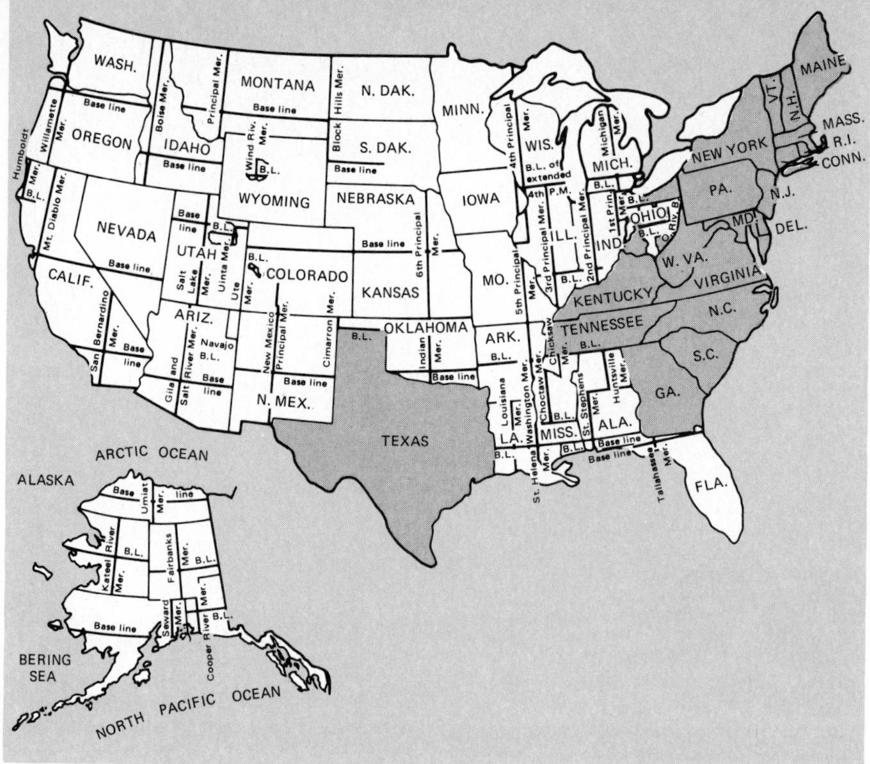

Figure 23-1. Areas covered by the public-lands surveys. Areas excluded are shaded. (Texas has a similar rectangular system.)

2. The need of the colonies for revenue from sale of the public lands. Monetary returns from their disposal were disappointing, but the planners' far-sighted vision of a grid system of subdivision deserves commendation.

23-2. INSTRUCTIONS FOR SURVEY OF THE PUBLIC LANDS. The U.S. system of public-lands surveys was inaugurated in 1784, with the territory northwest of the Ohio River as a test area. Sets of instructions for the surveys were issued in 1785 and 1796. Manuals of instructions were issued in 1855, 1881,1890, 1894, 1902, 1930, 1947, and 1973.

In 1796 a surveyor general was appointed, and the numbering of sections was changed to the system now in use (see Figure 23-7). Supplementary rules were promulgated by each local surveyor general "according to the dictates of his own judgment" until 1836, when the General Land Office was reorganized. Copies of changes and instructions for local use were not always preserved and sent to Washington. As a result, no office in the United States has a complete set of instructions under which the original surveys were supposed to have been made. Although the same general method of subdivision was followed, detailed procedures were altered in surveys made at different times in various areas of the country. As an example, instructions for New Mexico said only township lines were to be run where the land was deemed unfit for cultivation.

Most of the later public-lands surveys have been run by the procedures to be described, or variations of them. The task of present-day surveyors consists primarily of retracing the original lines and perhaps further subdividing sections. To do so, they must be thoroughly familiar with the rules, laws, equipment, and field conditions that governed their predecessors in a given area.

Basically, the rules of survey stated in the 1973 *Manual of Surveying Instructions* are as follows:[1]

> The public lands shall be divided by north and south lines run according to the true meridian, and by others crossing them at right angles, so as to form townships six miles square. . . .
>
> The corners of the townships must be marked with progressive numbers from the beginning; each distance of a mile between such corners must be also distinctly marked with marks different from those of the corners.
>
> The township shall be subdivided into sections, containing as nearly as may be, six hundred and forty acres each, by running parallel lines through the same from east to west and from south to north at the distance of one mile from each other (originally at the end of every two miles but amended in 1800), and marking corners at the distance of each half mile. The sections shall be numbered, respectively, beginning with the number one in the northeast section, and proceeding west and east alternately through the township with progressive numbers until the thirty-six be completed.

Additional rules of survey covering field books, subdivision of sections, adjustment for excess and deficiency, and other matters are given in the manuals. Private surveyors, on a contract basis, were paid $2/mi of line run until 1796 and $3/mi thereafter. Sometimes the amount was adjusted in accordance with the importance of a line, the terrain, location, and other factors. From this meager fee surveyors had to pay and feed a party of at least four while on the job and in transit to and from distant points. They had to brush out and blaze (mark trees by scarring the bark) the line, set corners and other marks, and provide satisfactory notes and one or more copies of completed plats. The contract system was completely discarded in 1910. Public-lands surveyors are now appointed.

Since meridians converge, it is evident the requirements that "lines shall conform to the true meridians and townships shall be 6 mi square" are mathematically impossible. An elaborate system of subdivision was therefore worked out as a practical solution.

Two principles furnished the legal background for stabilizing land lines:

1. Boundaries of public lands established and returned by duly appointed surveyors are unchangeable.
2. Original township and section corners established by surveyors must stand as the true corners which they were intended to represent, whether in the place shown by the field notes or not.

Expressed differently, the original surveyors had an official plan with detailed instructions for its layout, and presumably set corners to the best of

[1] Since new surveys (particularly in Alaska) as well as retracements are still being carried out, the 1973 manual employs more than one tense, and this chapter does also.

their ability. After title passed from the United States, their established corners (monuments), regardless of errors, became the lawful ones. Therefore, if monuments have disappeared, the purpose of resurveys is to determine where they were, not where they should have been. Correcting mistakes or errors now would disrupt too many accepted property lines and result in an unmanageable number of lawsuits.

In general, the procedure in surveying the public lands provides for the following subdivisions:

1. Division into quadrangles (tracts) approximately 24 mi on a side (after about 1840).
2. Division of tracts (quadrangles) into townships (16), approximately 6 mi on a side.
3. Division of townships into sections (36), approximately 1 mi square.
4. Subdivision of sections (usually by the local surveyor).

It will be helpful to keep in mind that the purpose of the grid system was to obtain sections 1 mi on a side. To this end, surveys proceeded from south to north and east to west, and all discrepancies were thrown into the sections bordering the north and west township boundaries to get as many *regular sections* as possible.

23-3. INITIAL POINT. Thomas Jefferson recognized the importance of surveys and served as chairman of a committee to develop a plan for locating and selling the western lands. His report to the Continental Congress in1784, adopted as an ordinance on May 20, 1785, called for survey lines to be run and marked before land sales. Many of today's property disputes would have been eliminated if all property lines were resurveyed and monuments checked and/or set before sales became final!

Subdivision of the public lands became necessary in any area as settlers moved in and mining or other land claims were filed. The early hope that surveys would precede settlement was not fulfilled.

In each area an initial point was established and located by astronomical observations. The manual of 1902 was the first to specify an indestructible monument, preferably a copper bolt, firmly set in a rock ledge if possible and witnessed by rock bearings. Thirty-seven initial points are available, five of them in Alaska. A principal meridian and a base line were passed through each initial point, such as the one in the center of Figure 23-2.

23-4. PRINCIPAL MERIDIAN. From each initial point, a true north-south line called a *principal meridian* (Prin. Mer. or PM) was run north and/or south to the limits of the area to be covered. Generally a solar attachment—a device for solving mechanically the mathematics of the astronomical triangle—was used. Monuments were set for section and quarter-section corners every 40 ch and at the intersections with all meanderable bodies of water (streams 3 ch or more in width, and lakes covering 25 acres or more).

The line was supposed to be within 3 min of the cardinal direction. Two independent sets of linear measurements were required to check within

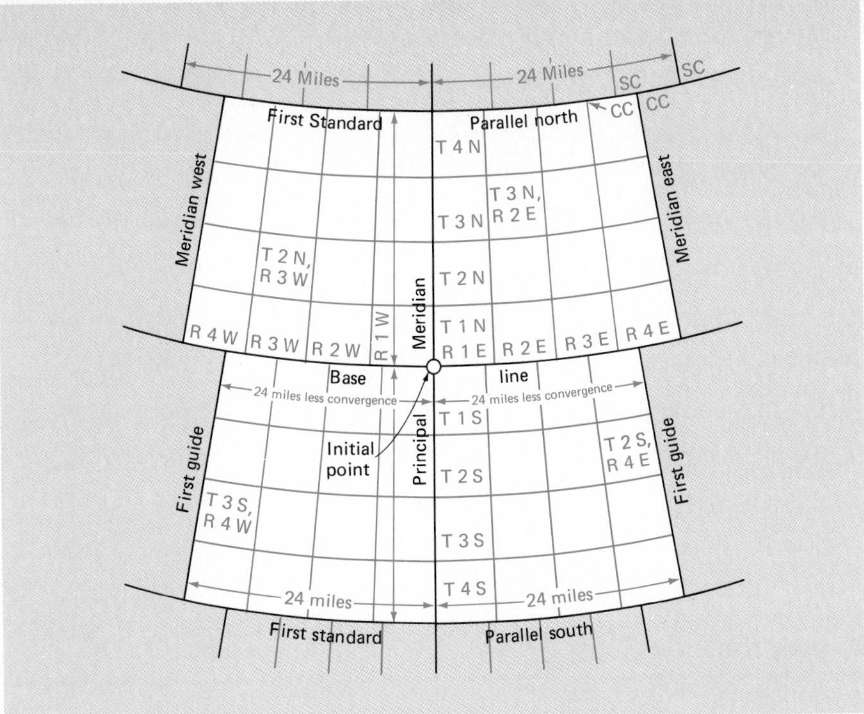

Figure 23-2. Survey of quadrangles. (Only a few of the standard corners and closing corners are identified.)

20 lk (13.2 ft)/80 ch, which corresponds to a precision ratio of only $\frac{1}{400}$. The allowable difference between sets of measurements is now limited to 7 lk/80 ch.

Areas within a PM system vary greatly as depicted in Figure 23-1.

23-5. BASE LINE. From the initial point, the base line was extended east and/or west as a true parallel of latitude to the limits of the area to be covered. As required on the principal meridian, monuments were set for section and quarter-section corners every 40 ch and at the intersections with all meanderable bodies of water. Permissible closures were the same as those for the principal meridian.

Base lines were run as circular curves with chords of 40 ch by the (a) solar method, (b) tangent method, or (c) secant method.

Solar Method. An observation is made with a solar attachment to determine the direction of true north. A right angle is then turned off and a line extended 40 ch, where the process is repeated. The series of lines so established, with a slight change in direction every half-mile, closely approaches a true parallel. Obviously if the sun is obscured, the method cannot be used.

Tangent Method. This method of laying out a true parallel is illustrated in Figure 23-3. A 90° angle is turned to the east or the west, as may be required from a true meridian, and corners set every 40 ch. At the same time, proper offsets are taken from tables and measured north from the tangent to the parallel. In the example shown, the offsets in links are 1, 2, 4, $6\frac{1}{2}$, . . . , 37. The error

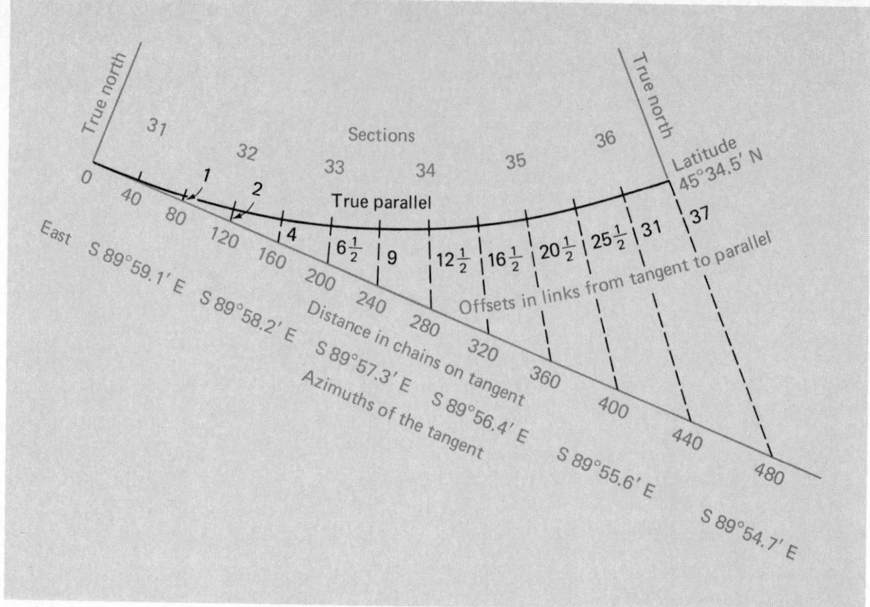

Figure 23-3. Layout of parallel by tangent method. (Adapted from 1973 *Manual of Surveying Instructions.*)

resulting from taking right-angle offsets instead of offsets along the converging lines is negligible. The main objection to the tangent method is that the parallel departs considerably from the tangent, so both the tangent and parallel must be brushed out.

Secant Method. This method of laying out a true parallel is shown in Figure 23-4. It actually is a modification of the tangent method in which a line parallel to the tangent at the 3-mi (center) point is passed through the 1- and 5-mi points to produce minimum offsets, as shown in Tables E-3 and E-4.

Field work includes establishing a point on the true meridian, south of the beginning corner, at a distance taken from a table for the latitude of a desired parallel. The proper bearing angle from the same table is turned to the east or west from the true meridian to define the secant, which is then projected 6 mi. Offsets are measured north or south from the secant to the parallel.

Advantages of the secant method are its simplicity, the offsets are small and can be measured perpendicular to the secant without appreciable error, and the amount of clearing is reduced.

23-6. STANDARD PARALLELS (CORRECTION LINES). After the principal meridian and base line have been run, standard parallels (Stan. Par. or SP), also called correction lines, are run as true parallels of latitude 24 mi apart in the same manner as was the base line. All 40-ch corners are marked. In some of the early surveys, standard parallels were placed at intervals of 30 or 36 mi.

Standard parallels are numbered consecutively north and south of the base line; examples are First Standard Parallel North and Third Standard Parallel South.

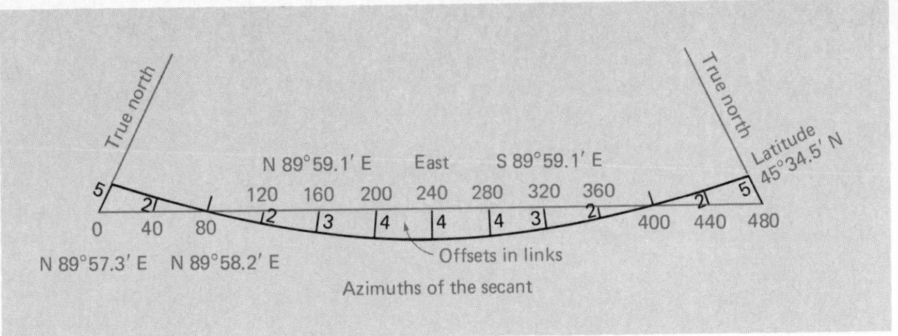

Figure 23-4. Layout of parallel by secant method. (Adapted from 1973 Manual of Surveying Instructions.)

23-7. GUIDE MERIDIANS. Guide meridians (GM) are run due north from the base line and standard parallels at intervals of 24 mi east and west of the principal meridian, in the same manner as was the principal meridian, and with the same limits of error. Before work is started, the chain or tape must be checked by measuring 1 mi on the base line or standard parallel. All 40-ch corners are marked.

Because meridians converge, a *closing corner* (CC) is set at the intersection of each guide meridian, standard parallel, or base line (see Figure 23-2). The distance from the closing corner to the *standard corner* (SC), which was set when the parallel was run, is measured and recorded in the notes as a check. Any error in the 24-mi-long guide meridian is put in the northernmost half-mile.

Guide meridians are numbered consecutively east and west of the principal meridian; examples are First Guide Meridian West and Fourth Guide Meridian East.

23-8. TOWNSHIP EXTERIORS, MERIDIONAL (RANGE) LINES, AND LATITUDINAL (TOWNSHIP) LINES. Division of a quadrangle, or tract, into townships is accomplished by running range (R) and township (T or Tp) lines.

Range lines are true meridians through the standard township corners previously established at intervals of 6 mi on the base line and standard parallels. They are extended north to intersect the next standard parallel or base line, and closing corners set (see Figures 23-2 and 23-5).

Formulas for convergence of meridians (derived in various texts on geodesy, with results given in Table E-2) are as follows:

$$\theta = 52.13d \tan \phi \tag{23-1}$$

$$c = \tfrac{4}{3}Ld \tan \phi \qquad \text{(slight approximation)} \tag{23-2}$$

where θ is the angle of convergence, in seconds; d the distance between meridians, in miles, on a parallel; ϕ the mean latitude; c the linear convergence, in feet; and L the length of meridians, in miles.

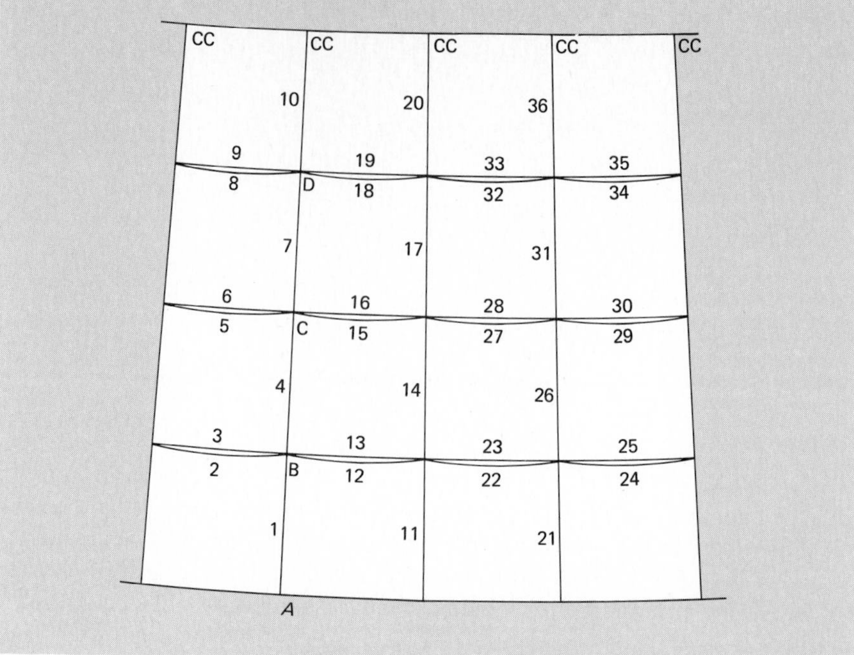

Figure 23-5. Order of running lines for the subdivision of a quadrangle into townships.

Township lines connect township corners previously established at intervals of 6 mi on the principal meridian, guide meridians, and range lines.

23-9. DESIGNATION OF TOWNSHIPS. A township is identified by a unique description based on the principal meridian governing it.

North and south rows of townships are called *ranges*, and numbered in consecutive order east and west of the principal meridian as indicated in Figure 23-2.

East and west rows of townships are named *tiers* and numbered in order north and south of the base line. By common practice, the term *tier* is usually replaced by township in designating the rows.

An individual township is identified by its number north or south of the base line, followed by the number east or west of the principal meridian. An example is Township 7 South, Range 19 East, of the Sixth Principal Meridian. Abbreviated, this becomes T 7 S, R 19 E, 6th PM.

23-10. SUBDIVISION OF A QUADRANGLE INTO TOWNSHIPS. The method to be used in subdividing a quadrangle into townships is fixed by regulations in the *Manual of Surveying Instructions*. Under the old regulations, township boundaries were required to be within 21 min of the cardinal direction. Later this was reduced to 14 min to keep interior lines within 21 min of the cardinal direction.

The detailed procedure for subdividing a quadrangle into townships can best be described as a series of steps designed to ultimately produce the maxi-

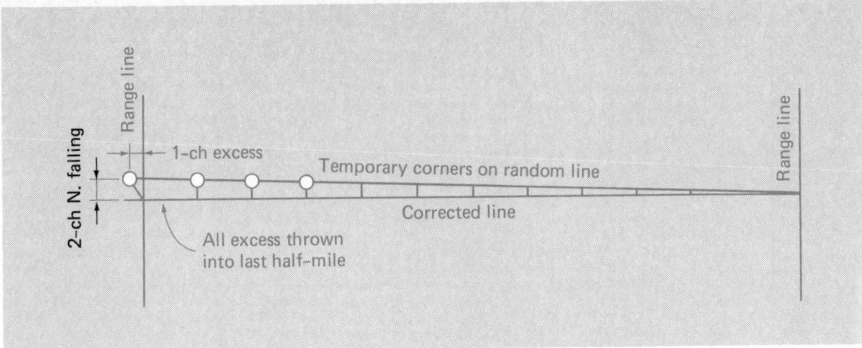

Figure 23-6. Double correction of random line for excess and falling.

mum number of regular sections with a minimum amount of unproductive travel by the field party. The order of running the lines is shown by consecutive numbers in Figure 23-5. Some details are described in the following steps:

1. Begin at the southeast corner of the southwest township, point *A*, after checking the chain or tape against a 1-mi measurement on the standard parallel.
2. Run north on the true meridian for 6 mi (line 1), setting alternate section and quarter-section corners every 40 ch. Set township corner *B*.
3. From *B*, run a random line (line 2) due west to intersect the principal meridian. Set temporary corners every 40 ch.
4. If the random line has an excess or deficiency of 3 ch or less (allowing for convergence), and a falling north or south of 3 ch or less, the line is accepted. It is then corrected back (line 3) and all corners are set in their proper positions. Any excess or deficiency is thrown into the most westerly half-mile. The method of correcting a random line having an excess of 1 ch and a north falling of 2 ch is shown in Figure 23-6.
5. If the random line misses the corner by more than the permissible 3 ch, all four sides of the township must be retraced.
6. The same procedure is followed until the southeast corner, *D*, of the most northerly township is reached. From *D*, range line 10 is continued as a true meridian to intersect the standard parallel or base line, where a closing corner is set. All excess or deficiency in the 24 mi is thrown into the most northerly half-mile.
7. The second and third ranges of townships are run the same way, beginning at the south line of the quadrangle.
8. While the third range is being run, random lines are also projected to the east and corrected back, and any excess or deficiency is thrown into the most westerly half-mile. (All points may have to be moved diagonally to the corrected line, instead of just the last point as in Figure 23-6.)

23-11. SUBDIVISION OF A TOWNSHIP INTO SECTIONS. Sections are now numbered from 1 to 36, beginning in the northeast corner of a township and ending in the southeast corner, as shown in Figure 23-7. The method used to

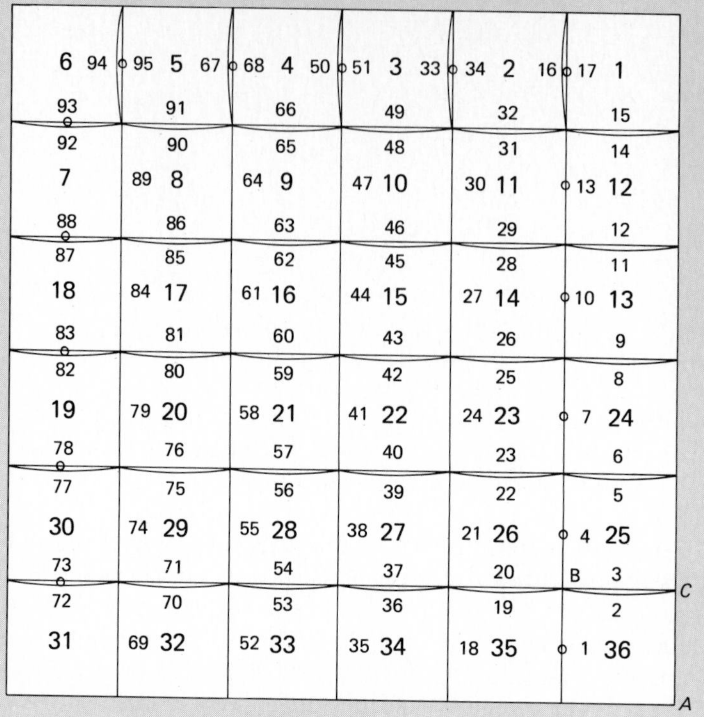

Figure 23-7. Order of running lines for the subdivision of a township into sections.

subdivide a township can be described most readily as a series of steps to produce the maximum number of regular sections 1 mi on a side. Lines were run in the following order:

1. Set up at the southeast corner of the township, point *A*, and observe the meridian. Retrace the range line northward and township line westward for 1 mi to compare the meridian, needle readings, and taped distances with those recorded.
2. From the southwest corner of section 36, run north *parallel* with the east boundary of the township. Set quarter-section and section corners on line 1 (Figure 23-7).
3. From the section corner just set, run a random line *parallel* with the south boundary of the township eastward to the range line. Set a temporary quarter-corner at 40 ch.
4. If the 80-ch distance on the random line is within 50 lk, falling or distance, the line is accepted. The correct line is calculated and a quarter-corner located at the *midpoint* of line *BC* connecting the previously established corner *C* and the new section corner *B*.
5. If the random line misses the corner by more than the permissible 50 lk, the township lines must be rechecked and the cause of the error determined.

6. The east range of sections is run in a similar manner until the southwest corner of section 1 is reached. From this point a random line is run northward to connect with the north township line section corner. A quarter-corner is set 40 ch from the south section corner (on line 17, corrected back by later manuals). All discrepancies in the 6 mi are thrown into the last half-mile.
7. Successive ranges of sections across the township are run until the first four have been completed. All north-south lines are parallel with the township east side. All east-west lines are run randomly parallel with the south boundary line and then corrected back (so parallelism may be destroyed).
8. When the fifth range is being run, random lines are projected to the west as well as to the east. Quarter-corners in the west range are set 40 ch from the east side of the section, with all excess or deficiency resulting from the errors and convergence being thrown into the most westerly half-mile.
9. If the north side of the township is a standard parallel, instead of running a random line to the north, lines parallel with the east township boundary are projected to the correction line and closing corners set. The distance to the nearest corner is measured and recorded.
10. True bearings of interior north-south section lines for any latitude can be obtained by applying corrections from tables for the convergence at a given distance from the east boundary.

By throwing the effect of meridian convergence into the westernmost half-mile of the township and all errors to the north and west, 25 regular sections nominally 1 mi^2 are obtained. Also, the south half of sections 1, 2, 3, 4, and 5; the east half of sections 7, 18, 19, 30, and 31; and the southeast quarter of section 6 are normal size.

23-12. SUBDIVISION OF SECTIONS. A section was the basic unit of the General Land Office system, but land was often patented in parcels smaller than a section. Subdivision of sections was performed by local surveyors and others as the owner took up the land. The BLM provides guidelines on the proper and intended way a section should be subdivided.

To divide a section into quarter-sections, straight lines are run between opposite quarter-section corners previously established or reestablished. This rule holds whether or not the quarter-section corners are equidistant from the adjacent section corners.

To divide a quarter-section into quarter-quarter-sections, straight lines are run between opposite quarter-quarter-section corners established at the midpoints of the four sides. The same procedure is followed to obtain smaller subdivisions.

If the quarter-sections are on the north or west side of a township, the quarter-quarter-section corners are placed 20 ch from the east or south quarter-section corners—or by single proportional measurement (see Sec. 23-19) on line if the total length on the ground is not equal to that on record.

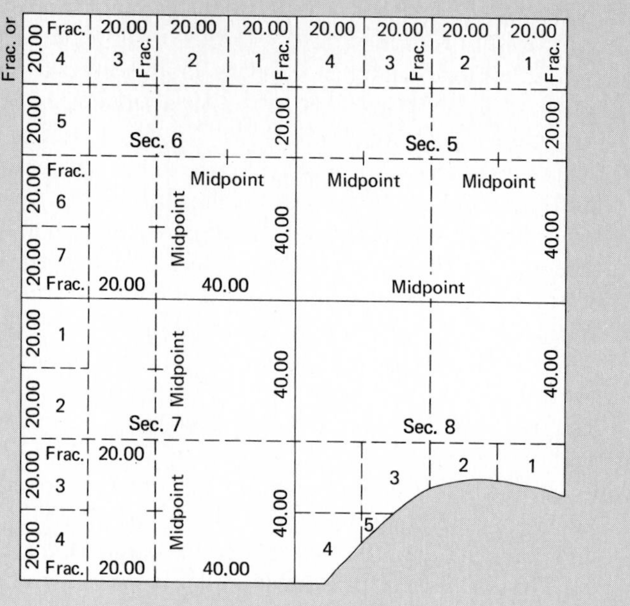

Figure 23-8. Subdivision of regular and fractional sections.

23-13. FRACTIONAL SECTIONS. In sections made fractional by rivers, lakes, or other bodies of water, lots are formed bordering on the body of water and numbered consecutively through the section (see section 8 in Figure 23-8). Boundaries of lots usually follow the quarter-section and quarter-quarter-section lines, but extreme lengths and narrow widths are avoided, as are areas of fewer than 5 acres or more than 45 acres.

Quarter-sections along the north and west boundaries of a township, made irregular by discrepancies of measurements and convergence of the range lines, are usually numbered and sold as lots (see Figure 23-8). One such quarter-section in a Wisconsin township contains 640 acres!

Lot lines are not actually run in the field. Like quarter-section lines, they are merely indicated on the plats by protraction (subdivisions of parcels on paper only). Lot areas are computed from plats. Many quarter-section lines are not indicated on plats.

23-14. NOTES. Specimen field notes for each of the several kinds of lines to be run are shown in various instruction manuals. Actual recording had to follow closely the model sets.

The original notes, or copies of them, are maintained in a land office in each state for the benefit of all interested persons.

23-15. OUTLINE OF SUBDIVISION STEPS. Pertinent points in the subdivision of quadrangles into townships, and townships into sections, are summarized in Table 23-1.

TABLE 23-1. **SUBDIVISION STEPS**

ITEM	SUBDIVISION OF A TRACT	SUBDIVISION OF A TOWNSHIP
Starting point	SE corner of SW township	SW corner of SE section (36)
Meridional lines		
Name	Range line	Section line
Direction	True north	North, parallel with east range line
Length	6 mi = 480 ch	1 mi = 80 ch
Corners set	Quarter-section and section corners at 40 and 80 ch alternately	Quarter-section corner at 40 ch; section corner at 80 ch
Latitudinal lines		
Name	Township line	Section line
Direction of random	True east-west parallel	East, parallel with south side of section
Length	6 mi less convergence	1 mi
Permissible error	3 ch, length or falling	50 lk, length or falling
Distribution of error		
Falling	Corners moved proportionately from random to true line	Corners moved proportionately from random to true line
Distance	All error thrown into west quarter-section	Error divided equally between quarter-sections

[Work repeated until north side of area is reached. Subdivision of last area on the north of the range of townships and sections follows.]

CASE I. WHEN LINE ON THE NORTH IS A STANDARD PARALLEL

ITEM	SUBDIVISION OF A TRACT	SUBDIVISION OF A TOWNSHIP
Direction of line	True north	North, parallel with east range line
Distribution of error in length	Placed in north quarter-section	Placed in north quarter-section
Corner placed at end	Closing corner	Closing corner
Permissible errors	Specified in *Manual of Surveying Instructions*	Specified in *Manual of Surveying Instructions*

CASE II. WHEN LINE ON THE NORTH IS NOT A STANDARD PARALLEL

ITEM	SUBDIVISION OF A TRACT	SUBDIVISION OF A TOWNSHIP
Direction of line	No case	Random north and correct back to section corner already established
Distribution of error in length		Same as case I

[Other ranges of townships and sections continued until all but two are laid out.]

ITEM	SUBDIVISION OF A TRACT	SUBDIVISION OF A TOWNSHIP
Location of last two ranges	On east side of tract	On west side of township
Next-to-last range subdivided	As before	As before
Last range		
Direction of random	True east	Westerly, parallel with south side of section
Nominal length	6 mi less convergence	1 mi less convergence
Correction for temporary corners	Corners moved proportionately from random to true line	Corners moved proportionately from random to true line
Distribution of error of closure	Corners moved westerly (or easterly) to place error in west quarter-section	Corner placed on the true line so that total error falls in west quarter-section

23-16. MARKING CORNERS. Various materials were approved and used for monuments in the original surveys. These included pits and mounds, stones, posts, charcoal, and broken bottles. A zinc-coated, alloyed iron pipe with $2\frac{1}{2}$ in outside diameter, 30 in long, is now standard except in rock outcrop, where a $3\frac{1}{4}$-in-diameter brass tablet with $3\frac{1}{2}$-in stem is specified.

Stones and posts were marked with one to six notches on one or two faces. The arrangements identify a monument as a particular section or township corner. Each notch represents 1 mi of distance to a township line or corner. Quarter-sections were marked with the fraction "$\frac{1}{4}$" on a single face.

In prairie country, where large stones and trees were scarce, a system of pits and mounds was used to mark corners. Different groupings of pits and mounds, 12 in deep and 18 in square, designated corners of the several classes. Unless perpetuated by some other type of mark, these corners were lost in the first plowing.

23-17. WITNESS CORNERS. Whenever possible, monuments were witnessed by two or three adjacent objects such as trees and rock outcrops. Bearing trees were blazed on the side facing the corner and marked with scribing tools.

When a regular corner fell in a creek, pond, swamp, or other place where it was impracticable to place a mark, *witness corners* (WC) were set on all lines leading to the corner. Letters WC were added to all other marks normally placed on the corner, and the WC then witnessed also.

23-18. MEANDER CORNERS. A meander corner (MC) was established on survey lines intersecting the bank of a stream having a width greater than 3 ch, or a lake, bayou, or other body of water of considerable extent. The distance to the nearest section corner or quarter-section corner was measured and recorded in the notes. A monument was set and marked MC on the side facing the water, and the usual witnesses noted. If practicable, the line was carried across the stream or other body of water by triangulation to another corner set in line on the farther bank.

A traverse joined successive meander corners along the banks of streams or lakes and followed as closely as practicable the sinuosities of the bank. The traverse was checked by calculating the position of the new meander corner and comparing it with its known position on a surveyed line.

Meander lines follow the mean high-water mark and are used for plotting and protraction of area only. They are *not* boundaries defining the limits of property adjacent to the water.

23-19. LOST AND OBLITERATED CORNERS. A common problem in resurveys of the public lands is replacment of lost or obliterated corners. This difficult task requires a combination of experience, hard work, and ample time to re-establish the location of a wooden stake or post monument incorrectly set, perhaps 100 years ago on an undependable section line, and with all witness trees long since cut or burned by apathetic owners.

An *obliterated corner* is one for which there are no remaining traces of the monument or its accessories, but whose location has been perpetuated or can be recovered beyond reasonable doubt. The corner may be restored from the acts or testimony of interested landowners, surveyors, qualified local authorities, witnesses, or from written evidence. Satisfactory evidence has value in the following order:

1. Evidence of the corner itself.
2. Bearing trees or other witness marks.
3. Fences, walls, or other evidence showing occupation of the property to the lines or corners.
4. Testimony of living persons.

A *lost corner* is one whose position cannot be determined, beyond reasonable doubt, either from traces of the original marks or from acceptable evidence or testimony that bears on the original position. It can be restored only by rerunning lines from one or more independent corners (existing corners that were established at the same time and with the same care as the lost corner). Proportionate measurements distribute the excess or deficiency between a recently measured distance d separating the nearest found monuments that straddle the lost point, and the record distance D given in the original survey notes between those monuments. Then the distance x from one of the found monuments required to set the lost point is calculated by proportion as: $x = X(d/D)$, where X is the record distance from that monument.

Single-proportionate measurement follows the procedure just described, and is used to relocate lost corners that have a specific alignment in one direction only. These include standard corners on base lines and standard parallels, intermediate section corners on township boundaries, all quarter-section corners, and meander corners established originally on lines carried across a meanderable body of water.

Double-proportionate measurements are used to establish lost corners located originally by specific alignment in two directions, such as interior section corners and corners common to four townships. The general procedure for single-proportionate measurement is used, but in two directions. It will establish two points: one on the north-south line and another on the east-west line. The lost corner is then located where lines from the two points, perpendicular to their respective north-south and east-west lines, intersect.

23-20. ACCURACY OF PUBLIC-LANDS SURVEYS. The accuracy required in the early surveys was a very low order. Frequently it fell below what the notes indicated. A small percentage of the surveys were made by drawing upon the imagination in the comparative comfort of a tent; no monuments were set and the notes serve only to confuse the situation for present-day surveyors and landowners. A few surveyors threw in an extra chain length at intervals to assure a full measure.

Many surveys in one California county were fraudulent. In another state, a meridian 108-mi long was run without including the chain handles in its 66-ft length. When discovered, it was rerun without additional payment.

The poor results obtained in various areas were due primarily to the following:

1. Lack of trained personnel; some contracts were given to persons with no technical training.
2. Very poor equipment by today's standards.
3. Work done by contract at low prices.
4. Surveys made in piecemeal fashion as the Indian titles and other claims were extinguished.
5. Marauding Indians, swarms of insects, dangerous animals and reptiles.
6. Lack of appreciation for the need to do accurate work.
7. Field inspection errratic or missing.
8. Magnitude of the problem.

In general, considering the handicaps listed, the work was reasonably well done in most cases.

23-21. DESCRIPTIONS BY TOWNSHIP, SECTION, AND SMALLER SUBDIVISION. Description by the sectional system offers a means of defining boundaries uniquely, clearly, and concisely. Several examples of acceptable descriptions are listed.

Sec. 6, T 7 S, R 19 E, 6th PM.
Frac. Sec. 34, T 2 N, R 5 W, Ute Prin. Mer.
The SE$\frac{1}{4}$, NE$\frac{1}{4}$, Sec. 14, Tp. 3 S, Range 22 W, SBM [San Bernardino Meridian].
E$\frac{1}{2}$ of NE$\frac{1}{4}$ of Sec. 20, T 15 N, R 10 E, Indian Prin. Mer.
E 80 acres of NE$\frac{1}{4}$ of Sec. 20, T 15 N, R 10 E, Indian Prin. Mer.

Note that the last two descriptions do not necessarily describe the same land. A California case in point occurred when the owner of a Southwest quarter-section, nominally 160 acres but actually 162.3 acres, deeded the westerly portion as "the West 80 acres" and the easterly portion as "the East $\frac{1}{2}$."

Sectional land which is privately owned may be partitioned in any legal manner at the option of the owner. The metes-and-bounds form is preferable for irregular parcels. In fact, metes and bounds are required to establish the boundaries of mineral claims and various grants and reservations.

Differences between the physical and legal (or record) ground locations, and areas, may result because of departures from accepted procedures in description writing, loose and ambiguous statements, or dependence on the accuracy of early surveys.

23-22. SOURCES OF ERROR. Some of the many sources of error in retracing the public-lands surveys follow:

1. Discrepancy between the length of the early surveyor's chain and a modern tape or an EDMI reading.
2. Change in magnetic declination, local attraction or both.

3. Lack of agreement between field notes and actual measurements.
4. Changes in water courses.
5. Nonpermanent objects used for corner marks.
6. Loss of witness corners.

23-23. MISTAKES. Some typical mistakes in the retracement of boundaries in public-land surveys are:

1. Failing to follow the general rules of procedure governing the original survey.
2. Neglecting to check the tape used against distances on record for marks in place
3. Resetting corners without exhausting every means of relocating the original corners.

PROBLEMS

23-1. What advantage would a stadia interval factor of 1:132 have in public-lands surveys?
23-2. Why are boundaries of public lands established by duly appointed surveyors unchangeable, even though incorrectly set in the original surveys?
23-3. What primary differences exist between the original and present instructions and practices in spacing corners and running section lines?
23-4. List the advantages and disadvantages of running base lines and standard parallels by each of the three methods.

Determine the offsets from, and azimuths of, a secant to lay out a parallel of latitude in Problems 23-5 and 23-6. Sketch the values.
23-5. At latitude 42°N.
23-6. At latitude 31°30′N.
23-7. At latitude 44°N, what is the maximum offset from secant to parallel, and the difference between the secant azimuth and cardinal direction?
23-8. What is the convergence in feet of two meridians (a) originally 12 mi apart, extended 12 mi, at latitude 36°N; and (b) 6 mi apart, extended 12 mi, at latitude 46°30′N?
23-9. Calculate the nominal distance in miles between the (a) First Guide Meridian East and West Range line of R 27 E, and (b) SE corner of Sec. 2, T 6 S, R 5 E, Choctaw PM, and the NW corner of Sec. 11, T 6 S, R 3 E.
23-10. How far from the PM will the quarter-quarter-section corner of Sec. 36, T 1 N, theoretically fall exactly opposite a section corner of Sec. 6, T 1 S, due to convergence at latitude 38°N?
23-11. Sketch locations of closing corners on township lines and areas they govern.
23-12. What steps in public-lands subdivision are left to the local surveyor?

Sketch and label pertinent lines and legal distances, and compute nominal areas of the parcels described in Problems 23-13 through 23-15.
23-13. $S\frac{1}{2}$, $NW\frac{1}{4}$, Sec. 26, T 3 N, R 5 W, Boise PM.
23-14. $NE\frac{1}{4}$, $SW\frac{1}{4}$, Sec. 21, T 4 N, R 2 W, 4th PM.
23-15. $NE\frac{1}{4}$, $SE\frac{1}{4}$, $NE\frac{1}{4}$, Sec. 8, T 1 N, R 2 W, Indian PM.
23-16. How many rods are required to enclose (a) the $SE\frac{1}{4}$, $SW\frac{1}{4}$, Sec. 27, T 12 N, R 5 W, and $NE\frac{1}{4}$, $SW\frac{1}{4}$ of Sec. 27? (b) Secs. 9, 16, 21, 22, 27?
23-17. What distance error is permitted in locating the NW corner T 1 N, R 1 E?
23-18. Name the line run as a random line, and those laid out parallel.

23-19. Corners of the NW$\frac{1}{4}$ of the SE$\frac{1}{4}$ of Sec. 10 are to be monumented. If all section and quarter-section corners originally set are in place, sketch all lines to be run and corners set.

23-20. The east line of the NE$\frac{1}{4}$ of Sec. 4 has a record distance of 40.12 ch and length measured in the field of 40.20 ch. Where should the SE corner of lot 1 be set? Explain.

23-21. The southern boundary of a township lies on a standard parallel at latitude 42°N. What is the theoretical length of its northern boundary?

23-22. The quarter section corner between Secs. 28 and 29 is found to be 40.28 ch from the common corner for Secs. 20, 21, 28, and 29. Where should the quarter-section corner be set in subdividing Sec. 29?

23-23. As shown in the figure, in a normal township the exterior dimensions of Sec. 6 on the west, north, east, and south sides are 81, 78, 82, and 79 ch respectively. Explain with a sketch how to divide the section into quarter-quarter sections.

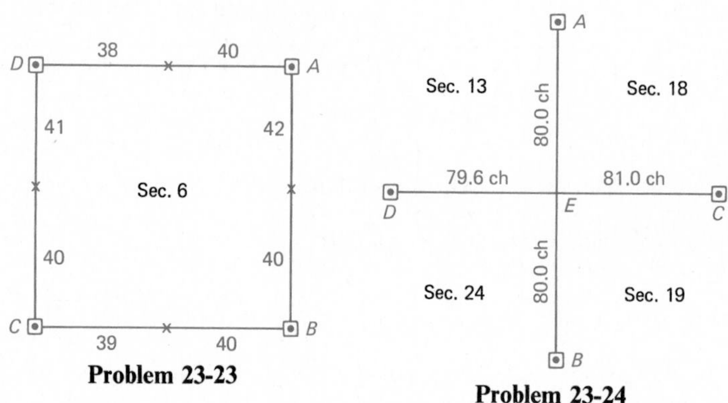

Problem 23-23

Problem 23-24

23-24. The problem figure shows original distances. Corners *A*, *B*, *C*, and *D* are found, but corner E is lost. Measured distances are *AB* = 160.4 ch and *CD* = 164.2 ch. Explain how to establish corner *E*.

To restore the corners in Problems 23-25 through 23-28, which method is used, single-proportion or double-proportion?

23-25. Township corners on guide meridians; section corners on range lines.

23-26. Section corners on section lines; township corners on township lines.

23-27. Quarter-section corners on range lines.

23-28. Quarter-quarter-section corners on section lines.

23-29. How are metes and bounds used in public-lands surveyed areas?

23-30. Why are meander lines not accepted as the boundaries defining ownership of lands adjacent to a stream or lake?

23-31. Where does a bend occur in the tier of a township?

23-32. Is an eighth-corner a standard term? Explain.

23-33. Are deeds and agreements involving land effective if not in writing?

23-34. Why are the areas of many public-lands sections smaller than the nominal size?

BIBLIOGRAPHY

Bowman, L. J. 1978. "Subdivision of Sections of Rectangular Surveys of the United States." *Surveying and Mapping* 33(no. 4):307.

Brown, C. M. 1969. *Boundary Control and Legal Principles*, 2nd ed. New York: Wiley.

Brown, C. M. 1976. "Identifying Monuments." *Surveving and Mapping* 36(no. 3):223.

Brown, C. M., W. G. Robillard, and D. A. Wilson. 1981. *Evidence and Procedures for Boundary Location*, 2nd ed. New York: Wiley.

Brown, N. D. 1975. "Corner Restoration in Missouri." *ASCE Journal of the Surveying and Mapping Division* 101(no. SU1):101.

Grady, B. J., and M. Stough. 1981. "Forestry Survey Jumble Touches off Boundary Disputes." *Surveying and Mapping* 74(no. 1):33.

Greulich, G. 1981. "Cancerous Survey Jumble — A Second Opinion." *Surveying and Mapping* 41(no. 3):289.

Lampert, L. M. 1980. "The Retracement." *Surveying and Mapping* 40(no. 3):315.

McCall, C. E. 1980. "Subdivision of Land Today." *Surveying and Mapping* 40(no. 4):415.

McEntyre, J. G. 1978. *Land Survey Systems.* New York: Wiley.

McEntyre, J. G. 1981. "Subdivision of a Section." *Surveying and Mapping* 41(no. 4):385.

Report to Congress. 1854. "The Surveyor and the Law." *Surveying and Mapping* 37(no. 2):161.

Shartle, S. M. 1979. "Re: Subdivision of Sections of Rectangular Surveys of the United States." *Surveying and Mapping* 38(no. 4):307.

U.S. Department of the Interior, Bureau of Land Management. 1973. *Manual of Surveying Instructions.* Washington, D.C.: U.S. Government Printing Office.

U.S. Department of the Interior, Bureau of Land Management. 1974. *Restoration of Lost or Obliterated Corners and Subdivision of Sections.* Washington, D.C.: U.S. Government Printing Office.

Whittlesey, C. 1977. "Origin of the American System of Land Surveys." *Surveying and Mapping* 37(no. 2):129.

William, M. G. 1981. "Sir William Petty, Thomas Jefferson, and the Down Survey: A Fresh Perspective on the U.S. Public Lands System." *Surveying and Mapping* 41(no. 1):77.

24 CONSTRUCTION SURVEYS

24-1. INTRODUCTION. Construction is one of the largest industries in the United States, so surveying, as the basis for it, is extremely important. It is estimated that 60% of all hours spent in surveying are on location-type work giving line and grade. Nevertheless, insufficient attention is frequently given to this type of survey.

An accurate topographic survey and site map are the first requirements in designing streets, sewer and water lines, and structures. Surveyors then lay out and position these facilities according to the design plan. A final "as-built" map, incorporating any modifications made to the design plans, is prepared during and after construction, and filed. Such maps are extremely important, especially where underground utilities are involved, to assure they can be located quickly if trouble develops and will not be disturbed by later improvements.

The surveyor in charge should receive copies of the plans well in advance of construction to become familiar with the job and have time to "tie-out" or "transfer" any established control points that might be destroyed during building operations. The methods of Figure 16-1, parts 6(a) and (b) particularly, can be used with intersection angles as close to 90° as possible. Differential levels are run to set new bench marks well out of the construction area but near enough for convenient reference.

A few common types of construction surveys, with which every surveyor, engineer, and architect should be familiar, will be described briefly in this chapter. Construction surveying is best learned on the job by adapting fundamental principles to the undertaking at hand. Since each project may involve individual problems, textbook coverage tends to be limited to introductory material.

24-2. EQUIPMENT FOR CONSTRUCTION SURVEYS. Layout of stakes for line and grade to guide construction operations has traditionally been accomplished using the surveyor's standard equipment — levels, theodolites, tapes, and, more recently, EDMIs. Basic surveying procedures for measuring horizontal and vertical distances and angles, as described in earlier chapters, have been applied. Although these techniques are still widely employed, recent advances in modern technology have produced some innovative equipment and procedures that have improved, simplified, and greatly increased the speed at which construction stakes can be set. These new devices include visible laser beam instruments, short-range electronic distance-measuring devices that can be operated in a "tracking mode," and "total-station" units.

The fundamental purpose of laser instruments is to create a visible line of known orientation, or a plane of known elevation, from which measurements for line and grade can be made. Two general types of lasers are described here.

1. *Single-beam lasers* project visible "string lines" or "plumb lines" utilized in linear and vertical alignment applications such as tunneling, sewer pipe placement, and building construction.

2. *Rotating-beam lasers* are merely single-beam lasers with spinning optics that rotate the beam in azimuth, thereby creating planes of reference. They expedite placement of grade stakes over large areas such as airports, parking lots, and subdivisions, but are also useful for topographic mapping. Laser beams are not readily visible to the human eye in bright sunlight so special detectors attached to a hand-held rod are used. Newer rotating lasers are self-leveling and quickly set up. Accurate readings can be taken at distances up to 1000 ft. If somehow bumped out of level, the laser beam shuts off and does not come back on until it relevels. Figure 24-1 shows one such model, a laser tracking level that automatically swings around to follow (track) a level rod equipped with a "Trak-Tronic system" and projects a red dot on it. After the HI has been established, the instrument can control elevations over a 1,000,000-ft^2 area from a single setup without requiring an operator, just a rodperson.

Theodolites combined with EDMIs that can automatically reduce measured slope distances to their horizontal and vertical components, and "total-station" instruments, are also very convenient for construction stakeout. In using these devices, coordinates of the occupied station and points to be staked are computed. From these coordinates, azimuths and horizontal distances to required stake positions are calculated. At the occupied station, a backsight is taken with the theodolite on a line of known azimuth, and the direction to a required stake turned. Operating the EDMI in its tracking mode, a range pole equipped with a prism is placed on-line and adjusted in position to read the required horizontal distance, where the stake is driven. The stake can also be

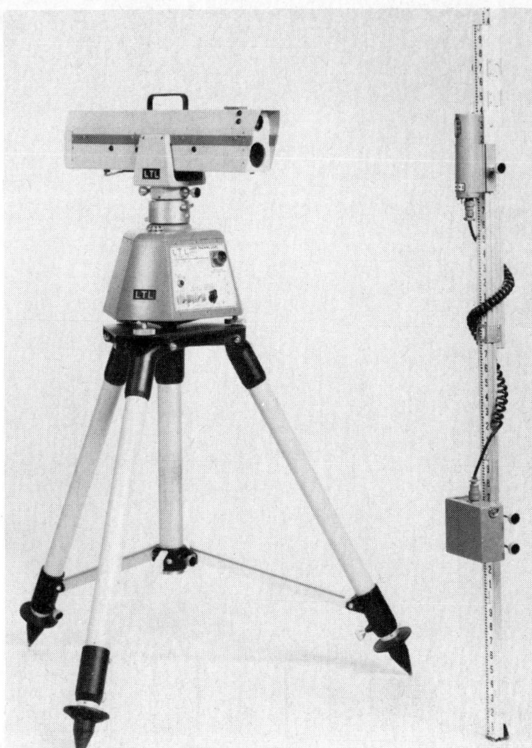

Figure 24-1. Laser tracking level and rod equipped with Trak-Tronic system. (Courtesy Construction Laser Div., Blount Industries, Inc.)

set at required elevation. With the prism attached to the range pole a distance above the bottom equal to the h.i. of the EDMI, the height of the stake can be adjusted until the vertical distance component obtained by the EDMI corresponds to the difference in elevation between the required stake elevation and that of the occupied station.

Special tracking prism systems have been developed that transmit signals to the range pole operator to assist in more quickly setting it at required locations. The tracking range pole-reflector combination shown in Figure 24-2, for example, produces a visible green light at the prism if it is left of line, red if right, and white when on-line. It also allows for voice communication from the EDMI operator, which greatly assists the stakeout process, especially on long sights.

With these total-station devices, since each point is set independently, it is important to check stake locations after they are set. This can usually be done by making quick measurements of distances between points, or by visual checks to see that points on a straight line are in fact straight, and curves are smooth.

Airplane and ship construction requires special equipment and methods as part of a unique branch of surveying, *optical tooling*. Precise location and erection of offshore oil drilling platforms many miles from a coast use new surveying technology such as inertial and satellite doppler systems (see Sections 20-14 and 20-15).

Figure 24-2. Range pole with prism and special tracking system for guiding placement of stakes. (Courtesy AGA Geodimeter, Inc.)

24-3. HORIZONTAL AND VERTICAL CONTROL. The importance of a good framework of horizontal and vertical control in a project area cannot be overemphasized. It provides the basis for positioning structures, utilities, roads, and so on, in both the design and construction stages. Too often, attempts have been made to skimp on establishing proper marks and monumenting them for preservation.

On most projects, additional control is required to supplement any already available in the job area. The points must be:

1. Convenient for use by the contractor's personnel — that is, located sufficiently close to the item being built so workers using relatively simple equipment such as carpenters' levels and string lines can accurately transfer alignment and grade.
2. Far enough from the actual construction to ensure working room for the contractor and freedom from possible destruction of stakes.
3. Clearly marked and understood by the contractor in the absence of a surveyor.
4. Supplemented by guard stakes to deter removal, and referenced to facilitate restoring them. Contracts usually require the owner to pay the cost of setting initial control points and the contractor to replace damaged or removed ones.
5. Suitable for securing the accuracy agreed upon for construction layout (which may be to only the nearest foot for a manhole, 0.01 ft for an anchor bolt, or 0.001 ft for a critical feature).

24-4. STAKING OUT A PIPELINE. Flow in water lines is generally under pressure, but most sewers have gravity flow. Alignment and grade must therefore be more accurate for the latter. Larger water lines also have definite grades because "blowoffs" are needed at low points and "air releases" at high spots. Grades are fixed by existing conditions, such as topography, which affect excavation depth of connecting lines, manholes, outfalls, and catch basins.

Construction stakes, sometimes set on the center line at 50-ft stations when the ground is reasonably uniform, disappear on the first pass of a ditcher or bulldozer, so parallel *offset lines* are necessary. Marks should be closer together on horizontal and vertical curves than on straight segments. For pipes of large diameter on horizontal curves, stakes may be placed for each pipe length — say, 6 or 8 ft.

On hard surfaces where stakes cannot be driven, points are marked by paint, spikes, shiners (tin can top with nail through it), drill holes, or other means. A surveyor should know in advance approximately how much material will be deposited, on which side of the trench, and adjust offset staking accordingly.

Figure 24-3 shows the arrangement of *batter boards* for a sewer line. Batter boards are usually 1 × 6-in boards nailed to 2 × 4-in posts which have been pointed and driven into the ground on either side of the trench. The top of the batter board is generally placed a full number of feet above the invert (flow line or lower inside surface) of the pipe. Nails are driven into the board tops so a string stretched tightly between them will define the pipe center line. A graduated pole or special rod is used to measure the required distance from the string to the pipe invert. Thus the string gives both line and grade. It can be kept taut by hanging a weight on each end after wrapping it around the nails.

In Figure 24-3, instead of a fixed batter board, a 2 × 4 carrying a level vial can be placed on top of the offset-line stake whose elevation is known.

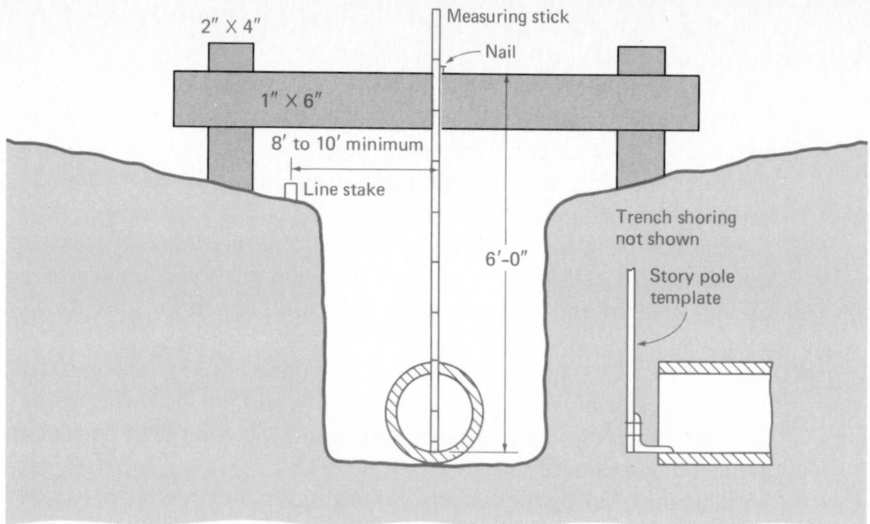

Figure 24-3. Batter board for sewer line.

Measurement is made from the underside of a leveled 2 × 4 with a tape or graduated pole to establish the flow line.

On some jobs that have a deep wide cut, a level or laser instrument is set up in the ditch to give line and grade. If the pipe is large enough, the laser device can be placed inside it.

Using a profile similar to that in Figure 7-8, grades for trenches are designed to avoid excessive cut and fill, and to permit connection to other facilities.

24-5. STAKING PIPELINE GRADES. Staking pipeline grades is essentially the reverse of running profiles, although in both operations the center line must first be marked and stationed in horizontal location. The actual profiling and staking are on an offset line.

Information conveyed to the contractor on stakes for laying pipelines usually consists of two parts: (1) giving the amount of cut (or fill), normally only to the nearest 0.1 ft, to enable a rough trench to be excavated; and (2) providing precise grade information, generally to the nearest 0.01 ft, to guide actual placement of the pipe invert at its planned elevation. Cut (of fill) values for the first part are vertical distances from ground elevation at the offset stakes to pipe invert. After the pipe's grade line has been computed and offset line run, cuts (or fills) can be determined by a leveling process illustrated in Figure 24-4, and in corresponding field notes in Plate D-12, and summarized as follows:

1. Stations staked on the pipeline are listed in column 1 of the field notes.
2. Compute flow line or invert elevation at each station (column 6).
3. Set up the level and get an HI by reading a plus sight on a BM; for example, HI = 2.11 + 100.65 = 102.76 (see Figure 24-4).

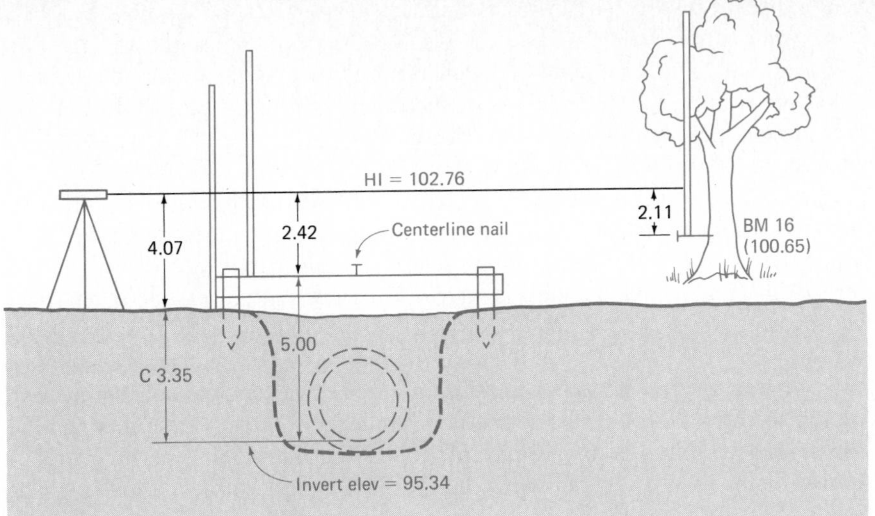

Figure 24-4. Leveling process to determine cut or fill and set batter boards for laying pipelines.

4. The elevation at each station is obtained from a rod reading on the ground at every stake (column 4)—for example, 4.07 at station 1 + 00 (see Figure 24-4)—and subtracting it from the HI (column 5); for example, 102.76 – 4.07 = 98.69 at station 1 + 00.
5. Ground elevation minus pipe elevation equals cut (+) or fill (–) (column 7); for example, 98.69 – 95.34 = C 3.35 (see Figure 24-4).
6. The cut or fill is marked with keel on an offset stake facing the center line; the station number is written on the other side.

In another variation, which produces the same results, *grade rod* difference between the HI and pipe invert) is computed, and *ground rod* (reading with rod held at the stake) is subtracted from it to get cut or fill. For station 1 + 00, grade rod = 102.76 – 95.34 = 7.42, and 7.42 – 4.07 = C 3.35.

After the trench has been excavated based on cuts and fills marked on the stakes, batter boards are set. Marks needed to place them can be made with a pencil on the offset stakes during the same leveling operation used to obtain cut and fill information. Figure 24-4 also illustrates the process. Suppose that at station 1 + 00, the batter board will be set so its top is exactly 5.00 ft above pipe invert. The rod reading necessary to set the batter board is obtained by subtracting pipe invert elevation plus 5.00 ft from the HI; thus, 102.76 – (95.34 + 5.00) = 2.42 ft (see Figure 24-4). The rod is held at the stake and adjusted in vertical position by direction from the level operator until a rod reading of 2.42 ft is obtained; then a mark is made at the rod's base on the stake. (To facilitate this process, a rod target or colored rubber band can be placed on the rod at the required reading.) The board is then fastened to the stake with its top at the mark using nails or C clamps, and a carpenter's level employed to align it horizontally across the trench. A nail marking the pipe center line is set by measuring the stake's offset distance along the board.

Staking lines having $\frac{1}{2}$ or 1% grades is easy. Since the usual stadia interval factor is 100 to 1, sight lines through the middle and upper, or middle and lower cross hairs of a level, transit, or theodolite differ by $\frac{1}{2}$ ft in 100 ft to define a $\frac{1}{2}$% grade. Between the upper and lower wires there is a 1% grade. This can be used to advantage if several consecutive batter boards are set at the same height above pipe invert. With the instrument placed on the offset line, the first batter board is set as just described. The lower cross hair then defines a $-\frac{1}{2}$% grade. By turning the leveling screws to set the upper hair on the rod reading for the first stake, the lower hair is on a − 1% grade. Successive stakes can then be set using the same rod reading as that at the first stake on the hair being used.

Surveying up and down mountain slopes can be time-consuming and costly, so often a "tilted plane" method similar to that just described can be used. The procedure is also applicable for laying out sloping parking lots. A transit or theodolite is set up with the h.i. a full number of feet above a grade stake, with two of the four screws parallel with the grade line (one screw on-line and two perpendicular to it for a three-screw instrument). A vertical circle reading of 0°00′ is placed on the instrument and retained. Using the parallel screws, the line of sight is tilted to produce the required grade—for example, a 2.00-ft drop in 100 ft for a grade of −2%. Likewise, with the transverse screws the line of sight can be tilted to set grade stakes on a line perpendicular to the center line.

24-6. STAKING OUT A BUILDING. The first task in staking out a building is to locate it properly on the correct lot by making measurements from the property lines. Most cities have an ordinance establishing setback lines from the street and between houses to improve appearance and provide fire protection.

Stakes may be set initially at the exact building corners as a visual check on positioning of the structure, but obviously such points are lost immediately when excavation is begun on the footings. A set of batter boards and reference stakes, placed as shown on Plate D-11, is therefore erected near each corner but out of the way of construction. The boards are nailed a full number of feet above the footing base, or at first-floor elevation. Corner stakes and batter-board points for rectangular buildings are checked by measuring the diagonals for comparison with each other and their computed values. A bench mark (two or more on large projects) beyond the construction area but within easy sight distance is necessary to control elevations.

Nails are driven into the batter-board tops so that strings stretched tightly between them define the outside wall or form line of the building. Again, the boards give line and grade.

Permanent foresights are helpful in establishing principal lines of the structure. Targets or marks on nearby existing buildings can be used if movement due to thermal effects or settlement is considered negligible. On formed concrete structures such as retaining walls, offset lines are necessary because the outside wall face is obstructed.

Plate D-11 shows the location of batter boards and steps to be followed in setting them for a small structure. Positions of such things as interior foot-

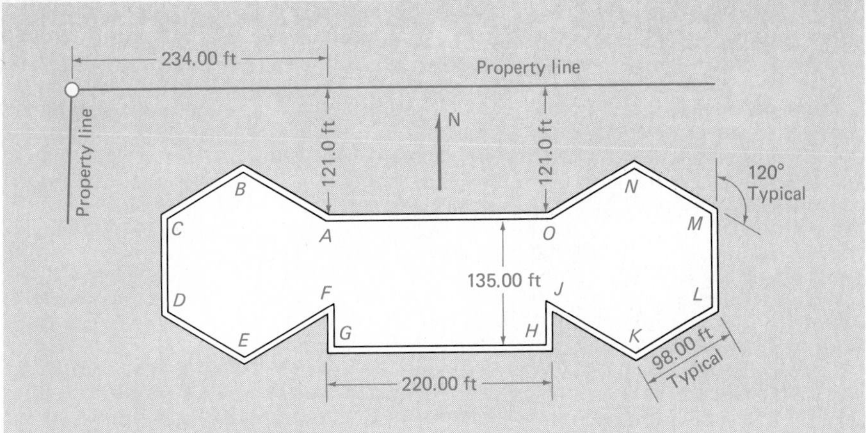

Figure 24-5. Building layout.

ings, anchor bolts for columns, and special piping or equipment can first be marked by 2 × 2 in stakes with tacks. Survey disks, scratches on bolts or concrete surfaces, and steel pins are also used. Batter boards set inside the building dimensions for column footings have to be removed as later construction develops.

Staking out a building can be a time-consuming and lengthy process if the surveyor does not give sufficient forethought to the basic control points required and the best method to establish them. The number of instrument setups should be minimized to conserve time, and calculations made in the office if possible, rather than in the field while a survey party waits.

One method of handling difficult jobs is to stake all (or many) points from a single instrument setup using precalculated angles and distances. Figure 24-5 shows an unusual building shape which was laid out rapidly using only two setups by choosing an initial one to reach half the corners, and having the same calculated angles and distances (Figure 24-6) for both ends of the structure. In using this method it is essential that enough building dimensions are checked by taping or an EDMI after marking the corners to ensure no large errors or mistakes were made.

24-7. STAKING OUT A HIGHWAY. After suitable control has been established, the construction area limits are staked so a contractor can clear to them. Next, some contractors want points set on the right-of-way with subgrade elevations showing cut or fill to a given elevation for use in performing rough grading and preliminary excavation of excess material. Location stakes are then placed on the center line or an offset line at full stations, the beginnings and ends of horizontal and vertical curves, and other critical points. A profile run on the center line determines the elevation at each stake.

To guide a contractor in making final excavations and embankments, *slope stakes* are driven at the *slope intercepts* (intersections of the original ground and each side slope), or offset a short distance, perhaps 4 ft, Figure 24-7.

BUILDING LAYOUT

Sta.	Angle	To the	Distance
	⊼ @ Point *A*		
O	0°00′	Rt	220.00 ft
F	90°00′	Rt	98.00 ft
G	90°00′	Rt	135.00 ft
E	120°00′	Rt	169.74 ft
D	150°00′	Rt	196.00 ft
C	180°00′	Rt	169.74 ft
B	210°00′	Rt	98.00 ft
	@ Point *O*		
A	0°00′	Lt	220.00 ft
J	90°00′	Lt	98.00 ft
H	90°00′	Lt	135.00 ft
K	120°00′	Lt	169.74 ft
L	150°00′	Lt	196.00 ft
M	180°00′	Lt	169.74 ft
N	210°00′	Lt	98.00 ft

Figure 24-6. Precalculated angles and distances for building stakeout.

The cut or fill at each location is marked on the slope stake. Note that actually there is *no* cut or fill at a slope stake—the figure given is the vertical distance from the ground elevation at the slope-stake to grade.

Slope stakes can be set at slope intercept locations predetermined in the office from cross-section data. (Methods for determining slope intercepts from cross sections are described in Chapter 27.) If predetermined slope intercepts are used, the ground elevation at each stake should still be checked in the field to verify its agreement with the cross section. If a significant discrepancy in

Figure 24-7. Slope stakes (shoulders and ditches not shown).

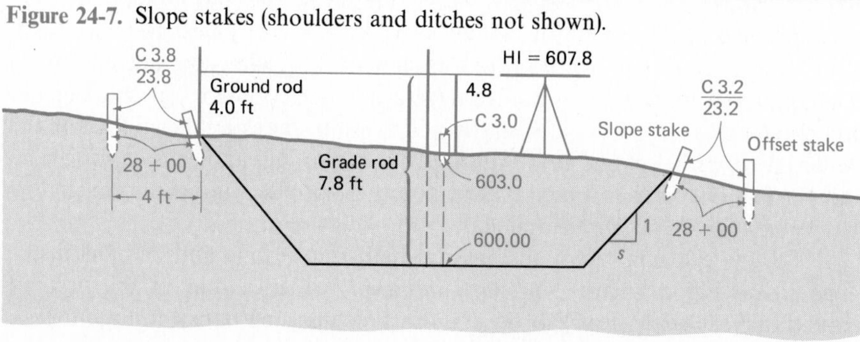

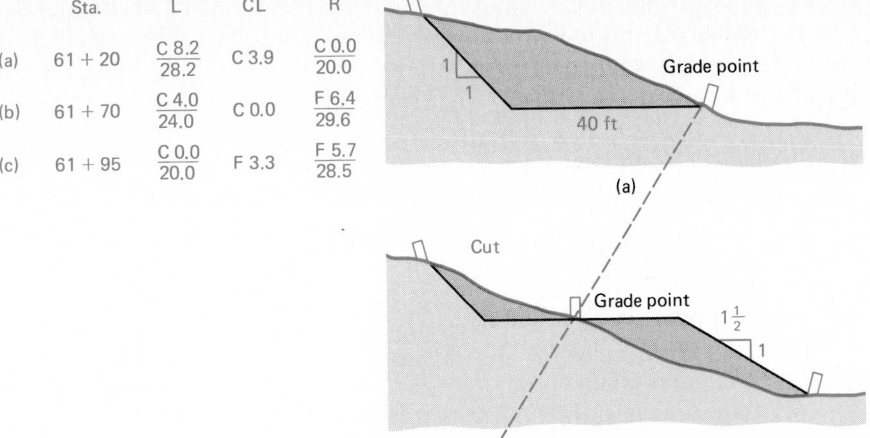

	Sta.	L	CL	R
(a)	61 + 20	C 8.2 / 28.2	C 3.9	C 0.0 / 20.0
(b)	61 + 70	C 4.0 / 24.0	C 0.0	F 6.4 / 29.6
(c)	61 + 95	C 0.0 / 20.0	F 3.3	F 5.7 / 28.5

Figure 24-8. Grade points at transition sections.

elevation exists, the stake's position must be adjusted by trial and error as described below. The amount of cut or fill marked on the stake is computed from the actual difference in elevation between slope stake ground elevation and grade elevation.

If slope intercepts have not been precalculated from cross-section data, slope stakes are located by a trial-and-error method based upon mental calculations involving the HI, grade rod, ground rod, half-roadway width, and side slopes. One or two trials are generally sufficient to fix the stake position within an allowable error of 0.3 to 0.5 ft for rough grading. The infinite number of ground variations prohibits use of a standard formula in slope staking. An experienced surveyor employs only mental arithmetic, without scratch paper or hand calculator. Whether using the method to be described, or any other, systematic procedures must be followed to avoid confusion and mistakes.

Grade stakes are set at points that have the same ground and grade elevation. Three transition sections normally occur in passing from cut to fill (or vice versa), and a grade stake is set at each one (see Figures 24-8 and 27-1.) A line connecting grade stakes, perhaps scratched out on the ground, defines the change from cut to fill, as line *ABC* of Figure 27-1.

Example 24-1 lists the steps to be taken in slope staking in sequence, *assuming for simplicity, academic conditions* of a level roadway. In practice, travel lanes and shoulders of modern highways have lateral slopes for drainage, then a steeper slope to a ditch in cut, and another slope up the hillside to the slope intercept. Transition sections may have half-roadway widths of cuts different from those in fills to accommodate ditches, and flatter side slopes for fills that tend to be less stable than cuts. But the same basic steps still apply and can be extended by students after learning the fundamental approach.

EXAMPLE 24-1

List the field procedures, including calculations, necessary to set slope stakes for a 40-ft wide level roadbed with side slopes of 1:1 in cut and fill (see Figures 24-7 and 24-8).

1. Compute the cut at the center line stake from profile and grade elevations (603.0 – 600.0 = C 3.0 in Figure 24-7). Check in field by grade rod minus ground rod = 7.8 – 4.8 = C 3.0 ft. Mark the stake C 3.0/0.0. (On some jobs the center stake is omitted and stakes set only at the slope intercepts.)
2. Estimate the difference in elevation between the left-side slope-stake point (20+ ft out) and the center stake. Apply the difference—say, +0.5 ft—to the center cut and get an estimated cut of 3.5 ft.
3. Mentally calculate the distance out to the slope stake, 20 + 1(3.5) = 23.5 ft, where 1 is the side slope.
4. Hold the zero end of a cloth tape at the center stake while the rodperson goes out at right angles (turned by prism or extended-arm method) with the other end and holds the rod at 23.5 ft.
5. *Forget all previous calculations to avoid confusion of too many numbers and remember only the grade-rod value.*
6. Read the rod with the level and get cut from grade rod minus ground rod, perhaps 7.8 – 4.0 = C 3.8 ft.
7. Compute required distance out for this cut, 20 + 1(3.8) = 23.8 ft.
8. Check the tape to see what is actually being held and find it is 23.5 ft.
9. Distance is within a few tenths of a foot and close enough. Move out to 23.8 ft if ground is level and drive stake. Move farther out if the ground slopes up, since a greater cut would result and thus the slope stake must be beyond the computed distance, or not so far if the ground has begun to slope down, which gives a smaller cut.
10. If the distance has been missed badly, make a better estimate of cut, compute a new distance out, and take a reading to repeat the procedure.
11. In going out on the other side, the rodperson lines up the center and left-hand slope stake to get the right-angle direction.
12. To locate grade stakes at the road edge, one person carries the zero end of the tape along the center line while the rodperson walks parallel holding the 20-ft mark until the required ground-rod reading is found by trial. Note that the grade rod changes during the movement but can be computed at 5- or 10-ft intervals. The notekeeper should have the grade rod listed in the fieldbook for quick reference at full stations and other points where slope stakes are to be set.
13. Grade points on the center line are located using a starting guess determined by comparing the cut and fill at back and forward stations.

Practice varies for different organizations, but often the slope stake is set 4 ft beyond the slope intercept. It is marked with the required cut or fill, distance out from the center line to the slope-stake point, side-slope ratio, and base half-width noted on the side facing the center line. Stationing is given on the back side. A reference stake having the same information on it may also be placed 6 ft or more farther out of the way of clearing and grading. On transition sections, grade-stake points are marked.

Slope staking using an EDMI and vertical angles is now being carried out, since the slope distance to a stake is readily measured. Theodolites with combined EDMIs, and total-station instruments which can automatically reduce measured slope distances to their horizontal and vertical components, greatly speed slope staking. They are especially convenient in rugged terrain where slope intercept elevations differ greatly from center-line grade. Stakeout procedures using them were briefly described in Section 24-2.

Slope staking should be done with the utmost of care, for once cut and fill embankments are started, it is difficult and expensive to reshape them if a mistake is discovered.

After rough grading has shaped cuts and embankments to near final elevation, finished grade is set more accurately from *blue tops* (stakes whose tops are marked with blue keel and driven to grade elevation). These are not normally offset, but rather driven directly on center line. A tight string line stretched through notches in the blue tops can provide close control of the grading.

Highway and railroad grades can often be rounded off to multiples of 0.05 or 0.10 % without appreciably increasing earthwork costs or sacrificing good drainage. Streets need a minimum 0.05 % grade for drainage from intersection to intersection, or from midblock both ways to the corners, and crowning for lateral flow to gutters.

Drainage profiles, prepared to verify or construct drainage cross sections, can be used to accurately locate drainage structures and easements. An experienced engineer, when asked a question regarding the three most important items in highway work, replied "drainage, drainage, and drainage"! This requirement must be satisfied by good surveying and design.

Utility relocation surveys may be necessary in connection with highway construction; for example, manhole or valve-box covers have to be set at correct grade before earthwork begins to fit the center-line finished grade and differential elevation resulting from transverse surface slope. They are located by center-line station and offset distance.

A three-person field crew can handle most construction staking jobs. Grade and other calculations should be prepared in the office, in advance if possible, to save field time and its greater cost.

Location staking for railways and canals follows the methods outlined for highways.

24-8. OTHER CONSTRUCTION SURVEYS. Hydrographic surveys for causeways, bridges, and offshore oil platforms add the problem of establishing points and depths where it may be difficult or impossible to hold a rod or reflector. Triangulation, trilateration, EDMIs, and sonar mapping devices are used to plot dredging cross sections for underwater trenching and pipe laying. Today more pipelines are crossing wider rivers than ever before. Mammoth pipeline projects now in progress to transport crude oil, natural gas, and water have introduced numerous new problems and solutions. Permafrost, extremely low temperature, and the need to provide animal crossings are examples of special problems encountered in the Alaska program.

Large earthwork projects—dams, levees, and superhighways—require widespread permanent control for quick setups and frequent replacement of slope stakes, all of which may disappear under fill in one day. Fixed signals for elevation and alignment painted or mounted on canyon walls or hillsides can mark important reference lines. Failures of some large structures, such as the Teton Dam, demonstrate the need for periodically monitoring them so any necessary remedial work can be done.

Underground surveys in tunnels and mines necessitate transferring line and elevations from the ground above, often down shafts. Directions of lines in mine tunnels can be most conveniently established using north-seeking gyros (see Section 19-1). In another, and still practiced method, two heavy plumb bobs hung on wires (and damped in oil or water) from opposite sides of the surface opening can be aligned by theodolite there and in the tunnel. (A vertical collimator will also provide two points on-line below ground.) A theodolite or laser is wiggled in on the short line defined by the two plumb-bob wires, a station mark set above the instrument, and the line extended. Later setups are made beneath *spads* (surveying nails with hooks) anchored in the roof. Elevations are brought down by taping or other means. Bench marks and instrument stations are set on the roof, out of the way of equipment.

Surveys are run at intervals on all large jobs to check progress for periodic payments to the contractor. And finally, an as-built survey is made to determine compliance with plans, note changes, and make terminal contract payment.

24-9. SOURCES OF ERROR. Important sources of error in construction surveys are:

1. Inadequate number and/or location of control points on the job site.
2. Errors in establishing control.
3. Measurement errors in layout.
4. Failure to double-center in laying out angles.
5. Careless referencing of key points.
6. Movement of stakes and marks.
7. Failure to use tacks for proper line where justified.

24-10. MISTAKES. Typical mistakes often made in construction surveys are:

1. Lack of foresight as to where construction will destroy points.
2. Notation for cut (or fill) and stationing on stake not checked.
3. Wrong datum for cuts, whether cut is to finished grade or subgrade.
4. Arithmetic mistakes, generally due to lack of checking.
5. Use of incorrect elevations, grades, and stations.
6. Failing to check the diagonals of a building.

7. Carrying out computed values beyond field accuracy possible (one good hundredth is worth all the bad thousandths).
8. Reading the rod on top of stakes instead of on the ground beside them in profiling and slope staking.
9. Failure to calculate midpoint grades by successive increments so that any error is evident if the last grade elevation does not check.

PROBLEMS

24-1. Describe in general, without discussing specific projects, the surveying work that normally must be performed in the construction industry.

24-2. List the types of construction projects where visible laser beam instruments are useful for stakeout.

24-3. Discuss how line and grade can be set with a total-station instrument operated in the "tracking" mode.

24-4. What orders of horizontal and vertical control should be used for construction of a (a) water line, (b) sewer line, and (c) long-span bridge?

24-5. How far apart should stakes be set on a sewer job having a (a) relatively flat grade, (b) steep grade, (c) sharp curve?

24-6. State two conflicting requirements which enter the decision on how far offset stakes should be set beyond the construction line.

24-7. A sewer pipe is to be laid from station 10 + 00 to station 13 + 35 on a −0.70% grade, starting with invert elevation 825.30 ft at 10 + 00. Calculate invert elevations at each 50-ft station along the line and the invert elevation at station 13 + 35.

24-8. A sewer pipe must be laid from a starting invert elevation of 1575.75 ft at station 9 + 50 to an ending invert elevation of 1569.10 ft at station 13 + 80. Determine the uniform grade needed, and calculate invert elevations at each 50-ft station.

24-9. Grade stakes for a pipeline running between stations 0 + 00 and 5 + 64 are to be set at each full station. Elevations of the pipe invert must be 1168.50 ft at station 0 + 00 and 1162.30 ft at 5 + 64, with a uniform grade between. After staking an offset centerline, an instrument is set up nearby, and a backsight of 3.06 taken on BMA (elevation 1173.25 ft). The following foresights are taken with the rod held on ground at each stake: (0 + 00, 3.51); (1 + 00, 3.67); (2 + 00, 4.03); (3 + 00, 5.16); (4 + 00, 5.92); (5 + 00, 6.80); and (5 + 64, 7.25). Prepare a set of suitable field notes for this project (see Plate D-12), and compute the cut required at each stake. Close the level circuit back to the bench mark.

24-10. If batter boards are to be set exactly 7.00 ft above pipe invert at each station on the project of Problem 24-9, calculate the necessary rod readings for placement of the batter boards. Assume the instrument has the HI as in Problem 24-9.

24-11. How are street grades arranged for drainage in a city with flat terrain?

24-12. List the most important things to consider in laying a grade line on a ground profile.

24-13. Describe situations where batter boards could be placed inside a building.

24-14. By means of a sketch, show how and where batter boards should be located for an I-shaped building. For an E-shaped structure.

24-15. A building in the shape of an L must be staked. The corners of the building *ABCDEF* all have right angles. Proceeding clockwise around the building, the required outside dimensions are $AB = 80.00$ ft, $BC = 40.00$ ft, $CD = 30.00$ ft, $DE = 30.00$ ft, $EF = 50.00$ ft, and $FA = 70.00$ ft. After staking the batter boards for this building and stretching string lines taut, check measurements of diagonals should be made. What should be the values of *AC*, *AD*, *AE*, *FB*, *FC*, *FD* and *BD*?

24-16. The design floor elevation for a building to be constructed is 975.50 ft. An instrument is set up nearby, leveled, and a backsight of 4.62 taken on BMA whose elevation is 974.89 ft. If batter boards are placed exactly 1.00 ft above floor elevation, what rod readings are necessary on the batter-board tops to set them properly?

24-17. Describe two or more ways to still the swing of a plumb-bob line extending several stories outside a building wall.

24-18. Can the corner of a building be plumbed by running the vertical line of a transit telescope up and down the wall line? Explain.

24-19. Compare the accuracy required on a survey for a long-span bridge with that for an ordinary property survey. List suitable tolerances in positioning anchor bolts for the bridge.

24-20. Which building line of a structure is shown on a record plat, the outside or center line of a wall?

24-21. Outline a suitable method for giving grade for a (a) parking lot, (b) reinforced-concrete culvert, and (c) footing for a bridge pier.

24-22. Discuss the suitability of a 0.00% grade for a street or highway.

24-23. A highway horizontal alignment survey is run by deflection angles. Which is the most important transit adjustment to check?

24-24. Discuss the importance of tying in and referencing critical center-line points on highway construction surveys.

24-25. Explain why slope stakes are placed at an offset distance from slope intercepts. What offset distance is recommended?

24-26. What information is normally given on slope stakes?

24-27. Discuss the procedure and advantages of using total-station instruments for slope staking.

24-28. Describe the field procedure in setting slope stakes.

24-29. How are directions of lines and elevations established in mines?

24-30. What rule is used in some cities regarding the responsibility of a contractor who breaks existing sewer or water lines during new construction?

24-31. On construction projects in your area, who is responsible for and pays the cost of setting initial control and survey stakes, the contractor or owner? For the replacement of stakes knocked out?

24-32. Compute the floor area for the building layout of Figure 24-5.

BIBLIOGRAPHY

Angeloni, W. A. 1975. "Planning and Execution of Surveys for Civil Works Projects." *Surveying and Mapping* 35(no. 4):341.

Barry B. A. 1973. *Construction Measurement*, New York Wiley.

Greulich, G. 1978. "Underground Utilities." *Bulletin, American Congress on Surveying and Mapping*, no. 60, p. 7.

Hole, S. W. 1979. "Pipelines—Positioning and Survey Problems." *ASCE Journal of the Surveying and Mapping Division* 105(no. SU1):15.

Hopper, A. G. 1982. "Laser Technology and the Inverted Plumb Bob." *Professional Surveyor* 2(no. 2):107.

Keene, D. F. 1974. "Precise Dam Surveys—Los Angeles County Flood Control District." *ASCE Journal of the Surveying and Mapping Division* 100(no. SU2):99.

Peterson, E. W., and P. Frobenius. 1973. "Tunnel Survey and Tunneling Machine Control." *ASCE Journal of the Surveying and Mapping Division* 99(no. SU1):21.

Posch, M. 1980. "Survey Work in Constructing the Arlberg Road Tunnel." *Bulletin, American Congress on Surveying and Mapping*, no. 69, p. 19.

Shewmon, D. C. 1981. "The Tilted Plane in Surveying." *Bulletin, American Congress on Surveying and Mapping*, no. 72, p. 23.

Thompson, B. J. 1974. "Planning Economical Tunnel Surveys." *ASCE Journal of the Surveying and Mapping Division* 100(no. SU2):95.

25 CIRCULAR CURVES

25-1. INTRODUCTION. Straight (tangent) sections of most types of transportation routes, such as highways, railroads, and pipelines, are connected by curves in both the horizontal and vertical planes. An exception is a transmission line, in which a series of straight lines is used with abrupt angular changes at tower locations if needed.

Two types of horizontal curves are employed: circular arcs and spirals. Both are readily laid out in the field with standard surveying equipment. A *simple curve* [Figure 25-1(a)] is a circular arc connecting two tangents. It is the type most often used. A *compound curve* [Figure 25-1(b)] is composed of two or more circular arcs of different radii tangent to each other, with their centers

Figure 25-1. Circular curves.

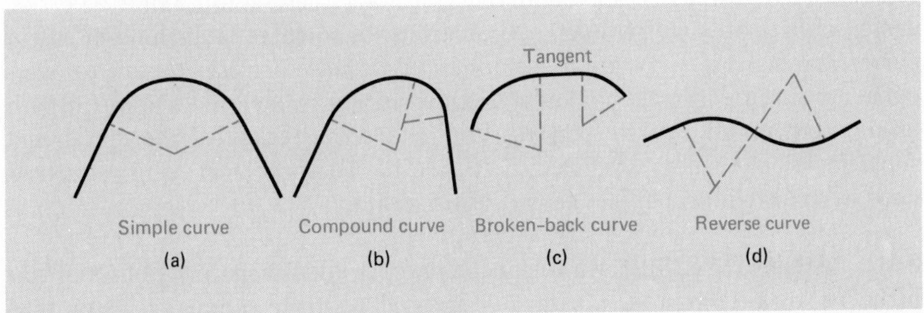

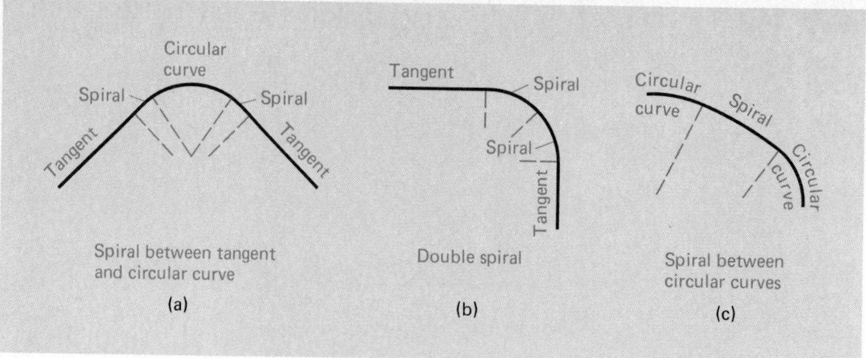

Figure 25-2. Use of spiral transition curves.

on the same side of the common tangent. The combination of a short length of tangent (less than 100 ft) connecting two circular arcs that have centers on the same side, as in Figure 25-1(c), is called a *broken-back curve.* A *reverse curve*, as in Figure 25-1(d), consists of two circular arcs tangent to each other, with their centers on opposite sides of the common tangent. Compound, broken-back, and reverse curves are unsuitable for modern high-speed highway, rapid-transit, and railroad traffic, and should be avoided if possible. They are sometimes necessary, however, in mountainous terrain to avoid excessive grades or very deep cuts and fills. Compound curves are often used on lower-speed exit and entrance ramps of interstate highways and expressways.

Easement curves are desirable, especially for railroads and rapid-transit systems, to lessen the sudden change in curvature at the junction of a tangent and circular curve. A *spiral* makes an excellent easement curve because its radius decreases uniformly from infinity at the tangent to that of the curve it meets. Spirals are used to connect a tangent with a circular curve, a tangent with a tangent (double spiral), and a circular curve with a circular curve. Figure 25-2 illustrates these arrangements.

The effect of centrifugal force on a vehicle passing around a curve can be balanced by *superelevation*, which raises the outer rail of a track or outer edge of a highway pavement. Correct transition into superelevation on a spiral increases uniformly with the distance from the beginning of the spiral, and is in inverse proportion to the radius at any point. Properly superelevated spirals ensure smooth and safe riding with less wear on equipment. As noted, spirals are used for railroads and rapid-transit systems. This is because trains are constrained to follow the tracks, and thus a smooth, safe, and comfortable ride can be assured only with properly constructed alignments that include easement curves. On highways, spirals are seldom used because drivers are able to overcome abrupt directional changes at circular curves by steering a spiraled path as they enter and exit the curves. For detailed coverage of the spiral and superelevation, readers are referred to one of the route surveying references cited in the Bibliography at the end of the chapter.

25-2. DEGREE OF CURVE. In European practice and the majority of American highway work, circular curves are designated by their radius—for example,

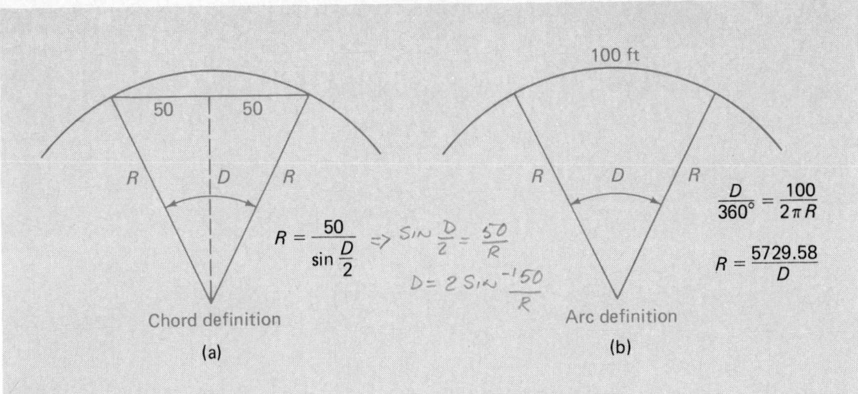

Figure 25-3. Degree of curve.

"1500-m curve" and "1000-ft curve." American railroads and some highway departments prefer to identify curves by their *degree*, using either chord or arc definition.

In railroad practice (and early highway construction), the degree of curve has been the angle at the center of a circular arc subtended by a *chord* of 100 ft. This is the *chord definition* shown in Figure 25-3(a). In most highway work, degree of a curve is the central angle subtended by a circular *arc* of 100 ft, the *arc definition* [Figure 25-3(b)]. Formulas relating radius R and degree D are shown beside the illustrations.

Radii of chord- and arc-definition curves for values of D from 1 to 4° are given in Table E-5. Although radius differences between the two definitions appear to to be small in this range, they have some significance in computations.

A chord-definition curve is consistent in using chords for computation and layout with 100.00 ft tape lengths for full stations on large-radius curves. Its disadvantages are (a) R is not directly proportional to the reciprocal of D, and (b) it is more difficult to check sharp curves.

The arc-definition curve has the disadvantage that most measurements between stations are less than a full tape length. Computations are facilitated, however, since an exact value for the radius is obtained by dividing the radius for a 1° curve by the degree D. Also, as will be shown later, the formula for length is exact, which is an advantage in preparing right-of-way descriptions.

Arc and chord definitions give practically the same result when applied to the flat curves common on modern highways and railroads.

25-3. DERIVATION OF FORMULAS. Circular curve elements are shown in Figure 25-4. The *point of intersection* of the tangents (PI) is also called the *vertex* (V). The beginning of the curve, or *point of curvature* (*PC*), and end of the curve or *point of tangency* (PT) are also sometimes called the BC and EC, respectively. Other expressions for these points are *tangent to curve* (*TC*) *and curve to tangent* (CT).

The distance from PC to PI and from PI to PT is the *tangent distance* (T). A line connecting the PC and PT is the *long chord* (LC). *Length of curve*

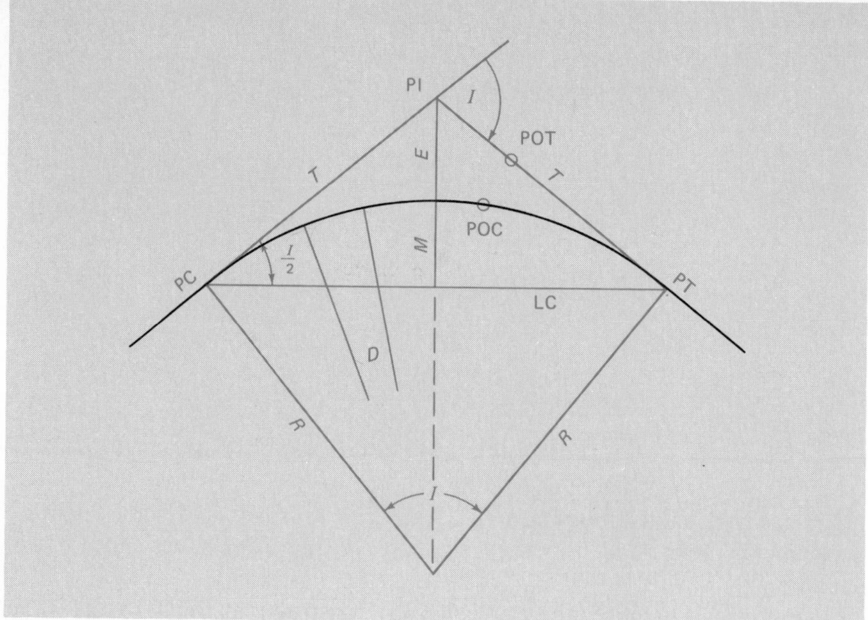

Figure 25-4. Circular curve elements.

(L) is the distance from PC to PT measured along the curve for arc definition, or by 100-ft chords for chord definition.

The *external distance* (E) is the length from vertex to curve on a radial line. *Middle ordinate* (M) is the (radial) distance from midpoint of the long chord to midpoint of the curve. Any *point on curve* is POC; any *point on tangent*, a POT. The degree of any curve is D_a (arc definition) or D_c (*chord definition*).

The change in direction of two tangents is the intersection angle I, which is also equal to the central angle subtended by the curve.

By definition, and from inspection of Figure 25-4, relations for the *arc definition* follow:

$$\frac{D^\circ}{360^\circ} = \frac{100}{2\pi R} \quad \text{and} \quad R = \frac{5729.58}{D} \text{ ft} \tag{25-1}$$

$$T = R \tan \frac{I}{2} \tag{25-2}$$

$$T_a = \frac{T_{1^\circ}}{D_a} \tag{25-3}$$

$$L = 100 \frac{I^\circ}{D^\circ} \tag{25-4}$$

also

$$L = RI \qquad (I \text{ in radians}) \tag{25-5}$$

$$LC = 2R \sin\frac{I}{2} \tag{25-6}$$

$$\frac{R}{R+E} = \cos\frac{I}{2} \quad \text{and} \quad E = R\left(\frac{1}{\cos I/2} - 1\right) \tag{25-7}$$

also

$$E = R \operatorname{exsec}\frac{I}{2} \qquad (25\text{-}7a)^1$$

$$E_a = \frac{E_{1^\circ}}{D_a} \tag{25-8}$$

$$\frac{R-M}{R} = \cos\frac{I}{2} \quad \text{and} \quad M = R\left(1 - \cos\frac{I}{2}\right) \tag{25-9}$$

also

$$M = R \operatorname{vers}\frac{I}{2} \qquad (25\text{-}9a)^1$$

In Eqs. (25-3) and (25-8), T_a and E_a are the tangent distance and external distance, respectively, for the given value of D_a; T_{1° and E_{1° are their lengths for a 1° curve and the given value of I.

The formulas for T, L, E, and M apply also to a chord-definition curve. Small corrections must be added to the values of T_a and E_a found by Eqs. (25-3) and (25-8), respectively.

The formula relating R and D for a chord-definition curve is as follows:

$$R = \frac{50}{\sin D/2} \tag{25-10}$$

25-4. CURVE STATIONING. Normally an initial route survey consists of establishing PIs, laying out tangents, and establishing continuous stationing along them from project beginning, through each PI, to project end. After this, I angles are measured and curves computed and staked. The stationing of points on any curve is based on that of its PI. To compute the PC station, tangent distance T is subtracted from the station of the PI, and to calculate the PT station, curve length L is added to the PC station.

EXAMPLE 25-1

Assume that $I = 8°24'$, the station of the PI is 64 + 27.46, and terrain conditions require the maximum degree of curve permitted by the specifications, which is, say, 2°00′ (arc definition). Calculate the stationing of the PC and PT and the external and middle ordinate distances for this curve.

[1] The functions exsec $I/2$ and vers $I/2$ are equal to $[(1/\cos I/2) - 1]$ and $(1 - \cos I/2)$, respectively. In the past they were frequently used, and given in tables for varying values of I. Now, with electronic calculators, Eqs. (25-7) and (25-9) are preferred.

SOLUTION
By Eq. (25-1):

$$R = \frac{5729.58}{2} = 2864.79 \text{ ft}$$

Check: From Table E-5; $R = 2864.79$.
By Eq. (25-2):

$$T = 2864.79 \times 0.73435 = 210.38$$

Check: By Eq. (25-3):

$$T = \frac{T_{1^\circ}}{D_a} = \frac{420.75}{2} = 210.38$$

where $T_{1^\circ} = 5729.58 \times 0.073435 = 420.75$.
By Eq. (25-4):

$$L = 100 \times \frac{8.40^\circ}{2^\circ} = 420.00 \text{ ft}$$

$$\begin{aligned} \text{PI station} &= 64 + 27.46 \\ -T &= \underline{\ 2 + 10.38} \\ \text{PC station} &= 62 + 17.08 \\ +L &= \underline{\ 4 + 20.00} \\ \text{PT station} &= 66 + 37.08 \end{aligned}$$

Also by Eq. (25-7):

$$E = 2864.79 \times 0.002693 = 7.71 \text{ ft}$$

Check: By Eq. (25-8):

$$E_a = \frac{E_{1^\circ}}{D_a} = \frac{15.42}{2} = 7.71$$

where $E_{1^\circ} = 5729.58\,[(1/0.997314) - 1] = 15.42$.
Also by Eq. (25-9):

$$M = 2864.79 \times 0.002686 = 7.69 \text{ ft}$$

Calculation for the stations of the PC and PT should be arranged as shown. Note that the station of the PT *cannot* be obtained by adding the tangent distance to the station of the PI, although the location of the PT on the ground is determined by measuring the tangent distance from the PI. Points representing the PC and PT must be carefully marked and placed exactly on the tangent lines at the correct distance from the PI, so other computed values will fit their fixed positions on the ground.

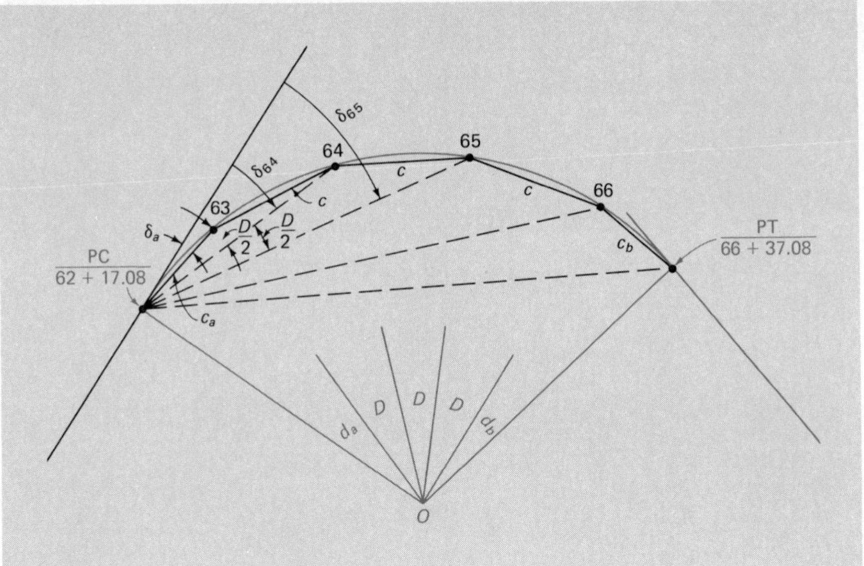

Figure 25-5. Curve layout by deflection angles.

Prior to laying out curves, a route survey is a series of tangents having continuous stationing, so an adjustment has to be made at each PT after curves have been inserted. Thus, for final stationing at the PT, there is a "station equation" or an *equation of chainage*, which relates the stationing back along the curve to that forward along the tangent. For Example 25-1, it would be 66 + 37.08 back = 66 + 37.84 ahead; that is, (PI = 64 + 27.46) + (T = 2 + 10.38) along the forward tangent at the PT before the curve was run in, which shortened the route.

The curve used in a particular situation is selected to fit ground conditions, and specification limitations on maximum D or minimum R. Normally the value of I and station of the PI are available from field measurements on the preliminary line. Then a value of D up to the maximum for a railroad line, or an R suitable for the highway type, is chosen. Sometimes the value of E or M required to miss a stream or steep slope outside or inside the PI is measured, and D or R computed. Tangent distance governs infrequently (one exception is to make a railroad, bus, or subway station fall on a tangent rather than a superelevated curve). Length of curve practically never governs.

25-5. GENERAL PROCEDURE OF CURVE LAYOUT BY DEFLECTION ANGLES WITH TRANSIT OR THEODOLITE AND TAPE. Except for unusual cases, radii of curves on route surveys are too large to permit swinging an arc from the curve center. Circular curves are therefore laid out by (a) deflection angles and chords, (b) tangent offsets, (c) chord offsets, (d) middle ordinates, and (e) other methods. Stakeout by deflection angles is the standard method.

Layout of a curve by deflection angles is illustrated in Figure 25-5. In the figure, assume that a transit or theodolite is set up over the PC (station 62 + 17.08 in Example 25-1). Each full station is to be marked along the curve,

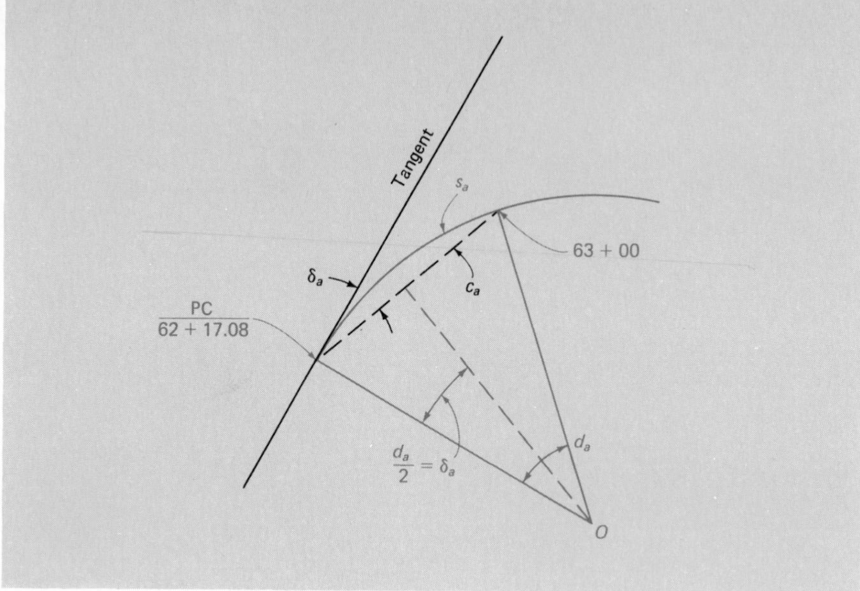

Figure 25-6. Subchords and subdeflections.

since cross sections are normally taken, construction stakes set, and computations of earthwork made at these points. (Half-stations or any other critical points can, of course, also be established.) The first station to be set in this example is 63 + 00. To mark that point from the PC, a backsight is taken on the PI with zero set on the instrument's horizontal circle. Deflection angle δ_a to station 63 + 00 is then turned; two tapepersons measure chord c_a from the PC and set 63 + 00 at the end of the chord on the instrument's line of sight. With station 63 + 00 set, the tapepersons measure chord length c from it, and stake station 64 + 00 where the line of sight of the instrument, now set to δ_{64}, intersects the end of that chord. This process is repeated until the entire curve is laid out.

25-6. COMPUTING DEFLECTION ANGLES AND CHORDS. From the preceding discussion it is clear that deflection angles and chords are important values which must be calculated if a curve is to be run by the deflection-angle method. To stake the first station, which is normally an odd distance from the PC (less than a full-station increment), *subdeflection* angle δ_a and *subchord* c_a are needed. These are detailed in Figure 25-6. In this figure, central angle d_a subtended by arc s_a from the PC to 63 + 00 is calculated by proportion according to the definition of D as:

$$\frac{d_a}{s_a} = \frac{D}{100} \quad \text{from which} \quad d_a = \frac{s_a D}{100} \tag{25-11}$$

where s_a is the difference in stationing between the two points. A fundamental theorem of geometry helpful in circular curve computation and stakeout is that *the angle at a point between a tangent and any chord is half the central angle subtended by the chord.* Thus the subdeflection angle δ_a needed to stake station 63 + 00 is $d_a/2$, or:

$$\delta_a = \frac{s_a D}{200} \quad \text{(degrees)} \tag{25-12}$$

Since subdeflection angles are usually small and are expressed in minutes, Eq. (25-12) becomes:

$$\delta_a = 0.3 s_a D \quad \text{(minutes)} \tag{25-13}$$

The length of the subchord, c_a, can be represented in terms of δ_a and the curve radius as:

$$\sin \delta_a = \frac{c_a}{2R} \qquad \text{from which} \quad c_a = 2R \sin \delta_a \tag{25-14}$$

Since the arc between full stations subtends an additional central angle of D, from the earlier stated geometric theorem, deflection angles to each full station beyond 63 + 00 are found by adding $D/2$ to the previous deflection angle. Full chords, c, are calculated using Eq. (25-14), except that $D/2$ is substituted for δ_a. Equations (25-13) and (25-14) are also used to compute the last subdeflection angle δ_b and subchord c_b but the difference in stationing between the last two stations replaces arc length s_a.

EXAMPLE 25-2

Compute subdeflection angles and subchords δ_a, c_a, δ_b, and c_b, and also calculate chord c of Example 25-1.

By Eq. (25-13):

$$\delta_a = 0.3(82.92)2^\circ = 49.75' = 0^\circ 49' 45''$$

$$\delta_b = 0.3(37.08)2^\circ = 22.25' = 0^\circ 22' 15''$$

By Eq. (25-14):[2]

$$c_a = 2(2864.79) \sin 0^\circ 49' 45'' = 82.92 \text{ ft}$$

$$c_b = 2(2864.79) \sin 0^\circ 22' 15'' = 37.08 \text{ ft}$$

$$c = 2(2864.79) \sin 1^\circ 00' 00'' = 99.99 \text{ ft}$$

An alternate method of computing deflection angles is to multiply the deflection angle per foot of arc or chord by the length of arc or chord being out.

[2] For up to 2°00′ (arc definition) curves, lengths of arcs and chords are nearly the same. On sharper curves chords are shorter than corresponding arc lengths. For example, the chord measurement to lay out a 100-ft arc for a 6°00′ curve is 2(5729.58/6) sin 3° = 99.95 ft. Table E-6 lists true chord lengths for full- and partial-station increments for varying values of D (both arc and chord definitions). True chords calculated in Example 25-2 can be checked by interpolating from this table.

TABLE 25-1. **DEFLECTION-ANGLE AND CHORD DATA FOR EXAMPLE CURVE**

STATION	CHORD	DEFLECTION INCREMENT	DEFLECTION ANGLE
66 + 37.08 (PT)	37.08	0°22′15″	4°12′00″
66 + 00	99.99	1°00′00″	3°49′45″
65 + 00	99.99	1°00′00″	2°49′45″
64 + 00	99.99	1°00′00″	1°49′45″
63 + 00	82.92	0°49′45″	0°49′45″
62 + 17.08 (PC)			

In Example 25-2 the deflection angle per foot of arc is $(D/2)/100 = 1°00'/100 = 0.60'/\text{ft}$. Then deflection angle δ_a from the PC to station 63 + 00 is $0.60' \times 82.92 = 49.75'$. Deflection angles per foot are listed in Table E-5 for varying values of D. Computations for deflection angles and chords on chord-definition curves use the same formulas; but R is calculated by Eq. (25-10). Note that for a given value of degree of curve, R is longer for chord-definition than arc-definition and true subchords for chord-definition curves are longer than their "nominal values" (differences in stationing). A check of Table E-6 verifies this fact.

25-7. CURVE NOTES. Based on principles discussed, deflection-angle and chord data for stakeout of the complete curve of Examples 25-1 and 25-2 have been computed and listed in Table 25-1. Normally, as has been done in this case, the data are prepared for stakeout from the PC, although field conditions may not allow the curve to be completely run from there. This problem is discussed in Section 25-9.

Values of deflection angles are normally carried out to several decimal places for checking purposes, and to avoid accumulating small errors when D is a noninteger number, such as, perhaps, 3°17.24′. In early railroad work with 1-min transits, D was generally rounded off to a multiple of 2 min. Note in Table 25-1 that the deflection angle to the PT is 4°12′, exactly half the I angle of 8°24′. This comparison affords an important check on the calculations of all deflection angles.

Field notes for the curve of this example are recorded in Plate D-13 as they would appear in a field book. Notes run up the page to simplify sketching while looking in a forward direction. Electronic computers of various types and sizes can conveniently perform all necessary computations for curve stakeout by deflection angles.

In many cases it is desirable to *back-in* a curve by setting up over the PT instead of the PC. One setup is thereby eliminated and the long sights are taken on the first measurements. In precise work it is better to run in the curve from both ends to the center, where small errors can be adjusted more readily. On long or very sharp circular curves, or if obstacles block sights from the PC or PT, setups on the curve (POCs) are necessary (see Section 25-9).

25-8. DETAILED PROCEDURES FOR CURVE LAYOUT BY DEFLECTION ANGLES WITH TRANSIT OR THEODOLITE AND TAPE. Regardless of the method used to stake intermediate curve points, the first steps in curve layout are: (1) establishing the PC and PT, normally by measuring tangent distance T from the PI along both the back and forward tangents, and (2) measuring the total deflection angle at the PC from PI to PT. This latter step should be performed whenever possible, since the measured angle must check $I/2$; if it doesn't, an error exists in either measurement or computation and time should not be wasted running an impossible curve.

It is good practice to also stake the midpoint of a curve before beginning to set intermediate points, especially on long curves. The midpoint can be set by bisecting angle $(180° - I)$ at the PI, and laying off the external distance from there. A check of the deflection angle from the PC to the midpoint should yield $I/4$. When staking intermediate points along the curve has reached the midpoint, a chord check measurement should be made to it.

The remaining steps in staking intermediate curve points by the deflection-angle method with transit (or theodolite) and tape are presented with reference to the curve of Examples 25-1 and 25-2. With the instrument set up and leveled over the PC, it is oriented by backsighting on the PI, or on a point along the back tangent, with 0°00′ on the circle. The subdeflection angle of 49′45″ is then turned. Meanwhile, the 17-ft mark of the tape is held on the PC. The 100-ft end of the (subtracting) tape is swung until the line of sight hits a point 0.08 ft back from the 100-ft mark. This is station 63 + 00. To stake station 64 + 00, the rear tapeperson next holds the zero mark on station 63, and the forward tapeperson sets station 64 at distance 99.99 ft by direction from the instrument operator, who has placed an angle of 1°49′45″ on the circle. An experienced forward tapeperson will walk along the extended first full chord, know or estimate the chord offset, and from an outside-the-chord position be holding the tape end and a stake within a foot of the correct location when the instrument operator has the deflection angle ready.

After placing the final full station (66 + 00 in this example), to determine any misclosure in staking a curve, the *closing* PT should be staked using the final deflection angle and subchord. This will rarely agree with the PT established by distance T along the forward tangent from the PI because of accumulated errors. This misclosure or "falling" should be measured; then field precision can be expressed as a numerical ratio like that used in traverse checks. Measured falling distance is the numerator, and $L + 2T$ the denominator. If the misclosure of this example was 0.25 ft, the precision would be $0.25/(420.00 + 2 \times 210.38) = 1/3360$.

25-9. SETUPS ON THE CURVE. Obstacles and extremely long sight distances sometimes make it necessary to set up on the curve. The simplest procedure to follow is one that permits use of the same notes computed for running the curve from the PC.

In this method the instrument is backsighted on any curve station with the telescope inverted and the circle set to *the deflection angle for that station from the PC*. The telescope is plunged to the normal position, and deflection angles previously computed for the various stations from the PC are used.

POC

In the example of the preceding sections, if a setup is required at station 65, place 0°00′ on the instrument and sight to the PC with the telescope inverted. Plunge, set the angle to read the deflection angle 3°49′45″, and stake station 66. Or, if the PC is not visible, set 0°49′45″ on the instrument, sight on station 63, plunge, set the angle to 3°49′45″, and locate station 66.

A simple sketch will make clear the geometry basic to this procedure.

25-10. CURVE LAYOUT BY DEFLECTION ANGLES WITH ELECTRONIC TACHEOMETERS. The new electronic tacheometers or total-station instruments described in Chapter 10 are extremely convenient for curve layout by deflection angles. With these instruments, field party size is reduced from three to two persons. Deflection angles are calculated and laid off as in the preceding example, but chords are all measured electronically from the PC or other station where the instrument is placed. If stakeout is planned from the PC, chords from there are calculated by Eq. (25-14), except that the deflection angle for each station is substituted for δ_a to obtain the corresponding chord. The chords necessary to stake the curve of the preceding example using an electronic tacheometer set up at the PC are 82.92 ft for 63 + 00, 182.89 ft for 64 + 00, 282.80 ft for 65 + 00, 382.63 ft for 66 + 00, and 419.62 ft to the PT, which is the long chord given by Eq. (25-6).

To stake curves using an electronic tacheometer, the instrument is placed in its tracking mode. The deflection angle to each station is turned and the required chord to that station entered into the instrument. The instrument operator directs the person with the reflector into the proper alignment. The reflector is then moved forward or back, as necessary, until the proper chord distance is achieved, where the stake is set. Although curves can be staked rapidly with this equipment, a danger associated with the procedure is that each stake is set individually and does not depend on previous stations. Thus a check at the end of the curve is not achieved as in the transit or theodolite and tape method described in Section 25-8, so mistakes in angles or distances could go undetected. Large blunders can usually be discovered by visual inspection of the curve stakes, but quickly taping the chords between adjacent stations gives a better check.

25-11. CURVE LAYOUT BY OFFSETS. For short curves, when a transit or theodolite is not available, or for checking purposes, one of four offset-type methods can be used for circular curves: *tangent offsets*, *chord offsets*, *middle ordinates*, and *ordinates from the long chord*. Figure 25-7 shows the relationship of chord offsets (CO), tangent offsets (TO), and middle ordinates (MO). Visually and by formula comparison, the chord offset for full stations is

$$\text{CO} = 2c \sin \frac{D}{2} = \frac{c^2}{R} = 2\text{TO} = \text{approximately } 8\text{MO} \qquad (25\text{-}15)$$

Since $\sin 1° = 0.0175$ (approx.), $\text{CO} = c(0.0175)D = 0.01\tfrac{3}{4}D$ (approx.), where D is in degrees and decimals.

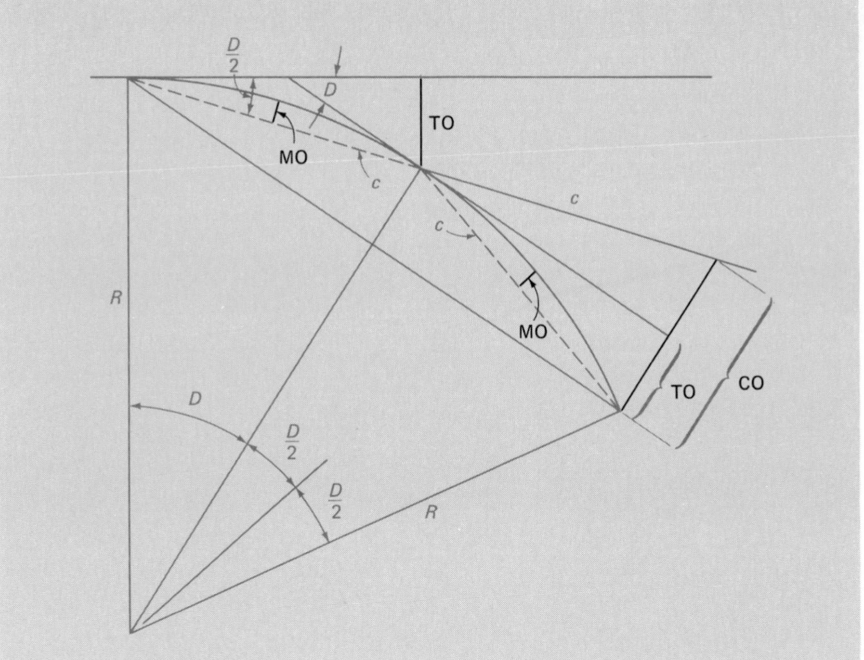

Figure 25-7. Curve offsets.

The middle ordinate m for any subchord is $R(1 - \cos\delta)$, with δ being the deflection angle for the chord. A useful equation in layout or checking curves in place is:

$$D\,(\text{in degrees}) = m\,(\text{in inches})\text{ for a 62-ft chord (approx.)} \qquad (25\text{-}16)$$

Geometry of the tangent offset method is shown in Figure 25-8. The figure illustrates that the curve is most conveniently laid out in both directions from the PC and PT to a common point near the middle of the curve. This procedure avoids long measurements and provides a check where small adjustments can more easily be made if necessary. To lay out a curve using this method, tangent distances are measured to established temporary points at A, B, and C of the figure. From these points, right-angle measurements (tangent offsets) are made to set the curve stakes. Tangent distances (TD) and tangent offsets (TO) are calculated using chords and δ angles in the following formulas:

$$\text{TD} = c\cos\delta \qquad (25\text{-}17)$$

$$\text{TO} = c\sin\delta \qquad (25\text{-}18)$$

In Eqs. (25-17) and (25-18), δ angles are calculated using either Eq. (25-12) or (25-13), and chords c are determined from Eq. (25-14).

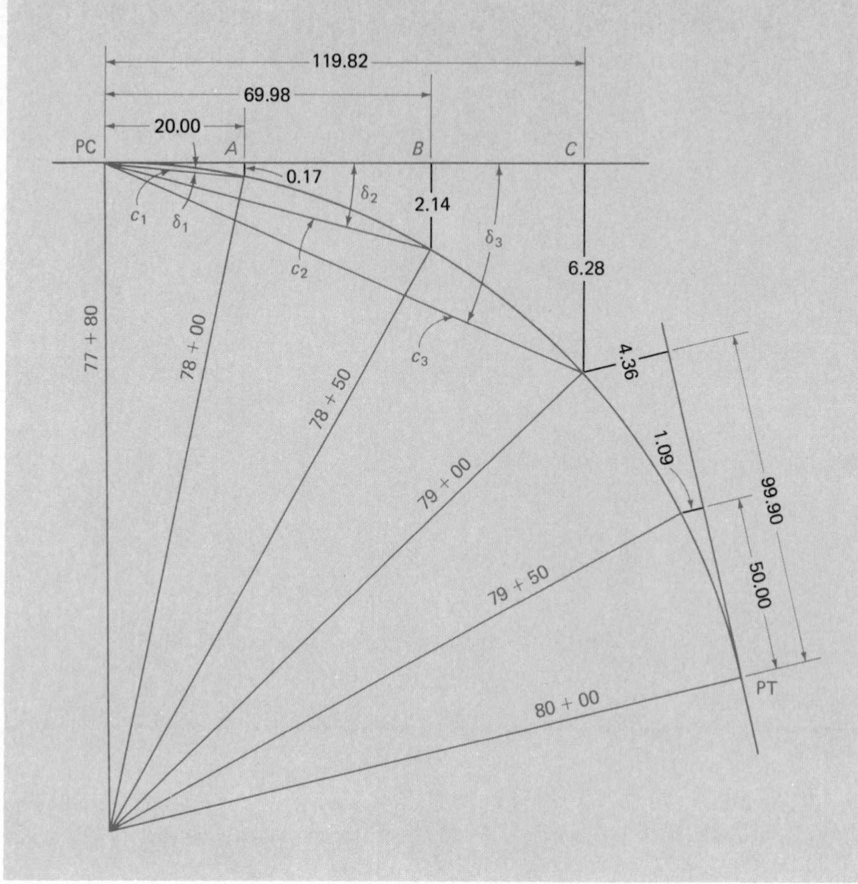

Figure 25-8. Curve stakeout by tangent offsets.

EXAMPLE 25-3

Compute and tabulate the data necessary to stake, by tangent offsets, the half-stations of a circular curve having $I = 11°00'$, $D_c = 5°00'$ (chord definition), and PC = 77 + 80.00.

SOLUTION

By Eq. (25-4), curve length is:

$$L = 100\left(\frac{11}{5}\right) = 220 \text{ ft}$$

Therefore the PT station is (77 + 80) + (2 + 20) = 80 + 00. Intermediate stations to be staked are 78 + 00, 78 + 50, 79 + 00, and 79 + 50 as shown in Figure 25-8.

TABLE 25-2. TANGENT OFFSET DATA FOR EXAMPLE 25-3.

STATION	DEFLECTION ANGLE δ	CHORD c	TANGENT DISTANCE ($c \cos \delta$)	TANGENT OFFSET ($c \sin \delta$)
80 + 00 (PT)				
79 + 50	1°15′	50.01	50.00	1.09
79 + 00	2°30′	100.00	99.90	4.36
79 + 00	3°00′	119.98	119.82	6.28
78 + 50	1°45′	70.01	69.98	2.14
78 + 00	0°30′	20.00	20.00	0.17
77 + 80 (PC)				

By Eq. (25-13), δ angles from the PC are:

$$\delta_1 = 0.3(20)5 = 0°30'$$
$$\delta_2 = 0.3(70)5 = 105' = 1°45'$$
$$\delta_3 = 0.3(120)5 = 180' = 3°00'$$

By Eq. (25-10) the radius is:

$$R = \frac{50}{\sin 2°30'} = 1146.28 \text{ ft} \qquad \text{(check by Table E-5)}$$

By Eq. (25-14), chords from the PC are:

$$c_1 = 2(1146.28) \sin 0°30' = 20.00 \text{ ft}$$
$$c_2 = 2(1146.28) \sin 1°45' = 70.01 \text{ ft}$$
$$c_3 = 2(1146.28) \sin 3°00' = 119.98 \text{ ft}$$

Now using Eqs. (25-17) and (25-18), tangent distances and tangent offsets are calculated. Chords, δ angles, tangent distances, and tangent offsets to stake points from the PT are computed in the same manner. All data for the problem are listed in Table 25-2. Tangent distances tabulated are lengths from the PC or PT that must be measured to establish points A, B, C, and so on, and tangent offsets are distances from these points needed to locate curve stakes.

25-12. SPECIAL CIRCULAR CURVE PROBLEMS. Many special problems arise in the design and computation of circular curves. Three of the more common ones are discussed here, and each can be solved using coordinate geometry formulas given in Appendix B.

25-12.1. PASSING A CIRCULAR CURVE THROUGH A FIXED POINT. One problem that often occurs in practice is to determine the radius of a curve connecting two established tangents and going through a fixed point such as

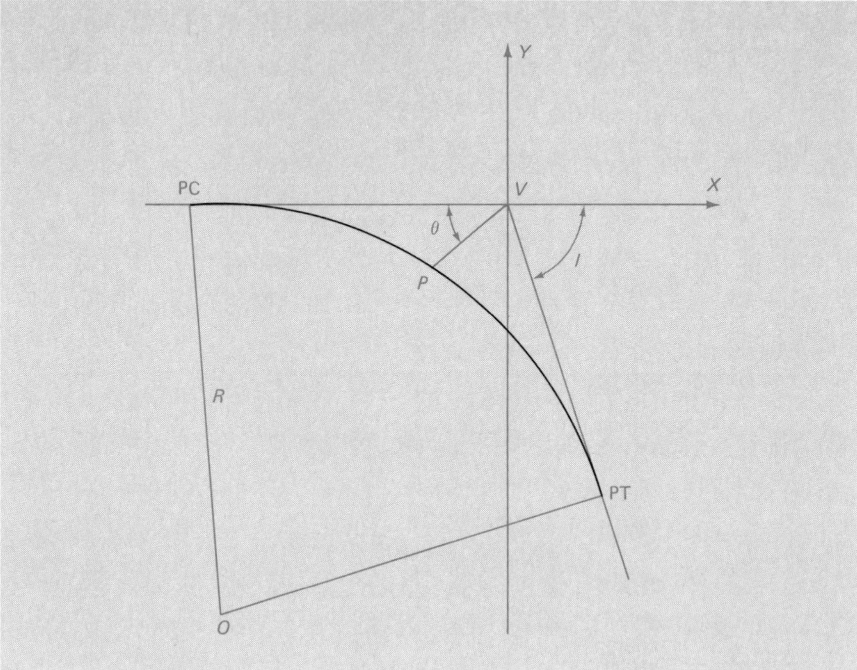

Figure 25-9. Passing a curve through a fixed point.

an underpass, overpass, or existing bridge. The problem can be solved by establishing an X-Y coordinate system as shown in Figure 25-9, where the origin occurs at the PI (V) and Y coincides with the back tangent. Coordinates of the radius point in this system are $X_O = -R \tan I/2$, and $Y_O = -R$. From measurements of distance PV and angle θ, coordinates X_p and Y_p of point P can be determined. Then the following equation for a circle, obtained by substitution into Eq. (B-6) of Appendix B, can be written:

$$R^2 = \left(X_p + R \tan\frac{I}{2}\right)^2 + (Y_p + R)^2 \tag{25-19}$$

With X_p, Y_p, and $I/2$ known in Eq. (25-19), a solution for R can be found. The equation is quadratic but can be solved using Eq. (B-8) from Section B-5, Appendix B.

25-12.2. INTERSECTION OF A CIRCULAR CURVE AND STRAIGHT LINE. Another frequently encountered problem involving curve computation is determination of the intersection point of a circular curve and straight line. An example is illustrated in Figure 25-10. The problem is solved by writing an equation for the straight line, Eq. (B-5), and one for the circle, Eq. (B-6), in which the unknown coordinates X_p and Y_p of the intersection point are included. The two equations are then solved simultaneously for X_p and Y_p. In typical cases,

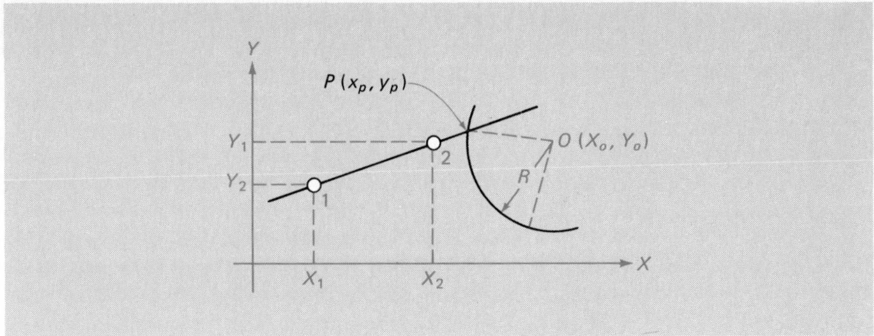

Figure 25-10. Intersection of circular curve and straight line.

coordinates X_1, Y_1, X_2, Y_2, X_O, and Y_O are known, as well as R. An example similar to that illustrated in Figure 25-10 is solved in Appendix B.

25-12.3. INTERSECTION OF TWO CIRCULAR CURVES. Another problem, illustrated in Figure 25-11, involves computing the intersection point of two circular curves. This is also best handled by coordinate geometry and requires writing equations in which unknown coordinates X_p and Y_p are included. The equations are solved to determine the unknowns. Coordinates X_{O_1}, Y_{O_1}, X_{O_2}, and Y_{O_2} are typically determined through survey, and R_1 and R_2 selected based on design or topographic constraints. An example similar to that illustrated in Figure 25-11 is given in Appendix B.

The problems described in this section and the preceding one arise most often in the design of subdivisions and interchanges, and in calculating right-of-way points along highways and railroads.

Figure 25-11. Intersection of two circular curves.

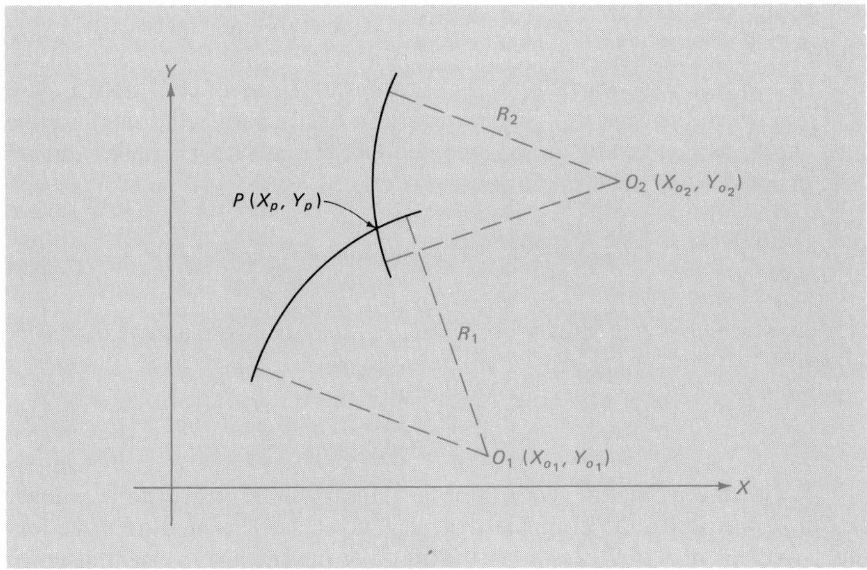

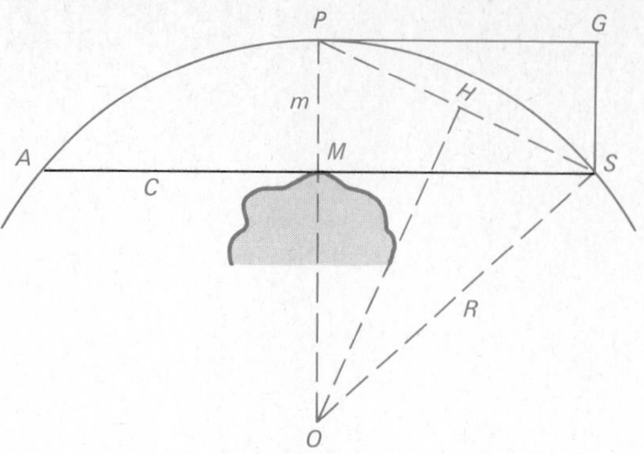

Figure 25-12. Sight distance.

25-13. COMPOUND AND REVERSE CURVES. Compound and reverse curves are combinations of two or more circular curves. They should be used only for low-speed trafficways, and in terrain where simple curves cannot be fitted to the ground without excessive construction costs.

Special formulas have been derived to facilitate computations for such curves and are demonstrated in texts on route surveying. A compound curve is run at the PC and PT, or perhaps by one setup at the *point of compound curvature* (CC). Reverse curves are handled in similar fashion.

25-14. SIGHT DISTANCE ON HORIZONTAL CURVES. Highway safety requires certain minimum sight distances in zones where passing is permitted, and in nonpassing areas to assure a reasonable stopping distance if there is an object on the roadway. Specifications and tables list suitable values based on vehicular speeds, the perception and reaction times of an average individual, braking distance for a given coefficient of friction during deceleration, and type and condition of the pavement.

A minimum sight distance of 600 ft is desirable for speeds as low as 30 mi/hr.

An approximate formula for sight distance can be derived from Figure 25-12, in which the clear sight distance past an obstruction is the length of the long chord *AS*, denoted by *C*; and the required clearance is the middle ordinate *PM*, denoted by *m*. Then in triangles *SPG* and SOH,

$$m : SP = \frac{SP}{2} : R \quad \text{and} \quad m = \frac{(SP)^2}{2R}$$

Usually *m* is small compared with *R*, and *SP* may be assumed equal to $C/2$. Then

$$m = \frac{C^2}{8R} \tag{25-20}$$

If distance *m* from the center line of a highway to the obstruction is known or can be measured, the available sight distance *C* is calculated from the formula. Actually cars travel on either the inside or outside lane, so sight distance

AS is not exactly the true stopping distance, but the computed length is on the safe side and satisfactory for practical use.

25-15. SOURCES OF ERROR. Some sources of error in curve computations and layout are:

1. Inability to set an instrument to the exact required part of a minute for deflection angles.
2. Poor intersections between some tape and sight lines (which may be almost parallel) on very flat curves.
3. Use of shorter than full 100.00-ft tape lengths on arc-definition curves.

25-16. MISTAKES. Typical mistakes that occur in laying out a curve in the field are the following:

1. Failure to take equal numbers of direct and reversed measurements of the deflection angle at the PI before computing or laying out the curve.
2. Adding the tangent distance to the PI station to get the PT station.
3. Using 100.00-ft chords to lay out arc-definition curves having D greater than 2°.
4. Taping subchords of nominal length for chord-definition curves having D greater than 5° (a *nominal* 50-ft subchord for a 6° curve requires a measurement of 50.02 ft).

PROBLEMS

25-1. For the following circular curves having a radius R, what is their degree of curve by (1) arc definition and (2) chord definition?
(a) 800 ft (b) 1400 ft (c) 3600 ft

25-2. Describe and sketch compound, broken-back, and reverse curves. Why are they objectionable for transportation route alignments?

25-3. List advantages and disadvantages of the chord and arc definitions for the degree of a circular curve in construction layout and land descriptions.

Compute T, L, E, M, R, or D, and stations of the PC and PT for the circular curves in Problems 25-4 through 25-7. Distances are in feet. Use the chord definition for railroad curves, arc definition for highways.

25-4. Railroad curve with $D = 3°00'$, $I = 18°00'$, and PI station = 95 + 53.90.

25-5. Highway curve with $R = 1875.00$, $I = 15°30'$, and PI station = 65 + 17.50.

25-6. Highway curve with $D = 4°36'$, $I = 24°20'$, and PI station = 70+ 28.64.

25-7. Railroad curve with $R = 2050.00$, $I = 19°00'$, and PI station = 123 + 40.28.

Tabulate the curve data R, D, T, L, E, M, PC, PT, and deflection angles and chords to lay out the curves at full stations in Problems 25-8 through 25-15. Use arc definition for highways and chord definition for railroads.

25-8. Highway curve with $R = 1455.00$, $I = 33°30'$, and PI station = 82 + 64.20.

25-9. Railroad curve with $D = 3°10'$, $I = 20°18'$, and PI station = 58 + 54.62.

25-10. Highway curve having $R = 2800$ ft, $I = 12°00'$, and PI station = 132 + 11.86.

25-11. Railroad curve with $D = 2°30'$, $I = 14°40'$, and PI station = 106 + 22.37.

25-12. Highway curve with $L = 550$ ft, $R = 2585.00$, and PI station = 39 + 29.50.

25-13. Railroad curve with $L = 400$ ft, $D = 4°15'$, and PI station = 62 + 35.00.

25-14. Highway curve having $T = 275.00$ ft, $R = 1270.00$, and PI station = 28 + 07.25.
25-15. Railroad curve with $T = 195.00$, $D = 2°50'$, and PI station = 106 + 52.91.

In problems 25-16 through 25-19 tabulate the curve data and deflection angles and chords to lay out the following curves at half-station increments.
25-16. The curve of Problem 25-8.
25-17. The curve of Problem 25-9.
25-18. The curve of Problem 25-10.
25-19. The curve of Problem 25-11.

25-20. A streetcar line on the center of an 80-ft street makes a 76°24′ turn into another street of equal width. The corner curb line has $R = 12$ ft. What is the largest R that can be given the track center line if the law requires it be at least 15 ft from the curb?

Tabulate all data required to lay out by deflection angles, at full-station increments, the simple circular curves of Problems 25-21 and 25-22.
25-21. The R for a highway curve will be rounded off to the nearest larger multiple of 100 ft. Field conditions require M to be approximately 20 ft to avoid an embankment. The PT = 49 + 18.70 and $I = 22°00'$.
25-22. The D for a highway curve will be rounded off to the nearest multiple of 20 ft. Field measurements show T should be approximately 190 ft to avoid an overpass. The PI = 36 + 64.50 and $I = 14°00'$.

25-23. A highway survey PI falls in a pond, so a cutoff line $AB = 280.00$ ft is run between the tangents. In the triangle formed by points A, B, and the PI, the angle at $A = 16°30'$ and at $B = 22°18'$. The station of A is 45 + 29.30. Calculate and tabulate curve notes to run a 4°30′ curve at half-station increments to connect the tangents.
25-24. A single circular highway curve will join tangents XV and VY and also be tangent to BC. Calculate R, L, and stations of the PC and PT in the figure.

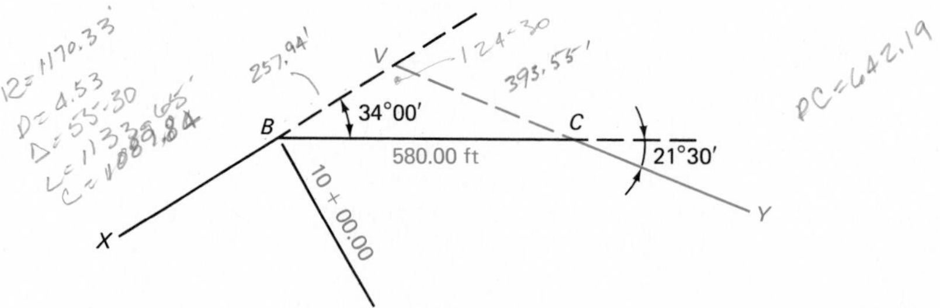

25-25. Compute R_x to fit requirements of the figure and make the tangent distances of the two curves equal.

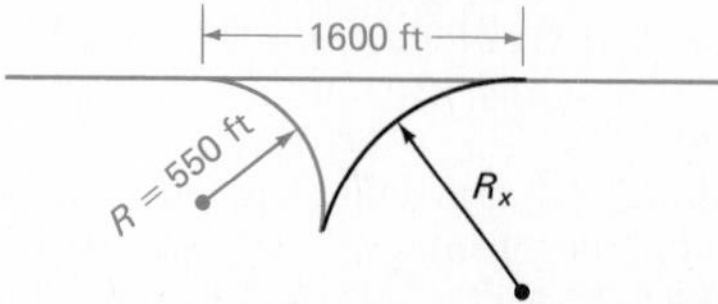

25-26. After a backsight on the PC with 0°00′ set on the instrument, what is the deflection angle to the following curve points?
(a) Setup at midpoint, deflection to the PT.
(b) Instrument at midpoint, deflection to the $\frac{3}{4}$ point.
(c) Setup at $\frac{1}{4}$ point of curve, deflection to $\frac{3}{4}$ point.
25-27. Why should the vertex angle be measured by repetition?

25-28. Assume the curve of Problem 25-8 will be staked at full-station increments from the PC using an electronic tacheometer. Compute and tabulate the curve data, deflection angles, and chords needed.
25-29. Similar to Problem 25-28, except for the curve of Problem 25-9.

For Problems 25-30 and 25-31 compute and tabulate coordinates to stake out a circular railroad curve by tangent offsets. Regular and 50-ft stations are to be set, and approximately one-half the curve laid out from each tangent.
25-30. $L = 240.00$ ft, $D = 4°40'$, and station of PC = 82 + 15.00.
25-31. Station of PI = 55 + 45.50, $D = 5°30'$, and $L = 460.00$ ft.

25-32. Under what field conditions is layout of a circular curve by offsets from the LC an appropriate method? What are its disadvantages?
25-33. A running track must be exactly 5000 ft along its center line with two semicircles and two tangents. The two curves are to constitute one-half its total length. Calculate L, R, and D for the curves.
25-34. In the figure the coordinates of points A and O are $X_A = 250.00$ and $Y_A = 400.00$; and $X_O = 630.00$ and $Y_O = 750.00$. If the azimuth of line AB is 39°28′ and the circular curve radius 180.00 ft, calculate the coordinates of intersection point P.

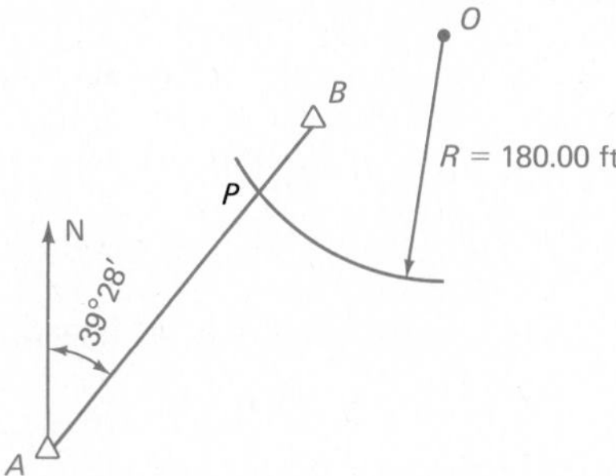

25-35. Coordinates of a circle center O_1 are $X_{O_1} = 1000.00$ and $Y_{O_1} = 1000.00$ ft; and for circle center O_2, $X_{O_2} = 1460.00$ and $Y_{O_2} = 590.00$ ft; $R_1 = 320$ ft and $R_2 = 350$ ft. Compute the coordinates of intersection point P shown in the figure.

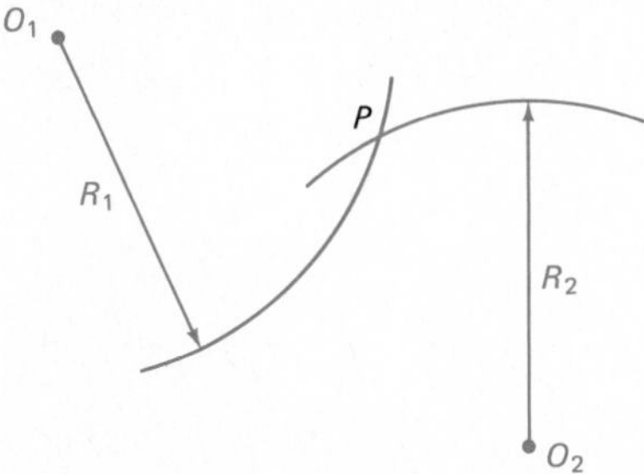

What sight distance is available if there is an obstruction on a radial line through the PI inside the curves in Problems 25-36 and 25-37?
25-36. For Problem 25-8, obstacle 20 ft from curve.
25-37. For Problem 25-13, obstacle 25 ft from curve.

25-38. If the misclosure for the curve of Problem 25-8 computed as described in Section 25-8 is 0.35 ft, what is the field layout precision? Do you think it would be acceptable for a farm road?
25-39. In laying out a park road, the center-line tangents are to be connected by the longest possible radius curve which will provide a 10-ft clearance between the

north edge of a 20-ft road and a tree of historic value (see figure). Determine the roadway centerline radius.

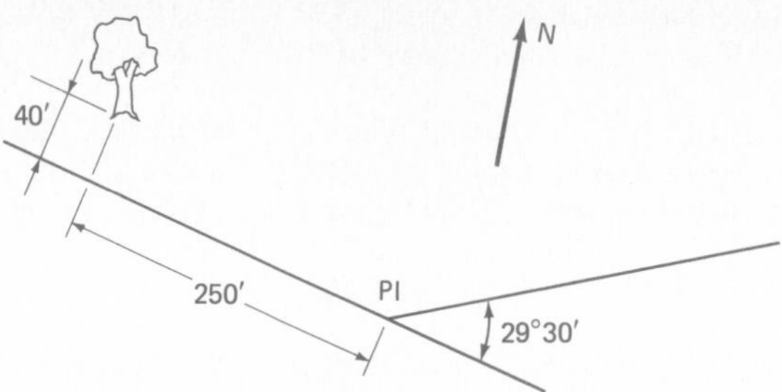

25-40. Compute the area bounded by the two arcs and tangent in Problem 25-25.

25-41. Write a computer program for calculating circular curve notes by the deflection angle method.

25-42. Write a computer program for calculating circular curve notes by the tangent offset method.

BIBLIOGRAPHY

Hikerson, T. F. 1967. *Route Location and Design*, 5th ed. New York: McGraw-Hill.

Hubbell, C. C. 1951. "Circular Curve Formulas." *Surveying and Mapping* 11(no. 3):287.

Meyer, C. F., and D. W. Gibson. 1980. *Route Surveying and Design*, 5th ed. New York: Harper & Row.

Pearson, F. 1983. "Non-Centerline Layout." *ASCE, Journal of Surveying Engineering* 109(no. 1):24.

Smirnoff, M. V. 1958. "About Finding the Radius of a Circular Curve." *Surveying and Mapping* 18(no. 3):321.

26 PARABOLIC CURVES

26-1. INTRODUCTION. Parabolic curves are normally used to provide a smooth transition between grade lines in a vertical plane for highways and railroads. An example is illustrated in Figure 26-1, which shows the profile view of a proposed section of highway to be constructed from A to B. A grade line consisting of three tangent sections is designed to fit the ground profile. Two vertical curves are needed; curve a to join tangents 1 and 2, and curve b to connect tangents 2 and 3. The function of these curves is to provide a gradual change in grade from the initial (*back*) tangent to the grade of the second (*forward*) tangent. Because parabolas provide a *constant rate of change of grade*, they are ideal for vertical alignments used by vehicular traffic.

Besides their use in vertical planes on highways and railroads, parabolas are also employed in horizontal alignments on landscaping work and for other projects requiring a curve possessing a pleasing appearance that can be laid out with a tape. The long parabolic curves used in route surveying practice differ only slightly on the ground from large-radius circular curves.

Two basic types of parabolic curves, *crest* and *sag*, are illustrated in Figure 26-1. Curve a is a crest type, which by definition undergoes a negative change in grade; that is, the curve turns downward. Curve b is a sag type, in which the change in grade is positive and the curve turns upward. Vertical curves must be designed to (a) fit the grade lines they connect, (b) have lengths sufficient to meet specifications covering maximum rate of change of grade (which affects the comfort of vehicle occupants), and (c) provide sufficient sight distance for safe vehicle operation (see Section 26-9).

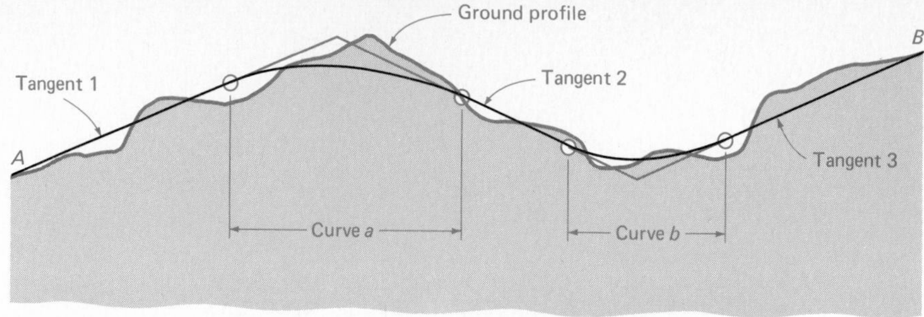

Figure 26-1. Grade line and ground profile of a proposed highway section.

Elevations at selected points (usually full or half-stations) along parabolic curves can be computed by either the *tangent-offset method* or the *chord-gradient method*. In this brief treatment, only the tangent-offset method will be demonstrated because it is simple, straightforward, conveniently performed with calculators and computers, and self-checking.

After elevations of curve points have been computed, they are staked in the field to guide construction operations so the route can be built according to plan.

26-2. THE GENERAL EQUATION OF A PARABOLIC CURVE. The general mathematical expression of a parabola, with respect to an XY rectangular coordinate system, is given by

$$Y_p = a + bX_p + cX_p^2 \tag{26-1}$$

Figure 26-2. Terms for a parabola.

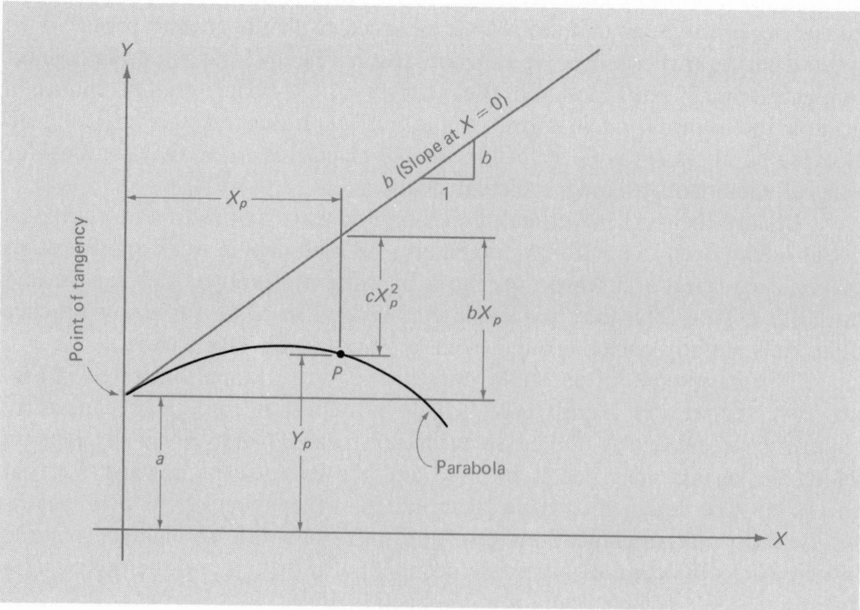

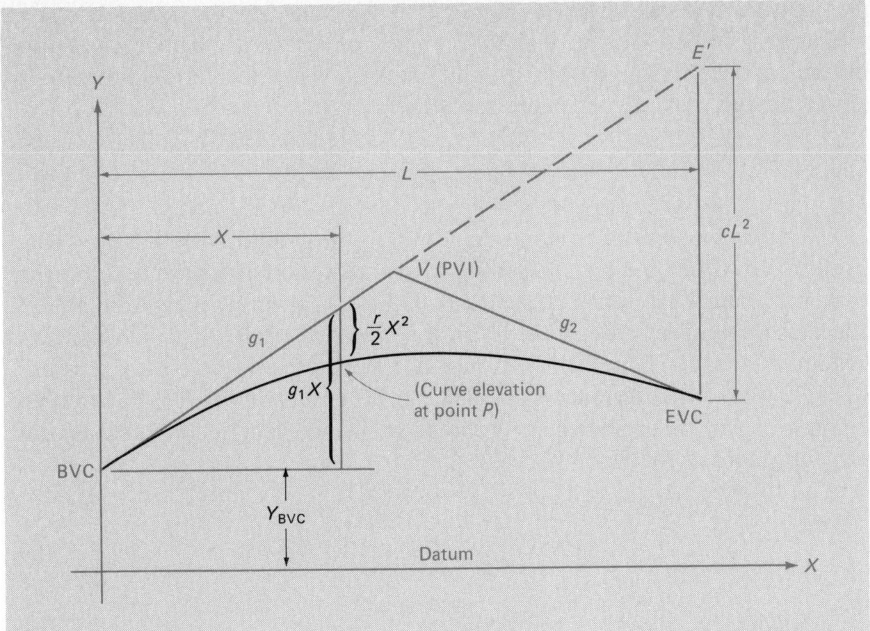

Figure 26-3. Parabolic curve relationships.

where Y_p is the ordinate at any point p on the parabola located a distance X_p from the origin of the curve, and a, b, and c are constants. Figure 26-2 shows a parabola in an XY rectangular coordinate system and illustrates the physical significance of terms in Eq. (26-1). Note from the figure that a is the ordinate at the beginning of the curve where $X = 0$; b the slope of a tangent to the curve at $X = 0$; bX_p the change in ordinate along the tangent over distance X_p, and cX_p^2 the parabola's departure from the tangent (*tangent offset*) in distance X_p. When the terms a, bX_p, and cX_p^2 are combined as in Eq. (26-1), and shown in Figure 26-2, they produce Y_p. For the crest curve of Figure 26-2, b has a positive algebraic sign and c is negative.

26-3. THE EQUATION OF AN EQUAL-TANGENT PARABOLIC CURVE IN SURVEYING TERMINOLOGY. Figure 26-3 shows a parabola that joins two intersecting tangents of a grade line. The parabola is essentially identical to that in Figure 26-2, except terms used are those commonly employed by surveyors and engineers. In the figure, BVC is the *beginning of vertical curve*, sometimes called the PVC (point of vertical curvature); V is the *vertex*, often called the PVI (point of vertical intersection); and EVC is the *end of vertical curve*, interchangeably called the PVT (point of vertical tangency). The percent grade of the back tangent is g_1, and that of the forward tangent is g_2. The curve's length, L, is the horizontal distance (in stations) from the BVC to the EVC. The curve of Figure 26-3 is called "equal tangent" because the horizontal distances from the BVC to V, and from V to the EVC are equal, each being L/2. Proof of this is given in Section 26-5.

On the XY axis system in Figure 26-3, X values are horizontal distances, in stations, measured from the BVC, and Y values are elevations, in feet (or meters), measured from datum, usually sea level. Using this terminology in Eq. (26-1), the parabola can be expressed as:

$$Y = Y_{BVC} + g_1X + cX^2 \tag{26-2}$$

Correspondence of terms in Eq. (26-2) to those of Eq. (26-1) are $a = Y_{BVC}$ (elevation of BVC) and $bX_p = g_1X$ (change in elevation along the back tangent with increasing X). To express constant c of Eq. (26-2) in surveying terminology, consider the tangent offset from E′ on the extended back tangent to the EVC in Figure 26-3. Its value (which is negative for the crest curve shown) is cL^2, where L (horizontal distance from BVC to E′) is substituted for X. From the figure, cL^2 can be expressed in terms of horizontal lengths (in stations) and percent grades as follows:

$$cL^2 = g_1\left(\frac{L}{2}\right) + g_2\left(\frac{L}{2}\right) - g_1L \tag{a}$$

Solving Eq. (a) for constant c gives:

$$c = \frac{g_2 - g_1}{2L} \tag{b}$$

Substituting Eq. (b) into Eq. (26-2) results in the following equation for an equal-tangent vertical curve in surveying terminology:

$$Y = Y_{BVC} + g_1X + \left(\frac{g_2 - g_1}{2L}\right)X^2 \tag{26-3}$$

The *rate of change of grade*, r, of an equal-tangent parabolic curve equals the total grade change from BVC to EVC, divided by the length L, in stations, over which the change occurs, or:

$$r = \frac{g_2 - g_1}{L} \tag{26-4}$$

As mentioned earlier, the value of r (which is negative for a crest curve and positive for a sag type) is an important design parameter because it controls rider comfort. To incorporate it in the equation for parabolic curves, Eq. (26-4) is substituted in Eq. (26-3):

$$Y = Y_{BVC} + g_1X + \left(\frac{r}{2}\right)X^2 \tag{26-5}$$

Figure 26-3 illustrates how the terms of Eq. (26-5) combine to give the curve elevation at point P.

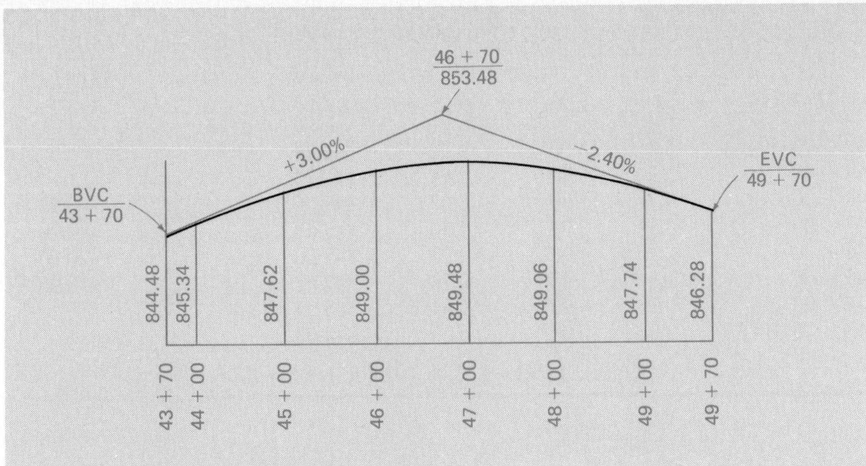

Figure 26-4. Crest curve of Example 26-1.

26-4. VERTICAL-CURVE COMPUTATION USING THE PARABOLIC EQUATION. Computations for vertical parabolic curves are normally done in tabular form. The following example for a crest curve illustrates these procedures.

EXAMPLE 26-1

A grade g_1 of $+3.00\%$ intersects grade g_2 of -2.40% at a vertex whose station and elevation are 46 + 70 and 853.48, respectively. An equal-tangent parabolic curve 600 ft long has been selected to join the two tangents. Compute and tabulate the curve for stakeout at full stations. (Figure 26-4 shows the curve.)

SOLUTION

By Eq. (26-4):

$$r = \frac{-2.40 - 3.00}{6} = -0.90\%/\text{station}$$

Stationing

$$\begin{aligned} V &= 46 + 70 \\ -L/2 &= \underline{\ 3 + 00\ } \\ \text{BVC} &= 43 + 70 \\ +L &= \underline{\ 6 + 00\ } \\ \text{EVC} &= 49 + 70 \end{aligned}$$

$$\text{elev}_{\text{BVC}} = 853.48 - 3.00(3) = 844.48 \text{ ft}$$

The remaining calculations utilize Eq. (26-5) and are listed in Table 26-1. A check on curve elevations is obtained by computing the first and second differences between the elevations of full stations, as shown in the right-hand

TABLE 26-1. **NOTES FOR THE CURVE OF EXAMPLE 26-1**

STATION	X	g_1X	$rX^2/2$	CURVE ELEVATION	FIRST DIFF.	SECOND DIFF.
49 + 70 (EVC)	6.0	18.00	−16.20	846.28		
49 + 00	5.3	15.90	−12.64	847.74		
48 + 00	4.3	12.90	− 8.32	849.06	−1.32	
47 + 00	3.3	9.90	− 4.90	849.48	−0.42	−0.90
46 + 00	2.3	6.90	− 2.38	849.00	0.48	−0.90
45 + 00	1.3	3.90	− 0.76	847.62	1.38	−0.90
44 + 00	0.3	0.90	− 0.04	845.34	2.28	−0.90
43 + 70 (BVC)	0.0	0.00	0.00	844.48		

Check: EVC = 853.48 − 2.40(3) = 846.28

columns of the table. Unless disturbed by rounding off, all second differences (rate of change) should be equal. For full-station increments the second differences equal r; for half-station increments they equal $r/4$.

It is sometimes desirable to calculate the elevation of the curve's center point. This can be done using $X = L/2$ in Eq. (26-5), and for Example 26-1 it is:

$$Y_{\text{center}} = 844.48 + 3.00(3) - \left(\frac{0.90}{2}\right)(3^2) = 849.43 \text{ ft}$$

This can be checked by employing the property of a parabolic curve that *the curve center falls halfway between the vertex and the midpoint of the long chord* (line from BVC to EVC). The elevation of the midpoint of the long chord (LC) is simply the average of the elevations of the BVC and EVC, and for Example 26-1 it is:

$$Y_{\text{midpoint}}\ \text{LC} = \frac{844.48 + 846.28}{2} = 845.38 \text{ ft}$$

By the property stated above, the elevation of the curve center for Example 26-1 is:

$$Y_{\text{center}} = \frac{845.38 + 853.48}{2} = 849.43 \text{ ft (Check)}$$

26-5. THE EQUAL-TANGENT PROPERTY OF A PARABOLA. The curve defined by Eqs. (26-3) and (26-5) has been called an *equal-tangent* parabolic curve, which means the vertex occurs a distance $X = L/2$ from the BVC. Proof of this property is readily made with reference to Figure 26-5, which illustrates a sag curve. In the figure assume the horizontal distance from BVC to V is an unknown val-

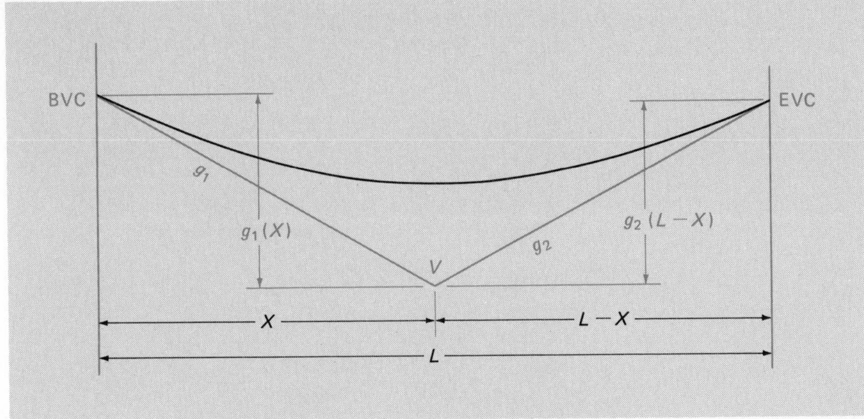

Figure 26-5. Proof of equal-tangent property of a parabola.

ue X; thus, the remaining distance from V to EVC is $(L - X)$. Two expressions can be written for the elevation of the EVC. The first, using Eq. (26-3) with $X = L$, yields:

$$Y_{\text{EVC}} = Y_{\text{BVC}} + g_1 L + \left(\frac{g_2 - g_1}{2L}\right) L^2 \qquad \text{(c)}$$

The second, using changes in elevation that occur along the tangents, gives:

$$Y_{\text{EVC}} = Y_{\text{BVC}} + g_1 X + g_2(L - X) \qquad \text{(d)}$$

Equating Eqs. (c) and (d) and solving, $X = L/2$. Thus, distances BVC to V, and V to EVC are equal—hence, the term *equal-tangent parabolic curve.*

26-6. COMPUTATIONS FOR AN UNEQUAL-TANGENT CURVE. An unequal-tangent vertical curve is simply a compounded pair of equal-tangent curves, where the EVC of the first is the BVC of the second. This point is called CVC, point of *compound vertical curvature.* In Figure 26-6, a 2.00% grade intersects a +1.60% grade at station 87 + 00 and elevation 743.24. A 400-ft vertical curve is to be extended back from the vertex, and a 600-ft curve run in forward to closely fit ground conditions.

To perform calculations for this type of curve, connect the midpoints of the tangents for the two curves, stations 85 + 00 and 90 + 00, to obtain line AB. Compute elevations for A and B and, using them, calculate the grade of AB by dividing the difference in elevation between B and A by the distance in stations separating these two points. From grade AB, determine the CVC elevation.

Now compute two vertical curves, one from BVC to CVC and another from CVC to EVC, by methods of Section 26-5. Since both curves are tangent to the same line AB at point CVC, they will be tangent to each other and form a smooth curve.

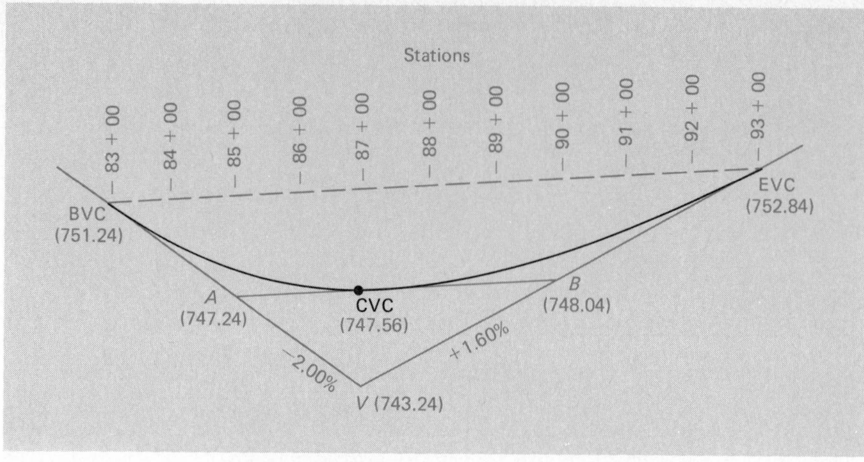

Figure 26-6. Unequal-tangent vertical curve.

EXAMPLE 26-2

For the configuration of Figure 26-6, compute and tabulate the notes necessary to stake the unequal-tangent vertical curve at full stations.

SOLUTION

1. Calculate elevations of BVC, EVC, A, B, and CVC, and grade AB.

$$\text{elev BVC} = 743.24 + 4(2.00) = 751.24 \text{ ft}$$

$$\text{elev } A = 743.24 + 2(2.00) = 747.24 \text{ ft}$$

$$\text{elev EVC} = 743.24 + 6(1.60) = 752.84 \text{ ft}$$

$$\text{elev } B = 743.24 + 3(1.60) = 748.04 \text{ ft}$$

$$\text{grade } AB = \left(\frac{748.04 - 747.24}{5}\right) = +0.16\%$$

$$\text{elev CVC} = 747.24 + 2(0.16) = 747.56 \text{ ft}$$

These elevations are shown in Figure 26-6.

2. In computing the first curve, the grade of AB will be g_2 in the formulas, and for the second curve it will be g_1. The rates of change of grade for the two curves, by Eq. (26-4), are:

$$r_1 = \frac{0.16 - (-2.00)}{4} = +0.54\%/\text{station}$$

$$r_2 = \frac{1.60 - 0.16}{6} = +0.24\%/\text{station}$$

3. Equation (26-5) is now solved in tabular form and the results listed in Table 26-2.

TABLE 26-2. NOTES FOR THE CURVE OF EXAMPLE 26-2

STATION	X	g_1X	$rX^2/2$	CURVE ELEVATION	FIRST DIFF.	SECOND DIFF.
93 + 00 (EVC)	6	0.96	4.32	752.84 (Check)		
					1.48	
92 + 00	5	0.80	3.00	751.36		0.24
					1.24	
91 + 00	4	0.64	1.92	750.12		0.24
					1.00	
90 + 00	3	0.48	1.08	749.12		0.24
					0.76	
89 + 00	2	0.32	0.48	748.36		0.24
					0.52	
88 + 00	1	0.16	0.12	747.84		0.24
					0.28	
87 + 00 (CVC)	4	−8.00	4.32	747.56 (Check)		
					−0.11	
86 + 00	3	−6.00	2.43	747.67		0.54
					−0.65	
85 + 00	2	−4.00	1.08	748.32		0.54
					−1.19	
84 + 00	1	−2.00	0.27	749.51		0.54
					−1.73	
83 + 00 (BVC)	0	0.00	0.00	751.24		

Vertical-curve computations by themselves are quite simple, hardly a challenge to an electronic computer. But when vertical curves are combined with horizontal curves, spirals, and superelevation on complex highway interchange coordinate calculations, programming can save time.

26-7. HIGH OR LOW POINT ON A VERTICAL CURVE. To investigate drainage conditions, clearance beneath overhead structures, cover over pipes, and sight distance, it may be necessary to determine the elevation and location of the low (or high) point on a vertical curve. At the low or high point, a tangent to the curve will be horizontal and its slope equal to zero. Based on this fact, by taking the derivative of Eq. (26-3) and setting it equal to zero, the following formula is readily derived:

$$X = \frac{g_1 L}{g_1 - g_2} \tag{26-6}$$

where X is the distance in stations from BVC to the high or low point of the curve, g_1 the tangent grade through the BVC, g_2 the tangent grade through the EVC, and L the curve length, in stations.

If g_2 is substituted for g_1 in the numerator of Eq. (26-6), distance X is measured back from the EVC.

EXAMPLE 26-3

Compute the station and elevation of the curve's high point in Example 26-1.

By Eq. (26-6):

$$X = \frac{3.00 \times 6}{3.00 - (-2.40)} = 3.3333 \text{ stations}$$

Then the station of the high point is:

$$\text{sta}_{\text{high}} = 43 + 70 + (3 + 33.33) = 47 + 03.33$$

By Eq. (26-3), elevation at this point is:

$$844.48 + 3.00(3.3333) + \frac{-2.40 - 3.00}{2(6)}(3.3333)^2 = 849.48$$

Note that in using Eq. (26-6) and all other equations of this chapter, correct algebraic signs must be applied to grades g_1 and g_2.

26-8. DESIGNING A CURVE TO PASS THROUGH A FIXED POINT. The problem of designing a parabolic curve to pass through a point of fixed station and elevation is frequently encountered in practice. It occurs, for example, where a new grade line must meet existing railroad or highway crossings, or a minimum vertical distance must be maintained between the grade line and underground utilities or drainage structures.

Given the station and elevation of the PVI, and grades g_1 and g_2 of the back and forward tangents, respectively, the problem consists of calculating the curve length required to meet the fixed condition. It is solved by substituting known quantities into Eq. (26-3) and reducing the equation to a quadratic containing only L as an unknown. Two values will satisfy the quadratic equation, but the correct one will be obvious.

EXAMPLE 26-4

In Figure 26-7, grades $g_1 = -4.00\%$ and $g_2 = +3.80\%$ meet at PVI station 52 + 00 and elevation 1261.50. Design a parabolic curve to meet a railroad crossing which exists at station 53 + 50 and elevation 1271.20.

SOLUTION

Referring to the figure, and substituting known quantities into Eq. (26-3), the following equation is obtained:

$$1271.20 = [1261.50 + 4.00(L/2)] + [-4.00(L/2 + 1.5)] + \left[\frac{3.80 + 4.00}{2L}(L/2 + 1.5)^2\right]$$

In the above expression, the value of X for the railroad crossing is $(\frac{L}{2} + 1.5)$, and the terms within successive brackets are Y_{BVC}, g_1X, and $(r/2)X^2$, respectively. Reducing the equation to quadratic form gives:

$$0.975L^2 - 9.75L + 8.775 = 0$$

Solving for L yields 9.1152 stations. To check the solution, $L = 9.1152$ stations, and $X = (9.1152/2 + 1.5)$ stations are used in Eq. (26-3) to calculate the elevation at station 53 + 50. A value of 1271.20 checks the computations.

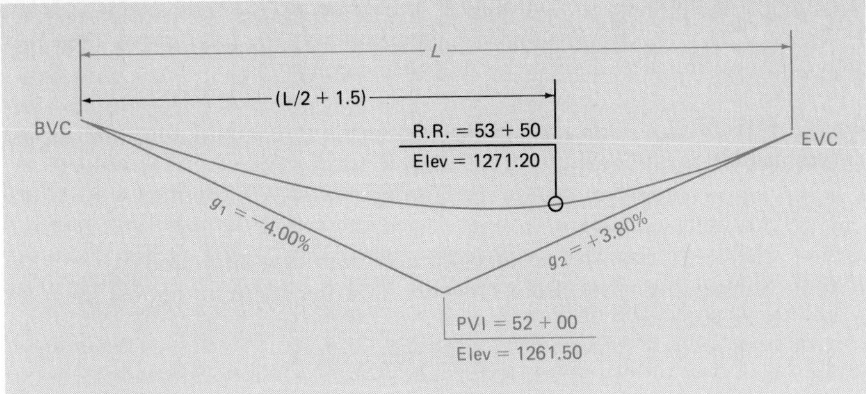

Figure 26-7. Designing a parabolic curve to pass through a fixed point.

26-9. SIGHT DISTANCE. The formula for sight distance S with the vehicle on a vertical curve and S less than the length L is

$$S^2 = \frac{8Lh}{g_1 - g_2} \tag{26-7}$$

where S is the sight distance, in stations; L the length of the curve, in stations; and h the height of the driver's eye and the object sighted above the roadway (by recommendation of the American Association of State Highway and Transportation Officials, 3.75 ft).

Then for a crest curve having a length of 800 ft, and grades of +2.00% and −1.60%, if $h = 3.75$ ft,

$$S = \sqrt{\frac{8 \times 8 \times 3.75}{2.00 - (-1.60)}} = 8.16 \text{ stations} = 816 \text{ ft}$$

Since this distance is greater than the length of curve and thus not in agreement with the assumption used in deriving the formula, a different expression must be employed.

If the vehicle is off the curve and on the tangent to it, S is greater than L, and the applicable sight distance formula is

$$S = \frac{L}{2} + \frac{4h}{g_1 - g_2} \tag{26-8}$$

Then in the preceding example, with $h = 3.75$ ft,

$$S = \frac{8}{2} + \frac{4 \times 3.75}{2.00 - (-1.60)} = 8.17 \text{ stations}$$

For a combined horizontal and vertical curve, the sight distance is the smaller of the two values computed independently for each curve. (See Section 25-14 for horizontal sight distance discussion.)

26-10. MISTAKES. Some typical mistakes made in computations for vertical curves include the following:

1. Arithmetic mistakes.
2. Failure to properly account for algebraic signs of g_1 and g_2.
3. Subtracting offsets from tangents for a sag curve, or adding them for a crest curve.
4. Failure to make the second-difference check.

PROBLEMS

26-1. Why is a parabola rather than a circular arc used for vertical curves?
26-2. What is meant by the rate of change of grade on vertical curves, and why is it important?

Tabulate station elevations for an equal-tangent parabolic curve for the data given in Problems 26-3 through 26-6. Check by second differences.

26-3. A +3.50% grade meets a −1.50% grade at station 60 + 50 and elevation 850.25 ft, 1000-ft curve, stakeout at full stations.
26-4. A −2.60% grade meets a +1.30% grade at station 36 + 00 and elevation 574.80, 1200-ft curve, stakeout at full stations.
26-5. A 350-ft curve, grades of $g_1 = -2.70\%$ and $g_2 = {}^{+}1.00\%$, PVI at station 98 + 70, elevation 310.00 ft, stakeout at half-stations.
26-6. An 800-ft curve, grades of −6.00% and −1.50%, PVI at station 50 + 00, elevation 900.00, stakeout at half-stations.

Field conditions require a highway curve to pass through a fixed point. Compute a suitable equal-tangent vertical curve and full-station elevations for Problems 26-7 through 26-9.

26-7. Grades of −3.50% and +0.50%, PVI elevation 800.00 at station 26 + 00. Fixed point elevation 803.00 ft at station 26 + 00.
26-8. Grades of $g_1 = -2.50\%$ and $g_2 = +1.50\%$, PVI elevation 1220.00 ft at station 15 + 00. Fixed point elevation 1227.00 ft at station 14 + 00.
26-9. Grades of +5.00% and +1.50%, PVI station 83 + 00 and elevation 730.00 ft. Fixed point elevation 729.00 ft at station 84 + 00.

26-10. A −1.00% grade meets a +0.50% grade at station 64 + 00 and elevation 900.00 ft. The +0.50% grade then joins a +2.50% grade at station 67 + 00. Compute and tabulate the notes for a vertical curve, at half stations, that passes through the midpoint of the 0.50% grade.

Compute and tabulate full-station elevations for an unequal-tangent vertical curve to fit the requirements in Problems 26-11 through 26-14.

26-11. A +4.00% grade meets a −2.00% grade at station 50 + 00 and elevation 1200.00 ft. Length of first curve 600 ft, second curve 400 ft.
26-12. Grade $g_1 = +1.50\%$, $g_2 = +5.00\%$, PVI at station 27 + 00 and elevation 1300.00 ft, $L_1 = 800$ ft, and $L_2 = 400$ ft.
26-13. The PVI of +4.00% and −3.00% grades at station 42 + 00 and elevation 822.24 ft. Lengths of curves are 400 and 600 ft.

26-14. A -0.50% grade meets a $+4.00\%$ grade at station 67 + 00 and elevation 466.80 ft. Length of first curve is 200 ft, second curve 600 ft.

26-15. Explain why the second differences of curve elevations are equal for a parabolic curve.
26-16. In laying out preliminary grades on a highway line, what consideration must be given to later introduction of vertical curves?
26-17. Why are parabolic curves not generally used for horizontal highway curves?
26-18. Write a computer program for vertical-curve computation.
26-19. When is it advantageous to use an unequal-tangent vertical curve instead of an equal-tangent one?
26-20. What factors influence the required minimal distances (tangent lengths) between a series of reversed vertical curves?
26-21. A manhole is 16 ft from the center line of a 40-ft-wide street that has a 4-in parabolic crown. The street center is at elevation 721.85 ft. What is the elevation of the manhole cover?
26-22. A 36-ft-wide street has an average parabolic crown from the center to each edge of $\frac{1}{4}$ in/ft. How much does the surface drop from the street center to a point 4 ft from the edge?
26-23. Determine the high point of the curve in Problem 26-3.
26-24. Calculate the low point of the curve in Problem 26-4.
26-25. Compute the sight distance available in Problem 26-3.
26-26. What sight distance does a driver have in Problem 26-9?
26-27. In determining sight distances on vertical curves, how does the designer determine whether the cars or objects are on the curve or tangent?

What is the minimum length of vertical curve to provide a required sight distance for the conditions given in Problems 26-28 through 26-30?
26-28. Grades of $+3.40\%$ and -2.40%. Sight distance of 700 ft.
26-29. A crest curve with grades of $+4.80\%$ and -3.40%. Sight distance 800 ft.
26-30. Sight distance of 1200 ft, grades of $+0.50\%$ and -1.20%.

BIBLIOGRAPHY

Colcord, J. E. 1962. "Vertical Curve Theory." *Surveying and Mapping* 22(no. 4):589.
Hickerson, T. F. 1967. *Route Location and Design*, 5th ed. New York: McGraw-Hill.
Lamont, C. M. 1960. "Self-Checking Method of Computing Curve Elevations." *ASCE Journal of the Surveying and Mapping Division* 86(no. SU1):1.
Meyer, C. F., and D. W. Gibson. 1980. *Route Surveying and Design*, 5th ed. New York: Harper & Row.
Vreeland, R. P. 1968. "The Unsymmetrical Vertical Parabola." *ASCE Journal of the Surveying and Mapping Division* 94(no. SU1):61.

27 VOLUMES

27-1. INTRODUCTION. Surveyors are often called upon to measure quantities of earthwork and concrete for various types of construction projects. Volume computations are also required to determine the capacity of bins, tanks, reservoirs, and buildings, and to check stockpiles of coal, gravel, and other material.

The unit of volume is a cube having edges of unit length. Cubic feet, cubic yards, and cubic meters are used in surveying calculations, although the cubic yard is most common for earthwork: 1 $yd^3 = 27\ ft^3$; 1 $m^3 = 35.315\ ft^3$.

27-2. METHODS OF VOLUME MEASUREMENT. Direct measurement of volumes is rarely made in surveying, since it is difficult to actually apply a unit of measure to the material involved. Instead indirect measurements are obtained by measuring lines and areas that have a relationship to the volume desired.

Three principal systems are used: (1) cross-section method, (2) unit-area, or borrow-pit, method, and (3) contour-area method.

27-3. THE CROSS-SECTION METHOD. The cross-section method is employed almost exclusively for computing volumes on linear construction projects such as highways, railroads, and canals. In this procedure, after the centerline has been staked, ground profiles called cross sections are taken (at right angles to the centerline), usually at intervals of 50 or 100 ft. Cross sectioning consists of measuring ground elevations and their corresponding distances orthogonally left and right from center line. Readings must be taken at the center line, high and low points, and locations where slope changes occur to determine accurately

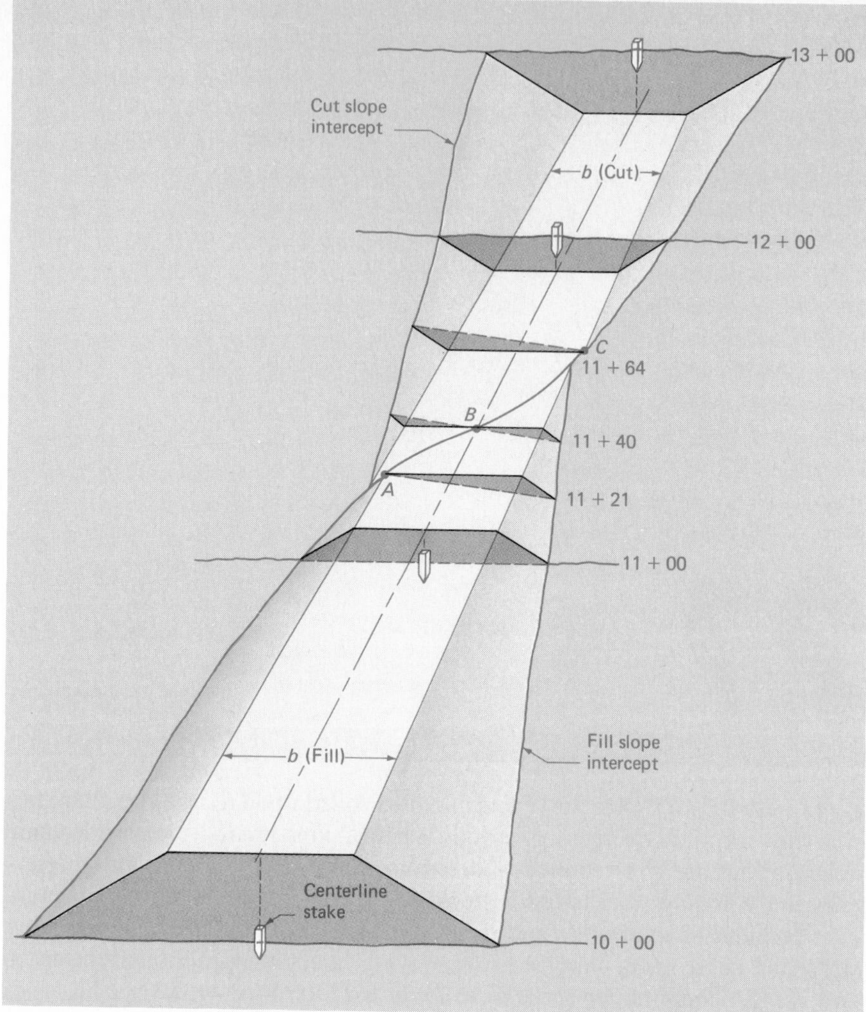

Figure 27-1. Section of roadway illustrating excavation (cut) and embankment (fill).

the ground profile. This can be done in the field using a level, level rod, and tape. Plate D-10 illustrates a set of field notes for cross sectioning.

Much of the field work formerly involved in running preliminary center lines, getting cross-section data, and making slope-stake and other measurements on long route surveys is now being done more efficiently by photogrammetry. It is not intended to discuss photogrammetric methods in this chapter; rather, basic field and office procedures for determining and calculating volumes will be presented briefly. Chapter 28 describes the subject of photogrammetry.

After the cross sections have been taken, they can be plotted, and *design templates* (outlines of planned excavation or embankment) superimposed on each plot. Templates cut from cardboard or plastic to form an outline of the required base width and side slopes, for cut and fill, can be used to aid in this work. This defines the section of excavation or embankment to be constructed at each station. Areas of these sections, called *end areas*, can be determined by computation or planimetering. As an alternate procedure, end areas can be computed directly from field cross-section data and design information. From the end areas, volumes are calculated by the *average-end-area* or *prismoidal* formula. These are discussed later in this chapter.

Figure 27-1 portrays a section of planned highway construction and illustrates some of the points just discussed. Center line stakes are shown in place, and mark locations where cross sections are taken, in this instance at full stations. End areas, based on the planned gradeline, size of roadway, and selected embankment and excavation slopes, are superimposed at each station shown shaded. Areas of these shaded sections are determined, whereupon volumes are computed using formulas given in Section 27-5 or 27-8. Note in the figure that embankment, or *fill*, is planned from stations 10 + 00 through 11 + 21, a transition from fill to excavation or *cut*, occurs from station 11 + 21 to 11 + 64, and cut is required from stations 11 + 64 through 13 + 00.

27-4. TYPES OF CROSS SECTIONS. The types of cross sections commonly used on route surveys are shown in Figure 27-2. In flat terrain the *level section* in (a) is suitable. The *three-level section* in (b) is generally used where ordinary ground conditions prevail. Rough topography may require a five-level section, (c), or more practically an irregular section, (d). A transition section, (e), and a side-hill section, (f), occur in passing from cut to fill, and on side-hill locations. In Figure 27-1, transition sections occur at stations 11 + 21 and 11 + 64, while a side-hill section exists at 11 + 40.

The width of base b or finished roadway is fixed by project requirements. As shown in Figure 27-1, it is usually wider in cuts than on fills to provide for drainage ditches. The side slope s [horizontal dimension required for a unit vertical rise and illustrated in Figure 27-2(a)] depends on the type of soil encountered. Side slopes in fill usually are flatter than those in cuts where the soil remains in its natural state.

Cut slopes of 1 horizontal to 1 vertical, and fill slopes of $1\frac{1}{2}$ to 1 might be satisfactory for ordinary loam soils, but $1\frac{1}{2}$ to 1 in excavation, and 2 to 1 in embankment are common. Even flatter proportions may be required—one cut in the Panama Canal area was 13 to 1—depending upon type of soil, rainfall, and other factors.

Formulas for areas of sections are readily derived and listed beside some of the sketches in Figure 27-2.

27-5. AVERAGE-END-AREA FORMULA. Figure 27-3 illustrates the concept of computing volumes by the average-end-area method. In the figure, A_1 and A_2 are end areas at two stations separated by a horizontal distance L. The volume between two stations is equal to the average of the end areas multiplied by the horizontal distance L between them. Thus,

$$V_e = \frac{A_1 + A_2}{2} \times \frac{L}{27} \text{ yd}^3 \qquad (27\text{-}1)$$

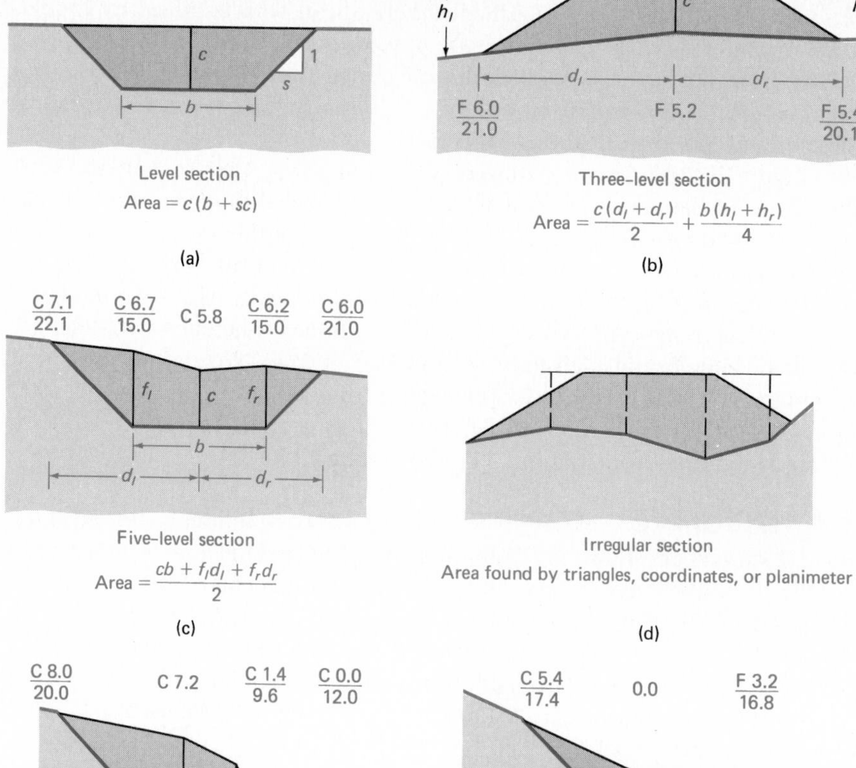

Figure 27-2. Earthwork sections.

Figure 27-3. Volume by average-end-area method.

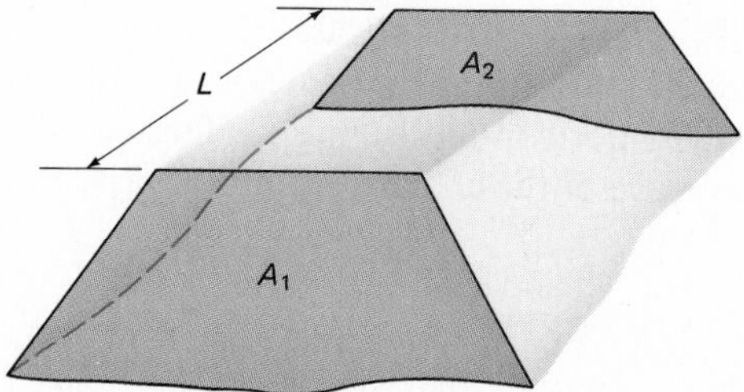

where V_e, the average-end-area volume, is in cubic yards, L in feet, and A_1 and A_2 in square feet. If L is 100 ft, as for full stations, Eq. (27-1) becomes:

$$V_e = 1.85(A_1 + A_2) \tag{27-2}$$

Equations (27-1) and (27-2) are approximate and give answers that generally are slightly larger than the true prismoidal volumes (see Section 27-8). They are used in practice because of their simplicity. Increased accuracy is obtained by decreasing the distance L between the sections. When the ground is irregular, cross sections must be taken closer together.

EXAMPLE 27-1

Compute the volume of excavation between station 24 + 00, with an end area of 711 ft^2, and station 25 + 00, having an end area of 515 ft^2.

By Eq. (27-2):

$$V = 1.85(A_1 + A_2) = 1.85(711 + 515) = 2268 \text{ yd}^3$$

27-6. DETERMINING END AREAS. End areas can be determined either graphically or by computation. In graphic methods, the cross section and template are plotted to scale on grid paper; then the number of small squares within the section can be counted and converted to area, or the area within the section measured using a planimeter. Computational procedures consist of either dividing the section into simple figures such as triangles and trapezoids and computing and summing these areas, or using the coordinate formula. These are discussed below.

27-6.1. END AREAS BY SIMPLE FIGURES. To illustrate the procedures of calculating end areas by simple figures such as triangles or trapezoids, assume the following excerpt of field notes applies to the cross section and end area shown in Figure 27-4.

Figure 27-4. End-area computation.

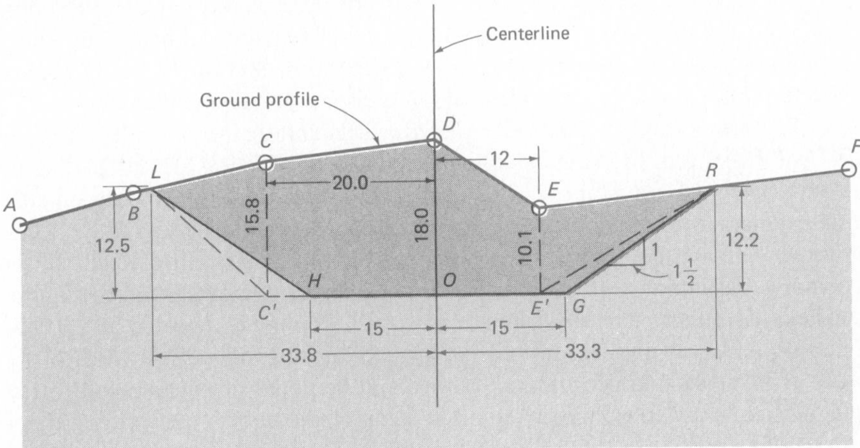

HI = 879.29

		867.3	870.9	874.7	876.9	869.0	872.8
24 + 00	Lt	12.0 / 50	8.4 / 36	4.6 / 20	2.4 / ℄	10.3 / 12	6.5 / 50

In this excerpt, the top numbers are elevations obtained by subtracting rod readings (middle numbers) from the HI of the leveling instrument. Bottom numbers are distances from center line, beginning from the left. Assume the design calls for a level roadbed of 30-ft width, cut slopes of $1\frac{1}{2}$:1, and a subgrade elevation at station 24 + 00 of 858.9. An appropriate design template is superimposed over the plotted cross section in Figure 27-4. Subtracting the subgrade elevation from cross-section elevations at *C*, *D*, and *E* yields the ordinates of cut required at those locations. Elevations and distances out from center line to the *slope intercepts* at *L* and *R* must be either scaled from the plot or computed. Assuming they have been scaled (methods for computing them are given in Section 27-7), the following tabulation of distances from center line and required cut ordinates at each point to subgrade elevation was made:

STATION	H	L	C	D	E	R	G
24 + 00	0 / 15	C 12.5 / 33.8	C 15.8 / 20	C 18.0 / 0	C 10.1 / 12	C 12.2 / 33.3	0 / 15

Numbers above the lines (preceded by the letter *C*) are cut ordinates in feet; those below the lines are distances out from the center line. Fills are denoted by the letter *F*. Using *C* instead of plus for cut, and *F* instead of minus, for fill, eliminates confusion.

From the cut ordinates and distances from center line shown, the area of the cross section in Figure 27-4 is computed by summing the individual areas of triangles and trapezoids. A list of the calculations is given in Table 27-1.

27-6.2. END AREAS BY COORDINATES. The coordinate method for computing end areas can be used for any type of section and has many engineering applications. The procedure was described in Section 14-6 as a means for determining the area contained within a closed polygon traverse.

To demonstrate the method in end area calculation, the example of Figure 27-4 will be solved. Coordinates of each point of the section are calculated, in an axis system having point *O* as its origin, using the earlier listed data on cuts and distances from center line. In computing coordinates, distances to the right of center line and cut values are considered plus; distances left and fill values are minus. Beginning with point *O* and proceeding clockwise around the figure, coordinates of each point are listed in sequence. Point *O* is repeated at the end. Then Eq. (14-7) is applied, in which products of diagonals downward to the right (solid arrows) are considered minus, and diagonal products down to the left (dashed arrows) are plus. Algebraic signs of the coordinates must be accounted for, and thus a positive product (dashed arrow) having a negative co-

TABLE 27-1. END AREA BY SIMPLE FIGURES

FIGURE	COMPUTATION	AREA
ODCC′	$\frac{1}{2}(18.0 + 15.8)20$	338
C′CL	$\frac{1}{2}(15.8)13.8$	109
HLC′	$-\frac{1}{2}(5)12.5$	−31
ODEE′	$\frac{1}{2}(18.0 + 10.1)12$	169
EE′R	$\frac{1}{2}(10.1)21.3$	108
E′RG	$\frac{1}{2}(3)12.2$	18
		Area = 711 ft^2

ordinate will actually be minus. Total area is obtained by dividing the absolute value of the algebraic summation of all products by 2. The calculations are illustrated in Table 27-2.

It is necessary to make separate computations for cut and fill end areas when they occur in the same section (as at station 11 + 40 of Figure 27-1), since they must always be tabulated independently for pay purposes. Payment is normally made only for excavation (its unit price includes making and shaping the fills) except on projects consisting primarily of embankment such as levees, earth dams, some military fortifications, and highways built up by continuous fill in flat areas.

27-7. COMPUTING SLOPE INTERCEPTS. Calculation of the elevations and distances out from center line of slope intercepts can be done using cross-section data and the cut or fill slope value. In Figure 27-4 for example, intercept *R* occurs between ground profile point *E* (distance 12 ft right and elevation 869.0) and point *F* (distance 50 ft right and elevation 872.8). The cut slope is $1\frac{1}{2}:1$, or 0.67 ft/ft. A more detailed diagram illustrating the geometry for calculating slope intercept *R* is given in Figure 27-5.

TABLE 27-2. END AREA BY COORDINATES

POINT	X	Y	PLUS $\pm$	MINUS $\mp$
O	0	0	↙	↘
H	−15	0	0	0
L	−33.8	12.5	0	+188
C	−20	15.8	−250	+534
D	0	18.0	0	+360
E	12	10.1	216	0
R	33.3	12.2	336	−146
G	15	0	183	0
O	0	0	0	0
			+485	+936
			+936	
			2) 1421	
			711 ft^2 (nearest ft^2)	

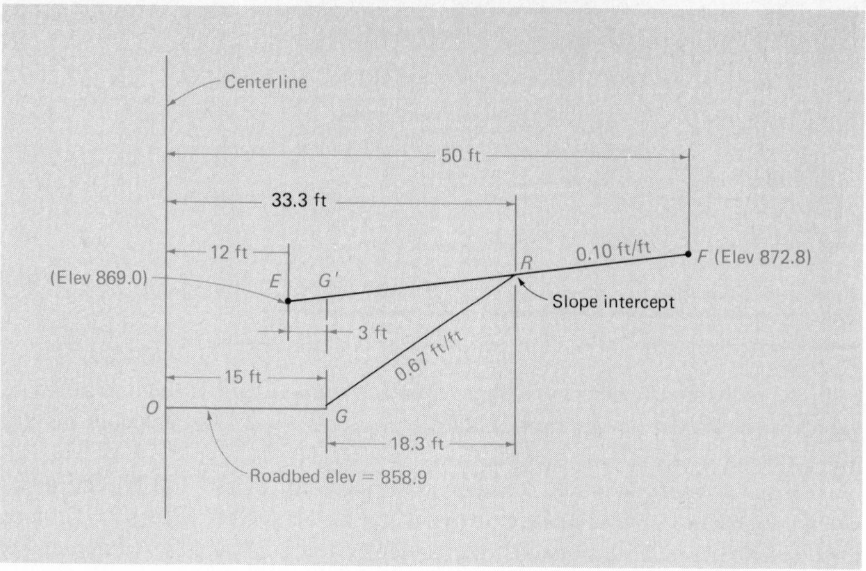

Figure 27-5. Computation of slope intercept *R* of Figure 27-4.

The slope along ground line *EF* is (872.8 – 869.0)/38 = 0.10 ft/ft, where 38 ft is the horizontal distance between the points. Elevation *G*′ (point vertically above *G*) is 869.0 + 0.10(3) = 869.3; thus, ordinate *GG*′ is (869.3 – 858.9) = 10.4 ft. Lines *EF* and *GR* converge at a rate equal to the difference in their slopes (because they are both sloping upward) or at (0.67 – 0.10) = 0.57 ft/ft. Dividing ordinate *GG*′ by this convergence yields horizontal distance *GR*, or 10.4/0.57 = 18.3 ft. Adding 18.3 to distance *OG* yields 18.3 + 15 = 33.3 ft, which is the distance from center line to slope intercept *R*. Finally, to obtain the elevation of *R*, the increase in elevation from *E* to *R* is added to the elevation of *E*, or 0.10(21.3) + 869.0 = 871.1. Cut ordinate at *R* equals 871.1 – 858.9 = 12.2 ft. Recall that 33.3 and 12.2 were the *X* and *Y* coordinates, respectively, used in the end-area calculations of Section 27-6.2.

The elevation and distance from center line of the slope intercept *L* of Figure 27-4 is calculated in a similar manner, except the rate of convergence of lines *CB* and *HL* is the sum of their slopes because *CB* slopes downward and *HL* upward. Calculations of slope intercepts are somewhat laborious, but routine when programmed for solution by electronic computer. If an electronic computer is not used for computing end areas and volumes, the usual procedure is to plot the cross sections and templates, determine the end area by planimeter and scale the slope intercepts from the plot. Slope intercepts are vital, since placement of slope stakes that guide construction operations is based on them.

27-8. PRISMOIDAL FORMULA. The prismoidal formula applies to volumes of all geometric solids which can be considered prismoids. Most earthwork volumes fit this classification, but relatively few of them warrant the precision of the prismoidal formula. The ground is not uniform from cross section to cross

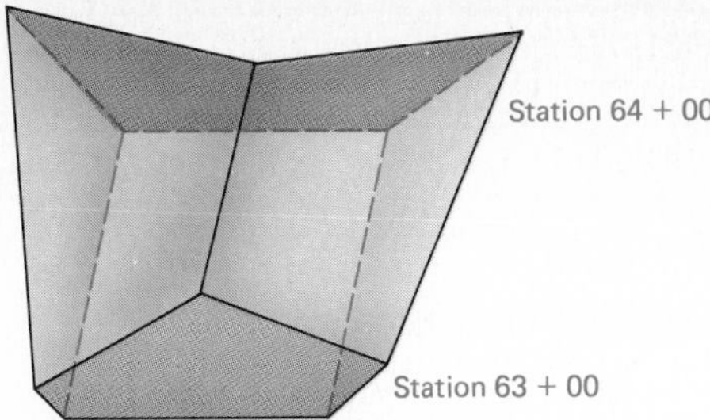

Figure 27-6. Sections for which the prismoidal correction is added to the end-area volume.

section, and right angles turned from the center line using a pentagonal prism or by the "arm" method (see Section 16-10.5) introduce errors.

One arrangement of the prismoidal formula is

$$V_p = \frac{L(A_1 + 4A_m + A_2)}{6 \times 27} \tag{27-3}$$

where V_p is the prismoidal volume in cy, A_1 and A_2 are areas of successive cross sections taken in the field, A_m is the area of a section midway between A_1 and A_2, and L is the horizontal distance between A_1 and A_2.

To use this formula it is necessary to know area A_m of the section half-way between stations. This is found by the usual computation *after averaging the heights and widths of the end sections.* Obviously the middle area is *not* the average of the end areas, since there would then be no difference between the results of the end-area formula and the prismoidal formula.

The prismoidal formula generally gives a volume less than that found by the average end-area formula. For example, the volume of a pyramid by the prismoidal formula is $Ah/3$, whereas by the average end-area method it is $A/2$. An exception occurs when the center height is great but the width narrow at one station, and the center height is small but the width large at the adjacent station. Figure 27-6 illustrates this condition. The difference between the volume obtained by the average end-area formula and that by the prismoidal formula is called the *prismoidal correction*, C_p.

Various books on route surveying give formulas and tables for computing prismoidal corrections which can be applied to average end-area volumes to get prismoidal volumes. Except in rock excavation and concrete work, use of the prismoidal formula is not normally justified by the low precision of field data. Frequently it is easier to compute the average end-area volume and the prismoidal correction than to calculate the prismoidal volume directly,

A prismoidal correction formula, accurate for three-level sections and close enough for most others, is

$$C_p = \frac{L}{12 \times 27}(c_1 - c_2)(w_1 - w_2) \tag{27-4}$$

TABLE 27-3. **TABULAR FORM OF VOLUME COMPUTATION.**

STATION (1)	END AREA (ft²) CUT (2)	FILL (3)	VOLUME (cy) CUT (4)	FILL (5)	FILL VOLUME +25% (cy) (6)	CUMULATIVE VOLUME (cy) (7)
10 + 00		892				0
				2614	3268	
11 + 00		421				−3268
				190	238	
11 + 21	0	68				−3506
			12	29	37	
11 + 40	34	31				−3531
			79	14	17	
11 + 64	144	0				−3469
			553			
12 + 00	686					−2916
			2967			
13 + 00	918					+51

where C_p is the volume of the prismoidal correction in cy, c_1 and c_2 are center heights in cut (or in fill), and w_1 and w_2 are widths of sections (from slope intercept to slope intercept) at adjacent sections.

If the product of $(c_1 - c_2)(w_1 - w_2)$ is minus, as in Figure 27-6, the prismoidal correction is added rather than subtracted from the end-area volume.

For projects with more than a few cross sections, computer programs are available and generally used, but a surveyor or engineer still must understand the method of computation. Earthwork calculations from photogrammetric data should check good field survey results within 1% in open country free of big trees or high brush. The method has received legal and contractor acceptance.

Except in rock excavation, average-end-area volumes are normally satisfactory and, unless otherwise noted, are legally accepted. Contractors are satisfied because pay quantities are usually greater than for the comparable prismoidal volumes.

27-9. VOLUME COMPUTATIONS. Volume calculations for route construction projects are often done by electronic computer, but if performed by hand-held calculator, they are usually arranged in tabular form. To illustrate this procedure, assume that end areas listed in columns (2) and (3) of Table 27-3 apply to the section of roadway illustrated in Figure 27-1. Using Eq. (27-1), cut and fill volumes are computed and tabulated in columns (4) and (5).

The volume computations illustrated in Table 27-3 include the transition sections of Figure 27-1. This is normally not done when preliminary earthwork volumes are being estimated (during design and prior to construction) because the exact locations of the transition sections and their configurations are usually unknown until slope staking occurs. Thus, for calculating preliminary earthwork quantities, an end area of zero would be used at the station of the centerline grade point (station 11 + 40 of Figure 27-1), and the the transition sections (stations 11 + 21 and 11 + 64 of Figure 27-1) would not appear in the computations. After slope staking (procedures for slope staking are described in Section 24-7) the locations and end areas of transition sections are known, and they should be included in final volume computations especially if they significantly affect quantities for which payment is made.

In highway and railroad construction, excavation or cut material is used to build embankments or fill sections. Unless there are other controlling factors, a well-designed grade line should nearly balance total cut volume against total fill volume. To accomplish a balance, either fill volumes must be expanded or cut volumes shrunk.[1] This is necessary because, except for rock cuts, embankments are compacted to a density greater than that of material excavated from its natural state, and to balance earthwork this must be considered. (Rock cut expands to occupy a greater fill volume; thus, either cut must be expanded or fill shrunk to obtain a balance.) The rate of expansion depends upon the type of material and can never be estimated exactly. However, samples and records of past projects in the immediate area are helpful in assigning reasonable factors. Column (6) of Table 27-3 lists expanded fills for the example of Figure 24-1, where a 25% factor was applied.

To investigate whether or not an earthwork balance is achieved, *cumulative volumes* are computed. This involves adding cut and expanded fill volumes algebraically from project beginning to end, with cuts considered positive and fills negative. Cumulative volumes are listed in column (7) of Table 27-3. In this example there is a cut volume excess of 51 cy between stations 10 + 00 and 13 + 00, or in other words, there is a surplus of that much excavation.

To analyze the movement of earthwork quantities on large projects, *mass diagrams* are constructed. These are plots of cumulative volumes for each station as the ordinate, versus the stations on the abscissa. Horizontal (balance) lines on the mass diagram then determine the limit of economic haul and direction of movement of material. Mass diagrams are described in more detail in books on route surveying.

If there is insufficient material from cuts to make the required fills, the difference must be *borrowed* (obtained from borrow pits or other sources such as by "daylighting" curves). If there is excess cut, it is *wasted* or perhaps used to extend and flatten the fills.

27-10. UNIT-AREA, OR BORROW-PIT, METHOD. The quantity of earth, gravel, rock or other material excavated or filled on a construction project can be determined by borrow-pit leveling. The quantities computed form the basis for payment to the contractor or materials supplier. The number of cubic yards of coal or other loose materials in stockpiles can be found in the same way.

As an example, assume the area shown in Figure 27-7 is to be graded to an elevation of 358.0 for a building site. Notes for the field work are shown in Plate D-6.

The area to be covered is staked in squares of 10, 20, 50, 100, or more feet, with the choice depending on project size and accuracy desired. A transit

[1] Expansion of fill volumes is generally preferred, since payment is usually based on actual volumes of material excavated.

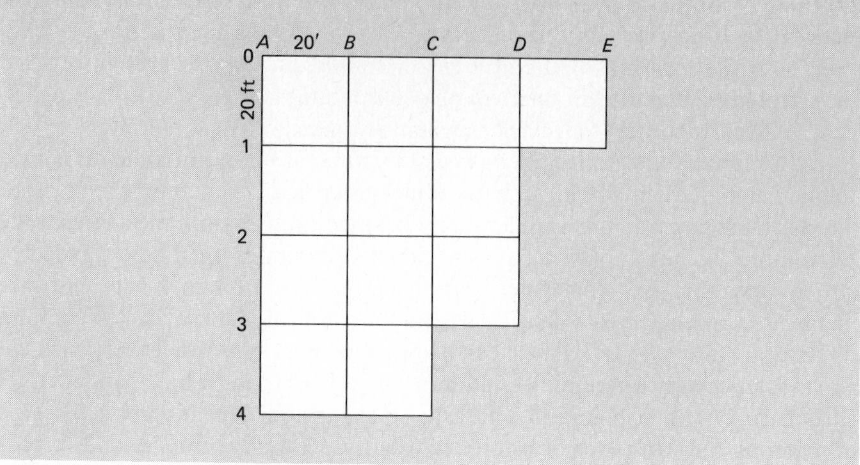

Figure 27-7. Borrow-pit leveling.

and/or tape may be used for the layout. A bench mark of known or assumed elevation is established outside the area in a place not likely to be disturbed.

A level is set up at any convenient location, a plus sight taken on the bench mark, and minus sights read on corners of the squares. If the terrain is not too rough, it may be possible to select a point near the area center and take sights on all corners from the same setup, as in this example.

Corners of the squares are designated by letters and numbers, such as *A*-1, *C*-4, and *D*-2. Since the site is to be graded to an elevation of 358.0, the amount of cut or fill at each corner can be obtained by subtracting 358.0 from its elevation. For each square, then, the average height of the four corners of the prism of cut or fill is determined and multiplied by the base area, 20×20 ft $= 400$ ft^2, to get the volume. The total volume is found by adding the individual values for each block and dividing by 27 to obtain a result in cubic yards.

As a simplification, the cut at each corner multiplied by the number of times it enters the volume computation can be shown in a separate column, a total secured, and divided by 4. The result multiplied by the base area of one block gives the volume. This procedure is shown in the sample noteform.

27-11. CONTOUR-AREA METHOD. Volumes based on contours can be obtained from contour maps by planimetering the area enclosed by each contour and multiplying the average of areas of adjacent contours by the contour interval, using Eq. (27-1). For example, if in the plan view of Figure 16-5, the area within the 10-ft contour is 19,650 ft^2 and that within the 20-ft contour is 12,720 ft^2, the average-end-area volume is $10(19{,}650 + 12{,}720)/2(27) = 5994$ yd^3. Use of the prismoidal formula is seldom, if ever, justified in this type of computation.

27-12. SOURCES OF ERROR. Some common errors in determining areas of sections and volumes of earthwork are:

1. Making errors in measuring cross sections.
2. Failing to use the prismoidal formula where it is justified.
3. Carrying out areas of cross sections beyond the nearest square foot, or beyond the limit justified by the field data.
4. Carrying out volumes beyond the nearest cubic yard.

27-13. MISTAKES. Some typical mistakes made in earthwork calculations are:

1. Confusing algebraic signs in end-area computations using the coordinate method.
2. Using Eq. (27-2) for full-station volume computation when partial stations are involved.
3. Using end-area volumes for pyramidal or wedge-shaped solids.
4. Mixing cut and fill quantities.

PROBLEMS

27-1. Why must cut and fill volumes be totaled separately?
27-2. Discuss comparative side slopes in cut and fill.
27-3. Why is a roadway in cut normally wider than the same roadway in fill?
27-4. Prepare a table of end areas versus depth of fill from 0 to 20 ft by increments of 2 ft for level sections, a 40-ft-wide level roadbed, and side slopes of 2 to 1.
27-5. Similar to Problem 27-4, except use side slopes of 4 to 1.

Draw the cross sections and compute V_e for the data given in Problems 27-6 through 27-9.

27-6. Two level sections of 100-ft stations with center heights of 5.1 and 6.3 ft in cut. Base width = 40 ft; side slopes are 2 to 1.
27-7. Two level sections at 70-ft stations with center heights 3.4 and 6.0 ft in fill. Base width = 30 ft; side slopes are $2\frac{1}{2}$ to 1.
27-8. The end area at station 36 + 00 is 232 ft^2. Notes giving distance from center line and cut ordinates for station 36 + 70 are C 3.8/15.7, C 4.9, C 5.8/18.7. Base = 20 ft.
27-9. An irrigation ditch with b = 16 ft and side slopes of 2 to 1. Notes giving distances from center line and cut ordinates for stations 52 + 00 and 53 + 00 are: C 2.4/12.8, C 3.0, C 3.7/15.4; and C 3.1/14.2, C 3.8, C 4.1/16.2.

27-10. For the data tabulated, calculate the volume of excavation in cubic yards between stations 10 + 00 and 15 + 00.

STATION	CUT END AREA (ft^2)	STATION	CUT END AREA (ft^2)
10 + 00	275	13 + 00	420
11 + 00	392	14 + 00	244
12 + 00	486	15 + 00	180

27-11. For the data listed, tabulate cut, fill, and cumulative volumes, in cubic yards, between stations 10 + 00 and 20 + 00. Use a shrinkage factor of 1.30 for cuts.

STATION	END AREA (ft^2) CUT	FILL	STATION	END AREA (ft^2) CUT	FILL
10 + 00	0		15 + 00		124
11 + 00	196		16 + 00		238
12 + 00	348		17 + 00		300
13 + 00	317		18 + 00		265
14 + 00	146		19 + 00		183
14 + 40	0	0	20 + 00		116

27-12. Compute the section areas in Problem 27-7 by the coordinate method.

27-13. Calculate the section area of station 36 + 70 in Problem 27-8 by the coordinate method.

27-14. Determine the section areas in Problem 27-9 by the coordinate method.

27-15. Compute C_p and V_p for Problem 27-6. Is C_p significant?

27-16. Calculate C_p and V_p for Problem 27-9. Would C_p be significant in rock cut?

27-17. From the following excerpt of field notes, plot the cross section on graph paper and superimpose upon it a design template for a 30-ft-wide level roadbed with fill slopes of 3:1 and a subgrade elevation at center line of 974.60 ft. Determine the end area graphically by counting squares.

HI = 972.31

20 + 00 Lt $\frac{4.8}{50}$ $\frac{5.2}{22}$ $\frac{5.9}{℄}$ $\frac{6.6}{12}$ $\frac{8.1}{30}$ $\frac{7.0}{50}$

27-18. For the data of Problem 27-17, determine the end area by planimeter.

27-19. For the data of Problem 27-17, calculate slope intercepts and determine the end area by the coordinate method. Check by computing areas of triangles and trapezoids.

27-20. From the following excerpt of field notes, plot the cross section on graph paper and superimpose upon it a design template for a 40-ft-wide level roadbed with cut slopes of $2\frac{1}{2}$:1 and a subgrade elevation of 1247.50 ft. Determine the end area graphically by counting squares. Check by planimeter.

HI = 1262.80

46 + 00 Lt $\frac{9.0}{50}$ $\frac{6.9}{27}$ $\frac{5.2}{10}$ $\frac{4.9}{℄}$ $\frac{5.6}{24}$ $\frac{3.8}{50}$

27-21. For the data of Problem 27-20, calculate slope intercepts and determine the end area by the coordinate method. Check by computing areas of triangles and trapezoids.

27-22. Complete the following notes and compute V_e and V_p. The roadbed is level, the base is 26 ft, and the side slopes are $2\frac{1}{2}$ to 1.

Station 88 + 00 $\frac{C\ 6.4}{\quad}$ C 3.6 $\frac{C\ 5.7}{\quad}$

Station 87 + 00 $\frac{C\ 3.1}{\quad}$ C 4.9 $\frac{C\ 4.3}{\quad}$

27-23. Similar to Problem 27-22, except the base is 36 ft and the side slopes are 3 to 1.

27-24. Calculate V_e and V_p for the following notes. Base in fill = 20 ft, base in cut = 30 ft, and side slopes are $1\frac{1}{2}$ to 1.

$$12 + 90 \quad \frac{C\ 2.4}{18.6} \quad \frac{C\ 1.0}{0} \quad \frac{0.0}{6.0} \quad \frac{F\ 2.0}{13.0}$$

$$12 + 40 \quad \frac{C\ 1.2}{16.8} \quad \frac{0.0}{0} \quad \frac{F\ 4.0}{16.0}$$

27-25. Calculate V_e, C_p, and V_p for the following notes. Base in cut = 36 ft, and side slopes are 2 to 1.

$$46 + 00 \quad \frac{C\ 4.2}{26.4} \quad \frac{C\ 2.0}{0} \quad \frac{C\ 3.6}{25.2}$$

$$45 + 00 \quad \frac{C\ 2.4}{22.8} \quad \frac{C\ 3.0}{0} \quad \frac{0.0}{18.0}$$

For Problems 27-26 and 27-27 compute the reservoir capacity (in acre-ft) between highest and lowest contours for planimetered areas on a topographic map.

27-26.

ELEVATION (ft)	960	970	980	990	1000	1010
AREA (ft²)	915	1106	1462	1967	2360	3070

27-27.

ELEVATION (ft)	415	420	425	430	435	440
AREA (ft²)	2017	2174	2222	2404	2596	2974

27-28. A calibrated polar planimeter gives an average reading of 1.476 revolutions of the roller over a 4-in-diameter circle. Twenty-foot contours planimetered on a reservoir site map to a scale of 1 in = 500 ft give the values tabulated. Calculate the reservoir volume in acre-ft.

CONTOUR	720	740	760	780	800
PLANIMETER READING	0.000	0.827	1.338	3.304	5.995

27-29. List several things that can be learned from a mass diagram.

27-30. State two situations where prismoidal corrections are most significant.

27-31. Do your state highway department construction contracts contain a provision covering free haul and overhaul? If so, list the details.

27-32. Write a computer program to calculate slope intercepts and end areas by the coordinate method, given cross-section notes and roadbed design information such as is given in Problem 27-17.

BIBLIOGRAPHY

Hickerson, T. F. 1967. *Route Location and Design*, 5th ed. New York: McGraw-Hill.

Meyer, C. F., and D. W. Gibson. 1980. *Route Surveying and Design*, 5th ed. New York: Harper & Row.

Wild, T. 1954. "Simplified Volume Measurement with the Polar Planimeter." *Surveying and Mapping* 14 (no. 2): 218.

28 PHOTOGRAMMETRY

28-1. INTRODUCTION. Photogrammetry may be defined as the science, art, and technology of obtaining reliable information from photographs. Photogrammetry encompasses two major areas of specialization: *metrical* and *interpretative.* The first area is of principal interest to surveyors, since it is applied to determine distances, elevations, areas, volumes, and cross sections, and for compiling topographic maps from measurements made on photographs. *Aerial* photographs (exposed from aircraft) are normally used, although for certain special work, *terrestrial* photos (taken from earth-based cameras) are employed.

Interpretative photogrammetry involves recognizing objects from their photographic images and judging their significance. Critical factors considered in identifying objects are the shapes, sizes, patterns, shadows, tones, and textures of their images. This area of photogrammetry was traditionally called *photographic interpretation* because initially it relied upon aerial photos. More recently other sensing and imaging devices such as multispectral scanners, thermal scanners, radiometers, and side-looking airborne radar have been developed which aid in interpretation. These instruments sense energy in wavelengths beyond those which the human eye can see or ordinary photographic film can record. They are often carried in aircraft as remote as satellites; hence, a new term, *remote sensing,* is now generally applied to the interpretative area of photogrammetry.

In this chapter, metrical photogrammetry using aerial photographs will be emphasized because it is the phase most frequently applied in surveying work.

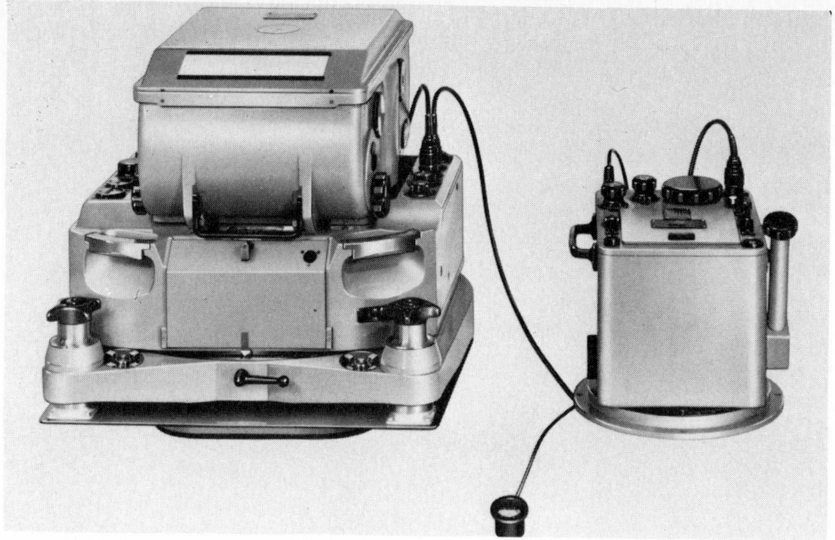

Figure 28-1. Aerial camera. (Courtesy Carl Zeiss, Oberkochen.)

28-2. USES OF PHOTOGRAMMETRY. Photography dates back to 1839, and the first attempt to use photogrammetry in preparing a topographic map occurred a year later. Photogrammetry is now the chief method of topographic mapping. The U.S. Geological Survey, for example, employs the procedure almost exclusively in compiling its quadrangle maps. Cameras, films, plotting instruments, and techniques have been continually improved so that photogrammetrically prepared maps today meet very high accuracy standards. Other advantages of this method of mapping are (a) speed of coverage of an area, (b) relatively low cost, (c) ease of obtaining topographic details, especially in inaccessible areas, and (d) reduced likelihood of omitting data due to the tremendous amount of detail shown in photographs.

Photogrammetry presently has many applications in surveying and engineering. It is used, for example, in land surveying to compute coordinates of section corners, boundary corners, or points of evidence that help locate these corners. Large-scale maps are made by photogrammetric procedures for many uses, one being subdivision design. Photogrammetry is used to map shorelines in hydrographic surveying, to determine precise ground coordinates of points in control surveying, and to develop maps and cross sections for route and engineering surveys. Photogrammetry is expected to play an important role in developing the necessary information for modern land data systems.

Photogrammetry is also being successfully applied in many nonengineering fields—for example, geology, archeology, forestry, agriculture, conservation, planning, military intelligence, traffic management, and accident investigation. It is beyond the scope of this chapter to describe all the varied applications of photogrammetry. Use of the science has increased dramatically in recent years, and its future growth for solving measurement and mapping problems seems assured.

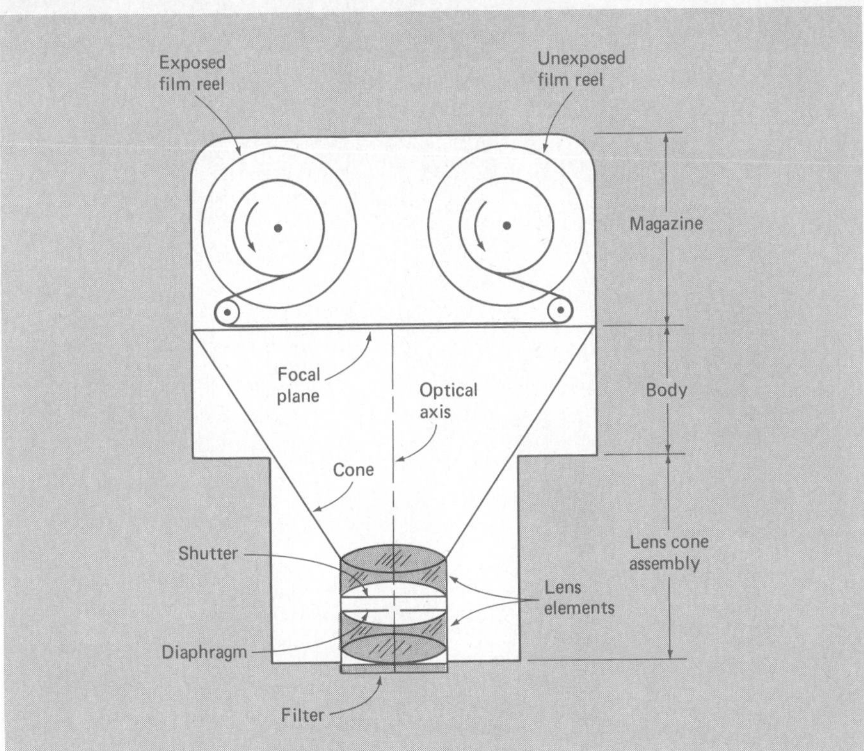

Figure 28-2. Principal components of a single-lens frame aerial camera.

28-3. AERIAL CAMERAS. Aerial mapping cameras are precision instruments designed to take photographs from aircraft. They must be capable of exposing a large number of photographs in rapid succession while moving in an aircraft at high speed, so a short cycling time, fast lens, efficient shutter, and large-capacity magazine are required.

Single-lens frame cameras are the type most often used in metric photogrammetry. These cameras expose the entire frame or format simultaneously through a lens held at a fixed distance from the focal plane. Generally they have a format size of 9 × 9 in and lenses with focal lengths of 6 in, although $3\frac{1}{2}$-, $8\frac{1}{4}$-, and 12-in lengths are also used. A single-lens frame camera is shown in Figure 28-1.

Principal components of a single-lens frame camera are shown in the diagram of Figure 28-2. These include the *lens* (most important part); *shutter* to control the interval of time that light passes through the lens; *diaphragm* to regulate the size of lens opening; *filter* to reduce the effect of haze and distribute light uniformly over the format; *camera cone* to support the lens-shutter-diaphragm assembly with respect to the focal plane and prevent stray light from striking the film; *focal plane*, surface upon which the film lies when exposed; *fiducial marks* (not shown in Figure 28-2), four or eight in number to define the photographic principal point; *camera body* to house the drive mechanism that

Figure 28-3. Vertical aerial photograph. (Courtesy Owen Ayres & Associates, Inc.)

cocks and trips the shutter, flattens the film, and advances it between exposures; and *magazine*, which holds the supply of exposed and unexposed film.

An aerial camera shutter can be operated manually by an operator or by an *intervalometer*, which automatically trips the shutter at a specified time. A level vial attached to the camera helps keep the optical axis of the camera lens (which is perpendicular to the focal plane) vertical in spite of any slight tip and tilt of the aircraft. More recently, gyroscopes have been developed to keep the camera axis approximately vertical. Polyester roll film is normally used with magazine capacities of 200 ft or more.

Images of the fiducial marks are printed on the photographs and lines joining opposite pairs intersect at or very near the *principal point*, defined as the point where a perpendicular from the emergent nodal point of the camera lens strikes the focal plane. Fiducial marks may be located in the corners, as shown in Figure 28-3, on the sides, as shown in Figure 28-4, or preferably in both places.

Aerial mapping cameras are laboratory calibrated to get precise values for the focal length and lens distortions. Flatness of the focal plane, relative position of the principal point with respect to the fiducial marks, and fiducial

Figure 28-4. Low oblique aerial photograph. (Courtesy Carl Zeiss, Oberkochen.)

mark locations are also specified. These calibration data are necessary for precise photogrammetric calculations.

28-4. TYPES OF AERIAL PHOTOGRAPHS. Aerial photographs exposed with single-lens frame cameras are classified as *vertical* (taken with the camera axis aimed vertically downward, or as nearly vertically as possible) and *oblique* (made with the camera axis intentionally inclined at an angle between the horizontal and vertical). Oblique photographs are further classified as *high* if the horizon shows on the picture, and *low* if it does not. Figures 28-3 and 28-4 show examples of vertical and low oblique photographs, respectively, which completely depict all natural and cultural features within the region covered, such as roads, railroads, buildings, rivers, bridges, trees, and cultivated lands.

28-5. VERTICAL AERIAL PHOTOGRAPHS. Vertical photographs are the principal mode of obtaining imagery for topographic mapping. A *truly vertical* photograph results if the axis of the camera is exactly vertical when the exposure is made. In spite of the precautions taken, small tilts, generally less than 1° and

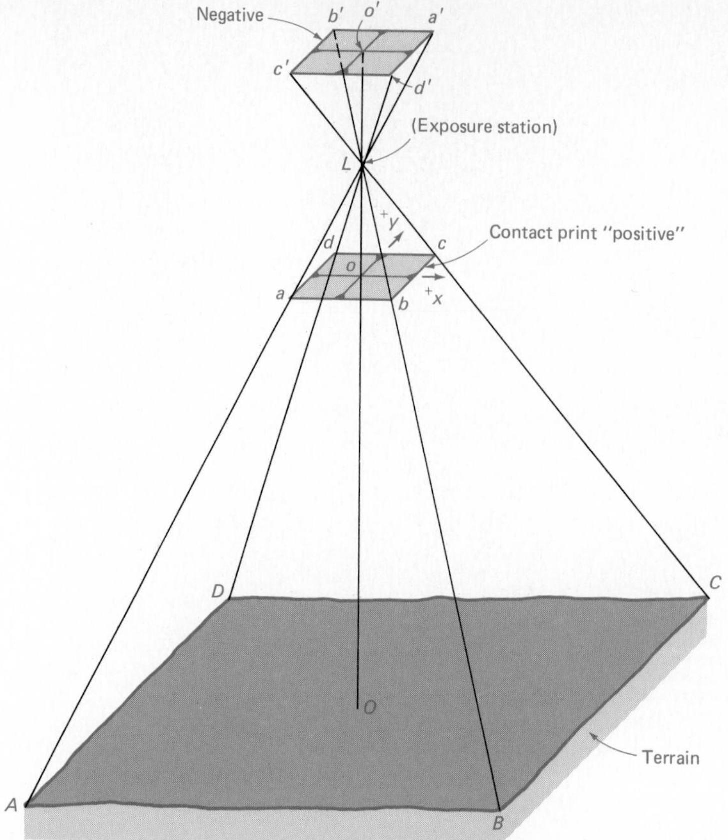

Figure 28-5. Geometry of the vertical aerial photograph.

rarely greater than 3°, are invariably present, and the resulting photos are called *near-vertical* or *tilted* photographs. Photogrammetric principles and practices have been developed to handle tilted photos and accuracy is not sacrificed in compiling maps from them.

Although vertical photographs look like maps to laypersons, they are not true orthographic projections of the earth's surface. Rather they are perspective views, and the principles of perspective geometry must be applied to prepare maps from them. Figure 28-5 illustrates the geometry of a vertical photograph taken at exposure station L. The photograph, considered as a contact print positive, is a 180° exact reversal of the negative. The positive shown on Figure 28-5 is used to develop photogrammetric equations in subsequent sections.

Distance oL (Figure 28-5) is the camera focal length. The x and y reference axis system for measuring photographic coordinates of images is defined by straight lines joining opposite-side fiducial marks shown on the positive of Figure 28-5. The x axis, arbitrarily designated as the line most nearly parallel with the direction of flight, is positive in the direction of flight. Positive y is 90° counterclockwise from positive x.

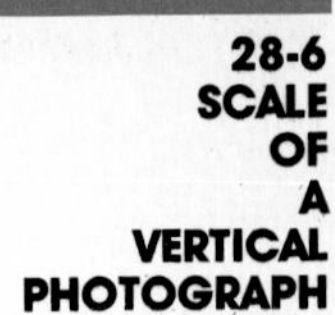

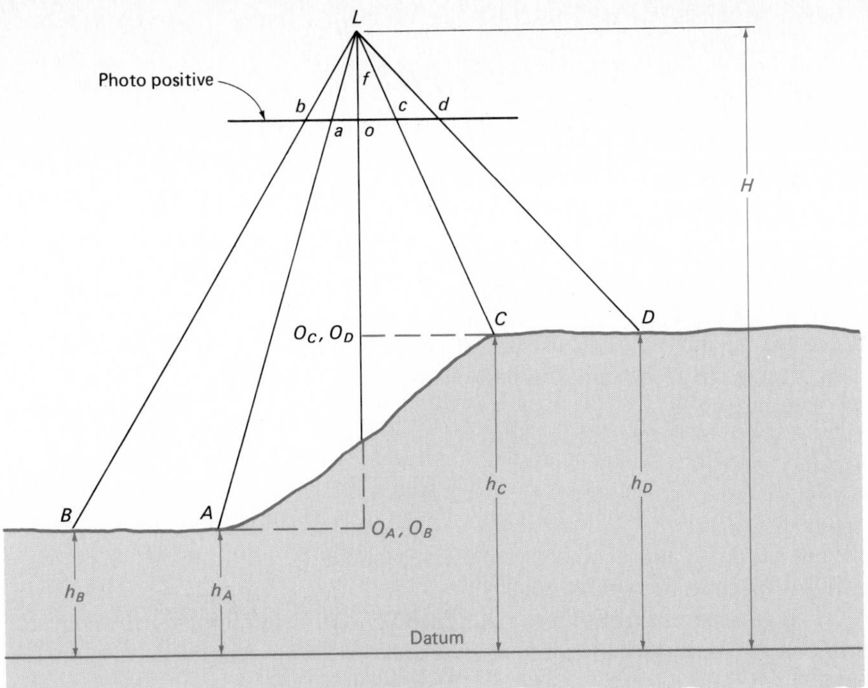

Figure 28-6. Scale of a vertical photograph.

Vertical photographs for topographic mapping are taken in strips which normally run lengthwise over the area to be covered. The strips or *flight lines* generally have a *sidelap* (overlap of adjacent flight lines) of 15 to 30%. *Endlap* (overlap of adjacent photographs in the same flight line) is usually about $60 \pm 5\%$. Figure 28-16(a) and (b) illustrate endlap and sidelap, respectively. If endlap is greater than 50%, all ground points will appear in at least two photographs and some show in three. Images common to three photographs permit extension of control through a strip of photographs using only minimal existing control.

28-6. SCALE OF A VERTICAL PHOTOGRAPH. Scale is ordinarily interpreted as the ratio of a distance on a map to that same distance on the ground, and is uniform throughout because a map is an orthographic projection. The scale of a vertical photograph is the ratio of photo distance to ground distance. Since a photograph is a perspective view, scale varies from point to point with variations in terrain elevation.

On Figure 28-6, L is the exposure station of a vertical photograph taken at an altitude H above the datum. The camera focal length is f, and o the photographic principal point. Points A, B, C, and D, which lie at elevations above datum of h_A, h_B, h_C, and h_D, respectively, are imaged on the photograph at a, b, c, and d. Scale at any point can be expressed in terms of its elevation, the camera focal length, and flying height above datum. From Figure 28-6, for similar triangles *Lab* and *LAB*, the following expression can be written:

$$\frac{ab}{AB} = \frac{La}{LA} \tag{a}$$

Also from similar triangles *Loa* and LO_AA, a similar expression results:

$$\frac{La}{LA} = \frac{f}{H - h_A} \tag{b}$$

Equating (a) and (b), recognizing that ab/AB equals photoscale at *A* and *B*, and considering *AB* to be infinitesimally short, the following equation for the scale at *A* results:

$$S_A = \frac{f}{H - h_A} \tag{c}$$

Scales at *B*, *C*, and *D* may be expressed similarly as: $S_B = f/(H - h_B)$, $S_C = f/(H - h_C)$, and $S_D = f/(H - h_D)$.

It is apparent from these relationships that scale increases at higher elevations and decreases at lower ones. This concept is seen graphically on Figure 28-6. Ground lengths *AB* and *CD* are equal but photo distances *ab* and *cd* are not, *cd* being longer and at larger scale than *ab* due to the higher elevation of *CD*. In general, by dropping subscripts, the scale *S* at any point whose elevation above datum is *h* may be expressed as:

$$S = \frac{f}{H - h} \tag{28-1}$$

where *S* is scale at any point on a vertical photo, *f* camera focal length, *H* flying height above datum, and *h* elevation of the point.

Use of an average photographic scale is frequently desirable but must be accepted with caution as an approximation. For any vertical photograph taken of terrain whose average elevation above datum is h_{avg}, the average scale S_{avg} is:

$$S_{avg} = \frac{f}{H - h_{avg}} \tag{28-2}$$

EXAMPLE 28-1

The photograph of Figure 28-6 was exposed with a 6-in focal length camera at a flying height above mean sea level of 10,000 ft. (a) What is the photo scale at point *a* if the elevation of point *A* on the ground is 2500 ft above mean sea level?

From Eq. (28-1):

$$S_A = \frac{f}{H - h_A} = \frac{6 \text{ in}}{10{,}000 - 2500} = \frac{1''}{1250 \text{ ft}} = 1{:}15{,}000$$

(b) For the same photograph, if average terrain is 4000 ft above mean sea level, what is the average photo scale?

From Eq. (28-2):

$$S_{avg} = \frac{f}{H - h_{avg}} = \frac{6 \text{ in}}{10{,}000 - 4000} = \frac{1''}{1000 \text{ ft}} = 1{:}12{,}000$$

Scale of a photograph can be determined from a map of the same area. This method does not require the focal length and flying height to be known. Rather it is necessary only to measure on the photograph a distance between two well-defined points identifiable on the map. Photo scale is then calculated from the equation

$$\text{photo scale} = \frac{\text{photo distance}}{\text{map distance}} \times \text{map scale} \tag{28-3}$$

In using Eq. (28-3), the distances must be in the same units, and the answer is the scale at average elevation of the two points used.

EXAMPLE 28-2

On a vertical photograph, the length of an airport runway measures 4.24 in. On a map plotted to a scale of 1:9600 it extends 7.92 in. What is the photo scale at the runway elevation?

From Eq. (28-3):

$$S = \frac{4.24}{7.92} \times \frac{1}{9600} = \frac{1}{17{,}900} \quad \text{or} \quad 1 \text{ in} = 1495 \text{ ft}$$

Scale of a photograph can also be computed readily if lines whose lengths are common knowledge appear in the photograph. Section lines, a football or baseball field, and so on, can be measured on the photograph and an approximate scale at that elevation ascertained as the ratio of measured photo distance to known ground length. With an approximate photographic scale known, rough determinations of the lengths of lines appearing in the photo can be determined.

EXAMPLE 28-3

On the photo of Example 28-2, a rectangular parcel of land measures 1.74 in by 0.83 in. Calculate the approximate ground dimensions of the parcel and its acreage. (Note: Photo scale was determined to be 1495 ft/in in Example 28-2.)

$$\text{Length} = 1495 \times 1.74 = 2600 \text{ ft}$$
$$\text{Width} = 1495 \times 0.83 = 1240 \text{ ft}$$
$$\text{Area} = \frac{2600 \times 1240}{43{,}560} = 74 \text{ acres}$$

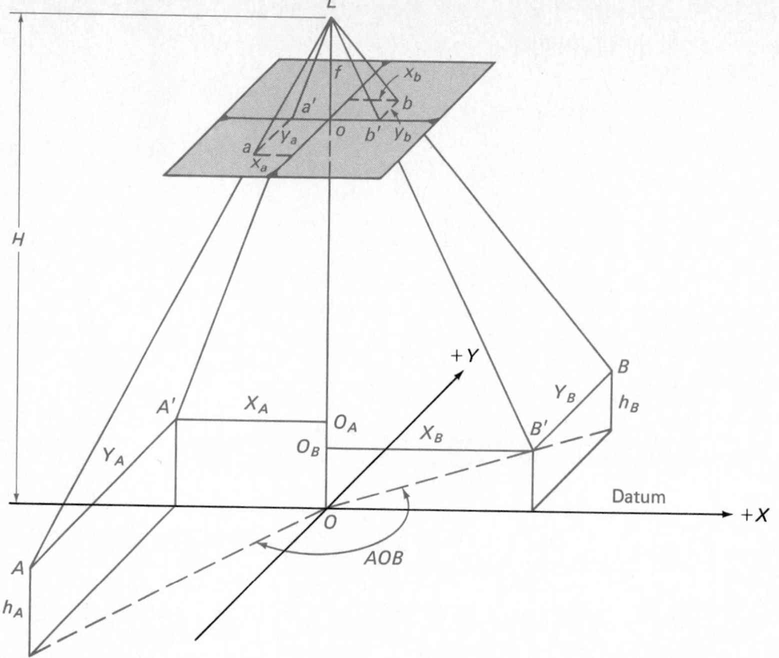

Figure 28-7. Ground coordinates from a vertical photograph.

28-7. GROUND COORDINATES FROM A SINGLE VERTICAL PHOTOGRAPH. Ground coordinates of points whose images appear in a vertical photograph can be determined with respect to an arbitrary ground axis system. The arbitrary X and Y ground axes are in the same vertical planes as photographic x and y, respectively, and the system's origin is in the datum plane vertically beneath the exposure station. Ground coordinates of points determined in this manner are used to calculate horizontal distances, horizontal angles, and areas.

Figure 28-7 illustrates a vertical photograph taken at flying height H above datum. Images a and b of the ground points A and B appear on the photograph. The measured photographic coordinates are x_a, y_a, x_b, and y_b; the ground coordinates, X_A, Y_A, X_B, and Y_B. From similar triangles LO_AA' and Loa',

$$\frac{oa'}{O_AA'} = \frac{f}{H - h_A} = \frac{x_a}{X_A}$$

then

$$X_A = \frac{(H - h_A)x_a}{f} \tag{28-4}$$

Also from similar triangles $LA'A$ and $La'a$,

$$\frac{a'a}{A'A} = \frac{f}{H - h_A} = \frac{y_a}{Y_A}$$

and

$$Y_A = \frac{(H - h_A)y_a}{f} \tag{28-5}$$

Similarly,

$$X_B = \frac{(H - h_B)x_b}{f} \tag{28-6}$$

$$Y_B = \frac{(H - h_B)y_b}{f} \tag{28-7}$$

From the X and Y coordinates of points A and B, the horizontal length of line AB can be calculated using the Pythagorean theorem:

$$AB = \sqrt{(X_B - X_A)^2 + (Y_B - Y_A)^2}$$

Areas are determined from X and Y coordinates by the method discussed in Chapter 14. The advantage of calculating lengths and areas by the coordinate formulas, rather than by average scale as in Example 28-3, is that greater accuracy results because differences in elevation are rigorously accounted for.

28-8. RELIEF DISPLACEMENT ON A VERTICAL PHOTOGRAPH. Relief displacement on a vertical photograph is the shift or movement of an image from its theoretical datum location caused by the object's relief—that is, its elevation above or below datum. Relief displacement on a vertical photograph occurs along radial lines from the principal point and increases in magnitude with greater distance of the image from it.

The concept of relief displacement in a vertical photograph taken from a flying height H above datum is illustrated in Figure 28-8. Camera focal length is f; the principal point o. Points B and C are the base and top, respectively, of a power pole with images at b and c on the photograph. A is an imaginary point on the datum plane vertically beneath B with corresponding position a on the photograph. Distance ab on the photograph is the image displacement due to h_B, the elevation of B above datum, and bc the image displacement due to height of the power pole.

From similar triangles LO_AA and Loa, an expression for relief displacement is formulated:

$$\frac{r_a}{R} = \frac{f}{H}$$

and rearranging,

$$r_aH = fR \tag{d}$$

Also from similar triangles LO_BB and Lob,

$$\frac{r_b}{R} = \frac{f}{H - h_B} \quad \text{or} \quad r_b(H - h_B) = fR \tag{e}$$

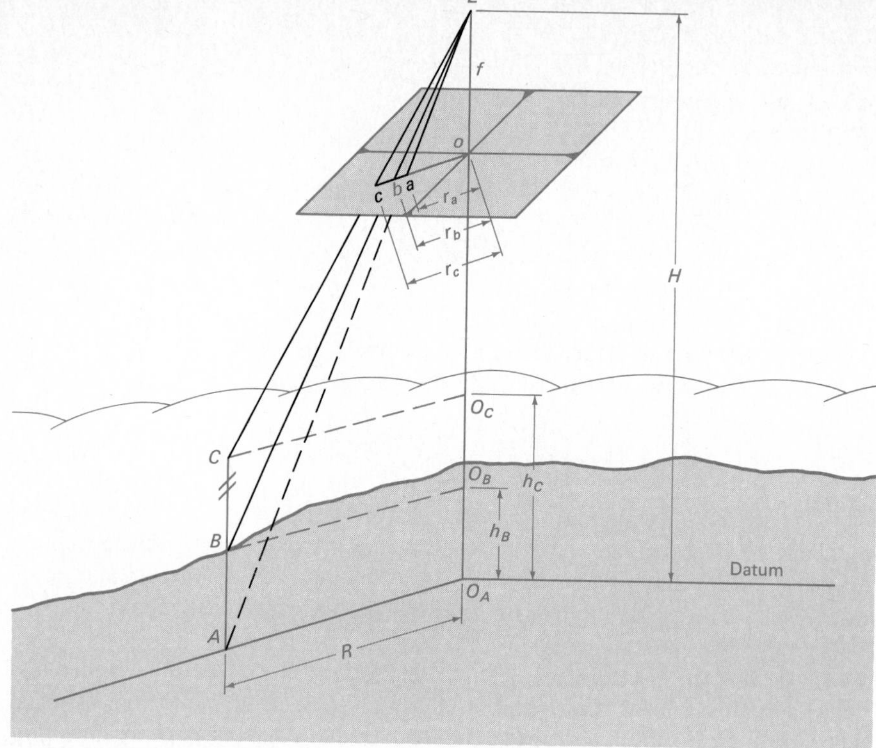

Figure 28-8. Relief displacement on a vertical photograph.

Equating (d) and (e):

$$r_a H = r_b(H - h_B)$$

and rearranging,

$$r_b - r_a = \frac{r_b h_B}{H}$$

If $d_b = r_b - r_a =$ the relief displacement of image b, then $d_b = r_b h_b/H$ can be written in general terms as

$$d = \frac{rh}{H'} \tag{28-8}$$

where d is the relief displacement, r the photo radial distance from principal point to image of the top or high point, h the height above datum of the top or high point, and H' the flying height above that same datum.

Equation (28-8) is used to locate the datum photographic positions of images on a vertical photograph. True horizontal angles may then be taken directly from the datum images, and if the photo scale at datum is known, true horizontal lengths of the lines are obtained directly. The datum position is located by scaling the calculated relief displacement d of a point along a radial line to the principal point (inward for a point whose elevation is above datum).

Equation (28-8) can also be applied in computing heights of vertical objects such as buildings, church steeples, radio towers, and power poles. To determine heights using the equation, images of both top and bottom of an object must be visible.

EXAMPLE 28-4

In Figure 28-8, the radial distance r_b to the image of a power pole base is 75.23 mm, and the radial distance r_c to the image of its top is 76.45 mm. Flying height H is 4000 ft above mean sea level, and the elevation of B is 450 ft. What is the height of the pole?

The relief displacement is $r_c - r_b = 76.45 - 75.23 = 1.22$ mm. Selecting a datum at the pole base and applying Eq. (28-8),

$$d = \frac{rh}{H'} \quad \text{so} \quad 1.22 = \frac{76.45h}{4000 - 450}$$

Then

$$h = \frac{3550(1.22)}{76.45} = 56.6 \text{ ft}$$

The relief displacement equation is particularly valuable to photo interpreters, who are usually interested in relative heights rather than absolute elevations.

Figure 28-3 vividly illustrates relief displacements. This vertical photo taken over the national capital shows the relief shift of Washington Monument in the upper right-hand portion of the format. This displacement, as well as that of other buildings throughout the photograph, occurs radially outward from the principal point.

28-9. FLYING HEIGHT OF A VERTICAL PHOTOGRAPH. From previous sections it is apparent that flying height above datum is an important parameter in solving basic photogrammetry equations. For rough computations, flying heights can be taken from altimeter readings if available. An approximate H can be obtained also by using Eq. (28-1) if a line of known length appears on a photograph.

EXAMPLE 28-5

The length of a section line is measured on a vertical photograph as 4.15 in. Find the approximate flying height above the terrain if $f = 6$ in. Assuming the datum at the section line elevation, Eq. (28-1) reduces to

$$\text{scale} = \frac{f}{H} \quad \text{and} \quad \frac{4.15}{5280} = \frac{6}{H}$$

from which

$$H = \frac{5280 \times 6}{4.15} = 7634 \text{ ft above the terrain}$$

If the images of two ground control points A and B appear on a vertical photograph, the flying height can be determined more precisely from the Pythagorean theorem:

$$L^2 = (X_B - X_A)^2 + (Y_B - Y_A)^2$$

Equations (28-4) through (28-7) are substituted in this expression:

$$L^2 = \left[\frac{(H - h_B)x_b - (H - h_A)x_a}{f}\right]^2 + \left[\frac{(H - h_B)y_b - (H - h_A)y_a}{f}\right]^2 \tag{28-9}$$

where L is the horizontal length of ground line AB, H flying height above datum, h the elevations of the control points above datum, and x and y measured photo coordinates of the control points.

In Eq. (28-9), all variables except H are known; hence, a direct solution can be found for the unknown flying height. The equation is a quadratic; so there are two solutions, but the correct one will be obvious.

28-10. STEREOSCOPIC PARALLAX. Parallax is defined as the apparent displacement of the position of an object with respect to a frame of reference due to a shift in the point of observation. For example, a person looking through the view finder of an aerial camera in an aircraft as it moves forward sees images of objects moving across the field of view. This apparent motion (parallax) is due to the changing location of the observer. Using the camera format as a frame of reference, parallax exists for all images appearing on successive photographs due to forward motion between exposures. Points closer to the camera (of higher elevation) will have greater parallaxes than lower ones. For 60% endlap, the parallax of images on successive photographs should average approximately 40% of the focal plane width.

Parallax of a point is a function of its relief and consequently a means of calculating elevations. It is also possible to compute X and Y ground coordinates from parallax.

Movement of an image across the focal plane between successive exposures takes place in a line parallel with the direction of flight. Thus, to measure parallax, that direction must first be established. For a pair of overlapping photos, this is done by locating positions of the principal points and *conjugate principal points* (principal points transferred to their places in the overlap area of the other photo). The line on each print ruled through these points defines the direction of flight. It also serves as the photographic x axis for parallax measurement. The y axis for making parallax measurements is drawn perpendicular to the flight line passing through the principal point of each photo. The x coordinate of a point is scaled on each photograph with respect to the axes so constructed, and parallax of the point is then calculated from the expression

$$p = x - x_1 \tag{28-10}$$

Photographic coordinates x and x_1 are measured on the left-hand and right-hand prints, respectively, with due regard given for algebraic signs.

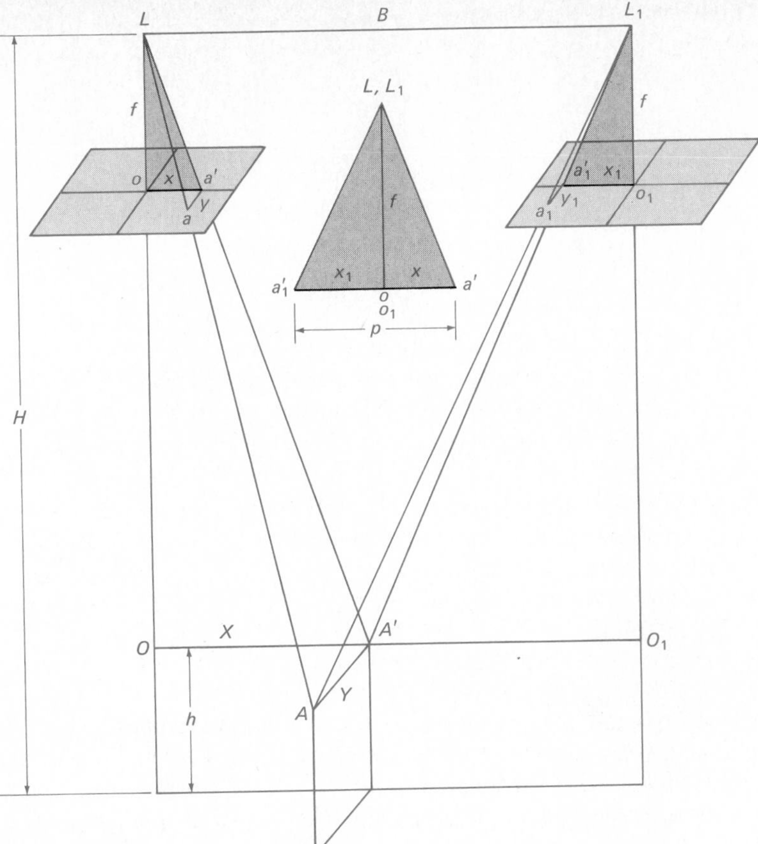

Figure 28-9. Parallax relationships.

Figure 28-9 illustrates an overlapping pair of vertical photographs exposed at equal flight heights H above datum. The distance between exposure stations L and L_1 is called B, the *air base*. The small inset figure shows two exposure stations L and L_1 in superposition to make the similarity of triangles La'_1a' and $LA'L_1$ more easily recognized. Equating these two similar triangles, there results

$$\frac{p}{f} = \frac{B}{H - h}$$

from which

$$H - h = \frac{Bf}{p} \qquad (28\text{-}11)$$

Also from similar triangles LOA' and Loa',

$$X = \frac{x}{f}(H - h)$$

Substituting Eq. (28-11) into that expression,

$$X = \frac{B}{p} x \tag{28-12}$$

And from triangles LAA' and Laa', with substitution of Eq. (28-11),

$$Y = \frac{B}{p} y \tag{28-13}$$

In these equations, X and Y are ground coordinates of a point with respect to an origin vertically beneath the exposure station of the left photograph, with positive X coinciding with the direction of flight and positive Y 90° counterclockwise to positive X. Parallax of the point is p, x and y photographic coordinates of the point on the left-hand print, H flying height above datum, h the elevation of the point above the same datum, and f the camera lens focal length.

Equations (28-11) through (28-13), commonly called the parallax equations, are useful in calculating horizontal lengths of lines and elevations of points. They also provide the fundamental basis for design and use of stereoscopic plotting instruments.

EXAMPLE 28-6

The length of line AB and elevations of points A and B from two vertical photographs which contain the images of a and b are needed. Flying height above mean sea level was 4050 ft and air base 2410 ft. The camera had a 6-in focal length. Measured photographic coordinates in inches on the left-hand print are $x_a = 2.10$, $x_b = 3.50$, $y_a = 2.00$, and $y_b = -1.05$; on the right-hand print, $x_{1a} = -2.25$ and $x_{1b} = -1.17$.

SOLUTION

From Eq. (28-10);

$$p_a = x_a - x_{1a} = 2.10 - (-2.25) = 4.35 \text{ in}$$
$$p_b = x_b - x_{1b} = 3.50 - (-1.17) = 4.67 \text{ in}$$

By Eqs. (28-12) and (28-13);

$$X_A = \frac{B}{p_a} x_a = \frac{2410 \times 2.10}{4.35} = 1163 \text{ ft}$$

$$X_B = \frac{2410 \times 3.50}{4.67} = 1806 \text{ ft}$$

$$Y_A = \frac{B}{p_a} y_a = \frac{2410 \times 2.00}{4.35} = 1108 \text{ ft}$$

$$Y_B = \frac{2410 \times (-1.05)}{4.67} = -542 \text{ ft}$$

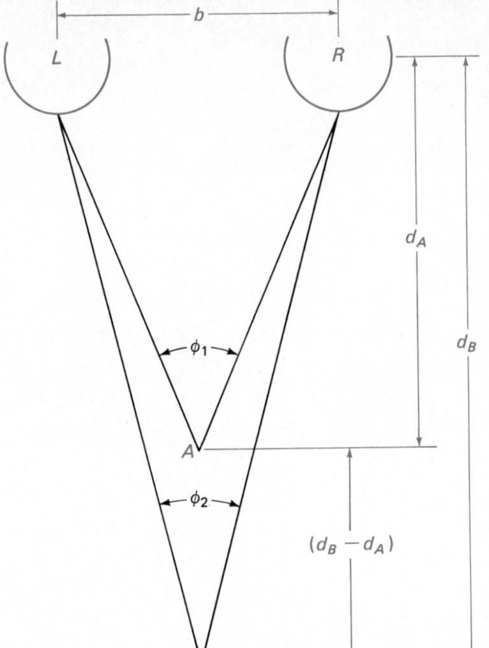

Figure 28-10. Parallactic angles in stereoscopic viewing.

From the Pythagorean theorem, length AB is

$$AB = \sqrt{(1806 - 1163)^2 + (-542 - 1108)^2} = 1771 \text{ ft}$$

By Eq. (28-11), the elevations of A and B are

$$h_A = H - \frac{Bf}{p_a} = 4050 - \frac{2410 \times 6}{4.35} = 726 \text{ ft}$$

$$h_B = 4050 - \frac{2410 \times 6}{4.67} = 954 \text{ ft}$$

28-11. STEREOSCOPIC VIEWING. The term *stereoscopic viewing* means seeing an object in three dimensions, a process requiring a person to have normal *binocular* (two-eyed) vision. In Figure 28-10 two *eyes* L and R are separated by a distance b called the *eye base.* When the eyes are focused on point A, their optical axes converge to form angle ϕ_1, and when sighting on B, ϕ_2 is produced. Angles ϕ_1 and ϕ_2 are called *parallactic angles* and the brain associates distances d_A and d_B with them. The depth $(d_B - d_A)$ of the object is perceived from unconscious comparison of these parallactic angles in the brain.

If two photographs of the same subject are taken from two different perspectives or camera stations, with the left print viewed with the left eye and

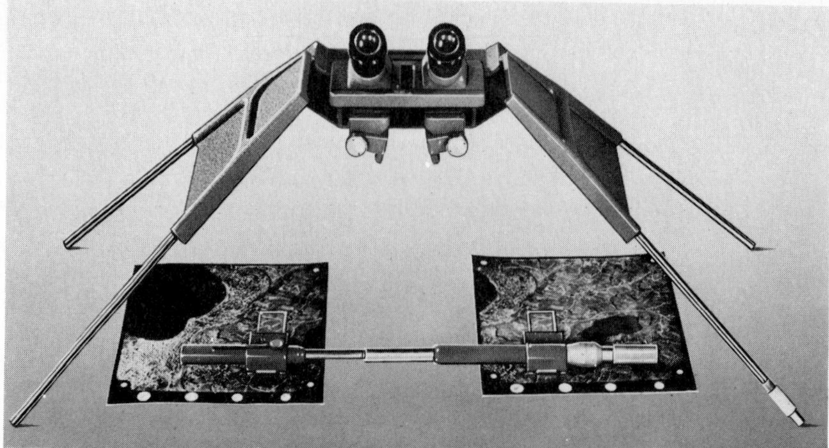

Figure 28-11. Folding-mirror stereoscope with parallax bar. (Courtesy Wild Heerbrugg Instruments, Inc.)

simultaneously the right print seen with the right eye, a mental impression of a three-dimensional model results. In normal stereoscopic viewing the *eye base* gives an impression of the parallactic angles. While looking at aerial photographs stereoscopically, the exposure station spacing simulates an eye base so the viewer actually sees parallactic angles comparable with having one eye at each of the two exposure stations.

The stereoscope shown in Figure 28-11 permits viewing photographs stereoscopically by enabling the left and right eyes to focus comfortably on the left and right prints, respectively, assuming proper orientation of the overlapping pair of photographs under the stereoscope. Correct orientation requires the two photographs to be laid out in the same order they were taken, with the stereoscope so set that the line joining the lens centers is parallel with the direction of flight. Spacing of the prints is varied, carefully maintaining this parallelism, until a clear stereoscopic model is obtained.

28-12. STEREOSCOPIC MEASUREMENT OF PARALLAX. The parallax of a point can be measured while viewing stereoscopically with the advantage of speed and, because binocular vision is used, greater accuracy. As the viewer looks through a stereoscope, two small identical marks etched on pieces of clear glass called *half-marks* are placed over each photograph. The viewer simultaneously sees one mark with the left eye and the other with the right eye; then the positions of the marks are shifted until they seem to fuse together as one mark which appears to lie at a certain elevation. The height of the mark will vary or "float" as the spacing of the half-marks is varied; hence, it is called the *floating mark*. Figure 28-12 demonstrates this principle and also illustrates that the mark can be set exactly on particular points such as A, B, and C by placing the half-marks at a and a', b and b', and c and c', respectively.

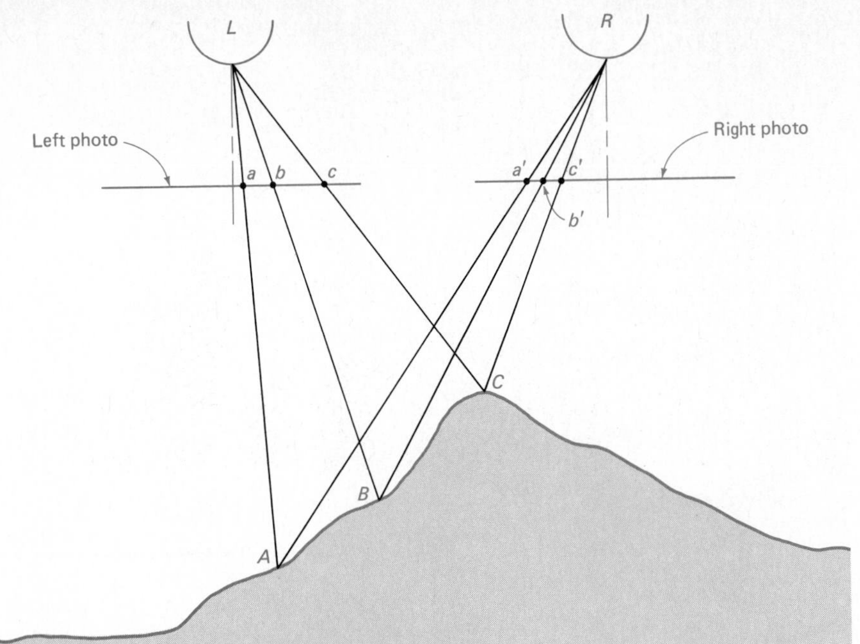

Figure 28-12. The principle of the floating mark.

Using the floating mark principle, parallax of points is measured stereoscopically with a parallax bar as shown beneath the stereoscope in Figure 28-11. It is simply a bar to which two half-marks are fastened. The right mark can be moved with respect to the left one by turning a micrometer screw and readings taken with the floating mark set stereoscopically on various points. The micrometer readings are added to the parallax bar *setup constant* to get parallax.

When a parallax bar is used, two overlapping photographs are oriented properly for viewing under a mirror stereoscope and fastened securely with respect to each other. The parallax bar constant for the setup is determined by measuring photo coordinates for a discrete point and applying Eq. (28-10) to obtain its parallax. The floating mark is placed on the same point, the micrometer read, and the constant for the setup found by

$$C = p - r \tag{28-14}$$

where C is the parallax bar setup constant, p the parallax of a point determined by Eq. (28-10), and r the micrometer reading obtained with the floating mark set on that same point.

Once the constant has been determined, the parallax of any other point can be computed by adding its micrometer reading to the constant. Thus a single measurement gives the parallax of a point. Each time another pair of photos is oriented for parallax measurements, a new parallax bar setup constant

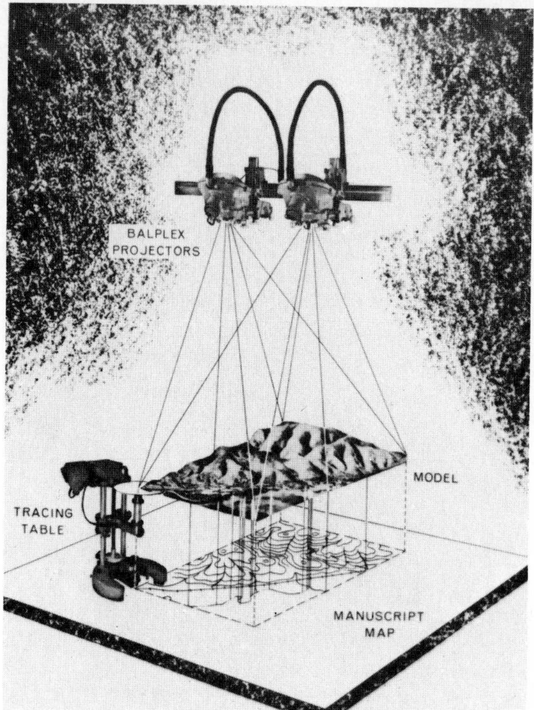

Figure 28-13. Balplex projectors and stereoscopic model. (Courtesy Bausch and Lomb, Inc.)

must be determined. A major advantage of the stereoscopic method is that parallaxes of nondiscrete points can be determined. Thus, elevations of hill tops, points in fields, etc. can be calculated using Eq. (28-11) even though their x coordinates cannot be measured for use in Eq. (28-10).

28-13. STEREOSCOPIC PLOTTERS. The primary use of stereoscopic plotters is to compile topographic maps from overlapping aerial photographs, but they are also used extensively for taking cross sections. There are two basic classifications of stereoscopic plotters: *optical projection* instruments and *mechanical projection* instruments. Each type consists of the following general components: (1) projection system (to create a stereomodel), (2) viewing system (which enables an operator to view the stereomodel), and (3) measuring/tracing system (for measuring or mapping, to scale, the stereomodel).

An optical projection type of stereoplotter is depicted in Figure 28-13. With this instrument, *diapositives* (positives developed on film or glass plates) of a pair of overlapping photographs are projected so that light rays carrying images common to both intersect to form a stereomodel. Projectors used in optical projection stereoscopic plotters resemble ordinary slide projectors; however, they are much more precise and can be adjusted in angular orientation and position to re-create the aerial camera's spatial location and attitude when the overlapping photos were exposed. This produces a "true" model of the terrain in the overlap area at greatly reduced scale.

Plotter viewing systems must be designed so the left and right eyes see only the projected images of the corresponding left and right diapositives. One method of accomplishing this is to place a blue filter in one projector and a red filter in the other. A pair of spectacles with corresponding blue and red lenses is worn by the operator. This system of viewing stereoscopically is called the *anaglyphic* method. Another system, called the *stereo image alternator* (SIA), operates by means of rapidly rotating shutters located on the projectors, and a viewing eyepiece. The shutters are synchronized so the left and right eye can see only the images from the corresponding left and right projector. SIA systems offer several advantages over the anaglyphic system, such as the capability of using color photographs, sharper projector images, and reduced light loss.

Various measuring and tracing systems have been devised for plotters. In the Balplex instrument (see Figure 28-13), light rays carrying corresponding images are intercepted on the *tracing table platen* (small white circular disk). A light emitted through a pinhole in the platen provides the floating mark which can be made to appear to rest exactly on a point of the model by raising or lowering the tracing table. A tracing pencil located directly below the floating mark permits positioning the point planimetrically. A counter linked to the tracing table responds to up-and-down motion so the elevation for any setting of the floating mark is read directly.

When diapositives are placed in the projectors and the lights turned on, corresponding rays will not intersect properly to form a clear model because of tilt in the photographs, unequal flying heights, and, of course, improper projector orientation. The projectors can be moved linearly along the X, Y, and Z axes and also rotated about each of them until the diapositives reproduce the relative conditions existing when the photographs were taken. This is called *relative orientation*, and when accomplished, corresponding rays will intersect to form a perfect three-dimensional model.

The model is brought to required scale by making the rays of at least two, but preferably three, ground control points intersect at their positions plotted on a manuscript map prepared at the desired scale. It is leveled by adjusting the projectors so the counter reads the correct elevations of each of a minimum of three, but preferably four, corner ground control points when the floating mark is set on them. *Absolute orientation* is a term applied to the processes of scaling and leveling the model.

When orientation is completed, measurements such as cross sections can be made from the model, or a map of it completed. Cross-section readings, used for highway design, borrow-pit and stockpile volume determination, and flood plain mapping have been greatly facilitated in recent years through the use of digital output systems interfaced with stereoplotters. In mapping, planimetric details are located first by bringing the floating mark into contact with objects in the model and tracing them. A pencil directly beneath the reference mark records their locations on the manuscript map below. Contours are traced by setting the elevation counter successively at each contour elevation and moving the reference mark over the model while keeping in contact with the terrain. Again, the pencil draws contours. When a manuscript is completed, it is examined for omissions and mistakes, and field-checked. The final map is drafted or scribed by tracing it.

Figure 28-14. Kelsh three-projector stereoplotting instrument. (Courtesy Kelsh Instrument Division, Danko Arlington, Inc.)

Figure 28-14 shows a Kelsh (optical projection) stereoplotter having three projectors. This has the advantage of enabling two adjacent stereomodels to be oriented simultaneously.

Mechanical projection stereoplotters use two precisely made metal space rods to simulate light rays. The diapositives are viewed through binoculars via an optical train of lenses and prisms. The floating mark, composed of a pair of half-marks superimposed in the optical train, is moved up or down by turning a hand screw or foot disk, and impelled in the *X* and *Y* directions, either manually or by means of hand wheels. Instruments equipped with a coordinatograph allow planimetry to be traced directly on the manuscript, and *X* and *Y* coordinates read. Figure 28-15 shows the Wild A-10 mechanical projection stereoplotter with attached coordinatograph on the right upon which the map is compiled.

Figure 28-15. A-10 Autograph, mechanical projection stereoplotter, with coordinatograph and other accessories. (Courtesy Wild Heerbrugg Instruments, Inc.)

28-14. ANALYTICAL PHOTOGRAMMETRY. In recent years, with the advent of high-speed electronic computers, much of the work done with stereoscopic plotters can now be performed economically by means of analytical photogrammetry. This science involves precise measurements of photographic coordinates of images, and construction of a mathematical model which can be solved by numerical methods. The procedures of analytical photogrammetry are highly accurate and particularly well adapted to establishing vertical and horizontal control through *aerotriangulation*. Accuracies of $\frac{1}{15,000}$th of flying height are readily obtained for computed X and Y ground coordinates, and $\frac{1}{10,000}$th of H for Z coordinates. Thus, for photos exposed from 6000 ft, X and Y can be calculated correctly to within approximately 0.4 ft, and Z to within about 0.6 ft.

28-15. ORTHOPHOTOS. As implied by their name, orthophotos are orthographic representations of the terrain in picture form. They are derived from aerial photos in a process called *differential rectification*, which removes scale variations and image displacements due to relief and tilt. Thus the imaged features are shown in their true planimetric positions.

Instruments used for differential rectification vary considerably in design, but basically they are modified stereoscopic plotters with either optical or mechanical projection. Using optical projection instruments, an orthophoto is derived by systematically *scanning* a stereomodel and photographing it in a series of adjacent narrow strips. *Rectification* (removal of tilt) is accomplished by leveling the model to ground control prior to scanning, and scale variations due to terrain relief are removed by varying the projection distance during scanning. As the instrument automatically traverses back and forth across the model, exposure is made through a narrow slit onto an orthonegative below. An operator, viewing the model in three dimensions, continually monitors the scans and adjusts the projection distance to keep the exposure slit in contact with the model. Because the model itself has uniform scale throughout, the resulting *orthonegative* (from which the orthophoto is made) is also of uniform scale.

Orthophotos combine the advantages of both aerial photos and line maps. Like photos, they show features by their actual images rather than as lines and symbols, thus making them more easily interpreted and understood. Like maps, they show the features in their true planimetric positions. Therefore true distances, angles, and areas can be scaled directly from them. Orthophotos can generally be prepared more rapidly and economically than line or symbol planimetric maps. With their many significant advantages, orthophotos have superseded conventional maps for many uses.

28-16. GROUND CONTROL FOR PHOTOGRAMMETRY. As pointed out in preceding sections, almost all phases of photogrammetry depend on ground control (points of known positions and elevations with identifiable images on the photograph). Ground control can be *basic control*—traverse, triangulation, or trilateration monuments and bench marks already in existence, and marked prior to photography to make them visible on the photos; or it can be *photo control*—natural points having images recognizable on the photographs and positions which are subsequently determined by ground surveys originating from basic control. Instruments and procedures used in the ground surveys were described in earlier chapters. Ordinarily, photo control points are selected after photography to ensure their satisfactory location and positive identification. Premarking points with artificial targets is sometimes necessary in areas that lack natural objects to provide definite images.

28-17. FLIGHT PLANNING. Certain factors, depending generally on the purpose of photography, must be specified to guide a flight crew in executing its mission of taking aerial photographs. Some of them are (a) boundaries of the area to be covered, (b) required scale of the photography, (c) camera focal length and format size, (d) endlap, and (e) sidelap. Having fixed these elements, it is possible to compute the entire flight plan and prepare a flight map upon which the required flight lines have been delineated. The pilot flies specified flight lines by choosing and correlating headings on existing natural features in the field shown on the flight map.

Purpose of the photography is the paramount consideration in flight planning. In taking aerial photos for topographic mapping with a stereoplotter, for example, endlap should optimally be 60% and sidelap between 25 and 35%. Required scale and contour interval of the final map must be evaluated to settle flying height. Enlargement capability from photo scale to map is restricted for all stereoplotters, and with many of them the optimum ratio is 5X. With such an instrument, if required map scale is 200 ft/in, photo scale becomes fixed at 1000 ft/in. If camera focal length is 6 in, by Eq. (28-2) flying height is established at $6 \times 1000 = 6000$ ft above average terrain.

C-factor (ratio of flight height above ground to the contour interval that is practical for any specific stereoplotter) is the criterion used to select flying height in relation to required contour interval. It has been determined for various plotters, and values range from approximately 800 to about 2000. Thus, if a plotter has a C-factor of, say, 1200, and a map is to be compiled with a 5-ft contour interval, a flight height of not more than 1200 × 5 = 6000 ft above the terrain should be sustained.

Information ordinarily calculated in flight planning includes (1) flying height above mean sea level; (2) distance between exposures; (3) number of photographs per flight line; (4) distance between flight lines; (5) number of flight lines; and (6) total number of photographs. A flight plan is prepared based on these items.

EXAMPLE 28-7

A flight plan for an area 10 mi wide and 15 mi long is required. Average terrain in the area is 1500 ft above mean sea level. The camera has a 6-in focal length with 9 × 9-in format. Endlap is to be 60%, sidelap 25%. Required scale of the photography is 1:12,000.

a. Flying height above datum from Eq. (28-2):

$$\text{scale} = \frac{f}{H - h_{avg}} \quad \text{so} \quad \frac{1}{12{,}000} = \frac{6/12}{H - 1500} \quad \text{and} \quad H = 7500 \text{ ft}$$

b. Distance between exposures: Endlap is 60% so the linear advance per photograph is 40% of the total coverage of 9 in × 1000 ft/in = 9000 ft. Thus the distance between exposures = 0.40 × 9000 = 3600 ft.

c. Total number of photographs per flight line:

$$\text{length of each flight line} = 15 \text{ mi} \times 5280 \text{ ft/mi} = 79{,}200 \text{ ft}$$

$$\text{number of photos per flight line} = \frac{79{,}200 \text{ ft}}{3600 \text{ ft/photo}} = 22$$

Add 2 photos on each end to ensure complete coverage, so the total is 22 + 2 + 2 = 26 photos per flight line.

d. Distance between flight lines: Sidelap is 25% so the lateral advance per flight line is 75% of the total photographic coverage.

$$\text{distance between flight lines} = 0.75 \times 9000 \text{ ft} = 6750 \text{ ft}$$

e. Number of flight lines:

$$\text{width of the area} = 10 \text{ mi} \times 5280 \text{ ft/mi} = 52{,}800 \text{ ft}$$

Number of spaces between flight lines is

$$\text{number} = \frac{52{,}800 \text{ ft}}{6750 \text{ ft/line}} = 7.8 \text{ (say 8)}$$

$$\text{total flight lines} = 8 + 1 = 9$$

$$\text{Planned spacing between flight lines} = \frac{52{,}800}{8} = 6600 \text{ ft}$$

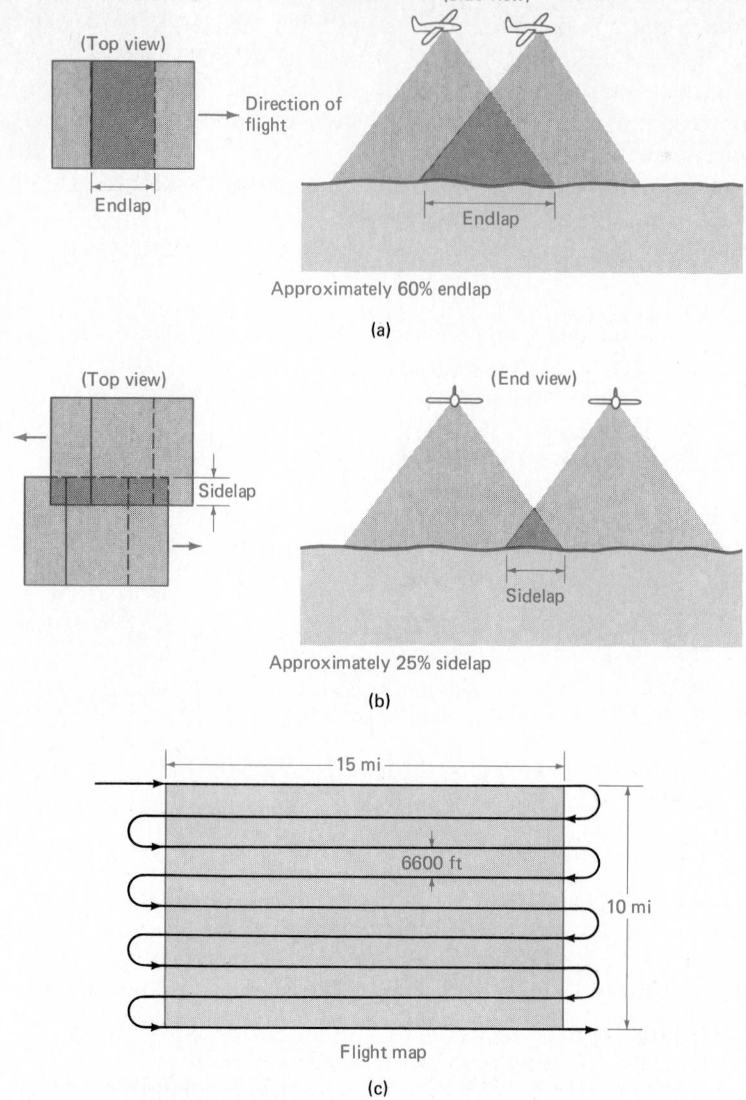

Figure 28-16. Endlap, sidelap, and flight map.

Note: The first and last flight lines should either coincide with or be near the edges of the area, thus providing a safety factor to ensure complete coverage.

f. Total number of photos required:

$$\text{total photos} = 26 \text{ per flight line} \times 9 \text{ flight lines} = 234 \text{ photos}$$

Figures 28-16(a) and (b) illustrate endlap and sidelap, and (c) shows the flight map.

28-18. SOURCES OF ERROR IN PHOTOGRAMMETRY. Some sources of error in photogrammetric work are:

1. Measuring scale not standard length.
2. Inaccurate location of principal and conjugate principal points.
3. Failure to use camera calibration data.
4. Assuming vertical exposures when photographs are actually tilted.
5. Presuming equal flying heights when they were unequal.
6. Not considering differential shrinkage or expansion of the photographic prints.
7. Incorrect orientation of photographs under a stereoscope or in a stereoscopic plotter.
8. Faulty setting of the floating mark on a point.

28-19. MISTAKES. Some mistakes that occur in photogrammetry are:

1. Reading measuring scales incorrectly.
2. Mistaking units—for example, inches and millimeters.
3. Confusion in identifying corresponding points on different photographs.
4. Neglect of relief displacement.
5. Failure to provide proper control or using erroneous control coordinates.
6. Attaching an incorrect sign (plus or minus) to a measured photographic coordinate.
7. Blunder in computations.
8. Misidentification of control-point images.

PROBLEMS

28-1. Define the terms (a) photogrammetry, (b) photo interpretation, and (c) remote sensing.

28-2. Discuss the advantages of preparing maps photogrammetrically.

28-3. List the principal components of a single-lens frame aerial camera.

28-4. The distance between two points on a vertical photograph is *ab* and the corresponding ground distance is *AB*. For the following data, compute photographic scale.
(a) $ab = 2.49$ in; $AB = 2910$ ft
(b) $ab = 3.86$ in; $AB = 4125$ ft
(c) $ab = 103.76$ mm; $AB = 3575$ m

28-5. On a vertical photograph of flat terrain, section corners appear a distance X apart. If the camera focal length was f, compute the flying height above the ground for the following data:
(a) $X = 5.86$ in; $f = 8\frac{1}{4}$ in
(b) $X = 65.85$ mm; $f = 152.4$ mm

28-6. On a vertical photograph of flat terrain, the scaled distance between two points is *ab*. In the following problems find the average photographic scale along *ab* if the measured length between the same line is *AB* on a map plotted at a scale of *S*.
(a) $ab = 2.84$ in; $AB = 3.76$ in; $S = 1{:}10{,}000$
(b) $ab = 110.64$ mm; $AB = 3.82$ in; $S = 1{:}24{,}000$

28-7. What are the average scales of vertical photographs for the following problems, given flying height above sea level H, camera focal length f, and average ground elevation h?
(a) $H = 7500$ ft; $f = 6$ in; $h = 1450$ ft
(b) $H = 5500$ ft; $f = 8\frac{1}{4}$ in; $h = 925$ ft
(c) $H = 3500$ m; $f = 88.90$ mm; $h = 645$ ft

28-8. The length of a football field from goal post to goal post scales 12.82 mm on a vertical photograph. Find the dimensions of a large rectangular building that also appears on this photo and whose sides measure 2.15 in by 1.40 in.

28-9. Compute the area in acres of a triangular parcel of land whose sides measure 53.27 mm, 48.05 mm, and 71.56 mm on a vertical photograph taken from 6000 ft above average ground with an $8\frac{1}{4}$-in focal length camera.

28-10. Calculate the flight height above average terrain that is required to obtain vertical photographs for constructing a mosaic at an average scale of S if the camera focal length is f, for the following data.
(a) $S = 1{:}5000$; $f = 8\frac{1}{4}$ in
(b) $S = 1{:}10{,}000$; $f = 152.4$ mm

28-11. Determine the horizontal distance between two points A and B whose elevations above datum are $h_A = 1560$ ft and $h_B = 1425$ ft, and whose images a and b have photographic coordinates $x_a = 2.59$ in, $y_a = 2.32$ in, $x_b = -1.46$ in, and $y_b = -2.66$ in on a vertical photograph. The camera focal length was 152.4 mm and the flying height above datum 9000 ft.

28-12. Similar to Problem 28-11, except that camera focal length was $3\frac{1}{2}$ in, flying height above datum 4500 ft, elevations h_A and h_B 955 ft and 1115 ft, respectively, and photocoordinates of images a and b were $x_a = 89.92$ mm, $y_a = -17.02$ mm, $x_b = -89.41$ mm, and $y_b = 57.40$ mm.

28-13. On the photograph of Problem 28-11, image c of a third point C appears. Its elevation $h_C = 1390$, and its photocoordinates are $x_c = 3.06$ in and $y_c = -2.53$ in. Compute the horizontal angles in the triangle ABC.

28-14. On the photograph of Problem 28-12, the image d of a third point D appears. Its elevation is $h_D = 1060$, and its photocoordinates are $x_d = 62.05$ mm and $y_d = 70.28$ mm. Calculate the area, in acres, of triangle ABD.

28-15. Determine the height of radio towers which appear on a vertical photograph for the following conditions of flying height above the tower base H', distance on the photograph from principal point to the tower base r_b, and distance from principal point to the tower top r_t.
(a) $H' = 6800$ ft; $r_b = 2.795$ in; $r_t = 3.030$ in
(b) $H' = 5500$ ft; $r_b = 86.57$ mm; $r_t = 95.82$ mm

28-16. An area has an average terrain elevation of 1250 ft above mean sea level and its highest point is 75 ft above average terrain. If the camera focal plane opening is 9 × 9 in, what flying height above mean sea level will limit relief displacement to a maximum of 0.05 in on a vertical photograph of this area?

28-17. On a vertical photograph, images a and b of ground points A and B have photographic coordinates $x_a = 3.47$ in, $y_a = 2.38$ in, $x_b = -1.85$ in, and $y_b = -2.42$ in. The horizontal distance between A and B is 5350 ft and the elevations of A and B above datum are 562 ft and 685 ft, respectively. Calculate the flying height above datum for a camera having a focal length of 152.4 mm.

28-18. Similar to Problem 28-17, except $x_a = -62.35$ mm, $y_a = 89.76$ mm, $x_b = -6.30$ mm, $y_b = -95.29$ mm, line length $AB = 4965$ ft, and elevations of points A and B are 825 and 727 ft, respectively.

28-19. An air base of 3250 ft exists for a pair of overlapping vertical photographs taken at a flying height of 8500 ft above MSL with a camera having a focal length of $8\frac{1}{4}$ in. Photocoordinates of points A and B on the left photograph are $x_a = 40.05$ mm, $y_a = 48.20$ mm, $x_b = 22.95$ mm, and $y_b = -29.15$ mm. The x photocoordinates on the right photograph are $x_a = -59.86$ mm and $x_b = -71.92$ mm. Calculate horizontal length AB.

28-20. Similar to Problem 28-19, except the air base is 7430 ft, flying height above mean sea level 15,250 ft, camera focal length 152.4 mm, x and y photocoordinates on the left photo $x_a = 39.78$ mm, $y_a = 70.45$ mm, $x_b = 26.43$ mm, and $y_b = -66.09$ mm, and x photocoordinates on the right photo $x_a = -51.27$ mm and $x_b = -61.78$ mm.
28-21. Calculate the elevations of points A and B of Problem 28-19.
28-22. Compute the elevations of points A and B of Problem 28-20.
28-23. A pair of overlapping vertical photographs is oriented and secured under a mirror stereoscope, and a parallax bar reading obtained on point a of $r_a = 14.32$ mm. The x photocoordinate of point a is 2.41 in on the left photograph and -1.17 in on the right photograph. The camera focal length is 152.4 mm, flying height 10,450 ft above datum, and the air base 4160 ft. Determine the difference in elevation between points B and C if their parallax bar readings are $r_b = 15.69$ mm and $r_c = 14.89$ mm.
28-24. Compare an orthophoto with a conventional line and symbol map.
28-25. How are orthophotos prepared?
28-26. Discuss the advantages of orthophotos as compared with maps and mosaics.
28-27. Discuss the various viewing systems used in optical projection stereoplotting instruments.

Aerial photography is to be taken of a tract that is X mi square. Flying height will be H' ft above average terrain, and the camera has a focal length f. If the focal lane opening is 9 × 9 in, and minimum sidelap is 30%, how many flight lines will be needed to cover the tract for the data given in Problems 28-28 and 28-29?
28-28. $X = 10$; $H' = 6000$; $f = 6$ in.
28-29. $X = 36$; $H' = 15{,}000$; $f = 8\frac{1}{4}$ in.

Aerial photography was taken at a flying height H' feet above average terrain. If the camera focal plane dimensions were 9 × 9 in, the focal length f, and the spacing between adjacent flight lines X ft, what is the percent sidelap for the data given in Problems 28-30 and 28-31?
28-30. $H' = 4100$; $f = 152.4$ mm; $X = 4500$.
28-31. $H' = 6200$; $f = 3\frac{1}{2}$ in; $X = 11{,}800$.

Photographs at a scale of S are required to cover an area X mi square. The camera has a focal length f and focal plane dimensions 9 × 9 in. If endlap is 60% and sidelap 30%, how many pictures will be required to cover the area for the data given in Problems 28-32 and 28-33?
28-32. $S = 1{:}12{,}000$; $X = 20$; $f = 152.4$ mm.
28-33. $S = 1{:}14{,}400$; $X = 48$; $f = 3\frac{1}{2}$ in.

BIBLIOGRAPHY

American Society of Photogrammetry. 1980. *Manual of Photogrammetry*, 4th ed. Falls Church, Va.: American Society of Photogrammetry.

Croom, C. H. 1973. "How Photogrammetry Aids the Surveyor." *Surveying and Mapping* 33 (no. 4):491.

Derenyi, E. E., and A. Maarek, 1974, "Photogrammetric Control Extension for Route Design." *ASCE Journal of the Surveying and Mapping Division* 100 (no. SU1):49.

Eldridge, W. H. 1967. "Photogrammetry for Property Surveying." *Surveying and Mapping* 27 (no. 1):63.

Moffitt, F. H., and E. Mikhail. 1980. *Photogrammetry*, 3rd ed. New York: Harper & Row.

Wolf, P. R. 1983. *Elements of Photogrammetry*, 2nd ed. New York: McGraw-Hill.

APPENDIX A

TESTING AND ADJUSTING INSTRUMENTS

A-1. INTRODUCTION. Surveying instruments are designed and constructed to give correct horizontal and vertical measurements. A good instrument, properly used and cared for, may stay in adjustment for months, or longer, and last a lifetime. Nevertheless, temperature changes, jarring, and improper handling can cause instruments to go out of adjustment; therefore, they should be tested periodically and adjusted when necessary to maintain accuracy. A level, for example, should be checked each day it is used on important work.

Proper field procedures, such as double centering and keeping backsights and foresights equal, permit accurate work to be done even though an instrument is out of adjustment. Frequently, however, a few minutes spent in making simple adjustments reduces the time and effort required to operate equipment efficiently. Furthermore, some errors may be introduced that can be eliminated only by adjusting the instrument.

Surveyors and engineers can perform tests to determine whether any surveying instrument is out of adjustment; and they should be capable of making routine adjustments of dumpy levels, tilting dumpy levels, and transits. Sections of this appendix describe standard techniques. Older wye levels, which are no longer made, may nevertheless still be in service and require testing and adjustment. Procedures for doing this were described in the 6th edition of this book.

Except for bull's-eye level vials, optical plummets, and some instrument plate bubbles, adjustment of automatic levels and theodolites should be left to experts. Dismantling either type of instrument can result in serious damage. If tests described in the following sections reveal that adjustments are required,

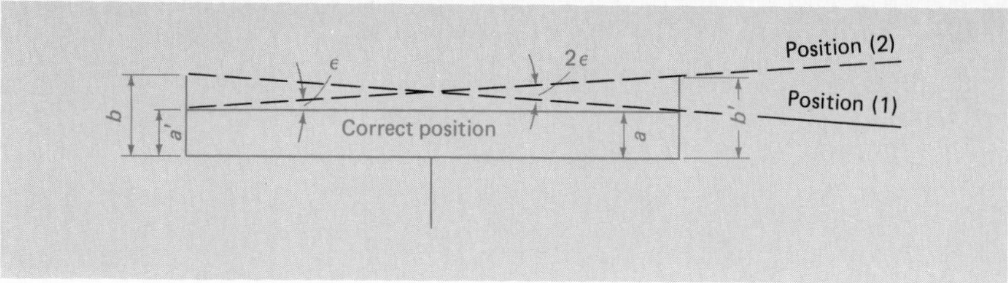

Figure A-1. Principle of reversion.

the equipment should be carefully packaged and shipped to the manufacturer or a laboratory where specialists are available.

A-2. METHODS OF TESTING INSTRUMENTS. Most tests used for checking surveying instruments fall into one of the following two categories:

1. *Comparisons with known values.* This method consists in making direct comparisons of measuring devices against a precisely calibrated standard. It is used in testing and calibrating tapes, EDMIs, and level rods.

2. *Principle of reversion.* This procedure consists of reversing an instrument in position to double the error and make it more apparent. It is used extensively in testing levels, transits, and theodolites. To illustrate, assume that in Figure A-1 the angular error between the correct and unadjusted line ε is caused by the difference in lengths a and b as shown in position (1). (An example would be the heights of the two ends of a level vial on a telescope.) After the telescope is turned 180° in azimuth, the unadjusted line occupies position (2) because a and b have changed places. Since the angle between positions (1) and (2) is 2ε, it follows that single reversion doubles the error. Specific methods of making reversion tests with a level, transit, or theodolite are described in this appendix.

A-3. REQUIREMENTS FOR TESTING AND ADJUSTING LEVELS, TRANSITS, AND THEODOLITES. Before making adjustments, *careful* tests should be made to ensure that any apparent lack of adjustment is actually caused by the instrument's condition, not by test deficiencies. To properly check and adjust levels, transits, and theodolites in the field, the following rules should be followed:

1. Choose terrain that permits solid setups in a nearly level area permitting sights of at least 200 ft in opposite directions. Three permanent points set approximately 200 ft apart in a straight line, on nearly level ground, and preferably at the same elevation expedite adjustments. Organizations having a number of instruments in use, and surveyors working in an area over a long period of time, find it profitable to set such permanent marks.
2. Perform adjustments when good atmospheric conditions prevail, preferably on cloudy days free of heat waves. No sight line should pass through alternate sun and shadow, or be directed into the sun.

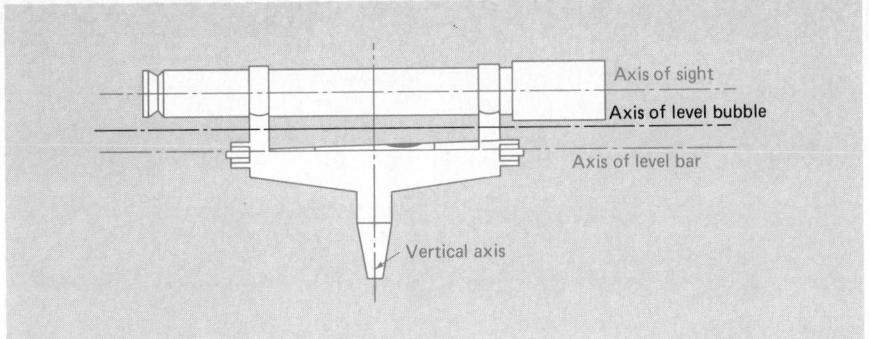

Figure A-2. Dumpy level.

3. Place the instrument in shade, or shield it from direct rays of the sun.
4. Make sure the tripod shoes are tight and the instrument firmly screwed onto the tripod. Spread the tripod legs well apart and position them so that the tripod plate is nearly level. Press the shoes firmly into the ground. With older conventional tripods, loosen the three tripod hinge screws to relieve stresses and then tighten again.

Standard methods and a prescribed order must be followed in adjusting levels, transits, and theodolites. Correct positioning of parts is attained by loosening or tightening the proper adjusting nuts and screws with special pins. Time is wasted if each adjustment is perfected on the first trial, since some adjustments affect others. The complete series of tests may have to be repeated several times if an instrument is badly off. A final check of all adjustments should be made to ensure that none has been disturbed. The simplest adjustment of all, removal of parallax by carefully focusing the objective lens and eyepiece, must be kept in mind at all times.

Straight, strong adjusting pins that fit the capstans should be used, and the capstans handled with care to avoid damaging the soft metal. Adjustment screws have been properly set when an instrument is shipped from the factory. Tightening them too much (or not enough) nullifies otherwise correct adjustment procedures and may leave the instrument in worse condition than it was before testing.

A-4. ADJUSTMENT OF A DUMPY LEVEL. A level in adjustment establishes a horizontal plane of sight when the telescope is revolved about a vertical axis. The principal lines of the dumpy level, as illustrated in Figure A-2, are (1) axis of sight, (2) axis of the level bubble, (3) axis of the level bar, and (4) vertical axis.

For perfect adjustment, it is necessary that the axis of sight, axis of the level bubble, and axis of the level bar be parallel to each other and perpendicular to the vertical axis. There are two adjustable parts: the cross hairs and the level vial.

A-4.1. ADJUSTMENT OF LEVEL VIAL

Purpose. To make the axis of the level bubble perpendicular to the vertical axis.

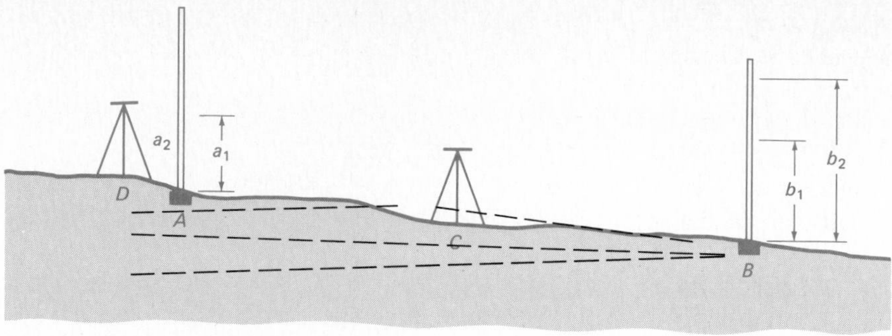

Figure A-3. Peg adjustment.

Test. Set up the level, center the bubble, and revolve the telescope 180° about the vertical axis. The distance the bubble moves off the central position is double the error.

Correction. Turn the capstan nuts at one end of the level vial to move the bubble halfway back to the centered position. Level the instrument using the leveling screws. Repeat the test until the bubble remains centered during a complete revolution of the telescope.

A-4.2. PRELIMINARY ADJUSTMENT OF HORIZONTAL CROSS HAIR

Purpose. To make the horizontal cross hair truly horizontal when the instrument is leveled.

Test. Sight a sharply defined point with one end of the horizontal cross hair. Turn the telescope slowly on its vertical axis so that the cross hair moves across the point. If it does not remain on the point for the cross hair's full length, the instrument is out of adjustment.

Correction. Loosen the four capstan screws holding the reticle. Rotate the reticle in the telescope tube until the horizontal hair remains on the point as the telescope is turned. The screws should be carefully tightened in their final position.

A-4.3. LINE OF SIGHT ADJUSTMENT

Purpose. To make the axis of sight perpendicular to the vertical axis and thus parallel to the axis of the level bubble. This adjustment is also called the *two-peg method* and also the *direct adjustment.*

Test. Level the instrument over a point *C* halfway between two stakes *A* and *B* about 200 ft apart. See Figure A-3. Determine the difference in rod readings a_1 and b_1 on *A* and *B*, respectively. Since the distance to the two points is equal, the true difference in elevation is obtained even though the axis of sight is not exactly horizontal.

Then set the instrument at *D* on line with the stakes and close to one of them—*A* in this case—and level. With the eyepiece only a few inches from the rod, a reading a_2 on A is taken by sighting through the objective lens end of the telescope. Usually a pencil is centered in the small field of view. A rod reading b_2 is also taken on B.

If the axis of sight is parallel to the axis of the level bubble (that is, horizontal), the rod reading b_2 should equal the rod reading at A plus the difference in elevation between A and B, or $(b_1 - a_1) + a_2$. The difference, if any, between the computed and actual readings is the error to be corrected by adjustment.

Correction. Loosen the top (or bottom) capstan screw holding the reticle, and tighten the bottom (or top) screw to move the horizontal hair up or down and give the required reading on the rod at B. Several trials may be necessary to get an exact setting. (*Caution:* One screw should be loosened before the other is tightened on older instruments to avoid breaking the cross hair!)

An alternative method of testing the adjustment is by reciprocal leveling. A setup is made close to A and readings taken on A and B. The level is moved to a position near B and similar sights taken. The difference in elevation is computed and the reticle shifted to give a reading on the distant rod equal to the reading on the near point plus the difference in elevation of the points.

If the difference in elevation of A and B is known, only one setup is required near either point for the adjustment.

A-5. ADJUSTMENT OF A TILTING DUMPY LEVEL. As described in Section 6-13, the vertical axis of a tilting dumpy level is oriented approximately vertical using a bull's-eye bubble, and the axis of sight is made horizontal for each individual backsight and foresight by carefully centering the main sensitive bubble. For a tilting dumpy level in good adjustment, the axis of its bull's-eye bubble should be approximately perpendicular to the vertical axis, and the line of sight must be parallel to the axis of the sensitive level bubble.

A-5-1. ADJUSTMENT OF THE BULL'S-EYE BUBBLE

Purpose. To make the axis of a bull's-eye bubble perpendicular to the vertical axis.

Test and Correction. Same as those described in Section A-8. Precise adjustment is not critical, however, because more accurate leveling of the telescope is accomplished by using the sensitive level bubble.

A-5.2. ADJUSTMENT OF THE HORIZONTAL CROSS HAIR

Purpose, Test, and Correction. Same as the second adjustment of the dumpy level (see Section A-4.2).

A-5.3. ADJUSTMENT OF THE SENSITIVE LEVEL BUBBLE

Purpose. To make the axis of the sensitive bubble parallel to the axis of sight.

Test. Perform the peg test as described in Section A-4.3. The telescope is brought to the rod reading required for the horizontal sight using the tilting knob. Maladjustment is indicated if the bubble is not centered in this position.

Correction. With the axis of sight horizontal, the adjusting screws on the sensitive bubble are raised or lowered as necessary to center the bubble, or make the two ends of the bubble match for coincidence bubbles.

A-6. ADJUSTMENT OF A TRANSIT.[1]

Transits are designed to measure the vertical and horizontal projections of angles and to serve as levels, so certain lines and axes must be precisely positioned. Principal lines of a transit, illustrated previously in Figure 11-15, are:

1. Axis of the plate bubble. The line tangent to the upper circular arc of the level vial at its midpoint.
2. Vertical axis. Same as for the level.
3. Horizontal axis. The line through the center of rotation of the telescope axle and its bearings in the standards.
4. Axis of sight. Same as for the level.
5. Axis of telescope bubble. Same as for the level.

For correct adjustment, the (1) axis of the plate bubble must be perpendicular to the vertical axis, (2) horizontal axis parallel to the plate bubble axis, (3) axis of sight perpendicular to the horizontal axis, and (4) axis of the telescope bubble parallel to the axis of sight. To maintain these relationships, plate-level vials, cross hairs, standards, telescope-level vial, and vertical-circle vernier are adjustable.

A-6.1. ADJUSTMENT OF PLATE-LEVEL VIALS

Purpose. To make the axis of each plate-level bubble perpendicular to the vertical axis.

Test. Set up the instrument, bring one plate-level vial over two opposite leveling screws, and center it. Revolve the instrument 180° about the vertical axis to place that level vial, turned end for end, over the same leveling screws. The distance the bubble moves from its central position is double the error.

Correction. Turn the capstan screws at one end of the level vial to move the bubble *halfway* back to the centered position. Level the instrument with the leveling screws. Repeat the test until the bubble remains centered during a complete revolution of the instrument.

Adjust the other bubble in the same manner.

A-6.2. PRELIMINARY ADJUSTMENT OF VERTICAL CROSS HAIR

Purpose. To place the vertical cross hair in a plane perpendicular to the horizontal axis of the instrument.

Test. Sight on a well-defined point with one end of the vertical cross hair. Turn the telescope on its horizontal axis so the cross hair moves along the point. If it departs, the cross hair is not perpendicular to the horizontal axis.

[1] The tests described in this section can be used also to check theodolites; however, as stated earlier, their adjustment is generally best left to experts.

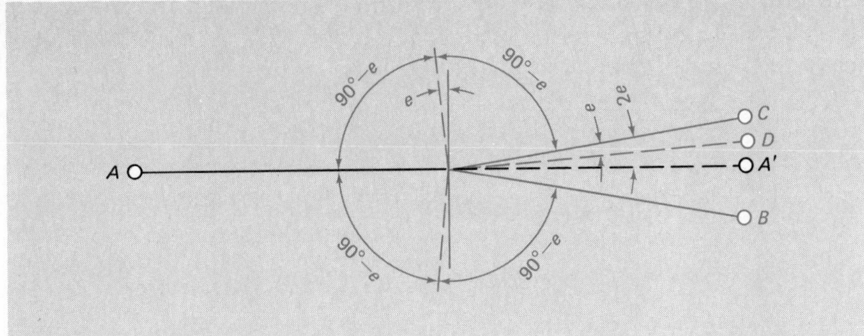

Figure A-4. Double reversion.

Correction. Loosen all four capstan screws holding the reticle, and turn the reticle slightly until the vertical hair remains on the fixed point during rotation of the telescope. Tighten the capstan screws and recheck the adjustment.

A-6.3. ADJUSTMENT OF VERTICAL CROSS HAIR

Purpose. To make the axis of sight perpendicular to the horizontal axis.

Test. This test applies the double-centering procedure for prolonging a straight line. Level the transit and backsight carefully on a well-defined distant point *A*, as in Figure A-4, clamping the plates. Plunge the telescope and set a foresight point *B* at approximately the same elevation as *A*, and at least 400 ft away if possible. With the telescope still in the inverted position, unclamp either plate, turn the instrument on the vertical axis, backsight on the first point *A* again, and clamp the plate. Plunge the telescope back to its normal position and set a point *C* beside the first foresight point *B*. The distance between *B* and *C* is *four times* the error of adjustment because of the double reversion.

Correction. Loosen one of the side capstan screws which hold the reticle to the telescope tube and tighten the opposite screw to move the vertical hair *one-fourth* of the distance *CB* to point *D*. Repeat the test until the telescope sights the same point *A′* after reversing from the backsight *A*.

A-6.4. ADJUSTMENT OF HORIZONTAL CROSS HAIR

Purpose. To bring the horizontal cross hair into the optical axis of the telescope. This is necessary if the transit is to be used for leveling or measuring vertical angles.

Test. Set up and level the transit over point *A*, as in Figure A-5. Line in two stakes *B* and *C* at approximately the same elevation. Stake *B* should be at minimum focusing distance from *A*, perhaps 5 or 10 ft, and stake *C* at least 300 ft away.

Take readings of the horizontal cross hair upon a rod held first on *B*, then on *C*. Plunge the telescope, turn the instrument about its vertical axis, and sight *B* again, setting the horizontal hair on the first rod reading. Then with vertical and

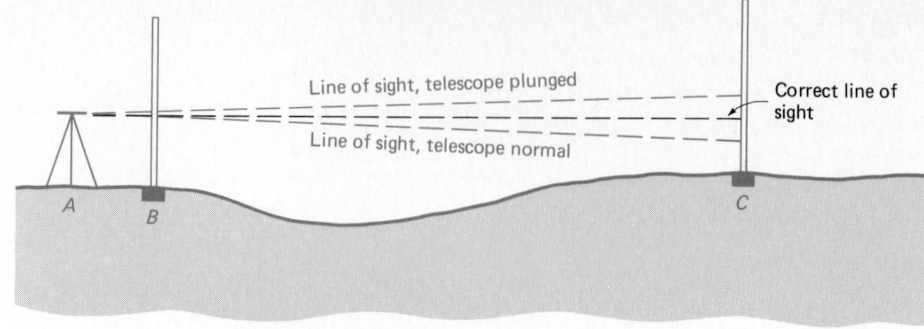

Figure A-5. Adjustment of horizontal cross hair.

horizontal axes clamped, read the rod held on far stake *C*. Any discrepancy between the two readings on the rod at *C* is approximately double the error.

Correction. By means of the top and bottom capstan screws holding the reticle, move the horizontal hair until it intercepts the rod halfway between the two readings on *C*. Repeat the test and adjustment until the horizontal-hair reading does not change for normal and plunged sights on the far point.

A-6.5. ADJUSTMENT OF STANDARDS

Purpose. To make the telescope's horizontal axis perpendicular to the transit's vertical axis.

Test. With the transit carefully leveled parallel to the horizontal axis, sight a well-defined high point *A*, as in Figure A-6, at a vertical angle of at least 30°, and

Figure A-6. Adjustment of standards.

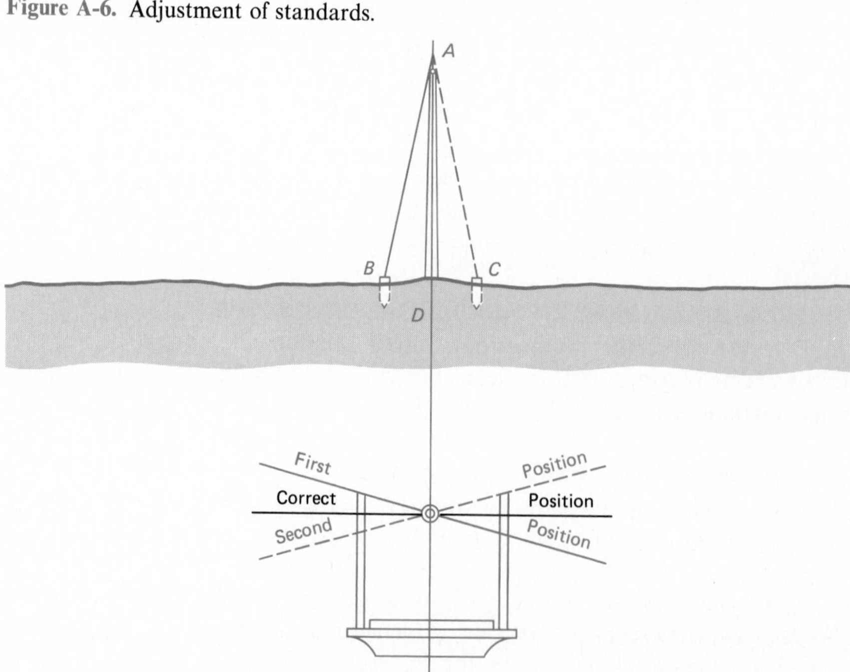

clamp the plates. Depress the telescope and mark a point B near the ground. Plunge the telescope, unclamp either plate, turn the instrument about the vertical axis, sight point A again, and clamp the plate. Now depress the telescope and set another point, C, near B. Any discrepancy between B and C is the result of unequal standard heights and represents approximately *twice* the error.

Correction. Set a point D approximately *halfway* between B and C, and sight on it. With the plates clamped, elevate the telescope and bring the line of sight on point A by raising or lowering the movable block in one standard. To raise the horizontal axis, first loosen the friction screws holding the trunnion cap on the standard, then tighten the capstan screw below the block. To lower the axis, reverse the procedure. The friction screws holding the trunnion cap must be set carefully to prevent the telescope from being too loose or binding.

Repeat the test and adjustment until the high and low points remain in the line of sight with the telescope in both normal and inverted positions.

A-6.6. ADJUSTMENT OF TELESCOPE-LEVEL VIAL

Purpose. To make the axis of the level bubble perpendicular to the vertical axis and parallel to the telescope axis of sight.

Test. Same as for the dumpy level peg adjustment (see Section A-4.3).

Correction. If rod reading b_2, as in Figure A-3, indicates an adjustment is necessary, the correct rod reading required to produce a horizontal axis of sight is set on the rod by means of the vertical-circle slow-motion screw. The telescope bubble is then centered by turning the capstan screws at one end of the level vial.

A-6.7. ADJUSTMENT OF VERTICAL-CIRCLE VERNIER

Purpose. To make the vernier of the vertical circle read zero when the telescope-level bubble is centered, thereby producing a zero index error.

Test. Level the instrument with both plate-level bubbles. Using the vertical-circle slow-motion screw, center the telescope bubble. Any nonzero angle on the vertical-circle vernier is an index error.

Correction. Loosen the capstan screws holding the vernier plate and move it so the vertical circle and vernier zero marks coincide. Tighten the capstan screws. Avoid leaving a gap between the vernier and vertical circle, since such a space introduces errors in reading angles.

A-7. ADJUSTMENT OF AN OPTICAL PLUMMET. The line of sight of an optical plummet should coincide with the instrument's vertical axis. To adjust, set the transit or theodolite over a fine point and aim the line of sight exactly at it by turning the leveling screws. Rotate the instrument 180° in azimuth. If the optical plummet reticle moves off the point, bring it *halfway* back by means of the adjusting screws provided. Center the reticle on the point again with the leveling screws, and repeat the test.

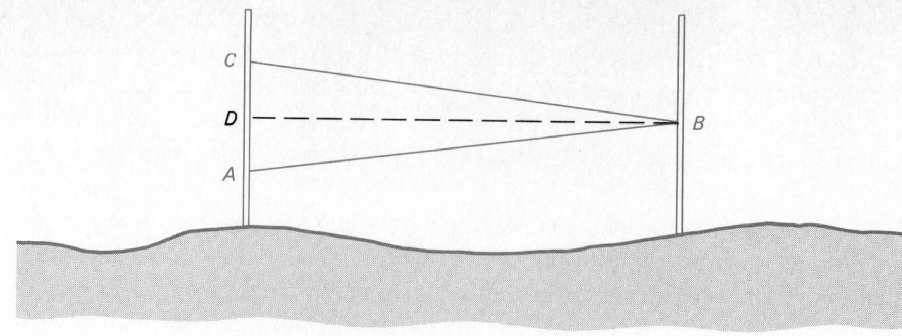

Figure A-7. Adjustment of hand level.

A-8. ADJUSTMENT OF BULL'S-EYE BUBBLES. If a bull's-eye bubble does not remain centered when an instrument is rotated in azimuth, an adjustment is required but need not be extremely precise because it does not control fine leveling of the sight axis. Carefully center the bubble using the leveling screws and turn the instrument 180° in azimuth. *Half* of the bubble run is corrected by manipulating the vial adjusting screws, the bubble centered by operating the leveling screws, and the test repeated.

A-9. ADJUSTMENT OF A HAND LEVEL. The only adjustable part of a Locke hand level is the horizontal cross hair.

Purpose. To make the axis of sight horizontal when the bubble is centered.

Test. With the hand level on a solid support at elevation *A*, and the bubble centered, mark a point *B* on a post or building corner as in Figure A-7. The distance *AB* should not be greater than 100 ft. Support the level at *B*, center the bubble, and note whether the line of sight strikes point *A*. If it does not, mark another point, *C*.

Correction. Bisect the distance *AC* and set point *D*. With the level at *B* and bubble centered, move the cross hair to *D* by means of the adjusting screws.

APPENDIX B

COORDINATE GEOMETRY IN SURVEYING CALCULATIONS

B-1. INTRODUCTION. Except for extensive geodetic control surveys, almost all other surveys are referenced to plane rectangular coordinate systems. State plane coordinates (see Chapter 21) are most frequently employed, although local arbitrary systems can be used. Advantages of referencing points in a rectangular coordinate system are (1) the relative positions of points are uniquely defined, (2) they can be conveniently plotted, (3) if lost in the field they can readily be recovered from other available points referenced to the same system, and (4) computations are greatly facilitated. The latter advantage is the subject of this appendix.

Computations involving coordinates are performed in a variety of surveying problems. Two situations were introduced in Chapter 13, where it was shown that the length and bearing (or azimuth) of a line can be calculated from the coordinates of its end points. Area computation using coordinates was discussed in Chapter 14. Additional problems that are conveniently solved using coordinates are determining the point of intersection of (a) two straight lines, (b) a line and circle, and (c) two circles. These problems are often encountered in route surveys where it is necessary to compute intersections of tangents and circular curves in horizontal alignments, and in boundary and subdivision work where parcels of land are often defined by straight lines and circular arcs.

Solutions to these problems can be obtained by writing equations for the lines and circles involved, which include unknown coordinates of the intersection points, then solving them simultaneously for the unknowns. The required

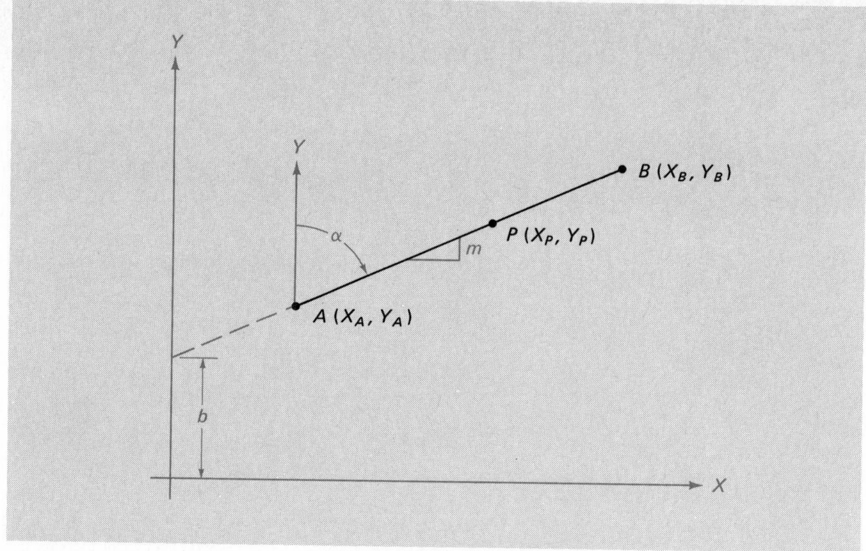

Figure B-1. Geometry of straight line in plane coordinate system.

equations, together with example problems, are presented in the following sections.

B-2. COORDINATE FORMS OF EQUATIONS FOR LINES. In Figure B-1, straight line *AB* is referenced in a plane rectangular coordinate system. Coordinates of end points *A* and *B* are X_A, Y_A, X_B and Y_B. Length *AB* and azimuth α of this line in terms of these coordinates are

$$AB = \sqrt{(X_B - X_A)^2 + (Y_B - Y_A)^2} \tag{B-1}$$

$$\alpha = \tan^{-1}\left(\frac{X_B - X_A}{Y_B - Y_A}\right) \tag{B-2}$$

The general mathematical expression for a straight line is:

$$Y_p = mX_p + b \tag{B-3}$$

where Y_p is the *Y* coordinate of any point *P* on the line whose *X* coordinate is X_p, *m* the slope of the line, and *b* the *Y* intercept of the line. Slope *m* can be expressed as:

$$m = \frac{Y_B - Y_A}{X_B - X_A} = \cot \alpha \tag{B-4}$$

For any straight line, the slope is constant; that is, in Figure B-1, the slope between *A* and *B* is the same as that between *A* and *P*. Thus the following equa-

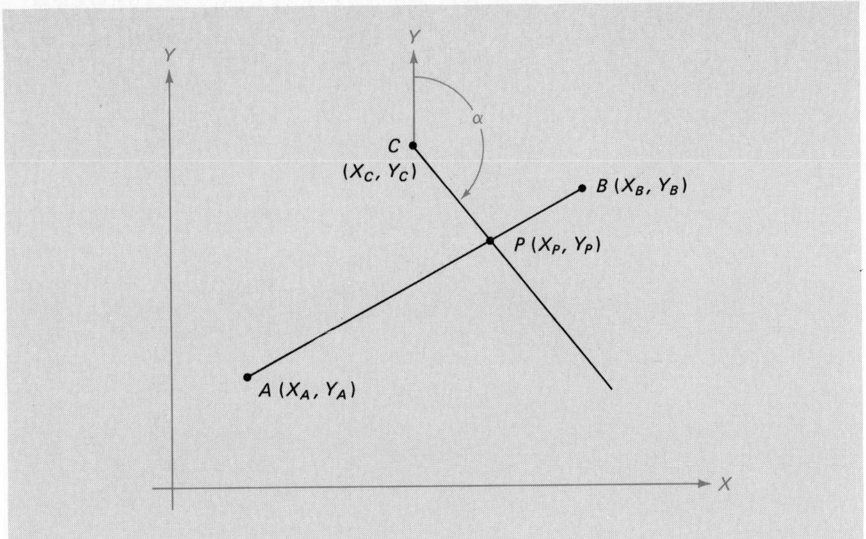

Figure B-2. Intersection of two lines.

tion can be written intuitively from Eq. (B-4):

$$\frac{Y_B - Y_A}{X_B - X_A} = \frac{Y_P - Y_A}{X_P - X_A} = \cot \alpha \tag{B-5}$$

B-3. INTERSECTION OF TWO LINES. Equation (B-5) is very useful in computing the point of intersection of two lines. Data generally known for this type of problem are coordinates of end points of lines, or fixed azimuths of lines, determined from survey or design data.

EXAMPLE B-1

In Figure B-2, assuming the following information is known for two lines, compute coordinates X_P and Y_P of the intersection point.

$X_A = 1425.07$	$X_B = 7484.80$	$X_C = 4497.96$	$\alpha = 141°30'$
$Y_A = 1971.28$	$Y_B = 5209.64$	$Y_C = 6062.00$	

By Eq. (B-5), an expression for line *AB* is:

$$\frac{5209.64 - 1971.28}{7484.80 - 1425.07} = \frac{Y_P - 1971.28}{X_P - 1425.07} \tag{a}$$

Also, by Eq. (B-5), an expression for the line at *C is:*

$$\frac{Y_P - 6062.00}{X_P - 4497.96} = \cot 141°30' \tag{b}$$

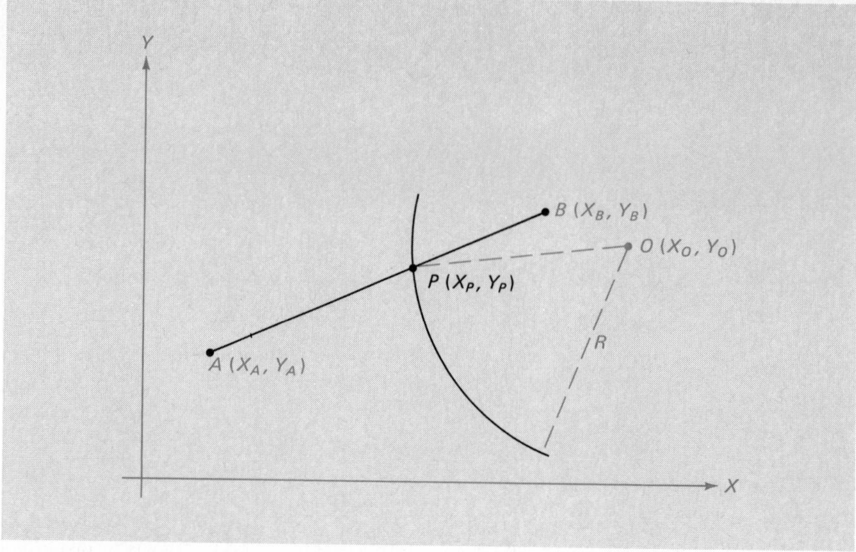

Figure B-3. Intersection of line and circle.

Reducing expressions (a) and (b):

$$0.53441X_P - Y_P = -1209.71 \tag{c}$$

$$1.25717X_P + Y_P = 11{,}716.43 \tag{d}$$

Solving Eqs. (c) and (d) simultaneously yields:

$$X_P = 5864.50 \text{ ft}$$

$$Y_P = 4343.76 \text{ ft}$$

B-4. COORDINATE FORM OF EQUATION FOR A CIRCLE. The general mathematical expression for a circle in rectangular coordinates is:

$$R^2 = (X_P - X_O)^2 + (Y_P - Y_O)^2 \tag{B-6}$$

In Eq. (B-6), and with reference to Figure B-3, R is the radius of the circle, X_O and Y_O coordinates of the radius point O, and X_P and Y_P coordinates of any point P on the circle. For most problems the radius and coordinates of the radius point are known, R having been selected on the basis of design specifications or geometric constraints, and X_O and Y_O having been computed as a result of survey measurements, or scaled from a design map.

B-5. INTERSECTION OF A LINE AND CIRCLE. Figure B-3 shows a line AB intersecting the circle at point P. The coordinates of points A, B, and O are known, as is the radius. To solve for the intersection point, an equation in the form of Eq. (B-5) can be written for the line, and one like Eq. (B-6) for the circle. These equations, when solved simultaneously, yield a quadratic expression for one of the unknowns, as:

$$aY_p^2 + bY_p + c = 0 \tag{B-7}$$

The solution for Y_p is then obtained from:

$$Y_p = \frac{-b \pm \sqrt{b^2 - 4ac}}{2a} \tag{B-8}$$

After calculating Y_p, its value can be substituted in one of the original equations to obtain X_p.

EXAMPLE B-2

In Figure B-3, assume the coordinates of the circle center are $X_O = 500.00$ and $Y_O = 200.00$; for points A and B, $X_A = 100.00$, $Y_A = 130.00$, $X_B = 300.00$, and $Y_B = 200.00$; and $R = 150$ ft. Determine the coordinates of intersection point P.

From Eq. (B-5):

$$\frac{X_P - 100.00}{Y_P - 130.00} = \frac{300.00 - 100.00}{200.00 - 130.00} \tag{e}$$

and by Eq. (B-6):

$$(X_P - 500.00)^2 + (Y_P - 200.00)^2 = (150.00)^2 \tag{f}$$

Reducing Eq. (e):

$$X_P = 2.8571\,Y_P - 271.43 \tag{g}$$

Substituting Eq. (g) into (f) and reducing:

$$Y_P^2 - 524.73\,Y_P + 66{,}854 = 0 \tag{h}$$

Solving Eq. (h) with Eq. (B-8):

$$Y_P = \frac{524.73 \pm \sqrt{(524.73)^2 - 4(66{,}856)}}{2} = 217.87$$

Then substituting $Y_P = 217.87$ into Eq. (g):

$$X_P = 2.8571(217.87) - 271.43 = 351.05$$

In solving quadratic Eq. (h), the decision to add or subtract the value under the radical can be made on the basis of experience or use of a carefully constructed

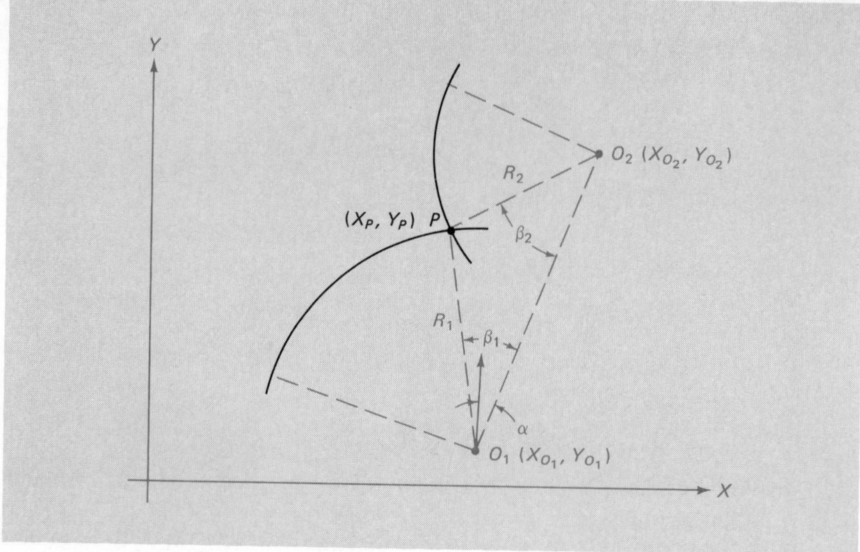

Figure B-4. Intersection of two circles.

scaled diagram, which also provides a check on the computations. One answer will be unreasonable and is discarded.

B-6. INTERSECTION OF TWO CIRCLES. Occasionally surveyors are required to compute the point of intersection of two circles having known radii and coordinates of radius points. The situation is illustrated in Figure B-4. The problem can be solved by writing equations in the form of Eq. (B-6), including unknown coordinates X_P and Y_P in both equations, and then solving simultaneously for the unknowns. The unknowns will both appear as second-degree terms in the equations, however; thus the solution is somewhat difficult to obtain.

In an alternative approach, the length and azimuth of O_1O_2 of Figure B-4 can be obtained from Eqs. (B-1) and (B-2), after which angles β_1 and β_2 are calculated using the law of cosines. With β_1 and β_2 known, azimuths O_1P and O_2P are computed, and the problem reduced to solving for the coordinates of point P given the length and direction from either known point O_1 or O_2.

EXAMPLE B-3

In Figure B-4, assume the following data are available, and X_P and Y_P at point P are required:

$$X_{O_1} = 2851.28 \qquad Y_{O_1} = 299.40 \qquad R_1 = 2000 \text{ ft}$$
$$X_{O_2} = 3898.72 \qquad Y_{O_2} = 2870.15 \qquad R_2 = 1500 \text{ ft}$$

By Eq. (B-1):

$$O_1O_2 = \sqrt{(3898.72 - 2851.28)^2 + (2870.15 - 299.40)^2} = 2775.95 \text{ ft}$$

By Eq. (B-2):

$$\alpha = \tan^{-1}\left(\frac{3898.72 - 2851.28}{2870.15 - 299.40}\right) = 22°10'05.5''$$

From the law of cosines:

$$\beta_1 = \cos^{-1}\left[\frac{(2000)^2 + (2775.95)^2 - (1500)^2}{2(2000)(2775.95)}\right] = 31°36'53.4''$$

$$\beta_2 = \cos^{-1}\left[\frac{(1500)^2 + (2775.95)^2 - (2000)^2}{2(1500)(2775.95)}\right] = 44°20'31.8''$$

$$\alpha O_1P = 22°10'05.5'' - 31°36'53.4'' = -9°26'47.9''$$

$$= 360° - 9°26'47.9'' = 350°33'12.1''$$

$$\alpha O_2P = 22°10'05.5'' + 180° + 44°20'31.8'' = 246°30'37.3''$$

With the coordinates of O_1 and the length and direction of O_1P known, the coordinates of point P can be obtained directly as:

$$X_P = 2851.28 + 2000 \times \sin 350°33'12.1'' = 2523.02 \text{ ft}$$

$$Y_P = 299.40 + 2000 \times \cos 350°33'12.1'' = 2272.28 \text{ ft}$$

These coordinates can be checked by a similar computation from point O_2 using the length and direction of O_2P.

B-7. TWO-DIMENSIONAL COORDINATE TRANSFORMATION. It is sometimes necessary to convert coordinates of points from one survey axis system to another. This happens, for example, if a survey is performed in some local-assumed or arbitrary coordinate system, and later it is desired to convert it to state plane coordinates. The process of making these conversions is called *coordinate transformation*, and if only planimetric coordinates (i.e., Xs and Ys) are involved, it is called *two-dimensional coordinate transformation.*

The geometry of a two-dimensional coordinate transformation is illustrated in Figure B-5. In the figure, X-Y represents a local assumed coordinate system, and E-N a state plane coordinate system. Coordinates of points A through D are known in the X-Y system, and those of A and B are also known in the E-N system. Points such as A and B, whose positions are known in both systems, are termed "control points." At least two control points are required in order to determine E-N coordinates of other points such as C and D.

If the scales of the two systems are equal (the usual case in surveying), only two steps are involved in coordinate transformation; (1) *rotation*, and (2) *translation.* Referring to Figure B-5, rotation consists of determining coordinates of points in the rotated X'-Y' axis system (shown dashed). The X'-Y' axes are parallel with E-N, but the origin of this system coincides with the origin of X-Y. In the figure, rotation angle θ, between the X-Y and X'-Y' axes systems, is:

$$\theta = \alpha - \beta \tag{B-9}$$

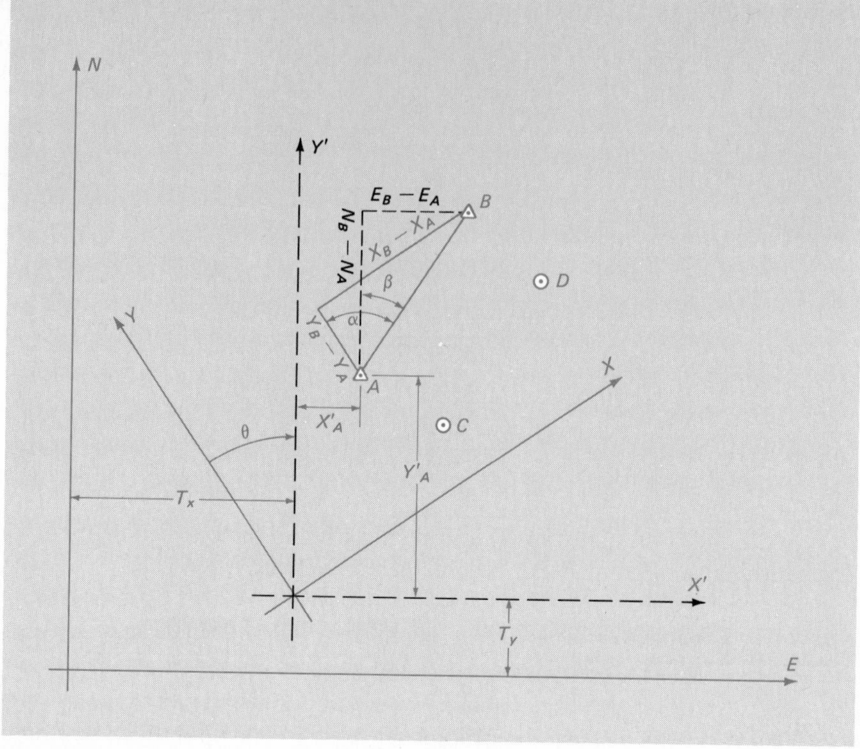

Figure B-5. Geometry of two-dimensional coordinate transformation.

In Eq. (B-9), α and β are calculated from the two sets of coordinates of control points A and B using Eq. (B-2) as follows:

$$\alpha = \tan^{-1}\left[\frac{X_B - X_A}{Y_B - Y_A}\right]$$

and

$$\beta = \tan^{-1}\left[\frac{E_B - E_A}{N_B - N_A}\right]$$

With θ known, X' and Y' coordinates of any point, for example A, can be calculated from:

$$\begin{aligned} X'_A &= X_A \cos\theta - Y_A \sin\theta \\ Y'_A &= X_A \sin\theta + Y_A \cos\theta \end{aligned} \qquad \text{(B-10)}$$

Individual parts of the rotation formulas [right-hand sides of Eqs. (B-10)], are detailed in Figure B-6.

Translation consists of shifting the origin of the X'-Y' axes to that of the E-N system. This is achieved by adding translation factors T_x and T_y (see

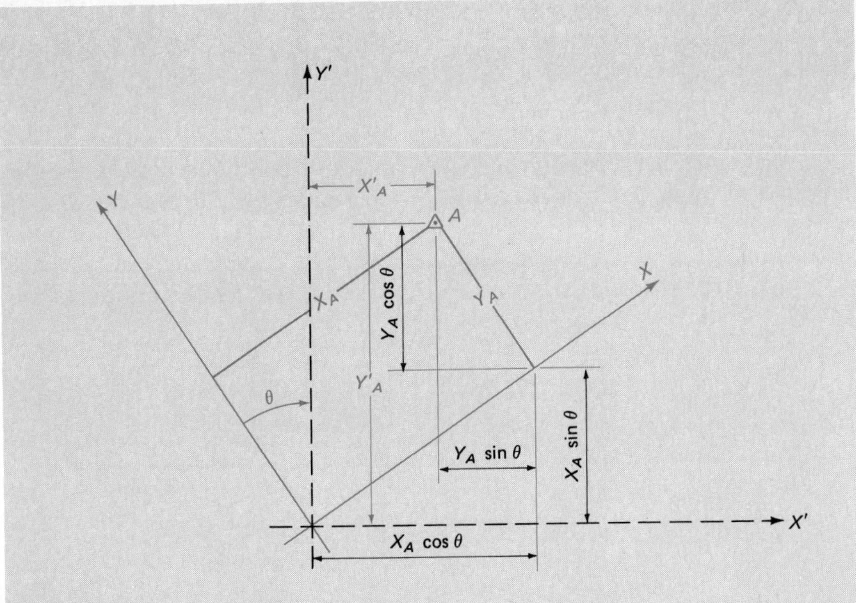

Figure B-6. Detail of rotation formulas in two-dimensional coordinate transformation.

Figure B-5) to X' and Y' coordinates to obtain E and N coordinates. Thus for point A:

$$\begin{aligned} E_A &= X'_A + T_x \\ N_A &= Y'_A + T_y \end{aligned} \tag{B-11}$$

Rearranging Eqs. (B-11), and using coordinates of one of the control points (such as A), numerical values for T_x and T_y can be obtained as:

$$\begin{aligned} T_x &= E_A - X'_A \\ T_y &= N_A - Y'_A \end{aligned} \tag{B-12}$$

The other control point (i.e., point B) should also be used in Eqs. (B-12) to calculate T_x and T_y and thus obtain a computational check.

Substituting Eqs. (B-10) into Eqs. (B-11), and dropping subscripts, the following equations are obtained for calculating E and N coordinates of noncontrol points (such as C and D) from their X and Y values:

$$\begin{aligned} E &= X \cos\theta - Y \sin\theta + T_x \\ N &= X \sin\theta + Y \cos\theta + T_y \end{aligned} \tag{B-13}$$

In summary, the procedure for performing two-dimensional coordinate transformations consists of (1) calculating rotation angle θ using two control points and Eqs. (B-2) and (B-9), (2) solving Eqs. (B-10) and (B-12) using one control point (check with the other) to obtain translation factors T_x and T_y, and (3) using θ and T_x and T_y in Eqs. (B-13) to transform all noncontrol points. If more than two control points are available, an improved solution can be obtained using least squares, but that topic is beyond the scope of this text.

EXAMPLE B-4

In Figure B-5, the following *E-N* and *X-Y* coordinates are known for points *A* through *D*. Compute *E* and *N* coordinates for points *C* and *D*.

POINT	STATE PLANE COORDINATES (ft)		ARBITRARY COORDINATES (ft)	
	E	N	X	Y
A	194,683.50	99,760.22	2848.28	2319.94
B	196,412.80	102,367.61	5720.05	3561.68
C			3541.72	897.03
D			6160.31	1941.26

SOLUTION

1. Determine α, β, and θ from Eqs. (B-2) and (B-9):

$$\alpha = \tan^{-1}\left[\frac{5720.05 - 2848.28}{3561.68 - 2319.94}\right] = 66°36'59.7''$$

$$\beta = \tan^{-1}\left[\frac{196{,}412.80 - 194{,}683.50}{102{,}367.61 - 99{,}760.22}\right] = 33°33'12.7''$$

$$\theta = 66°36'59.7'' - 33°33'12.7'' = 33°03'47''$$

2. Determine T_x and T_y from Eqs. (B-10) and (B-12) using point *A*:

$$X'_A = 2848.28 \cos 33°03'47'' - 2319.94 \sin 33°03'47'' = 1121.39$$
$$Y'_A = 2848.28 \sin 33°03'47'' + 2319.94 \cos 33°03'47'' = 3498.18$$
$$T_x = 194{,}683.50 - 1121.39 = 193{,}562.11$$
$$T_y = 99{,}760.22 - 3498.18 = 96{,}262.04$$

3. Check T_x and T_y using point *B*:

$$X'_B = 5720.05 \cos 33°03'47'' - 3561.68 \sin 33°03'47'' = 2850.69$$
$$Y'_B = 5720.05 \sin 33°03'47'' + 3561.68 \cos 33°03'47'' = 6105.58$$
$$T_x = 196{,}412.80 - 2850.69 = 193{,}562.11 \text{ (Check!)}$$
$$T_y = 102{,}367.61 - 6105.58 = 96{,}262.03 \text{ (Check!)}$$

4. Solve Eqs. (B-13) for *E* and *N* coordinates of points *C* and *D*:

$$E_C = 3541.72 \cos 33°03'47'' - 897.03 \sin 33°03'47'' + 193{,}562.11$$
$$= 196{,}040.93$$
$$N_C = 3541.72 \sin 33°03'47'' + 897.03 \cos 33°03'47'' + 96{,}262.04$$
$$= 98{,}946.04$$
$$E_D = 6160.31 \cos 33°03'47'' - 1941.26 \sin 33°03'47'' + 193{,}562.11$$
$$= 197{,}665.81$$
$$N_D = 6160.31 \sin 33°03'47'' + 1941.26 \cos 33°03'47'' + 96{,}262.04$$
$$= 101{,}249.78$$

APPENDIX C COMPUTER PROGRAMS

C-1. INTRODUCTION. This appendix contains listings of three computer programs, together with input/output examples illustrating their use. The programs are written in BASIC and are compatible with virtually all computing systems that support this language. Keyboard entry of these listings, if made exactly as shown, will produce the results in the examples and solve similar problems.

The programs, described in more detail in succeeding sections, perform (1) traverse computations, (2) azimuth from Polaris observations, and (3) azimuth from the sun. In each program, "prompts" indicated by question marks tell the user to enter data required for the solution.

Permission is granted for the unrestricted use of these programs; however, the publisher and authors assume no responsibility for problems that may arise as a result of employing them.

C-2. PROGRAM FOR TRAVERSE COMPUTATION. The following program performs traverse computations for both types of closed traverses—that is, polygons that close on their starting stations and those that close on another point of known position. As presently dimensioned, traverses having up to 40 courses can be handled. The computations include determining latitudes and departures, balancing them by the compass rule, and calculating linear misclosure and precision. From given coordinates of the starting station, coordinates of all traverse points are determined, and if the traverse is a polygon, area is calculated by the coordinate method.

Prior to using the program, traverse stations must be identified by consecutive numbers beginning with 1. Input to the program consists of (1) number of courses in the traverse; (2) entry of the number 1 if the traverse is a polygon, 2 if it is not; (3) length (in feet) and azimuth (in degrees, minutes, and seconds) of each successive course beginning with line 1-2; and (4) coordinates of station 1 if a polygon traverse, or the first and last stations if not. In numbering traverse stations, 1 must be assigned to a station whose coordinates are known. The program can accept lengths in meters; however, feet are assumed, and based on that assumption, area is calculated and listed in acres. If meters are used, lengths and coordinates will be correctly listed in the output in meters. However, the listed area must be multiplied by 4.3560 to convert it to hectares.

The following are the program listing and example input/output for solving Examples 13-2 and 14-5 of the text. Note that the results agree with those given in Tables 13-3 and 14-3.

Program Listing

```
100 REM    TRAVERSE COMPUTATION
110 DIM DS(40),DG(40),MN(40),SC(40),RA(40),XU(41),YU(41),XB(41),YB(41),
    X(41),Y(41)
120 RD=3.14159265/180
130 PRINT "ENTER NUMBER OF COURSES IN TRAVERSE";
140 INPUT NC
150 PRINT "ENTER '1' IF POLYGON TRAVERSE, '2' IF NOT";
160 INPUT TT
170 IF TT=1 THEN 200
180 IF TT=2 THEN 200
190 GOTO 150
200 DX=0
210 DY=0
220 PE=0
230 PRINT "ENTER DISTANCE AND AZIMUTH (deg,min,sec) OF EACH COURSE"
240 FOR I=1 TO NC
250 PRINT "COURSE";I;
260 INPUT DS(I),DG(I),MN(I),SC(I)
270 PE=PE+DS(I)
280 RA(I)=RD*(DG(I)+MN(I)/60+SC(I)/3600)
290 XU(I)=DS(I)*SIN(RA(I))
300 YU(I)=DS(I)*COS(RA(I))
310 DX=DX+XU(I)
320 DY=DY+YU(I)
330 NEXT I
340 PRINT "ENTER COORDINATES (X,Y) OF STARTING STATION";
350 INPUT X(1),Y(1)
360 IF TT=1 THEN 420
370 PRINT "ENTER COORDINATES (X,Y) OF ENDING STATION";
380 INPUT X(NC+1),Y(NC+1)
390 CX=DX-(X(NC+1)-X(1))
400 CY=DY-(Y(NC+1)-Y(1))
410 GOTO 460
420 X(NC+1)=X(1)
430 Y(NC+1)=Y(1)
440 CX=DX
450 CY=DY
460 CL=SQR(CX*CX+CY*CY)
470 PC=INT(PE/CL)
480 PRINT
490 PRINT "COURSE";TAB(11);"LENGTH";TAB(26);"AZIMUTH";TAB(43);"COS";
    TAB(55);"SIN"
500 FOR I=1 TO NC
510 J=0
520 IF TT=2 THEN 550
530 IF I<>NC THEN 550
540 J=NC
550 PRINT I;"-";I+1-J;TAB(11);DS(I);TAB(23);DG(I);"-";MN(I);"-";SC(I);
    TAB(41);COS(RA(I));TAB(53);SIN(RA(I))
560 XB(I)=XU(I)-CX*DS(I)/PE
570 YB(I)=YU(I)-CY*DS(I)/PE
580 X(I+1)=X(I)+XB(I)
```

```
590 Y(I+1)=Y(I)+YB(I)
600 NEXT I
610 PRINT TAB(11);PE
620 PRINT
630 PRINT
640 PRINT TAB(6);"UNBALANCED";TAB(26);"BALANCED";TAB(50);"COORDINATES"
650 PRINT TAB(3);"LAT";TAB(14);"DEP";TAB(23);"LAT";TAB(33);"DEP";
    TAB(43);"STA";TAB(49);"NORTH";TAB(60);"EAST"
660 XX=0
670 FOR I=1 TO NC
680 XU(I)=INT(XU(I)*1000+0.5)/1000
690 YU(I)=INT(YU(I)*1000+0.5)/1000
700 XB(I)=INT(XB(I)*1000+0.5)/1000
710 YB(I)=INT(YB(I)*1000+0.5)/1000
720 X(I)=INT(X(I)*1000+0.5)/1000
730 Y(I)=INT(Y(I)*1000+0.5)/1000
740 PRINT YU(I);TAB(12);XU(I);TAB(22);YB(I);TAB(32);XB(I);TAB(43);I;
    TAB(48);Y(I);TAB(59);X(I)
750 IF TT=2 THEN 770
760 XX=XX+X(I)*Y(I+1)-Y(I)*X(I+1)
770 NEXT I
780 DX=INT(DX*1000+0.5)/1000
790 DY=INT(DY*1000+0.5)/1000
800 PRINT DY;TAB(12);DX
810 PRINT
820 PRINT
830 PRINT "LINEAR MISCLOSURE = ";INT(1000*CL)/1000;" Feet"
840 PRINT "PRECISION =  1 in ";PC
850 IF TT=2 THEN 870
860 PRINT "AREA = ";INT(10000*ABS(XX/(2*43560)))/10000;" Acres"
870 END
```

Listing of Input/Output for Example Problem

```
ENTER NUMBER OF COURSES IN TRAVERSE? 5
ENTER '1' IF POLYGON TRAVERSE, '2' IF NOT? 1
ENTER DISTANCE AND AZIMUTH (deg,min,sec) OF EACH COURSE
COURSE 1 ? 285.10,26,10,00
COURSE 2 ? 610.45,104,35,00
COURSE 3 ? 720.48,195,30,00
COURSE 4 ? 203.00,358,18,00
COURSE 5 ? 647.02,306,54,00
ENTER COORDINATES (X,Y) OF STARTING STATION? 10000,10000

COURSE      LENGTH          AZIMUTH            COS          SIN
 1 - 2       285.1        26 - 10 - 0        .897515      .440984
 2 - 3       610.45      104 - 35 - 0      -.251788       .967783
 3 - 4       720.48      195 - 30 - 0      -.963631      -.267238
 4 - 5       203         358 - 18 - 0       .99956      -.0296663
 5 - 1       647.02      306 - 54 - 0       .60042      -.799685
             2466.05

        UNBALANCED            BALANCED                   COORDINATES
   LAT         DEP        LAT        DEP        STA    NORTH        EAST
 255.882     125.724    255.963    125.663      1     10000        10000
-153.704     590.783   -153.529    590.651      2     10255.963    10125.663
-694.276    -192.54    -694.071   -192.696      3     10102.434    10716.314
 202.911    -6.022      202.969   -6.066        4     9408.363     10523.618
 388.484    -517.412    388.669   -517.552      5     9611.331     10517.552
-.704        .533

LINEAR MISCLOSURE =  .883  Feet
PRECISION =  1 in  2791
AREA =  6.2582  Acres
```

C-3. AZIMUTH FROM POLARIS OBSERVATIONS. The following program solves Eq. (19-1) for the azimuth of Polaris (or other stars) at any hour angle. The horizontal angle from mark to star must then be applied manually to obtain the line's azimuth. As presently dimensioned, the program will accept any number of direct plus reversed observations up to a total of 16, and if more than one is entered, it performs the calculations independently for each observation and lists the mean azimuth.

Input consists of (1) total number of observations, direct plus reversed; (2) latitude of observation station (in degrees, minutes, and seconds); (3) longitude of station (in degrees, minutes, and seconds); (4) star's declination (in degrees, minutes, and seconds) at the most recent time listed in the ephemeris prior to the first observation; (5) star's declination (in degrees, minutes, and seconds) at the nearest time listed in the ephemeris following the observations; (6) elapsed time (in days to nearest 0.1 day) between the first entered declination time and observation time; (7) interval of time (in days to nearest 0.1 day) between observation time and the time of the second entered declination; (8) Greenwich hour angle (in degrees, minutes, and seconds) of the star at 0 hours GCT on the day of observation; and (9) Greenwich civil times (in hours, minutes, and seconds) of each observation.

The program listing is presented, followed by input/output for solving Example 19-1 of the text. In the example, all four sets of data are entered and the star's mean azimuth (359°31′24.8″) computed. To obtain the line's azimuth, the mean of the four horizontal angles (49°38.8′) is subtracted, yielding a line azimuth of 309°52.6′. This agrees with the value obtained in the text's example soultion, which was for the first observation only.

Program Listing

```
100 REM    AZIMUTH FROM POLARIS OBSERVATIONS
110 DIM A(16)
120 PI=3.14159265
130 RO=PI/180
140 PRINT "HOW MANY OBSERVATIONS (DIRECT + REVERSE)";
150 INPUT N
160 PRINT "LATITUDE (deg,min,sec)";
170 INPUT D,M,S
180 LA=RO*(D+M/60+S/3600)
190 PRINT "LONGITUDE (deg,min,sec)";
200 INPUT D,M,S
210 LO=RO*(D+M/60+S/3600)
220 PRINT "DECL. (deg,min,sec) AT MOST RECENT TIME BEFORE OBSERVATIONS";
230 INPUT D,M,S
240 DB=RO*(D+M/60+S/3600)
250 PRINT "DECL. (deg,min,sec) AT NEAREST TIME AFTER OBSERVATIONS";
260 INPUT D,M,S
270 DA=RO*(D+M/60+S/3600)
280 PRINT "INTERVAL FROM TIME OF DECL. #1 TO TIME OF OBS. (NEAREST 0.1 DAY)";
290 INPUT TB
300 PRINT "INTERVAL FROM TIME OF OBS. TO TIME OF DECL. #2 (NEAREST 0.1 DAY)";
310 INPUT TA
320 DO=DB+(DA-DB)*(TB/(TA+TB))
330 PRINT "GHA (deg,min,sec) AT 0hrs GCT ON DAY OF OBSERVATIONS";
340 INPUT D,M,S
350 GH=RO*(D+M/60+S/3600)
360 FOR I=1 TO N
370 PRINT "GCT (hr,min,sec) OF OBSERVATION #";I;
380 INPUT D,M,S
390 GC=RO*15*(D+M/60+S/3600)
400 GC=GH+GC*1.002737909
410 T=GC-LO
420 FT=1
430 IF T<=0 THEN 460
440 IF T>=PI THEN 460
450 FT=-1
460 T=ABS(T)
470 IF T<=PI THEN 490
480 T=2*PI-T
490 Z=ATN(SIN(T)/(COS(LA)*TAN(DO)-SIN(LA)*COS(T)))
500 Z=Z*FT/RO
510 A(I)=Z
```

```
520 PRINT
530 NEXT I
540 X=0
550 FOR I=1 TO N
560 X=X+A(I)
570 NEXT I
580 X=X/N
590 IF X>=0 THEN 610
600 X=X+360
610 XD=INT(X)
620 XM=INT(60*(X-XD))
630 XS=60*(60*(X-XD)-XM)
640 XS=INT(10*XS)/10
650 PRINT "AZIMUTH OF STAR (deg-min-sec):";XD;"-";XM;"-";XS
660 END
```

Listing of Input/Output for Example Problem

```
HOW MANY OBSERVATIONS (DIRECT + REVERSE)? 4
LATITUDE (deg,min,sec)? 43,5.4,00
LONGITUDE (deg,min,sec)? 89,26,00
DECL. (deg,min,sec) AT MOST RECENT TIME BEFORE OBSERVATIONS? 89,11.06,00
DECL. (deg,min,sec) AT NEAREST TIME AFTER OBSERVATIONS? 89,11.02,00
INTERVAL FROM TIME OF DECL. #1 TO TIME OF OBS. (NEAREST 0.1 DAY)? 4.1
INTERVAL FROM TIME OF OBS. TO TIME OF DECL. #2 (NEAREST 0.1 DAY)? 5.9
GHA (deg,min,sec) AT 0hrs GCT ON DAY OF OBSERVATIONS? 218,31.8,00
GCT (hr,min,sec) OF OBSERVATION # 1 ? 1,30,49

GCT (hr,min,sec) OF OBSERVATION # 2 ? 1,39,00

GCT (hr,min,sec) OF OBSERVATION # 3 ? 1,44,33

GCT (hr,min,sec) OF OBSERVATION # 4 ? 1,49,46

AZIMUTH OF STAR (deg,min,sec): 359 - 31 - 24.8
```

C-4. AZIMUTH FROM SUN OBSERVATIONS. The following program solves Eq. (19-4) for azimuth of the sun based on observed altitude. The horizontal angle from mark to sun must then be manually applied to the sun's azimuth to obtain the line's azimuth. Any number of direct and reversed observations can be handled, from which the sun's mean azimuth is calculated.

The program computes corrections to observed altitudes for refraction and parallax. Refraction corrections are made by interpolation from ephemeris values entered by the user for vertical angles that bracket the sun's observed altitudes. Parallax is calculated as a function of the sun's altitude and requires no special input. The program does not correct for index error or semidiameter. Thus it is assumed that equal numbers of direct and reversed observations will be entered, and either a Roelof prism was used or equal numbers of observations of the sun's limbs were made in diagonally opposite quadrants of the field of view.

Input consists of (1) atmospheric pressure at observation time (in inches of mercury); (2) temperature (in degrees Fahrenheit); (3) latitude of the occupied station (in degrees, minutes, and seconds); (4) sun's declination (in degrees, minutes, and seconds) at 0 hours Greenwich civil time the day of the observations; (5) sun's declination (in degrees, minutes, and seconds) at 0 hours GCT the day after observing; (6) total number of observations, direct plus reversed; (7) an entry of 1 if the observations were made in the morning and 2 if in the afternoon; (8) observed altitude (in degrees, minutes, and seconds) and Greenwich civil time (in hours, minutes, and seconds) for each observation; and (9) interpolation values taken from an ephemeris for refraction (prompted by the average of all observed altitude angles). This consists of an altitude angle (in degrees, minutes, and seconds) lower than the mean observed altitude angle, and its refraction correction (in minutes); and an altitude angle (in degrees, minutes, and seconds) higher than the mean observed altitude angle, and its refraction correction (in minutes).

The program listing is followed by input/output for solving Example 19-4 of the text. Note that the resulting sun's azimuth listed agrees with the value obtained in the text example.

Program Listing

```
100 REM    AZIMUTH FROM SUN OBSERVATIONS
110 PI=3.14159265
120 PRINT "PRESSURE (Inches of Hg)";
130 INPUT P
140 PRINT "TEMPERATURE (FARENHEIT)";
150 INPUT T
160 CP=.014+.0333*P
170 CT=1.108-.00241*T+4.44E-6*T*T
180 PRINT "LATITUDE (deg,min,sec)";
190 INPUT D,M,S
200 LT=(D+M/60+S/3600)*PI/180
210 PRINT "DECL (deg,min,sec) AT 0hrs GCT DAY OF OBSERVATION";
220 INPUT D,M,S
230 DB=D+M/60+S/3600
240 PRINT "DECL. (deg,min,sec) AT 0hrs GCT DAY AFTER OBSERVATION";
250 INPUT D,M,S
260 DA=D+M/60+S/3600
270 PRINT "NUMBER OF OBSERVATIONS";
280 INPUT N
290 PRINT "ENTER '1' FOR MORNING OBSERVATIONS, '2' FOR AFTERNOON";
300 INPUT MA
310 IF MA=1 THEN 340
320 IF MA=2 THEN 340
330 GOTO 290
340 GC=0
350 H=0
360 FOR I=1 TO N
370 PRINT"OBSERVATION #";I
380 PRINT "ALTITUDE (deg,min,sec)";
390 INPUT D,M,S
400 H=H+D+M/60+S/3600
410 PRINT "GCT (hr,min,sec)";
420 INPUT D,M,S
430 GC=GC+D+M/60+S/3600
440 NEXT I
450 H=H/N
460 GC=GC/N
470 TD=(DB+(DA-DB)*GC/24)*PI/180
480 D=INT(H)
490 M=INT(60*(H-D))
500 PRINT"ENTER INTERPOLATION VALUES FOR REFRACTION (H=";D;"deg";M;"min)"
510 PRINT "LOWER ANGLE (deg,min), REFRACTION VALUE (min)";
520 INPUT D,M,RL
530 HL=D+M/60
540 PRINT "HIGHER ANGLE (deg,min), REFRACTION VALUE (min)";
550 INPUT D,M,RH
560 HH=D+M/60
570 R=RL+(RH-RL)*(H-HL)/(HH-HL)
580 Z=4.2606E-5*SIN((H+90)*PI/180)
590 HT=(H-R*CP*CT/60)*PI/180+ATN(Z/SQR(1-Z*Z))
600 Z=SIN(TD)/COS(LT)/COS(HT)-TAN(LT)*TAN(HT)
610 Z=90-ATN(Z/SQR(1-Z*Z))*180/PI
620 IF MA=1 THEN 640
630 Z=360-Z
640 D=INT(Z)
650 M=INT(60*(Z-D))
660 S=INT(60*(60*(Z-D)-M))
670 PRINT
680 PRINT
690 PRINT"AZIMUTH OF SUN (deg-min-sec):";D;"-";M;"-";S
700 END
```

```
PRESSURE (Inches of Hg)? 28.7
TEMPERATURE (FARENHEIT)? 80
LATITUDE (deg,min,sec)? 42,45,00
DECL. (deg,min,sec) AT 0hrs GCT DAY OF OBSERVATION? 18,42.2,00
DECL. (deg,min,sec) AT 0hrs GCT DAY AFTER OBSERVATION? 18,27.8,00
NUMBER OF OBSERVATIONS? 4
ENTER '1' FOR MORNING OBSERVATIONS, '2' FOR AFTERNOON? 2
OBSERVATION # 1
ALTITUDE (deg,min,sec)? 49,44,00
GCT (hr,min,sec)? 19,33,10
OBSERVATION # 2
ALTITUDE (deg,min,sec)? 49,32,00
GCT (hr,min,sec)? 19,34,20
OBSERVATION # 3
ALTITUDE (deg,min,sec)? 49,52,00
GCT (hr,min,sec)? 19,35,37
OBSERVATION # 4
ALTITUDE (deg,min,sec)? 49,40,00
GCT (hr,min,sec)? 19,36,49
ENTER INTERPOLATION VALUES FOR REFRACTION (H= 49 deg 42 min)
LOWER ANGLE (deg,min), REFRACTION VALUE (min)? 48,00,0.86
HIGHER ANGLE (deg,min), REFRACTION VALUE (min)? 50,00,0.80


AZIMUTH OF SUN (deg-min-sec): 245 - 4 - 37
```

APPENDIX D EXAMPLE NOTEFORMS

PLATE D-1.

DISTANCES BY PACING

A. DETERMINING LENGTH OF PACE

No. of Paces	Direction	Taped Dist.
154	S	400′
155	N	400′
155	S	400′
156	N	400′
155	Average	

Length of pace = $\frac{400'}{155}$ = 2.58′

No. of paces per 100′ = 39

B. CHECKING LENGTH OF COURSE DE

No. of Paces	Direction
119	N
118	S
119	N
120	S
119	Average

Distance = 119 × 2.58 = 307′

Taped length = 306.83′ (Check)

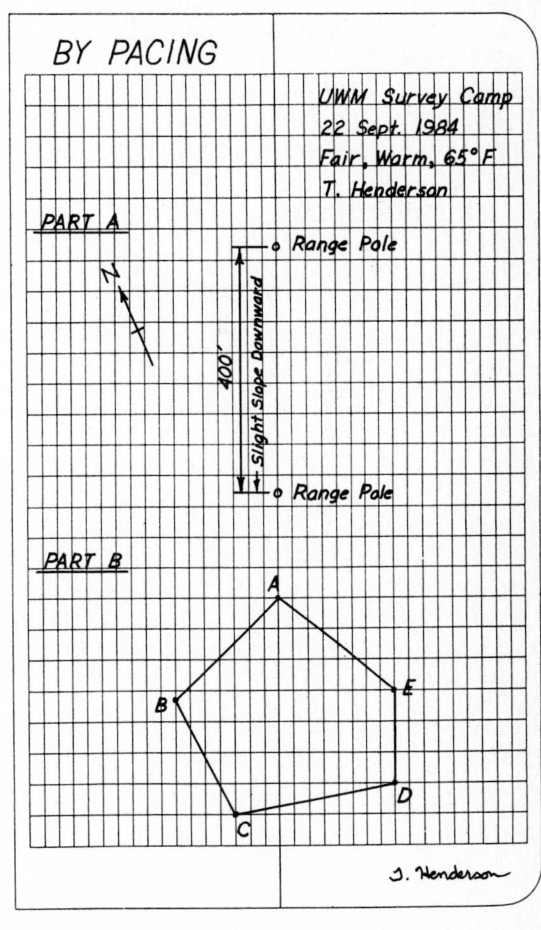

PLATE D-2.

MEASURING DISTANCES WITH

Hub	Sta.	Dist.
A	0+00	
		321.20′
B	3+21.42	
		276.57′
C	5+97.99	
		100.30′
D	6+98.29	
		306.79′
E	10+05.16	
		255.48′
A	12+60.64	
Σ	1260.64	1260.34

Ratio of error $= \frac{0.30}{1260.49} = \frac{1}{4200}$

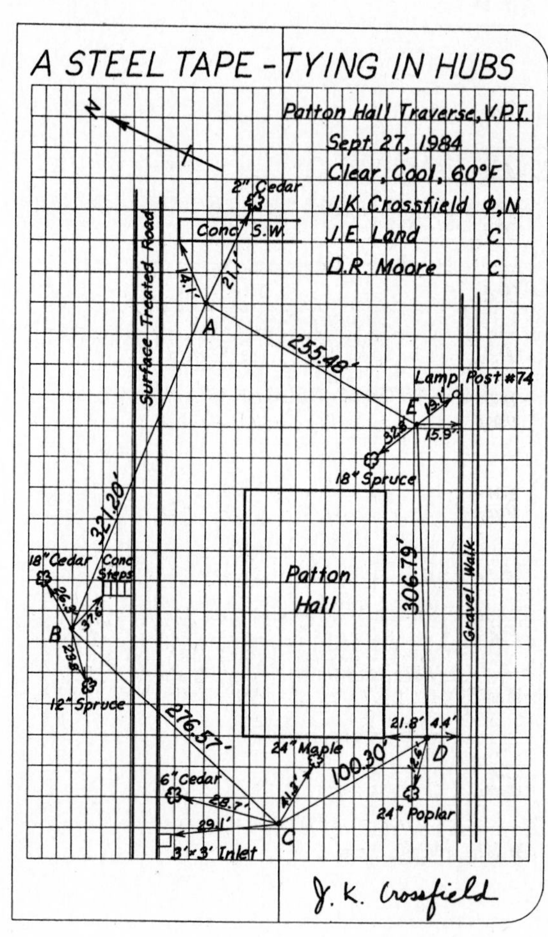

PLATE D-3.

DIFFERENTIAL LEVELS

Sta.	+ Sight	HI	− Sight	Elev.	Dist.
BM Mil.				100.00	
	1.33	101.33			150
TP 1			8.37	92.96	150
	0.22	93.18	7.91		135
TP 2			~~8.91~~	85.27	135
	0.96	86.23			160
TP 3			11.72	74.51	160
	0.46	74.97			160
BM Rutgers			8.71	66.26	160
	2.97		36.71		1210
BM Rutgers				66.26	
	11.95	78.21			180
TP 1			2.61	75.60	180
	12.55	88.15			180
TP 2			0.68	87.47	180
	12.77	100.24			155
BM Mil.			0.21	100.03	155
	37.27		3.50		1030
BM Rutgers	True elev. above MSL			2053.18	
	Elev. diff. (average)			33.75	
BM Mil.	MSL elev.			2086.93	

V.P.I. CAMPUS

BM Mil. to BM Rutgers
29 Sept. 1984
Clear, Warm, 70°F
M.M. Clark N
J.F. King φ
D.R. Moore ⊼
Gurley Level #6

BM Mil. on V.P.I. Campus SW of Old Military Bldg. 9.4 ft. north of sidewalk to instrument room and 1.6 ft. from bldg. Bronze disc in pipe flush with ground.

BM Rutgers SE of Patton Hall opposite main entrance and 8 ft. from curb around drill field. Bronze disc flush with ground set in 6" concrete cylinder and stamped "Rutgers."

Rod Sums		Elev. Checks	
−36.71	+37.27	+100.00	+66.26
+ 2.97	− 3.50	− 33.74	+33.77
−33.74	+33.77	+ 66.26 ✓	+100.03 ✓
	−33.74		

Loop misclosure 0.03

Permissible misclosure $= 0.05\sqrt{M}$

$= 0.05\sqrt{2240/5280} = 0.03'$

Marian M. Clark

PLATE D-4.

RECIPROCAL LEVELING

Station	+ Sight		− Sight	Elev. Diff.	Elev.
BM Rutgers	2.605				2053.182
BM Eggle			12.304		
			12.302		
			12.293		
			12.297		
BM Eggle		Avg.	12.299	9.694	2043.488
BM Rutgers	1.528				
	1.517				
	1.519				
	1.522				
	1.522	Avg.			
			11.203	9.681	2043.501
		Misclosure		0.013	
		Mean		9.688	
BM Eggle					2043.494

ACROSS DRILL FIELD

Description on page 3

V.P.I. Campus
04 Oct. 1984
Clear, Hot, 80°F
C.D. Ghilani N
H.W. Mills ϕ
M. Eidson ⊼
K&E Level #7

BM Eggle NE corner of 2nd step of entrance to Eggleston Hall. An "X" chiseled in concrete.

C. Ghilani

PLATE D-5.

PROFILE LEVELS

Station	+ Sight	HI	− Sight	Int. Sight	Elev.
BM Road	10.14	370.62			360.48
0+00				9.36	361.26
0+20				9.8	360.8
1+00				6.5	364.1
2+00				4.3	366.3
2+60				3.7	366.9
3+00				7.1	363.5
3+90				11.7	358.9
4+00				11.2	359.4
4+35				9.5	361.1
TP 1	7.33	366.48	11.47		359.15
5+00				8.4	358.1
5+54				11.08	355.40
5+74				10.66	355.82
5+94				11.06	355.42
6+00				10.5	356.0
7+00				4.4	362.1
TP 2	2.55	363.77	5.26		361.22
8+00				1.2	362.6
9+00				3.9	359.9
9+25.2				3.4	360.4
9+25.3				4.6	359.2
9+43.2				2.2	361.6
BM Store			0.76		363.01
Σ	20.02		17.49		

BM ROAD to BM STORE

BM Road 3 miles SW of Mpls. 200 yds. N of Pine St. overpass 40 ft. E of ℄ Hwy. 169. Top of RW conc. post No. 268.	SW Minneapolis on Hwy 169 6 Oct 1984
℄ Hwy 169, painted "X"	Cool, Sunny, 50°F
West drainage ditch	R.J. Hintz N
	N.R. Olson Φ
	R.C. Perry ⊼
Summit	Wild Level #3
Sag	
Summit	COPY
E gutter, Maple St.	+20.02
℄ Maple St.	−17.49
W gutter, Maple St.	+ 2.53
	360.48
	363.01
Summit	
Top of E curb, Elm St.	
Bottom of E curb, Elm St.	
℄ Elm St.	
BM Store. NE corner Elm St & 4th Ave. SE corner Store foundation wall. 3" brass disc set in grout.	

PLATE D-6.

BORROW-PIT LEVELING

Point	+ Sight	HI	− Sight	Elev.	Cut
BM Road	4.22	364.70		360.48	
A,0			5.2	359.5	1.5
B,0			5.4	359.3	1.3
C,0			5.7	359.0	1.0
D,0			5.9	358.8	0.8
E,0			6.2	358.5	0.5
A,1			4.7	360.0	2.0
B,1			4.8	359.9	1.9
C,1			5.2	359.5	1.5
D,1			5.5	359.2	1.2
E,1			5.8	358.9	0.9
A,2			4.2	360.5	2.5
B,2			4.7	360.0	2.0
C,2			4.8	359.9	1.9
D,2			5.0	359.7	1.7
A,3			3.8	360.9	2.9
B,3			4.0	360.7	2.7
C,3			4.6	360.1	2.1
D,3			4.6	360.1	2.1
A,4			3.4	361.3	3.3
B,4			3.7	361.0	3.0
C,4			4.2	360.5	2.5
BM Road	4.23				

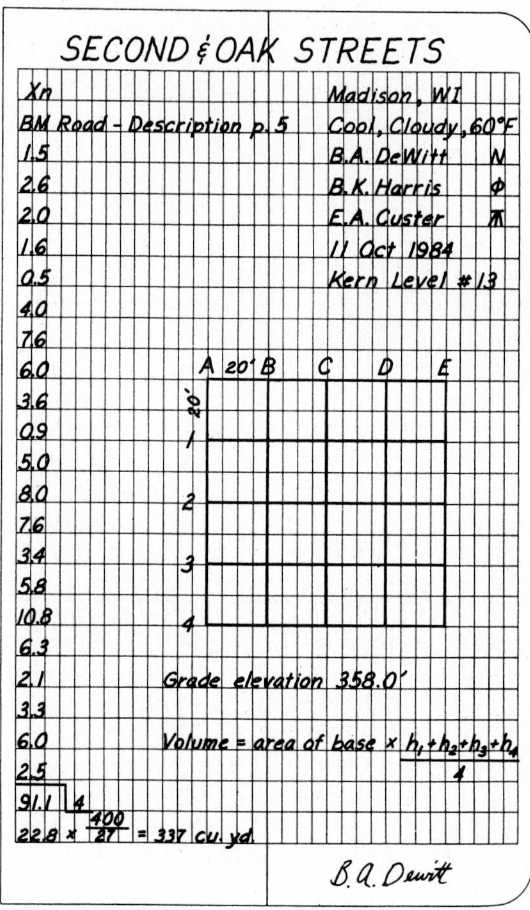

SECOND & OAK STREETS
Xn
Madison, WI
BM Road - Description p. 5
Cool, Cloudy, 60°F
B.A. DeWitt N
B.K. Harris Φ
E.A. Custer ⊼
11 Oct 1984
Kern Level # 13
1.5
2.6
2.0
1.6
0.5
4.0
7.6
6.0
3.6
0.9
5.0
8.0
7.6
3.4
5.8
10.8
6.3
2.1
3.3
6.0
2.5
A 20' B C D E
20'
1
2
3
4
Grade elevation 358.0'
Volume = area of base × (h1+h2+h3+h4)/4
91.1 | 4
22.8 × 400/27 = 337 cu. yd.
B.A. Dewitt

PLATE D-7.

CLOSING THE HORIZON

⊼ at ΔA

All angles read clockwise

Object	Vern. A	Vern. B	Mean	Unadj. Angle	Sta. Adj. Angle
	Reading ΔB to ΔC				
ΔB	0°00′00″	180°00′00″	0°00′00″		
3 Rep N	126°36′20″	306°36′20″	(Read for prelim. check)		
3 Rep P	253°13′00″	73°13′00″	253°13′00″	42°12′10″	42°12′09″
	Reading ΔC to ΔD				
3 Rep N	(Reading not required)				
3 Rep P	(612°) 252°53′40″	72°54′00″	252°53′50″	59°56′48″	59°56′47″
	Reading ΔD to ΔB				
3 Rep N	(Reading not required)				
3 Rep P	(2160°) 0°00′20″	180°00′20″	0°00′20″	257°51′05″	257°51′04″
				360°00′03″	360°00′00″
	Vernier misclosure			0°00′20″	
	Horizon misclosure			0°00′03″	
	Sta. adjustment			0°00′01″	

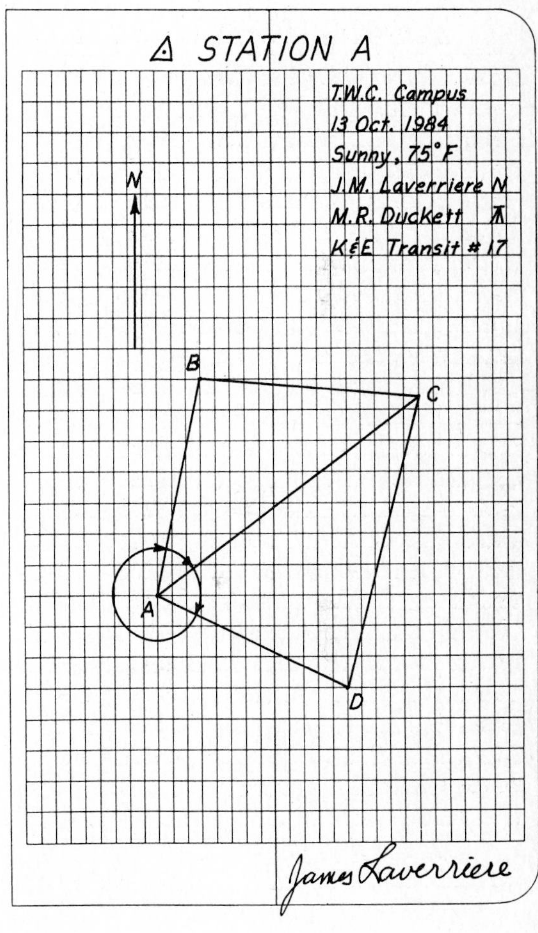

PLATE D-8.

DOUBLE DIRECT ANGLES

Hub	Dist.	Single ∡	Double ∡	Avg. ∡
A		38°58.0′	77°56.8′	38°58.4′
	321.31′			
B		148°53.6′	297°47.0′	148°53.5′
	276.57′			
C		84°28.1′	168°56.2′	84°28.1′
	100.30′			
D		114°40.3′	229°20.9′	114°40.4′
	306.83′			
E		152°59.4′	305°58.6′	152°59.3′
	255.48′			
A				
Σ	1260.49′			539°59.7′
			Misclosure	0°00.3′

Σ interior ∡s = (N-2) 180°
= (5-2) 180°
= 540° 00′

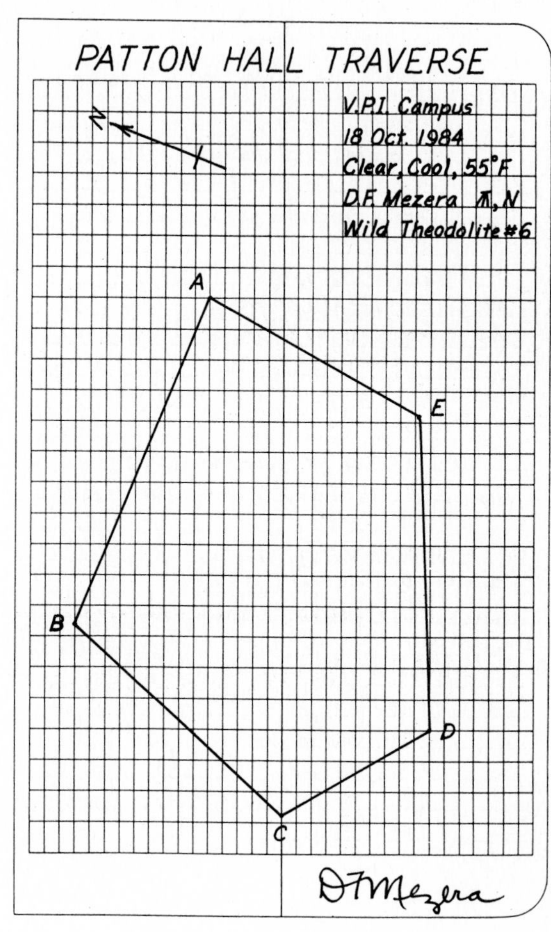

PLATE D-9.

STADIA SURVEY

Station	Stadia Interval	Azimuth	Vert.∠ or Rod R.	Horiz. Dist.	Elev.
	⊼ @ ⊡B,	Elev. 177.42,		h.i. = 5.0	
⊡A	6.74	148°04′	−0°34′	675	170.7
⊡C	4.21	60°12′	−1°35′	422	165.8
1	0.91	90°45′	8.3	92	174.1
2	1.66	120°20′	−2°12′	167	171.0
3	3.15	126°30′	−2°06′	316	165.9
4	4.60	143°45′	−1°23′	461	166.0
5	7.85	141°30′	−0°38′	786	168.7
6	2.47	167°20′	8.6	248	173.8
7	1.97	172°20′	9.6	198	172.8
8	4.99	181°15′	3.2	500	179.2
9	5.79	221°45′	+1°02′	580	187.8
10	3.47	256°00′	+1°50′	348	188.5
11	1.17	342°05′	2.4	118	180.0
12	1.71	350°15′	2.4	172	180.0
⊡D	4.49	6°10′	5.2	450	177.2
⊡A	6.74	148°04′	−0°34′	675	170.7
	⊼ @ ⊡C,	Elev. 165.77,		h.i. = 5.2	
⊡B	4.20	240°12′	+1°34′	421	177.3
13	3.21	286°00′	+2°01′	322	177.1
14	2.36	32°05′	8.5	237	162.5
15	2.60	41°50′	10.0	261	161.0
16	4.59	68°30′	−1°22′	460	154.8

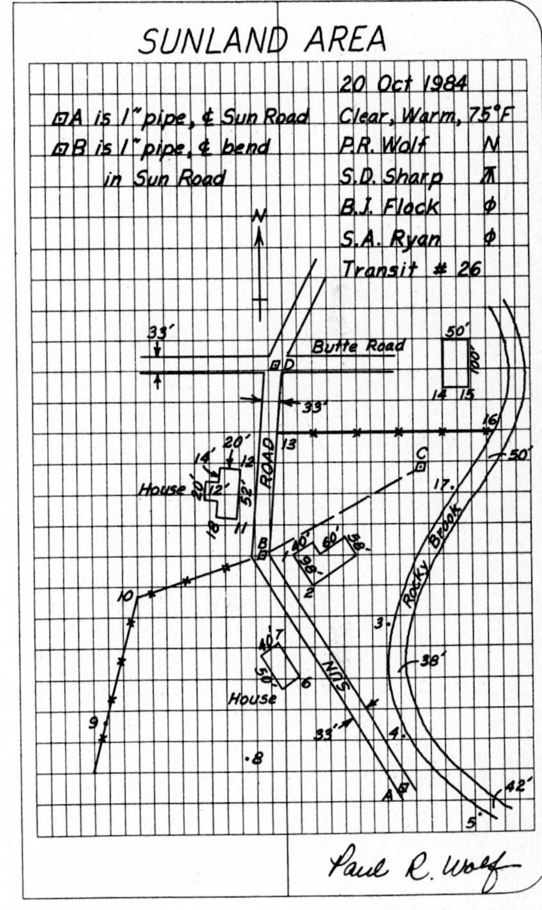

PLATE D-10.

CROSS-SECTION LEVELING

Sta.	+ Sight	HI	− Sight	Elev.
5+00			9.5	
4+00			12.6	
TP I	10.25	106.61	1.87	96.36
3+00			2.1	
2+50			5.8	
2+00			7.4	
1+35			9.7	
1+00			5.6	
0+50			7.6	
0+00			8.5	
BM Pod	8.51	98.23		89.72

HONOLULU-KAILUA HIGHWAY

99.2 / 7.4 / 52	101.5 / 5.1 / 30	97.4 / 9.2 / 10	97.1 / 9.5	95.8 / 10.8 / 12	97.0 / 9.6 / 28	103.8 / 2.8 / 45	
102.3 / 4.3 / 48	99.9 / 6.7 / 24	98.4 / 8.2 / 8	94.0 / 12.6	100.1 / 6.5 / 10	101.5 / 5.1 / 25	98.7 / 7.9 / 50	
	95.2 / 3.0 / 50	95.8 / 2.4 / 25	96.6 / 1.6 / 10	96.1 / 2.1	94.4 / 3.8 / 8	91.1 / 7.1 / 31	95.7 / 2.5 / 48
95.1 / 3.1 / 48	92.8 / 5.4 / 32	89.5 / 8.7 / 15	93.8 / 4.4 / 8	92.4 / 5.8	90.7 / 7.5 / 10	93.4 / 4.8 / 25	96.6 / 1.6 / 50
	92.3 / 5.9 / 54	90.0 / 8.2 / 30	90.8 / 7.4 / 10	90.8 / 7.4	91.3 / 6.9 / 9	93.2 / 5.0 / 25	95.9 / 2.3 / 40
	85.4 / 12.8 / 48	88.9 / 9.3 / 25	85.7 / 12.5 / 10	88.5 / 9.7	89.8 / 8.4 / 8	91.7 / 6.5 / 15	94.1 / 4.1 / 45
	88.6 / 9.6 / 52	97.2 / 1.0 / 28	92.2 / 6.0 / 12	92.6 / 5.6	95.8 / 2.4 / 10	93.6 / 4.6 / 28	95.4 / 2.8 / 50
	90.0 / 8.2 / 50	97.0 / 1.2 / 25	92.7 / 5.5 / 8	90.6 / 7.6	94.4 / 3.8 / 9	95.4 / 2.8 / 24	85.5 / 12.7 / 42
	88.6 / 9.6 / 50	96.1 / 2.1 / 25	92.0 / 6.2 / 10	89.7 / 8.5	93.5 / 4.7 / 8	97.0 / 1.2 / 25	91.5 / 6.7 / 50

Diamond Highway
25 Oct. 1984
Warm, Sunny, 70°
A.C. Chun ⊼
R.E. Neilan N
W.E. Grube Φ,C
M.L. Hagawa C
Lietz level # 10

BM Pod - Kalini Valley, Oahu, Ewa-makai corner Hibiscus and Kiawe Drives, Spike in 30" monkey pod tree, 2 ft. above ground.

Ruth E. Neilan

PLATE D-11.

STAKING OUT

Steps	
1	Set hubs A and B 5′ inside curb line, hub A 20′ from South property line, hub B 70.00′ from A.
2	Set ⊼ at hub A. BS on hub B. Turn 90°∡. Set batter board nails 1 and 2, stakes C and D.
3	Set ⊼ at hub B. BS on hub A. Turn 90°∡. Set batter board nails 3 and 4, stakes E and F.
4	Measure diagonals CF and DE Adjust error if small, restake if large.
5	Set ⊼ at C. BS on E. Set batter board nail 5. Plunge and set nail 6.
6	Set ⊼ at D. BS on F. Set nail 7. Plunge and set nail 8.
7	Set batter board nails 9, 10, 11, and 12 by measurements from established pts.

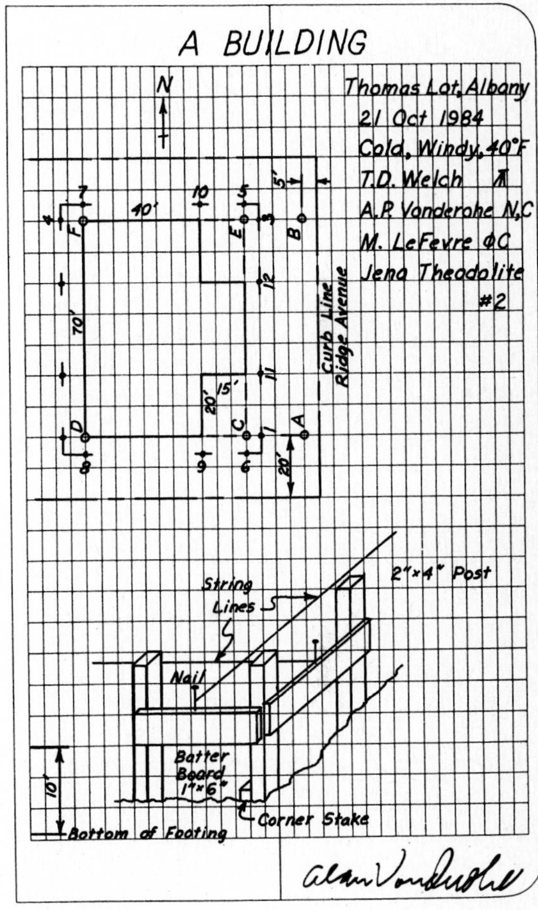

PLATE D-12.

8" SEWER STAKEOUT

(1) Station	(2) +Sight	(3) HI	(4) −Sight	(5) Ground Elev.	(6) Pipe Flow Line	(7) Cut/Fill
BM 16	2.11	102.76		100.65		
0+00			6.21	96.55	96.55	Floor of Existing Catch Basin
+00 ℄			3.20	99.56	96.55	C 3.01
+50 ℄			3.91	98.85	95.95	C 2.90
1+00 ℄			4.07	98.69	95.34	C 3.35
+31 ℄			8.22	94.54	94.97	F 0.43
+50 ℄			4.01	98.75	94.74	C 4.02
2+00 ℄			4.52	98.24	94.14	C 4.09
+33.7 ℄			5.03	97.73	93.73	C 4.00
+33.7			9.03	93.73	93.73	Floor of Existing Manhole
BM 16			2.11 ✓	100.65		

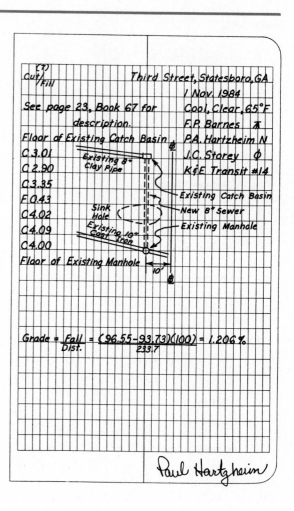

Third Street, Statesboro, GA
1 Nov. 1984
Cool, Clear, 65°F
F.P. Barnes ⊼
P.A. Hartzheim N
J.C. Storey ϕ
K&E Transit #14

See page 23, Book 67 for description.

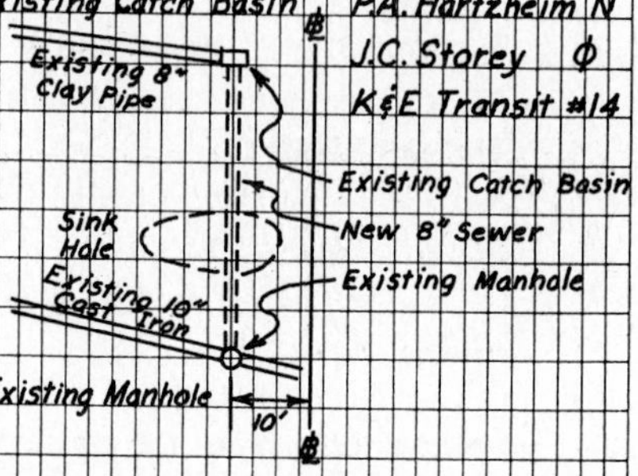

Flowline Calculations

Line drops 50' (1.206%) = 0.60' per 50'

Example

Sta. 0+50 = 96.55 − 0.60 = 95.95

1+00 = 96.55 − 1.21 = 95.34

1+31 = 96.55 − 131 (1.206%) = 94.97

$$\text{Grade} = \frac{\text{Fall}}{\text{Dist.}} = \frac{(96.55 - 93.73)(100)}{233.7} = 1.206\%$$

Paul Hartzheim

PLATE D-13.

ALIGNMENT OF LAFAYETTE HIGHWAY

Station	Chord	Total Def.	Calc. Bearing	Mag. Bearing	Curve Data
P.O.T. 68	100.00				
67	62.92				
			N24°42'E	N24°45'E	
P.T. 66+37.08	37.08	4°12'00"			
					Δ=8°24'
66	100.00	3°49'45" (48" corrected to 45")			R=2864.79'
					D=2°00'
P.O.C. 65	100.00	2°49'45"			L=420.00'
					T=210.38'
64	100.00	1°49'45"			E=7.71'
					M=7.69'
63	82.92	0°49'45"		Defl./ft.	=0.6 min.
P.C. 62+17.08	17.08	0°00'00"			
			N16°18'E	N16°30'E	
62	100.00				
61	100.00				
60	100.00				
P.O.T. 59					

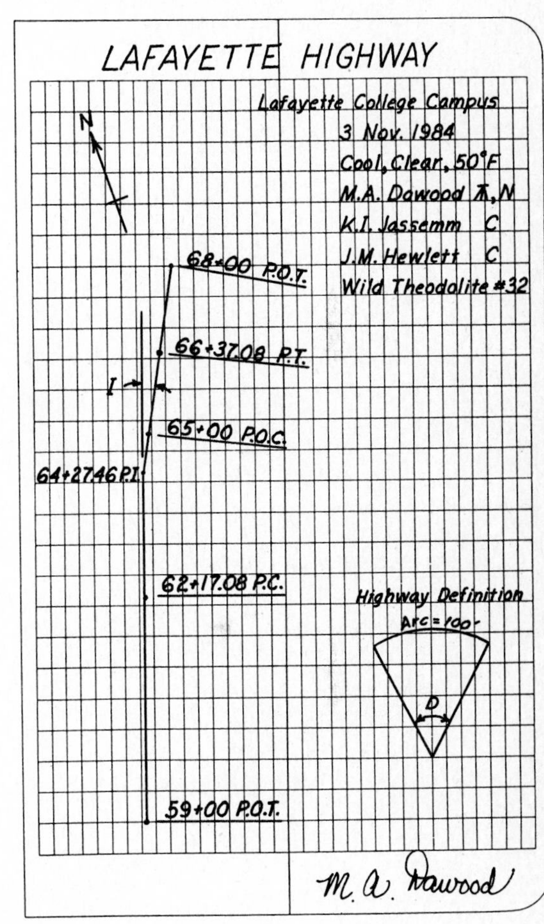

APPENDIX E TABLES

Δ Elevations are negative when z > 90°

TABLE E-1. STADIA REDUCTIONS

	$a = 0°$ $z = 90°$		$a = 1°$ $z = 91°$		$a = 2°$ $z = 92°$		$a = 3°$ $z = 93°$		
MINUTES ↓	**HOR. DIST.**	**DIFF. ELEV.**	**HOR. DIST.**	**DIFF. ELEV.**	**HOR. DIST.**	**DIFF. ELEV.**	**HOR. DIST.**	**DIFF. ELEV.**	
0	100.00	0.00	99.97	1.74	99.88	3.49	99.73	5.23	60′
2	100.00	0.06	99.97	1.80	99.87	3.55	99.72	5.28	
4	100.00	0.12	99.97	1.86	99.87	3.60	99.71	5.34	
6	100.00	0.17	99.96	1.92	99.87	3.66	99.71	5.40	
8	100.00	0.23	99.96	1.98	99.86	3.72	99.70	5.46	
10	100.00	0.29	99.96	2.04	99.86	3.78	99.69	5.52	50′
12	100.00	0.35	99.96	2.09	99.85	3.84	99.69	5.57	
14	100.00	0.41	99.95	2.15	99.85	3.89	99.68	5.63	
16	100.00	0.47	99.95	2.21	99.84	3.95	99.68	5.69	
18	100.00	0.52	99.95	2.27	99.84	4.01	99.67	5.75	
20	100.00	0.58	99.95	2.33	99.83	4.07	99.66	5.80	40′
22	100.00	0.64	99.94	2.38	99.83	4.13	99.66	5.86	
24	100.00	0.70	99.94	2.44	99.82	4.18	99.65	5.92	
26	99.99	0.76	99.94	2.50	99.82	4.24	99.64	5.98	
28	99.99	0.81	99.93	2.56	99.81	4.30	99.63	6.04	
30	99.99	0.87	99.93	2.62	99.81	4.36	99.63	6.09	30′
32	99.99	0.93	99.93	2.67	99.80	4.42	99.62	6.15	
34	99.99	0.99	99.93	2.73	99.80	4.47	99.61	6.21	
36	99.99	1.05	99.92	2.79	99.79	4.53	99.61	6.27	
38	99.99	1.11	99.92	2.85	99.79	4.59	99.60	6.32	
40	99.99	1.16	99.92	2.91	99.78	4.65	99.59	6.38	20′
42	99.99	1.22	99.91	2.97	99.78	4.71	99.58	6.44	
44	99.98	1.28	99.91	3.02	99.77	4.76	99.58	6.50	
46	99.98	1.34	99.90	3.08	99.77	4.82	99.57	6.56	
48	99.98	1.40	99.90	3.14	99.76	4.88	99.56	6.61	
50	99.98	1.45	99.90	3.20	99.76	4.94	99.55	6.67	10′
52	99.98	1.51	99.89	3.26	99.75	4.99	99.55	6.73	
54	99.98	1.57	99.89	3.31	99.74	5.05	99.54	6.79	
56	99.97	1.63	99.89	3.37	99.74	5.11	99.53	6.84	
58	99.97	1.69	99.88	3.43	99.73	5.17	99.52	6.90	
60	99.97	1.74	99.88	3.49	99.73	5.23	99.51	6.96	00′ ↑
$C = 0.75$	0.75	0.01	0.75	0.02	0.75	0.03	0.75	0.05	
$C = 1.00$	1.00	0.01	1.00	0.03	1.00	0.04	1.00	0.06	
$C = 1.25$	1.25	0.02	1.25	0.03	1.25	0.05	1.25	0.08	
		+		+		+		+	
	$z = 89°$		$z = 88°$		$z = 87°$		$z = 86°$		Minutes

Δ Elevations are positive when z < 90°

TABLE E-1. (Continued)

	$a = 4°$ $z = 94°$		$a = 5°$ $z = 95°$		$a = 6°$ $z = 96°$		$a = 7°$ $z = 97°$		
MINUTES	HOR. DIST.	DIFF. ELEV.	HOR. DIST.	DIFF. ELEV.	HOR. DIST.	DIFF. ELEV.	HOR. DIST.	DIFF. ELEV.	
0	99.51	6.96	99.24	8.68	98.91	10.40	98.51	12.10	60′
2	99.51	7.02	99.23	8.74	98.90	10.45	98.50	12.15	
4	99.50	7.07	99.22	8.80	98.88	10.51	98.49	12.21	
6	99.49	7.13	99.21	8.85	98.87	10.57	98.47	12.27	
8	99.48	7.19	99.20	8.91	98.86	10.62	98.46	12.32	
10	99.47	7.25	99.19	8.97	98.85	10.68	98.44	12.38	50′
12	99.46	7.30	99.18	9.03	98.83	10.74	98.43	12.43	
14	99.46	7.36	99.17	9.08	98.82	10.79	98.41	12.49	
16	99.45	7.42	99.16	9.14	98.81	10.85	98.40	12.55	
18	99.44	7.48	99.15	9.20	98.80	10.91	98.39	12.60	
20	99.43	7.53	99.14	9.25	98.78	10.96	98.37	12.66	40′
22	99.42	7.59	99.13	9.31	98.77	11.02	98.36	12.72	
24	99.41	7.65	99.11	9.37	98.76	11.08	98.34	12.77	
26	99.40	7.71	99.10	9.43	98.74	11.13	98.33	12.83	
28	99.39	7.76	99.09	9.48	98.73	11.19	98.31	12.88	
30	99.38	7.82	99.08	9.54	98.72	11.25	98.30	12.94	30′
32	99.38	7.88	99.07	9.60	98.71	11.30	98.28	13.00	
34	99.37	7.94	99.06	9.65	98.69	11.36	98.27	13.05	
36	99.36	7.99	99.05	9.71	98.68	11.42	98.25	13.11	
38	99.35	8.05	99.04	9.77	98.67	11.47	98.24	13.17	
40	99.34	8.11	99.03	9.83	98.65	11.53	98.22	13.22	20′
42	99.33	8.17	99.01	9.88	98.64	11.59	98.20	13.28	
44	99.32	8.22	99.00	9.94	98.63	11.64	98.19	13.33	
46	99.31	8.28	98.99	10.00	98.61	11.70	98.17	13.39	
48	99.30	8.34	98.98	10.05	98.60	11.76	98.16	13.45	
50	99.29	8.40	98.97	10.11	98.58	11.81	98.14	13.50	10′
52	99.28	8.45	98.96	10.17	98.57	11.87	98.13	13.56	
54	99.27	8.51	98.94	10.22	98.56	11.93	98.11	13.61	
56	99.26	8.57	98.93	10.28	98.54	11.98	98.10	13.67	
58	99.25	8.63	98.92	10.34	98.53	12.04	98.08	13.73	
60	99.24	8.68	98.91	10.40	98.51	12.10	98.06	13.78	00′
$C = 0.75$	0.75	0.06	0.75	0.07	0.75	0.08	0.74	0.10	
$C = 1.00$	1.00	0.08	1.00	0.10	0.99	0.11	0.99	0.13	
$C = 1.25$	1.25	0.10	1.24	0.12	1.24	0.14	1.24	0.16	
	$z = 85°$	+	$z = 84°$	+	$z = 83°$	+	$z = 82°$	+	Minutes

TABLE E-1. (Continued)

MINUTES	$a = 8°$ $z = 98°$ HOR. DIST.	DIFF. ELEV.	$a = 9°$ $z = 99°$ HOR. DIST.	DIFF. ELEV.	$a = 10°$ $z = 100°$ HOR. DIST.	DIFF. ELEV.	$a = 11°$ $z = 101°$ HOR. DIST.	DIFF. ELEV.	
0	98.06	13.78	97.55	15.45	96.98	17.10	96.36	18.73	60′
2	98.05	13.84	97.53	15.51	96.96	17.16	96.34	18.78	
4	98.03	13.89	97.52	15.56	96.94	17.21	96.32	18.84	
6	98.01	13.95	97.50	15.62	96.92	17.26	96.29	18.89	
8	98.00	14.01	97.48	15.67	96.90	17.32	96.27	18.95	
10	97.98	14.06	97.46	15.73	96.88	17.37	96.25	19.00	50′
12	97.97	14.12	97.44	15.78	96.86	17.43	96.23	19.05	
14	97.95	14.17	97.43	15.84	96.84	17.48	96.21	19.11	
16	97.93	14.23	97.41	15.89	96.82	17.54	96.18	19.16	
18	97.92	14.28	97.39	15.95	96.80	17.59	96.16	19.21	
20	97.90	14.34	97.37	16.00	96.78	17.65	96.14	19.27	40′
22	97.88	14.40	97.35	16.06	96.76	17.70	96.12	19.32	
24	97.87	14.45	97.33	16.11	96.74	17.76	96.09	19.38	
26	97.85	14.51	97.31	16.17	96.72	17.81	96.07	19.43	
28	97.83	14.56	97.29	16.22	96.70	17.86	96.05	19.48	
30	97.82	14.62	97.28	16.28	96.68	17.92	96.03	19.54	30′
32	97.80	14.67	97.26	16.33	96.66	17.97	96.00	19.59	
34	97.78	14.73	97.24	16.39	96.64	18.03	95.98	19.64	
36	97.76	14.79	97.22	16.44	96.62	18.08	95.96	19.70	
38	97.75	14.84	97.20	16.50	96.60	18.14	95.93	19.75	
40	97.73	14.90	97.18	16.55	96.57	18.19	95.91	19.80	20′
42	97.71	14.95	97.16	16.61	96.55	18.24	95.89	19.86	
44	97.69	15.01	97.14	16.66	96.53	18.30	95.86	19.91	
46	97.68	15.06	97.12	16.72	96.51	18.35	95.84	19.96	
48	97.66	15.12	97.10	16.77	96.49	18.41	95.82	20.02	
50	97.64	15.17	97.08	16.83	96.47	18.46	95.79	20.07	10′
52	97.62	15.23	97.06	16.88	96.45	18.51	95.77	20.12	
54	97.61	15.28	97.04	16.94	96.42	18.57	95.75	20.18	
56	97.59	15.34	97.02	16.99	96.40	18.62	95.72	20.23	
58	97.57	15.40	97.00	17.05	96.38	18.68	95.70	20.28	
60	97.55	15.45	96.98	17.10	96.36	18.73	95.68	20.34	00′
$C = 0.75$	0.74	0.11	0.74	0.12	0.74	0.14	0.73	0.15	
$C = 1.00$	0.99	0.15	0.99	0.17	0.98	0.18	0.98	0.20	
$C = 1.25$	1.24	0.18	1.23	0.21	1.23	0.23	1.22	0.25	
	$z = 81°$	+	$z = 80°$	+	$z = 79°$	+	$z = 78°$	+	Minutes

TABLE E-1. (Continued)

	$a = 12°$ $z = 102°$		$a = 13°$ $z = 103°$		$a = 14°$ $z = 104°$		$a = 15°$ $z = 105°$		
MINUTES ↓	HOR. DIST.	DIFF. ELEV.	HOR. DIST.	DIFF. ELEV.	HOR. DIST.	DIFF. ELEV.	HOR. DIST.	DIFF. ELEV.	
0	95.68	20.34	94.94	21.92	94.15	23.47	93.30	25.00	60′
2	95.65	20.39	94.91	21.97	94.12	23.52	93.27	25.05	
4	95.63	20.44	94.89	22.02	94.09	23.58	93.24	25.10	
6	95.61	20.50	94.86	22.08	94.07	23.63	93.21	25.15	
8	95.58	20.55	94.84	22.13	94.04	23.68	93.18	25.20	
10	95.56	20.60	94.81	22.18	94.01	23.73	93.16	25.25	50′
12	95.53	20.66	94.79	22.23	93.98	23.78	93.13	25.30	
14	95.51	20.71	94.76	22.28	93.95	23.83	93.10	25.35	
16	95.49	20.76	94.73	22.34	93.93	23.88	93.07	25.40	
18	95.46	20.81	94.71	22.39	93.90	23.93	93.04	25.45	
20	95.44	20.87	94.68	22.44	93.87	23.99	93.01	25.50	40′
22	95.41	20.92	94.66	22.49	93.84	24.04	92.98	25.55	
24	95.39	20.97	94.63	22.54	93.82	24.09	92.95	25.60	
26	95.36	21.03	94.60	22.60	93.79	24.14	92.92	25.65	
28	95.34	21.08	94.58	22.65	93.76	24.19	92.89	25.70	
30	95.32	21.13	94.55	22.70	93.73	24.24	92.86	25.75	30′
32	95.29	21.18	94.52	22.75	93.70	24.29	92.83	25.80	
34	95.27	21.24	94.50	22.80	93.67	24.34	92.80	25.85	
36	95.24	21.29	94.47	22.85	93.65	24.39	92.77	25.90	
38	95.22	21.34	94.44	22.91	93.62	24.44	92.74	25.95	
40	95.19	21.39	94.42	22.96	93.59	24.49	92.71	26.00	20′
42	95.17	21.45	94.39	23.01	93.56	24.55	92.68	26.05	
44	95.14	21.50	94.36	23.06	93.53	24.60	92.65	26.10	
46	95.12	21.55	94.34	23.11	93.50	24.65	92.62	26.15	
48	95.09	21.60	94.31	23.16	93.47	24.70	92.59	26.20	
50	95.07	21.66	94.28	23.22	93.45	24.75	92.56	26.25	10′
52	95.04	21.71	94.26	23.27	93.42	24.80	92.53	26.30	
54	95.02	21.76	94.23	23.32	93.39	24.85	92.49	26.35	
56	94.99	21.81	94.20	23.37	93.36	24.90	92.46	26.40	
58	94.97	21.87	94.17	23.42	93.33	24.95	92.43	26.45	
60	94.94	21.92	94.15	23.47	93.30	25.00	92.40	26.50	00′ ↑
$C = 0.75$	0.73	0.16	0.73	0.18	0.73	0.19	0.72	0.20	
$C = 1.00$	0.98	0.22	0.97	0.23	0.97	0.25	0.96	0.27	
$C = 1.25$	1.22	0.27	1.22	0.29	1.21	0.31	1.20	0.33	
		+		+		+		+	
	$z = 77°$		$z = 76°$		$z = 75°$		$z = 74°$		Minutes

TABLE E-2. CONVERGENCE OF RANGE LINES

LATITUDE (DEGREES)	DIFFERENCE BETWEEN SOUTH AND NORTH BOUNDARIES OF TOWNSHIP (LINKS)	ANGLE OF CONVERGENCY OF ADJACENT RANGE LINES (′ ″)	DIFFERENCE OF LONGITUDE PER RANGE: ARC (′ ″)	DIFFERENCE OF LONGITUDE PER RANGE: TIME (SEC)	DIFFERENCE OF LATITUDE, IN MINUTES OF ARC, FOR: 1mi	DIFFERENCE OF LATITUDE, IN MINUTES OF ARC, FOR: 6mi
25	33.9	2 25	5 44.34	22.96		
26	35.4	2 32	5 47.20	23.15		
27	37.0	2 39	5 50.22	23.35	0.871	5.229
28	38.6	2 46	5 53.40	23.56		
29	40.2	2 53	5 56.74	23.78		
30	41.9	3 00	6 00.36	24.02		
31	43.6	3 07	6 04.02	24.27		
32	45.4	3 15	6 07.93	24.53	0.871	5.225
33	47.2	3 23	6 12.00	24.80		
34	49.1	3 30	6 16.31	25.09		
35	50.9	3 38	6 20.95	25.40		
36	52.7	3 46	6 25.60	25.71		
37	54.7	3 55	6 30.59	26.04	0.870	5.221
38	56.8	4 04	6 35.81	26.39		
39	58.8	4 13	6 41.34	26.76		
40	60.9	4 22	6 47.13	27.14		
41	63.1	4 31	6 53.22	27.55		
42	65.4	4 41	6 59.62	27.97	0.869	5.217
43	67.7	4 51	7 06.27	28.42		
44	70.1	5 01	7 13.44	28.90		
45	72.6	5 12	7 20.93	29.39		
46	75.2	5 23	7 28.81	29.92		
47	77.8	5 34	7 37.10	30.47	0.869	5.212
48	80.6	5 46	7 45.79	31.05		
49	83.5	5 59	7 55.12	31.67		
50	86.4	6 12	8 04.83	32.32		
51	89.6	6 25	8 15.17	33.01		
52	92.8	6 39	8 26.13	33.74	0.868	5.207
53	96.2	6 54	8 37.75	34.52		
54	99.8	7 09	8 50.07	35.34		
55	103.5	7 25	9 03.18	36.22		
56	107.5	7 42	9 17.12	37.14		
57	111.6	8 00	9 31.97	38.13	0.867	5.202
58	116.0	8 19	9 47.83	39.19		
59	120.6	8 38	10 04.78	40.32		
60	125.5	8 59	10 22.94	41.52		
61	130.8	9 22	10 42.42	42.83		
62	136.3	9 46	11 03.38	44.22	0.866	5.198
63	142.2	10 11	11 25.97	45.73		
64	148.6	10 38	11 50.37	47.36		
65	155.0	11 08	12 16.82	49.12		
66	162.8	11 39	12 45.55	51.04		
67	170.7	12 13	13 16.88	53.12	0.866	5.195
68	179.3	12 51	13 51.15	55.41		
69	188.7	13 31	14 28.77	57.92		
70	199.1	14 15	15 10.26	60.68	0.866	5.193

TABLE E-3. AZIMUTHS OF THE SECANT*

LAT.	0 mi	1 mi	2 mi	3 mi	DEFLECTION ANGLE 6mi
°	° ′	° ′	° ′		° ′
25	89 58.8	89 59.2	89 59.6	90°	2 25
26	58.7	59.2	59.6	E or W.	2 32
27	58.7	59.1	59.6	" " "	2 39
28	58.6	59.1	59.5	" " "	2 46
29	58.6	59.0	59.5	" " "	2 53
30	58.5	59.0	59.5	" " "	3 00
31	58.4	59.0	59.5	" " "	3 07
32	58.4	58.9	59.5	" " "	3 15
33	58.3	58.9	59.4	" " "	3 23
34	58.2	58.8	59.4	" " "	3 30
35	58.2	58.8	59.4	" " "	3 38
36	58.1	58.7	59.4	" " "	3 46
37	58.0	58.7	59.3	" " "	3 55
38	58.0	58.6	59.3	" " "	4 04
39	57.9	58.6	59.3	" " "	4 13
40	57.8	58.5	59.3	" " "	4 22
41	57.7	58.5	59.2	" " "	4 31
42	57.7	58.4	59.2	" " "	4 41
43	57.6	58.4	59.2	" " "	4 51
44	57.5	58.3	59.2	" " "	5 01
45	57.4	58.3	59.1	" " "	5 12
46	57.3	58.2	59.1	" " "	5 23
47	57.2	58.1	59.1	" " "	5 34
48	57.1	58.1	59.0	" " "	5 46
49	57.0	58.0	59.0	" " "	5 59
50	56.9	57.9	59.0	" " "	6 12
51	56.8	57.9	58.9	" " "	6 25
52	56.7	57.8	58.9	" " "	6 39
53	56.6	57.7	58.8	" " "	6 54
54	56.4	57.6	58.8	" " "	7 09
55	56.3	57.5	58.8	" " "	7 25
56	56.2	57.4	58.7	" " "	7 42
57	56.0	57.3	58.7	" " "	8 00
58	55.8	57.2	58.6	" " "	8 19
59	55.7	57.1	58.6	" " "	8 38
60	55.5	57.0	58.5	" " "	8 59
61	55.3	56.9	58.4	" " "	9 22
62	55.1	56.7	58.4	" " "	9 46
63	54.9	56.6	58.3	" " "	10 11
64	54.7	56.5	58.2	" " "	10 38
65	54.4	56.3	58.1	" " "	11 08
66	54.2	56.1	58.1	" " "	11 39
67	53.9	55.9	58.0	" " "	12 13
68	53.6	55.7	57.9	" " "	12 51
69	53.2	55.5	57.8	" " "	13 31
70	89 52.9	89 55.3	89 57.6	" " "	14 15
	6 mi	5 mi	4 mi	3 mi	

* From U.S. Department of the Interior, *Standard Field Tables and Trigonometric Formulas*, 8th ed. Washington, D.C.: U.S. Government Printing Office, 1950.

TABLE E-4. OFFSETS, IN LINKS, FROM THE SECANT TO THE PARALLEL*

LAT.	0 mi	½ mi	1 mi	1½ mi	2 mi	2½ mi	3 mi
°							
25	2 N.	1 N.	0	1 S.	1 S.	2 S.	2 S.
26	2	1	0	1	1	2	2
27	3	1	0	1	2	2	2
28	3	1	0	1	2	2	2
29	3	1	0	1	2	2	2
30	3	1	0	1	2	2	2
31	3	1	0	1	2	2	2
32	3	1	0	1	2	2	3
33	3	1	0	1	2	2	3
34	3	2	0	1	2	3	3
35	4	2	0	1	2	3	3
36	4	2	0	1	2	3	3
37	4	2	0	1	2	3	3
38	4	2	0	1	2	3	3
39	4	2	0	1	2	3	3
40	4	2	0	1	3	3	3
41	4	2	0	2	3	3	4
42	5	2	0	2	3	3	4
43	5	2	0	2	3	4	4
44	5	2	0	2	3	4	4
45	5	2	0	2	3	4	4
46	5	2	0	2	3	4	4
47	5	2	0	2	3	4	4
48	6	3	0	2	3	4	4
49	6	3	0	2	3	4	5
50	6	3	0	2	4	4	5
51	6	3	0	2	4	5	5
52	6	3	0	2	4	5	5
53	7	3	0	2	4	5	5
54	7	3	0	2	4	5	6
55	7	3	0	3	4	5	6
56	7	3	0	3	4	6	6
57	8	3	0	3	5	6	6
58	8	4	0	3	5	6	6
59	8	4	0	3	5	6	7
60	9	4	0	3	5	7	7
61	9	4	0	3	5	7	7
62	9	4	0	3	6	7	8
63	10	4	0	3	6	7	8
64	10	5	0	4	6	8	8
65	11	5	0	4	6	8	9
66	11	5	0	4	7	8	9
67	12	5	0	4	7	9	9
68	12	6	0	4	7	9	10
69	13	6	0	5	8	10	10
70	14 N.	6 N.	0	5 S.	8 S.	10 S.	11 S.
	6 mi	5½ mi	5 mi	4½	4 mi	3½	3 mi

* From U.S. Department of the Interior, *Standard Field Tables and Trigonometric Formulas*, 8th ed. Washington, D.C.: U.S. Government Printing Office, 1950.

TABLE E-5. FUNCTIONS OF CIRCULAR CURVES

DEGREE OF CURVE D	DEFL. PER FT OF STA. (MIN)	CHORD DEFINITION		ARC DEFINITION	DEGREE OF CURVE D	DEFL. PER FT OF STA. (MIN)	CHORD DEFINITION		ARC DEFINITION
		RADIUS R	CO 1 STA.	RADIUS R			RADIUS R	CO 1 STA.	RADIUS R
0° 0′		Infinite		Infinite	1° 0′	0.30	5729.65	1.75	5729.58
1′	0.005	343775.	0.03	343775.	1′	0.305	5635.72	1.77	5635.65
2′	0.01	171887.	0.06	171887.	2′	0.31	5544.83	1.80	5544.75
3′	0.015	114592.	0.09	114592.	3′	0.315	5456.82	1.83	5456.74
4′	0.02	85943.7	0.12	85943.7	4′	0.32	5371.56	1.86	5371.48
5′	0.025	68754.9	0.15	68754.9	5′	0.325	5288.92	1.89	5288.84
6′	0.03	57295.8	0.17	57295.8	6′	0.33	5208.79	1.92	5208.71
7′	0.035	49110.7	0.20	49110.7	7′	0.335	5131.05	1.95	5130.97
8′	0.04	42971.8	0.23	42971.8	8′	0.34	5055.59	1.98	5055.51
9′	0.045	38197.2	0.26	38197.2	9′	0.345	4982.33	2.01	4982.24
10′	0.05	34377.5	0.29	34377.5	10′	0.35	4911.15	2.03	4911.07
11′	0.055	31252.3	0.32	31252.2	11′	0.355	4841.98	2.07	4841.90
12′	0.06	28647.8	0.35	28647.8	12′	0.36	4774.74	2.09	4774.65
13′	0.065	26444.2	0.38	26444.2	13′	0.365	4709.33	2.12	4709.24
14′	0.07	24555.4	0.41	24555.3	14′	0.37	4645.69	2.15	4645.60
15′	0.075	22918.3	0.44	22918.3	15′	0.375	4583.75	2.18	4583.66
16′	0.08	21485.9	0.47	21485.9	16′	0.38	4523.44	2.21	4523.35
17′	0.085	20222.1	0.49	20222.0	17′	0.385	4464.70	2.24	4464.61
18′	0.09	19098.6	0.52	19098.6	18′	0.39	4407.46	2.27	4407.37
19′	0.095	18093.4	0.55	18093.4	19′	0.395	4351.67	2.30	4351.58
20′	0.10	17188.8	0.58	17188.7	20′	0.40	4297.28	2.33	4297.18
21′	0.105	16370.2	0.61	16370.2	21′	0.405	4244.23	2.35	4244.13
22′	0.11	15626.1	0.64	15626.1	22′	0.41	4192.47	2.39	4192.37
23′	0.115	14946.8	0.67	14946.7	23′	0.415	4141.96	2.41	4141.86
24′	0.12	14324.0	0.70	14323.9	24′	0.42	4092.66	2.44	4092.56
25′	0.125	13751.0	0.73	13751.0	25′	0.425	4044.51	2.47	4044.41
26′	0.13	13222.1	0.76	13222.1	26′	0.43	3997.49	2.50	3997.38
27′	0.135	12732.4	0.79	12732.4	27′	0.435	3951.54	2.53	3951.43
28′	0.14	12277.7	0.81	12277.7	28′	0.44	3906.64	2.56	3906.53
29′	0.145	11854.3	0.84	11854.3	29′	0.445	3862.74	2.59	3862.64

30′	0.15	11459.2	0.87	11459.2	30′	0.45	3819.83	2.62	3819.71
31′	0.155	11089.6	0.90	11089.5	31′	0.455	3777.85	2.65	3777.74
32′	0.16	10743.0	0.93	10743.0	32′	0.46	3736.79	2.68	3736.68
33′	0.165	10417.5	0.96	10417.4	33′	0.465	3696.61	2.71	3696.50
34′	0.17	10111.1	0.99	10111.0	34′	0.47	3657.29	2.73	3657.18
35′	0.175	9822.18	1.02	9822.13	35′	0.475	3618.80	2.76	3618.68
36′	0.18	9549.34	1.05	9549.29	36′	0.48	3581.10	2.79	3580.99
37′	0.185	9291.25	1.07	9291.21	37′	0.485	3544.19	2.82	3544.07
38′	0.19	9046.75	1.11	9046.70	38′	0.49	3508.02	2.85	3507.91
39′	0.195	8814.78	1.13	8814.73	39′	0.495	3472.59	2.88	3472.47
40′	0.20	8594.42	1.16	8594.37	40′	0.50	3437.87	2.91	3437.75
41′	0.205	8384.80	1.19	8384.75	41′	0.505	3403.83	2.94	3403.71
42′	0.21	8185.16	1.21	8185.11	42′	0.51	3370.46	2.97	3370.34
43′	0.215	7994.81	1.25	7994.76	43′	0.515	3337.74	3.00	3337.62
44′	0.22	7813.11	1.28	7813.06	44′	0.52	3305.65	3.03	3305.53
45′	0.225	7639.49	1.31	7639.44	45′	0.525	3274.17	3.05	3274.04
46′	0.23	7473.42	1.34	7473.36	46′	0.53	3243.29	3.08	3243.16
47′	0.235	7314.41	1.37	7314.35	47′	0.535	3212.98	3.11	3212.85
48′	0.24	7162.03	1.39	7161.97	48′	0.54	3183.23	3.14	3183.10
49′	0.245	7015.87	1.43	7015.81	49′	0.545	3154.03	3.17	3153.90
50′	0.25	6875.55	1.45	6875.49	50′	0.55	3125.36	3.20	3125.22
51′	0.255	6740.74	1.48	6740.68	51′	0.555	3097.20	3.23	3097.07
52′	0.26	6611.12	1.51	6611.05	52′	0.56	3069.55	3.26	3069.42
53′	0.265	6486.38	1.54	6486.31	53′	0.565	3042.39	3.29	3042.25
54′	0.27	6366.26	1.57	6366.20	54′	0.57	3015.71	3.32	3015.57
55′	0.275	6250.51	1.60	6250.45	55′	0.575	2989.48	3.35	2989.34
56′	0.28	6138.90	1.63	6138.83	56′	0.58	2963.72	3.37	2963.58
57′	0.285	6031.20	1.66	6031.14	57′	0.585	2938.39	3.40	2938.25
58′	0.29	5927.22	1.69	5927.15	58′	0.59	2913.49	3.43	2913.34
59′	0.295	5826.76	1.72	5826.69	59′	0.595	2889.01	3.46	2888.86

TABLE E-5. (Continued)

DEGREE OF CURVE D	DEFL. PER FT OF STA. (MIN)	CHORD DEFINITION		ARC DEFINITION	DEGREE OF CURVE D	DEFL. PER FT OF STA. (MIN)	CHORD DEFINITION		ARC DEFINITION
		RADIUS R	CO 1 STA.	RADIUS R			RADIUS R	CO 1 STA.	RADIUS R
2° 0′	0.60	2864.93	3.49	2864.79	3° 0′	0.90	1910.08	5.24	1909.86
1′	0.605	2841.26	3.52	2841.11	1′	0.905	1899.53	5.26	1899.31
2′	0.61	2817.97	3.55	2817.83	2′	0.91	1889.09	5.29	1888.87
3′	0.615	2795.06	3.58	2794.92	3′	0.915	1878.77	5.32	1878.55
4′	0.62	2772.53	3.61	2772.38	4′	0.92	1868.56	5.35	1868.34
5′	0.625	2750.35	3.64	2750.20	5′	0.925	1858.47	5.38	1858.24
6′	0.63	2728.52	3.66	2728.37	6′	0.93	1848.48	5.41	1848.25
7′	0.635	2707.04	3.69	2706.89	7′	0.935	1838.59	5.44	1838.37
8′	0.64	2685.89	3.72	2685.74	8′	0.94	1828.82	5.47	1828.59
9′	0.645	2665.08	3.75	2664.92	9′	0.945	1819.14	5.50	1818.91
10′	0.65	2644.58	3.78	2644.42	10′	0.95	1809.57	5.53	1809.34
11′	0.655	2624.39	3.81	2624.23	11′	0.955	1800.10	5.56	1799.87
12′	0.66	2604.51	3.84	2604.35	12′	0.96	1790.73	5.58	1790.49
13′	0.665	2584.93	3.87	2584.77	13′	0.965	1781.45	5.61	1781.22
14′	0.67	2565.65	3.90	2565.48	14′	0.97	1772.27	5.64	1772.03
15′	0.675	2546.64	3.93	2546.48	15′	0.975	1763.18	5.67	1762.95
16′	0.68	2527.92	3.96	2527.75	16′	0.98	1754.19	5.70	1753.95
17′	0.685	2509.47	3.98	2509.30	17′	0.985	1745.29	5.73	1745.05
18′	0.69	2491.29	4.01	2491.12	18′	0.99	1736.48	5.76	1736.24
19′	0.695	2473.37	4.04	2473.20	19′	0.995	1727.75	5.79	1727.51
20′	0.70	2455.70	4.07	2455.53	20′	1.00	1719.12	5.82	1718.87
21′	0.705	2438.29	4.10	2438.12	21′	1.005	1710.57	5.85	1710.32
22′	0.71	2421.12	4.13	2420.95	22′	1.01	1702.10	5.88	1701.85
23′	0.715	2404.19	4.16	2404.02	23′	1.015	1693.72	5.90	1693.47
24′	0.72	2387.50	4.19	2387.32	24′	1.02	1685.42	5.93	1685.17
25′	0.725	2371.04	4.22	2370.86	25′	1.025	1677.20	5.96	1676.95
26′	0.73	2354.80	4.25	2354.62	26′	1.03	1669.06	5.99	1668.81
27′	0.735	2338.78	4.28	2338.60	27′	1.035	1661.00	6.02	1660.75
28′	0.74	2322.98	4.30	2322.80	28′	1.04	1653.01	6.05	1652.76
29′	0.745	2307.39	4.33	2307.21	29′	1.045	1645.11	6.08	1644.85

30′	0.75	2292.01	4.36	2291.83	30′	1.05	1637.28	6.11	1637.02
31′	0.755	2276.84	4.39	2276.65	31′	1.055	1629.52	6.14	1629.26
32′	0.76	2261.86	4.42	2261.68	32′	1.06	1621.84	6.17	1621.58
33′	0.765	2247.08	4.45	2246.89	33′	1.065	1614.22	6.19	1613.96
34′	0.77	2232.49	4.48	2232.30	34′	1.07	1606.68	6.22	1606.42
35′	0.775	2218.09	4.51	2217.90	35′	1.075	1599.21	6.25	1598.95
36′	0.78	2203.87	4.54	2203.68	36′	1.08	1591.81	6.28	1591.55
37′	0.785	2189.84	4.57	2189.65	37′	1.085	1584.48	6.31	1584.21
38′	0.79	2175.98	4.60	2175.79	38′	1.09	1577.21	6.34	1576.95
39′	0.795	2162.30	4.62	2162.10	39′	1.095	1570.01	6.37	1569.75
40′	0.80	2148.79	4.65	2148.59	40′	1.10	1562.88	6.40	1562.61
41′	0.805	2135.44	4.68	2135.25	41′	1.105	1555.81	6.43	1555.54
42′	0.81	2122.26	4.71	2122.07	42′	1.11	1548.80	6.46	1548.53
43′	0.815	2109.24	4.74	2109.05	43′	1.115	1541.86	6.49	1541.59
44′	0.82	2096.39	4.77	2096.19	44′	1.12	1534.98	6.51	1534.71
45′	0.825	2083.68	4.80	2083.48	45′	1.125	1528.16	6.54	1527.89
46′	0.83	2071.13	4.83	2070.93	46′	1.13	1521.40	6.57	1521.13
47′	0.835	2058.73	4.86	2058.53	47′	1.135	1514.17	6.60	1514.43
48′	0.84	2046.48	4.89	2046.28	48′	1.14	1508.06	6.63	1507.78
49′	0.845	2034.37	4.92	2034.17	49′	1.145	1501.48	6.66	1501.20
50′	0.85	2022.41	4.94	2022.20	50′	1.15	1494.95	6.69	1494.67
51′	0.855	2010.59	4.97	2010.38	51′	1.155	1488.48	6.72	1488.20
52′	0.86	1998.90	5.00	1998.69	52′	1.16	1482.07	6.75	1481.79
53′	0.865	1987.35	5.03	1987.14	53′	1.165	1475.71	6.78	1475.43
54′	0.87	1975.93	5.06	1975.72	54′	1.17	1469.41	6.81	1469.12
55′	0.875	1964.64	5.09	1964.43	55′	1.175	1463.16	6.83	1462.87
56′	0.88	1953.48	5.12	1953.27	56′	1.18	1456.96	6.86	1456.67
57′	0.885	1942.44	5.15	1942.23	57′	1.185	1450.81	6.89	1450.53
58′	0.89	1931.53	5.18	1931.32	58′	1.19	1444.72	6.92	1444.43
59′	0.895	1920.75	5.21	1920.53	59′	1.195	1438.68	6.95	1438.39

TABLE E-6. LENGTHS OF ARCS AND TRUE CHORDS*

D	CHORD DEFINITION OF D				ARC DEFINITION OF D		
	ARC FOR 1 STA.	TRUE CHORDS			TRUE CHORDS		
		$\frac{1}{10}$ STA.	$\frac{1}{4}$ STA.	$\frac{1}{2}$ STA.	$\frac{1}{4}$ STA.	$\frac{1}{2}$ STA.	1 STA.
1°	100.001	10	25	50	25	50	100
2°	100.005	10	25	50	25	50	99.99
3°	100.011	10	25	50	25	50	99.99
4°	100.020	10	25	50.01	25	50	99.98
5°	100.032	10	25.01	50.01	25	50	99.97
6°	100.046	10.01	25.01	50.02	25	50	99.95
7°	100.062	10.01	25.02	50.02	25	50	99.94
8°	100.081	10.01	25.02	50.03	25	49.99	99.92
9°	100.103	10.01	25.02	50.04	25	49.99	99.90
10°	100.127	10.01	25.03	50.05	25	49.98	99.88
11°	100.154	10.02	25.04	50.06	25	49.98	99.85
12°	100.183	10.02	25.04	50.07	25	49.98	99.82
13°	100.215	10.02	25.05	50.08	25	49.97	99.79
14°	100.249	10.02	25.06	50.09	25	49.97	99.75
15°	100.286	10.03	25.07	50.11	25	49.96	99.72
16°	100.326	10.03	25.08	50.12	25	49.96	99.68
17°	100.368	10.04	25.09	50.14	25	49.95	99.63
18°	100.412	10.04	25.10	50.16	24.99	49.95	99.59
19°	100.460	10.04	25.11	50.17	24.99	49.94	99.54
20°	100.510	10.05	25.12	50.19	24.99	49.94	99.49
21°	100.562	10.06	25.13	50.21	24.99	49.93	99.44
22°	100.617	10.06	25.14	50.23	24.99	49.92	99.39
23°	100.675	10.07	25.16	50.25	24.99	49.92	99.33
24°	100.735	10.07	25.17	50.27	24.99	49.91	99.27
25°	100.798	10.08	25.19	50.30	24.99	49.90	99.21
26°	100.863	10.08	25.20	50.32	24.99	49.89	99.14
27°	100.931	10.09	25.22	50.35	24.99	49.88	99.08
28°	101.002	10.10	25.23	50.38	24.98	49.88	99.01
29°	101.075	10.11	25.25	50.40	24.98	49.87	98.94
30°	101.152	10.11	25.27	50.43	24.98	49.86	98.86

TABLE E-6. (Continued)

D	CHORD DEFINITION OF D				ARC DEFINITION OF D		
	ARC FOR 1 STA.	TRUE CHORDS			TRUE CHORDS		
		$\frac{1}{10}$ STA.	$\frac{1}{4}$ STA.	$\frac{1}{2}$ STA.	$\frac{1}{4}$ STA.	$\frac{1}{2}$ STA.	1 STA.
32°	101.312	10.13	25.31	50.49	24.98	49.84	98.71
34°	101.482	10.15	25.35	50.56	24.98	49.82	98.54
36°	101.664	10.16	25.39	50.62	24.97	49.79	98.36
38°	101.857	10.18	25.43	50.69	24.97	49.77	98.18
41°	102.166	10.21	25.51	50.81	24.97	49.73	97.88
44°	102.500	10.25	25.59	50.94	24.96	49.69	97.56
48°	102.986	10.30	25.70	51.12	24.95	49.63	97.10
52°	103.516	10.35	25.82	51.31	24.95	49.57	96.60
57°	104.246	10.42	25.98	51.59	24.94	49.49	95.93
64°	105.394	10.53	26.26	52.01	24.92	49.35	94.88
72°	106.896	10.68	26.61	52.57	24.90	49.18	93.55
82°	109.073	10.90	27.12	53.38	24.87	48.94	91.68
95°	112.445	11.23	27.91	54.63	24.82	48.58	88.93
115°	118.992	11.88	29.44	57.03	24.74	47.93	84.04

Chord definition of D

For degrees of curve not listed obtain excess arc per station approximately by interpolation, or exactly to three decimal places (up to $D = 15°$) from:

$$\text{excess} = 0.00127D^2$$

Arc definition of D

For degrees of curve not listed, obtain chord deficiency per station approximately by interpolation, or exactly to two decimal places (up to $D = 30°$) from:

$$\text{deficiency} = 0.00127D^2$$

*From Meyer & Gibson, *Route Surveying and Design* p. 328, 5th ed.

TABLE E-7. TRIGONOMETRIC FORMULAS FOR THE SOLUTION OF RIGHT TRIANGLES

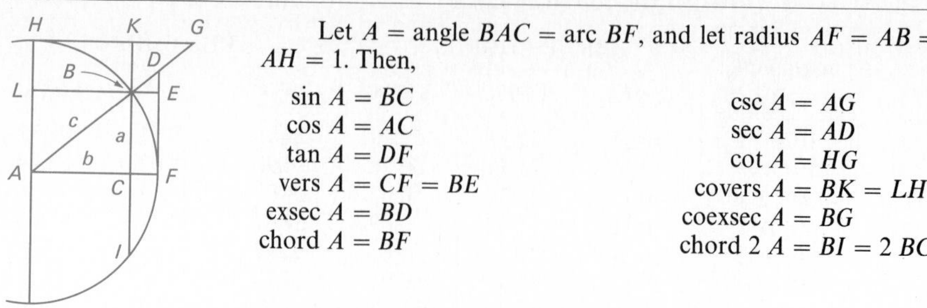

Let A = angle BAC = arc BF, and let radius $AF = AB = AH = 1$. Then,

$\sin A = BC$	$\csc A = AG$
$\cos A = AC$	$\sec A = AD$
$\tan A = DF$	$\cot A = HG$
$\text{vers } A = CF = BE$	$\text{covers } A = BK = LH$
$\text{exsec } A = BD$	$\text{coexsec } A = BG$
$\text{chord } A = BF$	$\text{chord } 2A = BI = 2\,BC$

In the right-angled triangle ABC, let $AB = c$, $BC = a$, $CA = b$. Then

1. $\sin A = \dfrac{a}{c}$

2. $\cos A = \dfrac{b}{c}$

3. $\tan A = \dfrac{a}{b}$

4. $\cot A = \dfrac{b}{a}$

5. $\sec A = \dfrac{c}{b}$

6. $\csc A = \dfrac{c}{a}$

7. $\text{vers } A = 1 - \cos A = \dfrac{c - b}{c}$
$= \text{covers } B$

8. $\text{exsec } A = \sec A - 1 = \dfrac{c - b}{b} = \text{coexsec } B$

9. $\text{covers } A = \dfrac{c - a}{c} = \text{vers } B$

10. $\text{coexsec } A = \dfrac{c - a}{a} = \text{exsec } B$

11. $a = c \sin A = b \tan A$

12. $b = c \cos A = a \cot A$

13. $c = \dfrac{a}{\sin A} = \dfrac{b}{\cos A}$

14. $a = c \cos B = b \cot B$

15. $b = c \sin B = a \tan B$

16. $c = \dfrac{a}{\cos B} = \dfrac{b}{\sin B}$

17. $a = \sqrt{c^2 - b^2}$
$= \sqrt{(c - b)(c + b)}$

18. $b = \sqrt{c^2 - a^2}$
$= \sqrt{(c - a)(c + a)}$

19. $c = \sqrt{a^2 + b^2}$

20. $C = 90^\circ = A + B$

21. $\text{area} = \frac{1}{2}ab$

TABLE E-8. TRIGONOMETRIC FORMULAS FOR THE SOLUTION OF OBLIQUE TRIANGLES

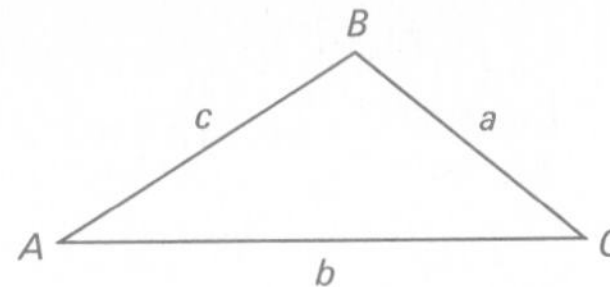

NO.	GIVEN	SOUGHT	FORMULA
22	A, B, a	C, b, c	$C = 180° - (A + B)$
			$b = \dfrac{a}{\sin A} \times \sin B$
			$c = \dfrac{a}{\sin A} \times \sin(A + B) = \dfrac{a}{\sin A} \times \sin C$
		Area	$\text{Area} = \frac{1}{2}ab \sin C = \dfrac{a^2 \sin B \sin C}{2 \sin A}$
23	A, a, b	B, C, c	$\sin B = \dfrac{\sin A}{a} \times b$
			$C = 180° - (A + B)$
			$C = \dfrac{a}{\sin A} \times \sin C$
		Area	$\text{Area} = \frac{1}{2}ab \sin C$
24	$C, a, b,$	c	$c = \sqrt{a^2 + b^2 - 2ab \cos C}$
25		$\frac{1}{2}(A + B)$	$\frac{1}{2}(A + B) = 90° - \frac{1}{2}C$
26		$\frac{1}{2}(A - B)$	$\tan \frac{1}{2}(A - B) = \dfrac{a - b}{a + b} \times \tan \frac{1}{2}(A + B)$
27		A, B	$A = \frac{1}{2}(A + B) + \frac{1}{2}(A - B)$ $B = \frac{1}{2}(A + B) - \frac{1}{2}(A - B)$
28		c	$c = (a + b) \times \dfrac{\cos \frac{1}{2}(A + B)}{\cos \frac{1}{2}(A - B)} = (a - b) \times \dfrac{\sin \frac{1}{2}(A + B)}{\sin \frac{1}{2}(A - B)}$
29		Area	$\text{Area} = \frac{1}{2}ab \sin C$
30	a, b, c	A	$\text{Let } s = \dfrac{a + b + c}{2}$
31			$\sin \frac{1}{2}A = \sqrt{\dfrac{(s - b)(s - c)}{bc}}$
			$\cos \frac{1}{2}A = \sqrt{\dfrac{s(s - a)}{bc}}$
			$\tan \frac{1}{2}A = \sqrt{\dfrac{(s - b)(s - c)}{s(s - a)}}$
32			$\sin A = \dfrac{2\sqrt{s(s - a)(s - b)(s - c)}}{bc}$
			$\cos A = \dfrac{b^2 + c^2 - a^2}{2bc}$
33		Area	$\text{Area} = \sqrt{s(s - a)(s - b)(s - c)}$

TABLE E-9. RELATIONS BETWEEN LINEAR AND ANGULAR ERRORS

ALLOWABLE ANGULAR ERROR FOR GIVEN LINEAR PRECISION		ALLOWABLE LINEAR ERROR FOR GIVEN ANGULAR PRECISION					
PRECISION OF LINEAR MEASUREMENTS	ALLOWABLE ANGULAR ERROR	LEAST READING IN ANGULAR MEASUREMENTS	ALLOWABLE LINEAR ERROR IN 100′	500′	1000′	5000′	RATIO
$\frac{1}{500}$	6′53″	5′	0.145	0.727	1.454	7.272	$\frac{1}{688}$
$\frac{1}{1000}$	3′26″	1′	0.029	0.145	0.291	1.454	$\frac{1}{3440}$
$\frac{1}{5000}$	0′41″	30″	0.015	0.073	0.145	0.727	$\frac{1}{6880}$
$\frac{1}{10{,}000}$	0′21″	20″	0.010	0.049	0.097	0.485	$\frac{1}{10{,}300}$
$\frac{1}{50{,}000}$	0′04″	10″	0.005	0.024	0.049	0.242	$\frac{1}{20{,}600}$
$\frac{1}{100{,}000}$	0′02″	5″	0.002	0.012	0.024	0.121	$\frac{1}{41{,}200}$
$\frac{1}{1{,}000{,}000}$	0′00.2″	2″	0.001	0.005	0.010	0.048	$\frac{1}{103{,}100}$
		1″		0.002	0.005	0.024	$\frac{1}{206{,}300}$

TABLE E-10. PRECISION OF COMPUTED VALUES

SIZE OF ANGLE AND FUNCTION		ANGULAR ERROR				
		1′	30″	20″	10″	5″
		PRECISION OF COMPUTED VALUE USING SINE OR COSINE				
sin 5° or cos 85°		1/300	1/600	1/900	1/1800	1/3600
10	80	1/610	1/1210	1/1820	1/3640	1/7280
20	70	1/1250	1/2500	1/3750	1/7500	1/15,000
30	60	1/1990	1/3970	1/5960	1/11,970	1/23,940
40	50	1/2890	1/5770	1/8660	1/17,310	1/34,620
50	40	1/4100	1/8190	1/12,290	1/24,580	1/49,160
60	30	1/5950	1/11,900	1/17,860	1/35,720	1/71,440
70	20	1/9450	1/18,900	1/28,330	1/56,670	1/113,340
80	10	1/19,500	1/39,000	1/58,500	1/117,000	1/234,000
		PRECISION OF COMPUTED VALUE USING TAN OR COT				
tan or cot 5°		1/300	1/600	1/900	1/1790	1/3580
10		1/590	1/1180	1/1760	1/3530	1/7050
20		1/1100	1/2210	1/3310	1/6620	1/13,250
30		1/1490	1/2980	1/4470	1/8930	1/17,870
40		1/1690	1/3390	1/5080	1/10,160	1/20,320
45		1/1720	1/3440	1/5160	1/10,310	1/20,630
50		1/1690	1/3390	1/5080	1/10,160	1/20,320
60		1/1490	1/2980	1/4470	1/8930	1/17,870
70		1/1100	1/2210	1/3310	1/6620	1/13,250
80		1/590	1/1180	1/1760	1/3530	1/7050
85		1/300	1/600	1/900	1/1790	1/3580

APPENDIX F ANSWERS TO SELECTED PROBLEMS

CHAPTER 2

2-1(c). 3,705.01ft
2-2(c). 1,259.7m
2-3(c). 2,709.3ft
2-4(c). 40.068sq.Gunterchains
2-6(b). 52.15ft
2-8(b). 0.910acres
2-10(c). 147°20′20″
2-11(b). 1615
2-12(c). 1.2749×10^2
2-17. $\bar{M} = 728.890$ ft, $\sigma = \pm 0.046$ ft, $\sigma_m = \pm 0.015$ ft
2-21. 50%, 728.859 to 728.921 (7)
90%, 728.814 to 728.966 (8)
2-25. $M = 49°23'11''$, $\sigma = \pm 12.7''$, $\sigma_m = \pm 4.5''$
2-29. $\sigma = \pm 0.016$ ft
2-31(b). $\sigma = \pm 0.044$ m
2-33(b). $\bar{M}_w = 65°38'20.9''$
2-35(b). Area = 12,144 m^2, ± 2.4 m^2
2-38(b). Elev. diff. = -80.362 m, ± 0.032 m

CHAPTER 4

4-3(b). 1661.66 ft
4-4(c). 650.79 ft

4-9. 669.648 ft
4-12. 576.073 ft
4-15. 87.705 ft
4-18. 623.688 ft
4-21. 114.076 m
4-23. 156.689 m
4-26. 570.937 ft
4-30. $P_1 = 27.3$ lb
4-32(c). 0.026 ft
4-34. Triangle *CDG*; $C = 90°29.1'$, $D = 46°08.5'$, $G = 43°22.4'$
4-37. ± 0.140 ft
4-40. $+0.123$ ft

CHAPTER 5

5-5. 5.1×10^{-8} sec
5-8(c). ± 8.2 mm
5-9(c). $\frac{1}{107,000}$
5-13(b). Approx. 7°F
5-14(b). Approx. 0.6 in
5-16(b). Approx. 4°C
5-19. 937.13 m
5-21. 241.71 m
5-23. 76.73 m

CHAPTER 6

6-3. 7.23 mi
6-7. Net = -0.00172 ft
6-10. 29.8 ft, 57.8″
6-14. 20.0″
6-17. 0.036 ft
6-24. 147.0 ft
6-26. Eyepiece and/or objective lens not properly focused.
6-29. $n = 5$, $n = 20$
6-32. Erecting telescope requires an extra lens, suffers some light loss; inverting telescope is awkward for beginners.
6-33. Vial can be read when telescope is plunged.

CHAPTER 7

7-3. 0.51 ft
7-5. 0.0035 ft
7-6. Just below third order
7-9. 2.24 ft, 5.24 ft
7-11. 17 ft
7-12. elev BM $B = 272.28$ ft
7-15. 36.40%
7-18. elevs BMs B, C, D = 637.579, 670.033, 708.243
7-20. 0.188 ft
7-23. ± 6.9 mm

CHAPTER 8

8-1. 1.4355 rad, 83°23′06″, 236°15′
8-11. CD = 27°54′
8-14. $A = B = C = 90°$, D = 143°12′, E = 126°48′
8-16. Bearing FG = S58°02′E
8-17. Bearing DE = N85°00′E
8-21. Bearing EA = S21°50′W
8-23. Bearing to 25 + 48.0 = S43°39′E
8-27. Azimuth HI = 284°06′
8-30. Bearing EF = N36°14′E

CHAPTER 9

9-9. 9°15′E
9-11. At equator; at poles
9-14. Natural error, systematic
9-15. South Pole
9-20. N52°25′W
9-22. N58°45′E
9-24. N84°31′W
9-29. 5°45′E
9-33. No distances given; traverse lines do not cross, so they must meet somewhere to exactly close the traverse.

CHAPTER 10

10-8. Make 8 vernier divisions span 7 main scale divisions
10-9(c). 12′

CHAPTER 11

11-1(c). 24.8″
11-2(d). 43.0″
11-10. 49°36.5′
11-12. X = 136°53.25′; Y = 223°06.90′; misclosure = 0.15′
11-14. BAC = 79°41′32″; CAD = 102°09′50″; DAB = 178°08′38″
11-16. ±10.4″
11-18. 0.019 ft
11-22(c). 1′01.2″
11-25. Index error = −0°03′, true angle = 16°20′
11-27(c). 1.1″
11-30(c). 8.6′, 3.4′, 1.9′, 52″, 17″

CHAPTER 12

12-7(c). 1980°
12-8. 107°43′
12-9(b). $\sum$ left = $\sum$ right
12-15. ±37″
12-22. BearingFA

CHAPTER 13

13-4. Adjusted angle $C = 101°23'$, bearing of CD = S74°28′W
13-5. Latitude and departure of CD are −111.38 and −400.72 ft, respectively
13-6. Adjusted latitude and departure of CD are −111.38 and −400.79 ft, respectively. $X_C = 9531.18$ ft, $Y_C = 11{,}688.94$ ft
13-10. Bearing DE = S46°43′W, latitude and departure of DE are −734.14 and −779.51, respectively, linear misclosure = 1.03 ft and precision = $\frac{1}{5000}$
13-12. $X_D = 6456.42$ ft, $Y_D = 4283.73$ ft
13-15. Balanced latitudes and departures of DA, by Compass Rule are −777.27 and −398.12 ft, respectively; by Transit Rule they are −777.25 and −398.11 ft, respectively
13-21. Distance CD = 306.40 ft, bearing DE = S70°28′W
13-24. CD is probably 10 ft too long
13-31. Length = 1933.66 ft, bearing = S55°51.8′W
13-33. Adjusted angle $D = 106°27.6'$

CHAPTER 14

14-2. 5.952 acres
14-5. 1.331 acres
14-8. 5.418 acres
14-10. 0.853 acres
14-13. 3.302 acres
14-17. 1.729 acres
14-19. 4.749 acres
14-25. Area = 27.893 acres
14-29. 1.262 acres
14-32. Length of partition line = 269.50 ft
14-35. 30.028 acres

CHAPTER 15

15-7. 0.000025″
15-12. Index mark may be 0, 30, 50, or 100.
15-14. No. See Section 15-12.
15-17. Read middle wire on point, move upper or lower hair to point. Angle change should be 0°17′.
15-20. $H = 545.6$ ft, $V = 39.8$ ft
15-25. Approximately 2°00′
15-28. 422.6 ft
15-30. About 6 min.
15-33. EF = 449 ft, elev. F = 612.3 ft

CHAPTER 16

16-9. 2.5% grade
16-11. $6\frac{2}{3}\%$
16-16. 12.5 ft maximum for 10% of elevations tested
16-18. 53
16-23. All contour points have the same elevation.
16-28. Select topo point No. 1 and continue consecutively clockwise around 360°.
16-29. Probably not

CHAPTER 17

17-1. Moving in USGS and NGS, slower in highway surveying, and more opposition in property surveys
17-3. 3 ft
17-6. So bearing directions will not be misunderstood (See error on plat in Figure 17-6.)
17-9. 1:100, 1:200, 1:500, 1:1000, 1:2000, 1:10,000
17-16. Culture, relief, vegetation, drainage
17-21. 5 ft
17-26. Key, or explanation
17-32. Map covers predominantly land areas, a chart mostly water areas.
17-35. Depending on map area and scale, etc., a few 0.001 acre
17-36. Not exactly

CHAPTER 18

18-3. Scale of map; lengths of lines on ground and sheet; method used
18-7. A prism eyepiece is necessary to get an eye between telescope and table.
18-12. Table can be set anywhere three known points can be seen to determine position occupied.
18-14. Fill in questionable areas, field checking and plotting
18-19. 230 ft, 0.115″
18-24. 0.0067″ (insignificant)
18-26. $\frac{1}{130}$
18-28. 802 ft

CHAPTER 19

19-4. Declination and either Greenwich Hour Angle or Right Ascension
19-12. Civil time is constant, apparent time is not.
19-13(d). 4^h00^m A.M., PST
19-14(d). $5^h25^m9.5^s$
19-17. $11^h57^m35.6^s$, EST, 5 Mar 1983
19-19(b). 6^h57^m P.M., PST
19-23. 64°49′
19-25. For 1983, altitude = 42°49.8′
19-27(b). N86°23′W
19-28(c). For 1983, decliniation = N10°43.9′
19-34. 45°30′N
19-37. 40°17.2′N
19-39. 87°26′03.0″W

CHAPTER 20

20-22. m = 0.68 (First Order, Class II)
20-24. Elev BM Y = 860.652 ft

CHAPTER 21

21-3. For elev 3000 ft, factor = 0.99985652
21-4. For elev 2000 m, factor = 0.99968623
21-8. Grid length = 5,438.88 ft, grid bearing = S53°33′46.9″W

21-9. $X = 2{,}066{,}508.92$ ft, $Y = 70{,}252.41$ ft
21-15(b). 2486.07 ft
21-18. $BC = 2400.00$ ft
21-19. $BC = 2399.85$ ft
21-20. Azimuth $BC = 326°26'32''$
21-21. Precision = $\frac{1}{7600}$
21-22. For C, $X = 2{,}060{,}395.20$ ft, $Y = 564{,}264.71$ ft
21-23. $CD = 523.722$ m
21-24. Bearing CD = N20°02′38″E
21-25. For D, $X = 712{,}284.319$ m, $Y = 180{,}191.720$ m

CHAPTER 22

22-3. About 1870
22-9. Check records in local courthouse
22-13. Coordinates. Replacing lost corners accomplished with more certainty
22-23. Varies in different states but 20 years most common
22-27. One found corner enables location of others.
22-29. 10,595 ft^2
22-32. 10,530 ft^2
22-34. $Bx = 155.2$ ft
22-35. $HG = 260.8$ ft
22-36. $PQ = 127.07$ ft, $KQ = 51.60$
22-38. No. The land may have been sold but not recorded.

CHAPTER 23

23-1. A 1-ft rod reading represents 2 chains on a stadia sight.
23-7. 5 links, 0°02.5 min
23-9. 132 mi, 13 mi
23-12. Subdivision of sections into smaller units
23-16. 3840 rods
23-22. Set corner at midpoint, 20.14 ch, on section line
23-26. Double, double
23-31. On north and south township sides at section corners
23-33. No
23-34. Convergence of meridians, surveyor errors

CHAPTER 24

24-8. Pipe grade = -1.547%, elev at sta 11 + 50 = 1572.66 ft
24-9. Cut at 3 + 00 = 5.95 ft
24-10. Rod at 3 + 00 = 4.11 ft
24-15. $AC = 89.44$ ft, $BF = 106.30$ ft
24-16. 3.01 ft

CHAPTER 25

25-1(b). $D_a = 4°05'33''$, $D_c = 4°05'36''$
25-5. $T = 255.18$ ft, $L = 507.24$ ft, $E = 17.28$ ft, $M = 17.13$ ft, $D_a = 3°03'21''$, PC = 62 + 62.32 ft, PT = 67 + 69.56 ft
25-9. $R = 1809.57$ ft, $D_c = 3°10'$, $T = 323.96$ ft, $L = 641.05$ ft, $E = 28.77$ ft, $M = 28.32$ ft, PC = 55 + 30.66 ft, PT = 61 + 71.71 ft. For station 59 + 00, chord = 100.00 ft, deflection = 5°50.88′

25-14. $R = 1270.00$ ft, $D_a = 4°30'41''$, $T = 275.00$ ft, $L = 541.64$ ft, $E = 29.43$ ft, $M = 28.77$ ft, PC = 25 + 32.25 ft, PT = 30 + 73.89 ft. For station 29 + 00, chord = 99.97 ft, deflection = 8°17.73'.

25-19. $R = 2292.01$ ft, $D_c = 2°30'$, $T = 294.97$ ft, $L = 586.67$ ft, $E = 18.90$ ft, $M = 18.75$ ft, PC = 103 + 27.40 ft, PT = 109 + 14.07 ft. For station 106 + 50, chord = 50.00, deflection = 4°01.95'

25-22. $D_a = 3°40'$, PC = 34 + 72.64. For station 38 + 00, chord = 99.98 ft, deflection = 6°00.1'

25-24. $R = 1170.33$ ft, $L = 1133.65$ ft, PC = 6 + 42.19 ft, PT = 17 + 75.84 ft

25-28. For station 84 + 00, chord = 570.00 ft, deflection = 11°17.75'

25-31. For staking 55 + 00 from the PC, tangent distance = 187.45 ft, tangent offset = 17.00 ft

25-35. $X_p = 1305.24$ ft, $Y_p = 903.92$ ft

25-37. 519.3 ft

25-39. $R = 1333.49$ ft

CHAPTER 26

26-4. Elev of 35 + 00 = 581.46 ft, elev of sta 40 + 00 = 580.65 ft

26-8. $L = 1268.47$ ft, elev of 17 + 00 = 1225.97 ft

26-12. $g = +2.667\%$, elev of 24 + 00 = 1297.32 ft, elev of 27 + 00 = 1304.67 ft, elev of 29 + 00 = 1311.17 ft

26-21. 721.64 ft

26-24. Station 38 + 00 = elev 580.00 ft

26-26. $S = 892$ ft

26-30. $L = 635$ ft

CHAPTER 27

27-4. For fill of 8 ft, end area = 448 ft^2

27-5. For fill of 8 ft, end area = 576 ft^2

27-8. $V = 472$ cy

27-11. Cumulative vol. = −1490 cy

27-14. At 52 + 00, end area = 63.9 ft^2

27-16. $C_p = 0.5$ cy, not significant

27-21. Left intercept = 38.4 ft at elev = 1254.9 ft, end area = 599.2 ft^2

27-23. $V_e = 830$ cy, $V_p = 835$ cy

27-27. Vol = 1.36 acre feet

CHAPTER 28

28-4(c). 1:3445

28-6(b). 1:21,050

28-9. Area = 24.07 acres

28-12. $AB = 7554$ ft

28-14. 233.67 acres

28-18. 4693 ft

28-20. 11,360 ft

28-22. Elev $B = 2413$ ft

28-29. 18

28-31. 26%

28-33. 35 lines at 63 photos each = 2205

INDEX

U

V

84 85 86 7 6 5 4 3 2 1

ABBREVIATIONS

Construction Surveys

b.b.	batterboards
B.L.	building line
C.B.	catch basin
C.G.	center line of grade
℄	center line
const.	construction
C	cut
esmt.	easement
F	fill
F.G.	finish grade, Fin. Gr.
F.H.	fire hydrant
FL	fence line
F.L.	flow line (invert)
F.S.	finished surface
G.C.	grade change
G.P.	grade point
G.R.	grade rod (s.s. notes)
L	left (x-sect. notes)
M.H.	manhole
PL	property line
P.P.	power pole
pvmt.	pavement
R	right (x-sect. notes)
R/W	right-of-way
S.D.	storm drain
S.G.	subgrade
S.L.	spring line
spec.	specifications
Sq.	square
s.s.	slope stake, side slope
Std.	standard
Str. Gr.	straight grade
X-sect.	cross-section

Property Surveys

A	area
C.F.	curb face
ch "X"	chiseled X cross
C.I.	cast iron
diam.	diameter
Dr.	drive
ER	end of return
Ex.	existing
H & T	hub and tack
H.C.	house connection sewer
I.B.	iron bolt (bar)
I.P.	iron pipe; iron pin (confusing)
L & T	lead and tack
max.	maximum
min.	minimum
M.H.W.	mean high water
M.L.L.W.	mean low low water
M.L.W.	mean low water
Mon.	monument
No.	number
P	pipe; pin (confusing)
Rec.	record
St.	street
Std. Surv. Mon.	standard survey monument
Std. Trav. Mon.	standard traverse monument
2″ × 2″	two- × two-inch stake
X	cross cut in stone
yd	yard

Public Lands Surveys

ac	acres
AMC	auxiliary meander corner
bdy., bdys.	boundary, boundaries
BT	bearing tree
CC	closing corner
ch, chs	chain, chains
cor., cors.	corner, corners
corr.	correction
decl.	declination
dist.	distance
frac.	fractional (sec., etc.)
Gr.	Greenwich
G.M.	guide meridian
lk, lks	link, links
meas.	measurement
mer.	meridian
mkd.	marked
Mi. Cor.	mile corner
MC	meander corner
M.S.	mineral survey